TUMOR ANGIOGENESIS REGULATORS

T0138639

Tumor Angiogenesis Regulators

Editors

Ruben R. Gonzalez-Perez
Department of Microbiology
Biochemistry and Immunology
Morehouse School of Medicine
Atlanta, GA
USA

Bo R. Rueda
Director
Vincent Center for Reproductive Biology
Vincent Department of Obstetrics and Gynecology
Massachusetts General Hospital
Boston, MA
USA

and

Associate Professor
Obstetrics, Gynecology and Reproductive Biology
Harvard Medical School
Boston, MA
USA

CRC Press
Taylor & Francis Group
Boca Raton   London   New York

CRC Press is an imprint of the
Taylor & Francis Group, an **informa** business

A SCIENCE PUBLISHERS BOOK

CRC Press
Taylor & Francis Group
6000 Broken Sound Parkway NW, Suite 300
Boca Raton, FL 33487-2742

© 2013 Copyright reserved
CRC Press is an imprint of Taylor & Francis Group, an Informa business

Cover Illustrations: Reproduced by kind courtesy of Dr. Graham Pidgeon (author of Chapter 7), Drs. Manu O. Platt and Jerald E. Dumas (authors of Chapter 8), Drs. João Incio and Raquel Soares (authors of Chapter 12).

No claim to original U.S. Government works

Printed in the United States of America on acid-free paper

International Standard Book Number: 978-1-4665-8097-8 (Hardback)

This book contains information obtained from authentic and highly regarded sources. Reasonable efforts have been made to publish reliable data and information, but the author and publisher cannot assume responsibility for the validity of all materials or the consequences of their use. The authors and publishers have attempted to trace the copyright holders of all material reproduced in this publication and apologize to copyright holders if permission to publish in this form has not been obtained. If any copyright material has not been acknowledged please write and let us know so we may rectify in any future reprint.

Except as permitted under U.S. Copyright Law, no part of this book may be reprinted, reproduced, transmitted, or utilized in any form by any electronic, mechanical, or other means, now known or hereafter invented, including photocopying, microfilming, and recording, or in any information storage or retrieval system, without written permission from the publishers.

For permission to photocopy or use material electronically from this work, please access www.copyright.com (http://www.copyright.com/) or contact the Copyright Clearance Center, Inc. (CCC), 222 Rosewood Drive, Danvers, MA 01923, 978-750-8400. CCC is a not-for-profit organization that provides licenses and registration for a variety of users. For organizations that have been granted a photocopy license by the CCC, a separate system of payment has been arranged.

Trademark Notice: Product or corporate names may be trademarks or registered trademarks, and are used only for identification and explanation without intent to infringe.

Library of Congress Cataloging-in-Publication Data

Tumor angiogenesis regulators / editors, Ruben R. Gonzalez-Perez, Bo R. Rueda.
       p. ; cm.
   Includes bibliographical references and index.
   ISBN 978-1-4665-8097-8 (hardcover : alk. paper)
   I. Gonzalez-Perez, Ruben R. II. Rueda, Bo R.
   [DNLM: 1. Angiogenesis Inducing Agents. 2. Neoplasms--metabolism. 3. Angiogenesis Inhibitors. 4. Neoplasms--drug therapy. 5. Neovascularization, Pathologic--drug therapy.   QZ 202]
   616.99'4061--dc23

                                                                2012049252

Visit the Taylor & Francis Web site at
http://www.taylorandfrancis.com

CRC Press Web site at
http://www.crcpress.com

Science Publishers Web site at
http://www.scipub.net

Preface

It has long been recognized that once a critical mass of tumor cells is achieved, the cells can secrete and ever growing list of factors that can reprogram the surrounding microenvironment making it more receptive to tumor establishment, maintenance and growth. These processes are highly reliant on angiogenesis. There is a number of prominent pro-angiogenic factors involved in tumor angiogenesis that have been studied extensively. However, there is also an increasing number of lesser known factors and conditions that are contributing to tumor angiogenesis. Moreover, accumulating data provides evidence to suggest angiogenesis is controlled by opposing actions of an increasing number of pro- and anti-angiogenic factors emphasizing the challenges facing scientists and clinicians.

The book provides a comprehensive update of classical and non-classical factors modulating the angiogenesis process: VEGF, cancer and endothelial stem cells, vasculogenic mimicry, tumor-activated macrophages, proteins, proteases with special emphasis on cathepsins; Notch and crosstalk to cytokines, Krüppel-Like factor, inflammatory and bioactive lipids and steroids, with special emphasis on cyclooxygenase (COX) and lipoxygenase (LOX) pathways.

Several chapters are devoted to the mechanisms/pathways regulating the expression and actions of classical and novel tumor angiogenesis regulators. In addition, anti-angiogenic drugs currently in use to combat several types of cancers are discussed. The book also provides relevant and up to date information on the potential influence of pandemic overweight and obesity as well as diabetes and metabolic syndrome on the regulation of tumor angiogenesis. Additional, current data on genetic polymorphisms in several molecules potentially responsible for health disparity and differential tumor angiogenesis outcome among ethnic groups is also presented.

In summary, the book addresses several aspects of the, biochemical, biological and physiological actions of well-characterized and novel regulators of the tumor angiogenesis process. The information provided can serve as the basis for the further development of old and new areas of basic and translational and clinical research on the regulation of tumor angiogenesis. Therefore, the book is intended to serve as a source for updated scientific information on angiogenesis to researchers, scholars and clinicians from multiple disciplines that have an interest in tumor biology.

Editors

Ruben R. Gonzalez-Perez
Bo R. Rueda

Contents

Chapter 1

VEGF Signaling in Normal and Tumor Angiogenesis

Evangelia Pardali, Rinesh Godfrey and Johannes Waltenberger*

Department of Cardiovascular Medicine, University of Münster, Münster, Germany.

CHAPTER OUTLINE

- ▶ Introduction
- ▶ Vascular endothelial growth factor family members
- ▶ VEGF receptor signaling
- ▶ Neuropilin I and Neuropilin 2 (NRPs)
- ▶ VEGF signaling and regulation of angiogenesis
- ▶ Functional role of VEGFRs in regulation of angiogenesis
- ▶ VEGF signaling and tumor angiogenesis
- ▶ Conclusions and perspectives
- ▶ Acknowledgements
- ▶ References

ABSTRACT

The proper expression of VEGF-A is crucial for the initiation of vascular development and vascular growth. Likewise, the function of VEGF-A is dependent on the availability and function of VEGF-receptor 1 (*fms*-like tyrosine kinase, Flt-1) VEGF-receptor 2 (fetal live kinase, Flk-1; kinase-insert domain containing receptor, KDR) and VEGF-receptor 3. This is true both during embryonic development, in inflammation-related angiogenesis as well as in tumor angiogenesis. This

Correspondence Author: Department of Cardiology and Angiology, University of Münster, Albert-Schweitzer-Campus 1, Building A1, 48149 Münster, Germany. Email: Waltenberger@ukmuenster.de

chapter reviews and highlights the role of VEGF-A and other gene products of the VEGF family in normal and pathological angiogenesis, the role of the different VEGF-receptors in this context, the molecular basis of VEGF receptor signaling and function, and – finally – the impact of pathological conditions (metabolic, tumor environment) on the induction of angiogenesis and on the fate of newly developed vessels.

Key words*:*   VEGF, angiogenesis, endothelial, tumor

1   Introduction

Vertebrates have a highly organized vascular system consisting of a branching network of arteries, capillaries and veins which penetrate all body tissues besides cartilage. The vascular system enables the efficient transport of oxygen and nutrients as well as waste removal from the tissue. The development of the vascular system is based on two highly regulated and controlled processes: vasculogenesis and angiogenesis. Vasculogenesis is the de novo formation of a vascular plexus; this involves differentiation of mesoderm cells into angioblasts and endothelial cell precursor cells (Carmeliet et al. 1996; Risau 1997; Coultas et al. 2005). Angioblasts are then organized and form lumens/tubes which develop into the primary vascular plexus which will give rise to the vessels and capillaries. This primitive vascular network is then expanded via the process of angiogenesis. Angiogenesis involves sprouting, bridging and branching of preexisting vessels. As a result the primary vascular network develops into a highly branched and organized vascular tree consisting of arteries and veins. Endothelial cells (ECs) have a remarkably plastic capacity to give rise to vessels with diverse functional, morphological and molecular characteristics. After blood vessel assembly, ECs undergo specification to either arterial or venous fate in response to both genetic factors as well as haemodynamic stimuli (Adams and Alitalo 2007; Rocha and Adams 2009). Later during development, vessels that arise by angiogenesis may also adopt diverse functional specializations depending on locally derived signals (Adams and Alitalo 2007, Rocha and Adams 2009). In addition, the venous endothelium gives rise to the lymphatic vessel network, which collects and returns lymph fluid back to the vasculature (Tammela and Alitalo 2010). Besides endothelial cells mural cells (pericytes in medium-sized vessels and smooth muscle cells in large vessels) play also important role in the formation of the vascular network by supporting the new vessels.

Angiogenesis is characterized by two phases: the activation and the resolution phase of angiogenesis. During the activation phase an angiogenic stimulus induces EC proliferation and migration as well as degradation of the basement membrane (BM). During the resolution phase, EC stop to proliferate and migrate, mural cells are recruited to support the newly formed vessel and BM is synthesized. Angiogenesis is a highly regulated process and its perturbation may contribute to various vascular diseases. Vasculogenesis takes place predominantly at early phases of development while angiogenesis occurs both during development as well as later in life.

Nevertheless, it is known today that vasculogenesis can also occur later after birth at sites of neovascularization by recruited EC progenitor cells (Ribatti 2007). Angiogenesis takes place not only at the early stages of life but also throughout life.

As mentioned earlier angiogenesis is a tightly regulated process. Recent advances in vascular biology suggested that specialized ECs with distinct cellular specifications and functions contribute to new blood vessel formation (Fig. 1A). In stable vessels ECs are quiescent cells (phalanx cells). ECs remain quiescent until they receive an angiogenic stimulus. This leads to fundamental changes in EC phenotype and function. Induction of angiogenesis leads to loss of EC junctions, induced proliferation and migration of ECs, activation of proteases and degradation of extracellular matrix (Gerhardt et al. 2003; De Smet et al. 2009; Carmeliet and Jain 2011a). Some of the EC acquire a more invasive and motile phenotype and drive blood vessel sprouting. These ECs are known as 'tip cells' (Gerhardt et al. 2003; De Smet et al. 2009; Carmeliet and Jain 2011a). Tip cells are migratory cells which do not proliferate but invade the surrounding tissue by extending numerous filopodia. Tip cells have lost their polarity and do not form a lumen. Tip cell function relies on the "stalk cells'. Stalk ECs tail behind the tip cells, proliferate, are fully polarized and form lumens. However, they do not form so many filopodia. Stalk cell proliferation ensures elongation of sprouting vessel and recruitment of support cells such as pericytes (Kamei et al. 2006; Iruela-Arispe and Davis 2009). The newly formed branch connects with another branch via tip cell fusion (anastomosis). Tip ECs stop migrating and form new EC-EC junctions and a new vascular lumen is formed which allows blood flow. Both genetic and environmental factors regulate tip/stalk EC specialization. Finally, EC acquire a quiescent phenotype and become phalanx EC. Phalanx cells do not migrate and proliferate but by depositing a basement membrane contribute to vessel stabilization (Fig. 1A).

Several factors have been characterized for their role in regulating angiogenesis. One of the indispensable ones is vascular endothelial growth factor (VEGF). Direct or indirect perturbation of VEGF signaling leads to vascular disorders. In addition, increased VEGF signaling leads to vascular dysfunction and to the development of abnormal vasculature. Due to its critical role in angiogenesis, VEGF is considered to be one of the major targets for the development of anti-angiogenic drugs. This chapter will focus on the role of VEGF signaling in angiogenesis and pathological angiogenesis.

2   Vascular Endothelial Growth Factor Family Members

Vascular endothelial growth factor (VEGF) is a potent cytokine which is important for the normal developmental angiogenesis and for the revascularization associated with various diseases. VEGF had initially been identified as vascular permeability factor (VPF) (Senger et al. 1983). VEGF was independently identified as an *in vivo* inducer of microvascular

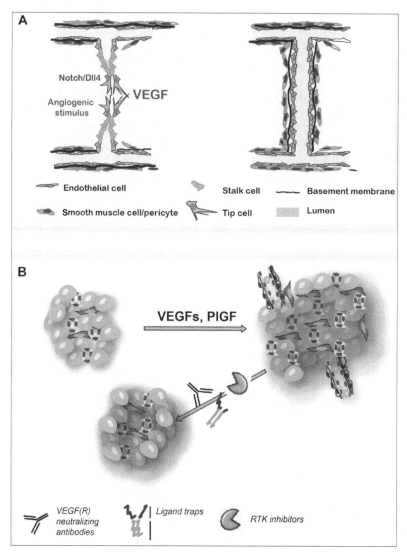

Figure 1 Angiogenesis and tumor angiogenesis. A) Regulation of angiogenesis. Angiogenesis comprises of an activation and a resolution phase. Following EC activation by an angiogenic stimulus (VEGF, bFGF, TGFβ) BM is degraded and the tip EC at the forefront of the sprout invades the surrounding tissue by extending numerous filopodia. The new sprouts elongate through proliferation of the stalk ECs and the new branches connect through tip cell fusion. Finally, ECs stop to proliferate, and sprout maturation occurs by reconstitution of BM and pericyte/SMC recruitment and acquire a quiescent phenotype (phalanx EC). B) Targeting VEGF signaling in tumor angiogenesis. Tumors growth depends on angiogenesis which is regulated by a variety of pro- and anti-angiogenic factors such as VEGF. Inhibition of VEGF signaling by neutralizing antibodies, ligand traps or receptor kinase inhibitors result in reduced tumor angiogenesis and tumor growth.

Color image of this figure appears in the color plate section at the end of the book.

hyperpermeability and also as a strong *in vitro* mitogen for endothelial cells (Ferrara and Henzel 1989). Later, it was discovered that the factor responsible for inducing vascular permeability and mitogenic activities are the same protein which is related to platelet-derived growth factor; namely VEGF (Keck et al. 1989; Leung et al. 1989; Conn et al. 1990).

The mammalian VEGF family consists of five related members: VEGFA (firstly identified and the prototype), VEGFB, VEGFC, VEGFD and placental growth factor (PIGF). In general, VEGFs are homodimeric polypeptides. However, heterodimers of VEGFA and PIGF have been reported (DiSalvo et al. 1995). VEGFs also exist in various isoforms generated through alternative splicing and proteolytic processing (Ferrara 2010). The crystal structure of VEGFA indicated that VEGFA monomer resembles that of platelet-derived growth factor (PDGF) with one major difference; the N-terminal segment of VEGFA is helical rather than extended (Muller et al. 1997). This chapter primarily focuses on VEGFA and the importance of this growth factor will be described in detail.

Human VEGFA gene is localized in chromosome 6p21.3 (Vincenti et al. 1996). VEGFA is alternatively spliced to generate VEGFA-121, VEGFA-145, VEGFA-165, VEGFA-189 and VEGF-206 (Houck et al. 1992). VEGFA-165 is the predominant isoform produced by normal and transformed cells with a molecular weight of 45 kDa. Splice variants, which occur rarely, have also been reported, including VEGFA-183, VEGFA-162 and VEGFA-165b (Jingjing et al. 1999, Lange et al. 2003) In contrast, VEGFA-165b is a variant which has been reported to have a negative effect on VEGFA-induced mitogenesis (Bates et al. 2002). Transcripts encoding VEGFA-121 and VEGFA-189 have been detected in most of cells and tissues expressing VEGF gene. The primary functional differences between the different VEGFA isoforms appear to be due to the differences in their affinities for heparin (Neufeld et al. 1996). While VEGFA-121 has little or no affinity for heparin, VEGFA-165 binds heparin with moderate affinity and VEGFA-189 and VEGFA-206 binds heparin with highest affinity. This property of VEGFA-189 and VEGFA-206 makes them retained at the cell surface and extracellular matrix (ECM) through the complex interactions with heparin sulfate proteoglycans (Park et al. 1993). VEGFA-121, due to its inability to bind to heparin, is present in the cell as a freely diffusible protein. VEGFA-165 is available within the cell as a diffusible protein but a significant portion is bound to the cell surface and extracellular matrix (Park et al. 1993). Cell bound VEGFA can be released into freely diffusible form by the action of heparinase. Plasminogen activation and generation of plasmin can also potentially aid in the release of cell bound VEGFA isoforms by cleaving at its COOH terminus. This leads to the generation of a bioactive proteolytic fragment with a molecular weight of approximately 34 kDa (Houck et al. 1992). Matrix metalloproteinases (MMPs) are reported to generate diffusible bioactive fragments of VEGFA, thereby regulating the bioavailability of VEGFA (Lee et al. 2005). Thus alternative splicing and proteolytic processing appears to be the two mechanisms by which different

VEGFA isoforms are generated within endothelial cells. Interestingly, loss of heparin binding leads to reduced mitogenic activity for vascular endothelial cells, probably due to the loss of stability of VEGF-VEGFR complex in the absence of heparin sulfate (Keyt et al. 1996).

Oxygen levels play a pivotal role in the regulation of VEGFA gene expression, both *in vitro* and *in vivo*. VEGF mRNA expression was found to be increased by low partial pressures of oxygen (pO_2) in cultured cells (Minchenko et al. 1994). In pig myocardium, induction of ischemia resulted in the increase in VEGFA mRNA expression, indicating that VEGFA plays an immediate role in the revascularization following myocardial ischemia (Banai et al. 1994). Hypoxia is now considered to be an inducer of VEGFA expression. Extensive promoter deletion analysis carried out in VEGFA mRNA, identified a 28-base sequence in the 5′ promoter region that can bind hypoxia-inducible factor-1 (HIF-1) and mediate hypoxia-induced transcription (Liu et al. 1995). In ECs, reactive oxygen species (ROS) mediated upregulation of VEGFA has also been reported (Yamagishi et al. 2003). ROS seems to modulate the expression of HIF-1 thereby indirectly control the expression patterns of VEGFA (Ushio-Fukai and Nakamura 2008). Hypoxia not only transactivates the expression of VEGFA, but also it has been reported to induce increased mRNA stability of VEGFA. Regions which mediate this posttranscriptional stability of VEGFA mRNA have been mapped to be present in the 3′ untranslated region (UTR) of VEGFA mRNA (Ikeda et al. 1995).

Cytokines and growth factors can upregulate the expression of VEGFA mRNA. Epidermal growth factor (EGF), transforming growth factor- α (TGF-α), transforming growth factor-β (TGF-β), insulin-like growth factor-1 (IGF-1), fibroblast growth factor (FGF) and platelet-derived growth factor (PDGF) has been reported to significantly induce VEGFA mRNA expression (Frank et al. 1995; Warren et al. 1996). Epithelial and fibroblastic cells upon treatment with TGF-β resulted in the increase the expression of VEGFA mRNA and also release of VEGFA protein into the culture medium. Thus, it was proposed that VEGF may act as a paracrine mediator for indirectly acting pro-angiogenic agents like TGF-β (Pertovaara et al. 1994). Inflammatory cytokines such as interleukin (IL)-1α, IL-6 induce expression of VEGFA in cultured synovial fibroblasts, indicating the role of VEGFA induction in mediating inflammatory angiogenesis (Ben-Av et al. 1995). Various hormones like TSH, ACTH have been reported to stimulate VEGF expression indicating the role played by VEGF in inducing adrenal cortical and thyroid angiogenesis (Soh et al. 1996; Shifren et al. 1998). Specific transforming events are also correlated with the induction of VEGFA mRNA. Mutated form of p53, onocogenic RAS, mutations in the WNT-signaling pathway are known to positively upregulate VEGFA (Kieser et al. 1994; Okada et al. 1998; Zhang et al. 2001). Recently, two forkhead box (FOX) family proteins, FOXO3a and FOXM1 were found to regulate VEGFA expression in a non-redundant fashion in breast cancer cells. FOXO3a was able to repress the expression of VEGFA, whereas FOXM1 was able to induce the expression of VEGFA (Karadedou et al. 2011).

3   VEGF Receptor Signaling

VEGF binding receptors (VEGFRs) were initially described on the surface of ECs *in vitro* and *in vivo*. Later, VEGFR was identified on the surface of hematopoietic cells such as monocytes and macrophages. In humans, three structurally related VEGFRs have been identified: VEGFR1, VEGFR2 and VEGFR3. In general, VEGFR1 is expressed on ECs and other cells such as monocytes and macrophages; whereas, VEGFR2 is expressed mainly on vascular ECs and VEGFR3 on lymphatic ECs. VEGFRs consist of an extracellular ligand-binding region made up of seven immunoglobulin-like domains. This extracellular region is connected to the cytoplasmic catalytic domain via a short transmembrane helix. VEGFA, VEGFB, and PlGF bind to VEGFR1. VEGFR2 is stimulated by VEGFA, VEGFC, and VEGFD whereas VEGFR3 is specific for VEGFC and VEGFD. VEGFs can also interact with co-receptors such as heparan sulfate proteoglycans and neuropilin-1 and neuropilin-2 (Stuttfeld and Ballmer-Hofer 2009).

Quite similar to other receptor tyrosine kinases, signaling by VEGFRs is initiated upon binding of a covalently linked ligand dimer to the extracellular receptor domain. This can happen *in cis* or *in trans*. This interaction leads to receptor homo- and heterodimerization. Dimerization is followed by changes in the intracellular domain conformation, resulting in the exposure of ATP binding site in the intracellular kinase domain leading to phosphorylation of specific tyrosine residues located in the intracellular juxtamembrane domain, the kinase insert domain, and the carboxy-terminal region of the receptor. As a result, a variety of signaling molecules are recruited to VEGFR dimers giving rise to the assembly of large molecular complexes leading to the activation of distinct cellular pathways. The interaction between VEGFRs and downstream signaling effectors are mainly mediated through Src homology-2 (SH-2) and phosphotyrosine-binding (PTB) domains. Phosphorylation of the receptor is tightly controlled by the internalization and degradation and by the action of tyrosine dephosphorylating enzymes; the protein tyrosine phosphatases (PTPs) like Protein tyrosine phosphate 1B (PTP1B), (vascular endothelial PTP) VE-PTP, SH2-domian containing PTP (SHP-1 and SHP-2) and density-enhanced phosphatase (DEP-1) (Kappert et al. 2005).

VEGFR1 is a 180-185 kDa glycoprotein that gets activated upon binding with VEGFA, VEGFB and PlGF. This receptor is expressed on vascular ECs at considerably higher levels during development and also in adults (Jakeman et al. 1992; Peters et al. 1993). VEGFR1 binds VEGFA with very high affinity (K_d=15 pM) than VEGFR2 (K_d=750 pM) (Shinkai et al. 1998). However, VEGFR1 tyrosine kinase activity is only weakly induced by its ligands, possibly due to the presence of a repressor sequence in the juxtamembrane domain of VEGFR1 (Meyer et al. 2006). One interesting aspect of the phosphorylation pattern of VEGFR1 is that it is ligand dependent. For example, Tyr1309 is phosphorylated only upon PlGF stimulation but not with VEGFA stimulation (Autiero et al. 2003). The

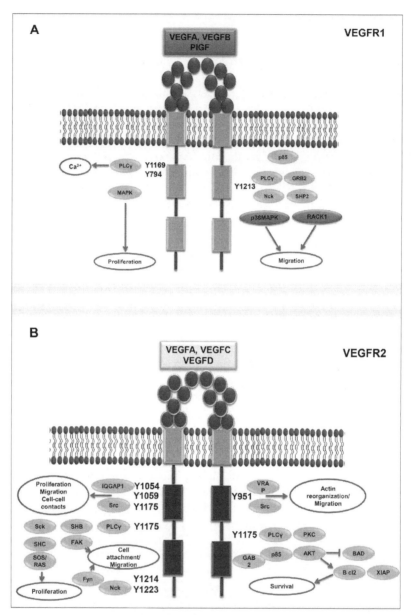

Figure 2 VEGF signal transduction. A) Schematic representation of activated and dimerized VEGFR1. Upon ligand binding, docking of adaptor molecules (shown in yellow) to specific phosphor sites occurs, leading to activation of subsequent molecules. The biological outcome of the signaling process is shown inside red oval boxes. B) Schematic representation of activated and dimerized VEGFR1. Upon ligand binding, docking of adaptor molecules (shown in yellow) to specific phosphor sites occurs, leading to activation of subsequent molecules. The biological outcome of the signaling process is shown inside red oval boxes.

Color image of this figure appears in the color plate section at the end of the book.

signaling cascades activated downstream of VEGFR1 is well characterized. A schematic representation of the VEGFR1 signaling cascades are depicted in Fig. 2A. Activation of Tyr^{794} and Tyr^{1169} are responsible for the binding and activation of phospholipase C- γ (PLCγ), which in turn results in Ca^{2+} release and the generation of inositol 1,4,5- triphosphate (Sawano et al. 1997). Tyr^{1213} binds various SH2-domain containing proteins such as growth factor receptor bound protein 2 (GRB2), SH2-containing protein tyrosine phosphatase (SHP-2), non-catalytic region of tyrosine kinase adaptor protein (Nck), PLCγ and p85 subunit of the phosphoinositide 3-kinase (PI3K) (Igarashi et al. 1998; Ito et al. 1998; Yu et al. 2001).

VEGFR2 is a 210-230 kDa glycoprotein, which binds to VEGFA and also proteolytically processed VEGFC and VEGFD. An alternatively spliced soluble VEGFR2 has been reported, and this sVEGFR2 is able to bind to VEGFC and prevents its binding to VEGFR3, thereby blocking lymphatic endothelial cell proliferation (Albuquerque et al. 2009). VEGFR2 is expressed in vascular ECs with the peak of its expression during embryonic vasculogenesis and angiogenesis (Millauer et al. 1993). Its expression is also present in non-endothelial cells, such as hematopoetic cells and megakaryocytes. VEGFR2 expression is upregulated during active physiological angiogenesis, e.g. in uterus during reproductive cycle (Tasaki et al. 2010) as well as during neovascularization during pathological processes such as cancer (Millauer et al. 1994).

VEGFR2 signal transduction mediates endothelia cell survival, proliferation, migration and formation of vascular tube. Upon ligand binding, major phosphorylation sites including Tyr^{951}, Tyr^{1054} and Tyr^{1059} in kinase domain, and Tyr^{1175} and Tyr^{1214} in the C-terminal domain get phosphorylated (Matsumoto et al. 2005). A schematic representation of the VEGFR2 signaling cascades are depicted in Fig. 2B. Phosphorylated Tyr^{951} results in the docking of SH2-domain containing VEGFR-associated protein (VRAP), which mediates VEGFA dependent actin reorganization and migration through the activation of Src (Matsumoto et al. 2005). Phosphorylation of Tyr^{1059} leads to the activation Src and subsequent phosphorylation of other residues in VEGFR2, mainly Tyr^{1175} and actin-binding protein IQ (isoleucine/glutamine)-motif-containing GTPase-activating protein (IQGAP1). Activation of IQGAP1 is responsible for the regulation of cell-cell contacts, proliferation and migration (Yamaoka-Tojo et al. 2006). Tyr^{1175} phosphorylation is critical for the binding of PLCγ, but also this site allows docking of adaptor proteins SHB (SH2-domain containing adaptor protein B) and SHC-related adaptor protein (Sck). Focal adhesion kinase (FAK) is recruited through SHB and mediates endothelial cell adhesion and migration (Warner et al. 2000; Abu-Ghazaleh et al. 2001; Meyer et al. 2008). Tyr^{1175} phosphorylation also allows binding of growth factor receptor-bound protein 2 (GRB2), which in turn mobilizes nucleotide-exchange factor son of sevenless (SOS) to VEGFR2 and activates the mitogenic RAS-MAPK pathway. PLCγ activation leads to the activation of three different protein kinase C (PKC) isoforms, PKCγ, PKCγ and PKCγ and along with downstream activation of ERK, mediates cell proliferation (Takahashi et al. 1999). Non-catalytic region of tyrosine kinase adaptor protein 1(Nck) and the soluble Src family tyrosine kinase Fyn is recruited through phosphorylated Tyr^{1214} and Nck-Fyn complex thus formed allows activation of p21-activated protein kinase (PAK2), cell division cycle 42 (cdc42) and p38MAPK (Lamalice et al. 2006). Lipid mediators are generated through the docking protein GRB2-binding protein (GAB1) GAB1 recruitment leads to the activation of Rac through PIP3. Rac-1 activation is important for the cell motility (Laramee et al. 2007). Anti-apoptotic proteins B-cell lymphoma 2 (Bcl2) and inhibitors of apoptosis family, the X-linked inhibitor of apoptosis (XIAP) and survivins are also expressed upon VEGFA mediated activation of VEGFR2 (Deveraux et al. 1998).

4   Neuropilin 1 and Neuropilin 2 (NRPs)

Some endothelial cells express VEGFA binding sites which are not VEGFR1 or VEGFR2. Lack of ability of the VEGFA-121 to bind to these sites indicated that exon-7 encoded sequences are mandatory for the binding. The receptor was later identified to be neuropilin-1 (NRP1) (Soker et al. 1996). They are 130 kDa transmenbrane proteins with a cytoplasmic tail lacking catalytic function. NRP1 binds to exon 8 encoded regions of VEGFA isoforms such as VEGFA-165 with the formation of a VEGFR2-NRP1 complex (Vander Kooi et al. 2007). Binding partners of NRP2 are different. It associates with VEGFR3 in a VEGFC or VEGFD-dependent fashion (Karpanen et al. 2006).

NRP1 modulates VEGFR signaling, resulting in increased migration and survival of endothelial cells (Favier et al. 2006, Karpanen et al. 2006). NRP1 is also implicated in the VEGFR2 mediated permeability and VEGFA-induced vessel sprouting and branching (Kawamura et al. 2008). NRP1 is expressed mainly in arteries and NRP2 in veins and lymphatic cells.

5   VEGF Signaling and Regulation of Angiogenesis

Angiogenesis plays crucial role in tissue homeostasis and perturbation of angiogenesis leads to pathological situations. Due to the important role of angiogenesis several studies have focused on the factors involved in the regulation of angiogenesis. Genetic studies have provided evidence for the indispensable role VEGFA in vasculogenesis and angiogenesis. VEGFA is the principal master regulator of vasculogenesis and angiogenesis in development, growth and disease. Deletion of one only VEGFA allele results in embryonic lethality between day 11 and 12 due to defects in vascularization in several organs, a reduced number of nucleated red blood cells within the blood islands in the yolk sac (Carmeliet et al. 1996; Ferrara et al. 1996). These results suggest that during development VEGFA regulates both vasculogenesis and early hematopoiesis concentration/gene-dosage dependent manner. It has been shown that VEGF-A is not only critical for the development of ECs during development but is also required for the survival of ECs. Interestingly specific deletion of the VEGFA gene in ECs in mice results also in embryonic lethality (Lee et al. 2007a), suggesting that VEGFA produced by adjacent cells cannot compensate for endothelial VEGFA and that VEGFA produced by endothelial cells may act in an autocrine manner to stimulate vessel survival (Lee et al. 2007a). VEGFA also plays an important role in postnatal life. Inhibition of VEGFA signaling by administration of a soluble VEGFR1 protein results in growth arrest and lethality due to inhibition of glomerular development and kidney failure when the treatment is initiated at day 1 or day 8 postnatally (Gerber et al. 1999a; Eremina et al. 2003). VEGFA is also important for directional growth and cartilage invasion by blood vessels (Gerber et al. 1999b; Carlevaro et al. 2000). In addition, it has been shown that VEGF inhibitors delay follicular development (Zimmermann et al. 2001b) and suppresses luteal angiogenesis in rodents (Ferrara and Henzel 1989; Zimmermann et al. 2001a) as well as in primates suggesting that VEGFA is a key regulator of reproductive angiogenesis.

Other studies have focused on the role of other VEGF ligands. VEGFB was shown to promote fatty acid uptake in endothelial cells, which is important in organs with high metabolic stress such as the heart (Hagberg et al. 2010). VEGFB-deficient animals display no vascular defects and survive development, but display reduced heart size and impaired recovery after cardiac ischemia (Bellomo et al. 2000; Aase et al. 2001). VEGFC was also shown to play important role in development since its inactivation results in embryonic lethality following defective lymphatic development and fluid accumulation in tissues (Karkkainen et al. 2004). In addition, deletion of both VEGFC and VEGFD in mice results in defects primarily in lymphatic vessels, while blood vessels remain unaffected (Haiko et al. 2008). Genetic studies in mice have demonstrated that the VEGFA homologue PlGF is dispensable for embryonic as well as adult angiogenesis (Carmeliet et al. 2001). Interestingly, PlGF expression is very low under physiological conditions but it is upregulated under certain pathological conditions (Failla et al. 2000; Green et al. 2001; Larcher et al. 2003).

6   Functional Role of VEGFRs in Regulation of Angiogenesis

Several studies have focused on the characterization of the role of VEGF receptors in angiogenesis. Genetic deletion of *vegfr1* in mice leads to death at embryonic day 9 (E9.0) due to increased proliferation of endothelial cells and as a result disorganization of the vascular system (Fong et al. 1995). Similarly, *vegfr1* knockdown enhances angiogenic EC behavior in zebrafish (Krueger et al. 2011). VEGFR1 was found to be dispensable for the proliferation and migration of endothelial cells *in vitro* (Waltenberger et al. 1994; Rahimi et al. 2000). However, EC differentiation and organization into vascular tubes require VEGFR1 induced activation of PI3K pathway (Cai et al. 2003). VEGFR1 is also implicated in the endothelial cell actin reorganization through receptor for activated C-kinase (RACK1) (Wang et al. 2011). Interestingly, although VEGFR1 has a higher affinity for VEGFA it possesses a weak Tyr kinase activity. In addition, deletion of the VEGFR1 tyrosine kinase domain in mice does not affect vascular development (Hiratsuka et al. 2001; Hiratsuka et al. 2005a; Hiratsuka et al. 2005b), suggesting that expression of VEGFR1 in ECs, may act as VEGFA decoy and by this way controls physiological angiogenesis (Kappas et al. 2008). VEGFR1 cross talks with VEGFR2 through various ways including dimerization, transphosphorylation and regulation of expression levels of receptors. VEGFR1-VEGFR2 heterodimers comprise of 25–50% of the active signaling VEGF signaling complexes and these heterodimers are formed as a result of VEGFA binding and not due to VEGFB or PIGF (Mac Gabhann and Popel 2007). In ECs, VEGFR1 may act to sequester VEGFA, thus reducing the activation of VEGFR2 and formation of angiogenic sprouts (Hiratsuka et al. 2005b). The negative regulatory role of VEGFR1 is in part mediated by soluble form of VEGFR1 (solVEGFR1) was shown to act as a negative regulator of angiogenesis probably by sequestering VEGF-A (Kendall et al. 1994), which sequesters VEGFA and by forming nonfunctional heterodimers with VEGFR2. solVEGFR1 is expressed during gestation in placenta in a regulated manner (Clark et al. 1998; Maynard et al. 2003) its excessive expression during pregnancy has been associated with hypertension in pregnant women in the pathogenesis of pre-eclampsia. Elevated levels of sVEGFR1 are implicated in various diseases such as cancer (Toi et al. 2002).

Additionally, it was shown that expression of VEGFR1 in non-EC cells types affects also angiogenesis. It was demonstrated that VEGFR1 expression in retinal myeloid cells suppresses angiogenesis in the mouse retina (Stefater et al. 2011). In other cells such as CD14+ monocytes, VEGFR1 is considered to be important for angiogenesis and arteriogenesis processes (Waltenberger et al. 2000; Waltenberger, 2009). In monocytes, VEGFR1 specific ligands VEGFB and PIGF induce Ras-MAPK pathway, PI3K/AKT pathway and the p38MAPK pathway and are implicated in the regulation of monocyte chemotaxis (Tchaikovski et al. 2009). VEGFR1 was shown to be also involved in inflammation-induced angiogenesis and it was shown that PIGF plays an important role in VEGFR1 induced inflammation associated angiogenesis (Carmeliet et al. 2001). VEGFR1 tyrosine kinase domain-deficient mice (VEGFR1 TK-/-) show impaired angiogenesis and inflammation in different disease models such as atherosclerosis (Zhao et al. 2004), rheumatoid arthritis (Murakami et al. 2006), chorioidal neo-vascularization (Kami et al. 2008), cerebral ischemia (Beck et al. 2010). It has been suggested that VEGFR1 exerts these effects by regulating the recruitment of bone marrow-derived monocytes/macrophages, which in turn regulate angiogenesis by secreting various angiogenic growth factors.

VEGFR2 is expressed in almost all adult vascular endothelial cells as well as in circulating endothelial progenitor cells (Millauer et al. 1993; Oelrichs et al. 1993; Quinn et al. 1993). In addition VEGFR2 expression is also induced in the uterus during the reproductive cycle (Tasaki et al. 2010). VEGFR2 also plays an important role in angiogenesis. VEGFR2 deficient mice die at E8.5 due to defects in the development of hematopoietic and endothelial cells (Shalaby et al. 1995). The *vegfr2-/-* mice have a phenotype highly reminiscing that of the *vegfa-/-* mice. Therefore it has been suggested that VEGFR2 is the main signal transducer of VEGFA in endothelial cells and regulates EC proliferation, migration, tube formation and sprouting. Interestingly, alternative splicing of VEGFR2 results in a soluble form of VEGFR2 (solVEGFR2), which is present in various tissues. It was demonstrated that solVEGFR2 binds VEGFC and in this way prevents activation of VEGFR3 and lymphatic endothelial cell proliferation (Albuquerque et al. 2009). In addition, solVEGFR2 may also contribute to vessel maturation by regulating mural cell migration and vessel coverage (Lorquet et al. 2010).

VEGFR3 (FLT4) plays critical role during lymphangiogenesis (Tammela and Alitalo, 2010) as well as in regulation of blood vascular development (Tammela et al. 2008). VEGFR3 regulates during development the formation of the primary vascular plexus (Kukk et al. 1996) while at later stages it is expressed in venous ECs of the cardinal vein, which subsequently gives rise to the VEGFR3-expressing lymphatics (Kaipainen et al. 1995). VEGFR3 deficient mice die at E10.5 (before the formation of lymphatics) due to impaired organization of the vasculature and cardiovascular defects (Dumont et al. 1998). These data show a central role of VEGFR3 in angiogenesis during embryonic development. Postnatally, VEGFR3 has an essential role in lymphatic endothelial cells (Tammela and Alitalo 2010; Tammela et al. 2011). Interestingly, VEGFR3 expression is also induced in endothelial cells at sites of active angiogenesis (Carmeliet et al. 2009) such as in chronic inflammatory wounds and in tumor vessels (Laakkonen et al. 2007; Paavonen et al. 2000). In particular, VEGFR3 is highly expressed in tip ECs and is required for EC sprouting in both mice and zebrafish (Siekmann and Lawson 2007; Tammela et al. 2008). In addition, it was shown that VEGFR3 is also expressed in other cells such as osteoblasts (Orlandini et al. 2006) and macrophages (Schmeisser et al. 2006), which indirectly support angiogenesis (Chung et al. 2009). Recent studies have demonstrated that there is molecular communication (receptor heterodimerization) between VEGFR3 and VEGFR2 in the angiogenic process (Hamada et al. 2000; Tammela et al. 2008; Nilsson et al. 2010). Studies in transgenic mice expressing mutated or truncated forms of VEGFR3 have shown that although the VEGFR3 kinase activity is required for lymphatic development is not important for blood vessel development (Zhang et al. 2010). These data further supported the notion that VEGFR3 on ECs regulates VEGFR2 signaling through receptor heterodimerization (Nilsson et al. 2010). It was suggested that VEGFC promotes the assembly of VEGFR2–VEGFR3 heterodimers, in tip ECs and by this way positively regulates EC sprouting and angiogenesis (Nilsson et al. 2010). Thus, VEGFR3 expression may be an important determinant of tip EC function.

As mentioned earlier sprouting ECs are categorized into leading tip-ECs and trailing stalk-ECs that exhibit very distinct and specialized cell actions. During sprouting angiogenesis a feedback loop between VEGF and Notch signaling regulates EC specification to tip or stalk-ECs. VEGF gradients and increased VEGFR2/VEGFR3 signaling induces the invasive, motile tip-EC phenotype which drives angiogenesis (Gerhardt et al. 2003; Carmeliet et al. 2009). VEGFR2 induces high expression of Deltalike ligand 4 (Dll4) on the tip-ECs and increased Notch signaling in the stalk-ECs and consequently in reduced VEGFR2 and VEGFR3 expression. In addition, Notch signaling in stalk-ECs induces the expression of VEGFR1 and solVEGFR1, which represses VEGFR2 function and blocks tip-EC function and promotes EC proliferation and stalk-EC behavior (Gerhardt et al. 2003; Carmeliet et al. 2009).

NRP1 and NRP2 co-receptors form complexes with various VEGFRs and by binding to VEGFA165 enhance VEGFR-mediated signaling and modulate angiogenic sprouting (Neufeld et al. 2002; Pan et al. 2007b). Genetic studies in mice clearly demonstrated the critical role of NRPs in vascular development. Nrp1-defficient mice display disrupted EC remodeling and vessel branching during development (Kawasaki et al. 1999; Gerhardt et al. 2004). Nevertheless, vascular defects in Nrp1 knockout mice are not severe, suggesting that the role of NRP during tip-EC specification and sprouting might not be crucial. Disruption of NRP1 leads to embryonic lethality at E12.5 to E13.5 due to excess of vessel formation (Fujisawa and Kitsukawa 1998) Recent reports have indicated a pro-tumor activity for NRPs as NRPs are expressed on the cell surface of several tumor cell lines (Klagsbrun and Eichmann 2005).

7   VEGF Signaling and Tumor Angiogenesis

Several anti-tumor therapeutic approaches, such as chemotherapy, rely on the expectation that the drugs will preferentially kill only the rapidly dividing cancer cells rather than the normal cells. In addition, tumor therapies using chemical kinase inhibitors are designed to selectively inhibit tumor cells. However, most of the current anti-tumor therapies lack selectivity due to the lack of specific tumor markers. These approaches lead to high toxicity in normal tissues with high proliferation rates. Tumors are characterized by high interstitial pressure due to the irregular tumor vasculature. This impairs the delivery of active agents to tumor sites and by this way further decreases the efficiency of the conventional therapies (Bosslet et al. 1998). Since early 70's it was suggested that inhibition of tumor angiogenesis may be an effective treatment for all types of cancers. This concept was based on the realization that growth of solid tumors depends on their capacity to acquire a blood supply (Folkman 1971) and that EC, compared to tumor cells, are genetically stable (Denekamp 1990). Subsequently, several studies have focused on the development of therapeutic interventions which will target EC function and angiogenesis (Table 1 and Fig. 1B).

Table 1. Anti-VEGF angiogenesis inhibitors in preclinical and clinical development.

	Specific targets	Clinical status	References
Monoclonal antibodies			
Bevacizumab/Avastin	hVEGF	Approved for CC (2004), NSCLC (2006), BC (2008), RCC and glioblastoma (2009)	(Van Meter and Kim 2010)
IMC-18F1/ MF-1	VEGFR1	Phase I	(Wu et al. 2006)
Ramucirumab/IMC-1121B	VEGFR2	Phase II/III	(Spratlin 2011)
CDP791	VEGFR2	Phase II	(Youssoufian et al. 2007)
Anti-VEGFR3 ab/mF4-31C1	VEGFR3	Preclinical	(Burton et al. 2008, Roberts et al. 2006)
Anti-NRP1 ab	NRP1	Preclinical	(Pan et al. 2007a)
Anti-NRP2 ab	NRP2	Preclinical	(Caunt et al. 2008)
Humanized rabbit monoclonal anti-VEGF abs	VEGF	Preclinical	(Yu et al. 2010)
Fusion proteins			
VEGF trap/aflibercept	VEGF, PlGF	Phase II/III	(Teng et al. 2010)
VEGFC/D trap	VEGFC, VEGFD	Preclinical	(Burton et al. 2008, Lin et al. 2005)
Oligonucleotide			
VEGF-AS/Veglin	VEGFA, VEGFC, VEGFD	Phase I	(Levine et al. 2006)
RNA interference			
VEGF shRNA	VEGFA mRNA	Preclinical	(Burton et al. 2008)
VEGFR2 siRNA	VEGFR2 mRNA	Preclinical	(Wang et al. 2009)
VEGFC siRNA	VEGFC mRNA	Preclinical	(Shibata et al. 2008)
RTKI			
Sunitinib/ SU11248 (Sutent)	VEGFRs, PDGFRs, c-Kit, Flt-3	Approved for RCC and GIST (2004)	(Sulkes 2010)
Sorafenib/ Bay43–9006 (Nexavar)	VEGFR2–3, PDGFR, Raf-1, Flt-3, c-Kit	Approved for RCC (2005) and GIST (2008)	(Sulkes 2010)
Axitinib/ AG013736	VEGFRs, PDGFR, c-Kit	Phase II/III	(Kelly et al. 2010)
pazopanib (Votrient)	VEGFRs, PDGFR, Flt-3, c-Kit	Approved for RCC (2009)	(Sternberg et al. 2010)
Cediranib/AZD2171 (Resentin)	VEGFRs, c-Kit	Phase II/III	(Lindsay et al. 2009)
Vandetanib/ZD6474 (Zactima)	VEGFR2–3, EGFR, RET	Phase II/III	(Morabito et al. 2009)
AMG706/motesanib	VEGFRs, PDGFR, c-Kit	Phase II/III	(Blumenschein et al. 2010)
Brivanib alanitate	VEGFR-2, FGFR-1	Phase II/III	(Diaz-Padilla and Siu 2011)

Linifanib/ABT869	VEGFRs, PDGFR, Flt-3	Phase II/III	(None 2010)
Tivozanib/AV-951	VEGFRs, PDGFR	Phase II/III	(De Luca and Normanno 2010)
Vatalanib/PTK787	VEGFRs, PDGFRs, c-Kit	Phase II	(Scott et al. 2007)
TKI258 (Dovitinib)	VEGFRs, PDGFR, FGFRs	Phase II	(Kim et al. 2011)
CEP-11981	VEGFRs	Phase I	www.cancer.gov

Tumor vessels are disorganized and morphologically abnormal and they are characterized by decreased pericyte coverage (Baluk et al. 2005) and poor blood flow (Fukumura and Jain 2007), as a result tumor blood vessels are leaky. This is also due to increased expression of VEGFA in the growing tumor. Tumor derived VEGFA promotes tumor EC proliferation, migration and permeability, while inhibiting vessel maturation (Greenberg et al. 2008; Carmeliet and Jain 2011b). In addition, VEGFA inhibits dendritic cell maturation and T cell development (Gabrilovich et al. 1996, Ohm et al. 2003). Furthermore, VEGFA recruits VEGFR1 expressing myeloid cells, which can further promote tumor vascularization (Rafii et al. 2002). Recent evidence has shown that tumor cells may also express VEGFRs, suggesting that VEGFA, by acting directly on tumor cells, could promote tumor growth and invasion (Lee et al. 2007b). In addition, VEGFA expression on cancer stem cells promotes cancer stemness and renewal through NRP1 (Beck et al. 2011). Therefore, it was suggested that targeting VEGFA in the tumor microenvironment might deteriorate tumor growth and inhibit metastatic spread and as a result improve treatment efficacy. Indeed, it was first demonstrated in animal models that VEGFA neutralizing antibodies inhibited tumor angiogenesis and tumor growth (Kim et al. 1993). This led to the development of humanized antibody Bevacizumab (Avastin), which specifically inhibits human VEGFA. Bevacizumab is the most successfully marketed compounds within the anti-angiogenic therapy field. Several other VEGFA neutralizing antibodies have been developed which are now in clinical trials. Another approach to inhibit VEGF and tumor angiogenesis is the use of soluble receptors which will bind, sequester and inhibit VEGF ligands from binding to their receptors. Since VEGFR1 binds VEGF with higher affinity, several studies have developed solVEGFR1 ligand-traps to inhibit tumor angiogenesis, tumor growth and metastasis (Mahasreshti et al. 2001; Bagley et al. 2011). In addition, a ligand trap containing extracellular parts of VEGFR1 and VEGFR2 fused to the Fc region of human immunoglobulin G1 antibody suppressed tumor vascularization and growth of tumors from different tissues and species (Holash et al. 2002). The clinical use of VEGF-trap in tumor development is still under evaluation.

PlGF was also shown to play an important role in tumor angiogenesis and tumor progression (Failla et al. 2000; Carmeliet et al. 2001; Green et al. 2001; Larcher et al. 2003; Fischer et al. 2007). Additionally certain types of human tumors produce high levels of PlGF, such as breast cancer (Parr et al. 2005). Overexpression of PlGF in a mouse melanoma model resulted in increased tumor growth and metastasis (Li et al. 2006). However, its overexpression in tumors which produce high VEGFA levels result in inactive heterodimers and decreased tumor growth (Xu et al. 2006). It has been shown that PlGF acts as a chemotactic and survival factor on endothelial cells, mural cells and VEGFR1-expressing tumor cells (Ziche et al. 1997; Yonekura et al. 1999; Fischer et al. 2007). In addition, it was shown that PlGF modulates immune response by recruiting VEGFR1-positive bone marrow cells to sites of neovascularization (Hattori et al. 2002; Dikov et al. 2005). Based on these observations, PlGF has been regarded as an attractive candidate for anti-angiogenic therapy. Indeed, PlGF neutralizing antibodies inhibit tumor growth as well as metastasis in different mouse tumor models (Fischer et al. 2007, van de Veire et al. 2010). Interestingly, anti-PlGF therapy enhanced the efficacy of chemotherapy and anti-VEGFR2 treatment. In contrast other studies provided evidence that although neutralization of PlGF inhibited metastasis, it had no significant effects on primary tumor growth in a number of tumor models (Bais et al. 2010). VEGFB, another VEGFR1 ligand, is highly expressed in a wide range of tumors and in some cases correlates with diseases expression (Eggert et al. 2000; Gunningham et al. 2001; Kawakami et al. 2003; Kanda et al. 2008), nevertheless its role in cancer has not yet been conclusively delineated.

Several studies suggested the involvement of VEGFR1 in tumor growth. VEGFR1 TK deficient mice show impaired tumor growth (Hiratsuka et al. 2001; Kerber et al. 2008; Muramatsu et al. 2010) and metastasis (Hiratsuka et al. 2002). Moreover, enhanced lung metastasis was associated with infiltration of VEGFR1-expressing monocytes/macrophages in the lungs (Hiratsuka et al. 2002). In addition a population of VEGFR1-expressing bone marrow-derived cells, referred to as "premetastatic niche", are mobilized to the lungs to provide a metastatic site for the original tumor (Kaplan et al. 2005). However, other studies have suggested that, although VEGFR1 receptor activity is important for metastatic dissemination, it is not required for premetastatic priming (Dawson et al. 2009a; Dawson et al. 2009b). Moreover, it was suggested that expression of VEGFR1 on tumor cells might increase tumor invasiveness (Seto et al. 2006; Mylona et al. 2007). Based on these results VEGFR1 is an attractive potential target for anti-tumor angiogenesis target, to inhibit tumor growth. Neutralizing antibodies against VEGFR1 inhibit tumor growth in preclinical models (Luttun et al. 2002; Wu et al. 2006; Lee et al. 2010). Moreover, peptides inhibiting binding of PlGF or VEGF to VEGFR1 resulted in reduced tumor growth in different xenograft models (Bae et al. 2005; Taylor and Goldenberg 2007; Giordano et al. 2010). Nevertheless, inhibition of VEGFR1 alone is not always sufficient to inhibit tumor growth. Thus, blocking simultaneously both VEGFR1 and VEGFR2 may result in more efficient tumor suppression. VEGFR1 is also found to be responsible for the initiating signals for migration and invasion of several cancer cell lines through the activation of Src and ERK pathways (Taylor et al. 2010).

VEGFR2 expression, often in combination with VEGFR3, is significantly upregulated in tumor vasculature in most common human solid cancers (Plate et al. 1993; Millauer et al. 1994; Smith et al. 2010). Although tumor cells typically express VEGFR1, tumor cells such as melanoma and hematological malignancies may also express VEGFR2 (Hicklin and Ellis 2005; Podar and Anderson 2005; Youssoufian et al. 2007). Downregulation of sVEGFR2 in advanced metastatic neuroblastoma, may promote lymphogenic spread of metastases (Becker et al. 2010). One of the essential biological roles of VEGFR2 *in vivo* is to promote survival of endothelial cells. Thus, blocking VEGFR2 kinase activity using small molecular weight inhibitors leads to regression of tumor capillaries and reduced tumor growth (Kamba et al. 2006). Preclinical studies in mice using monoclonal antibodies that specifically block the interaction of VEGF ligands to VEGFR2 has validated the importance of VEGFR2 as a therapeutic target to inhibit tumor angiogenesis and tumor growth (Prewett et al. 1999; Bruns et al. 2000; Tong et al. 2004). Humanized anti-VEGFR2 antibodies such as IMC-1121B (Ramucirumab) and CDP791 have been developed for clinical use (Youssoufian et al. 2007; Krupitskaya and Wakelee 2009). Ramucirumab is currently tested in several clinical trials, (Spratlin 2011) and it shows activity in patients previously treated with other anti-angiogenic agents, suggesting that direct targeting of VEGFR2 has more efficient anti-tumor effects. In addition, small chemical inhibitors, known as Receptor Tyrosine Kinase Inhibitors (RTKIs), which block VEGFR2 kinase activity, have been extensively evaluated. RTKs inhibitors have in general a broad spectrum of action due to the high degree of homology of the ATP binding site in the kinase domain of tyrosine kinases (Gotink and Verheul 2010). Thus RTK inhibitors target not only VEGFRs but also PDGF, EGF, c-kit and other receptors and in this way they can disrupt several independent signaling cascades at once (Kerbel 2000; Adams et al. 2002; Gotink and Verheul 2010). Although, in general this ability has been considered as a positive characteristic, studies have shown that these molecules might have side effects associated with their "off-target" effects on other kinases (Adams et al. 2002). Of the most advanced RTK inhibitors used in clinical trials are pazopanib (Votrient, GlaxoSmithKline), sorafenib (Nexavar, Bayer) and Sunitinib (Sutent, Pfizer), which inhibit not only VEGFR2, but also PDGFRs, c-Kit, and FLT3 (Sulkes 2005; Kumar et al. 2009). Other RTK inhibitors such as axitinib, cediranib, vatalanib, linifanib, tivozanib and motesanib have shown antitumor activity in several tumor types, and are used in clinical trials (Kiselyov et al. 2007).

VEGFR3 plays also an important role in the tumor microenvironment (He et al. 2004). Tumor cells produce VEGFC, which induces lymphatic endothelial destabilization, endothelial sprouting, leakage and enlargement of the vessels. This facilitates entry of tumor cells into the lymphatics and dissemination of metastatic cells to sites of metastasis (He et al. 2005; Achen and Stacker 2008). As VEGFR3 is expressed both in blood and lymphatic endothelial cells, it comprises an attractive therapeutic target. Preclinical studies have demonstrated that the use of VEGFR3 neutralizing antibodies or soluble VEGFR3 result in reduced lymphatic metastasis up to 50–70% (Tammela and Alitalo 2010;, Tvorogov et al. 2010). Additionally, small molecular weight chemical inhibitors that block VEGFR3 signaling by competing for the ATP-binding site in the kinase domain are available, although they do not primarily target VEGFR3 (Wood et al. 2000; Heckman et al. 2008; Luo et al. 2011) they may have potential in cancer treatment by targeting lymphatic sprouting.

8 Conclusions and Perspectives

Since the realization that antiangiogenic therapies targeting the non-malignant and genetically stable tumor endothelial cells may be more efficient in targeting tumor growth, several studies have focused on the development of anti angiogenic drugs. Many of them have focused on targeting VEGF signaling the master regulator of angiogenesis (Table 1). Nevertheless, despite the positive results from preclinical studies in mice, the initial results from clinical studies with drugs selectively targeting the tumor neovasculature, such as bevacizumab, sunitinib and sorafenib, has been disappointing since their effects are not great and the prolongation of progression-free survival is typically only 2–6 months. In addition, some of them resulted in the development of evasive resistance. The approval of some of them such as bevacizumab was recently revoked by the United States Food and Drug Administration (Twombly 2011). Studies in mice revealed that anti-angiogenic treatments targeting the VEGF signaling lead to therapy induced resistance and the development of tumors which are refractory to anti-angiogenic therapy (Bergers and Hanahan 2008). It was suggested that adaptation to anti-VEGF therapy is due substitution of VEGF by other compensatory pro-angiogenic factors such as bFGF (Casanovas et al. 2005), induced neovascularization by bone marrow derived EC progenitors, increased pericyte coverage of the blood vessels as well as increased invasiveness (Pietras and Hanahan 2005) and metastatic spread due to the hypoxic environment in the tumor (Ebos et al. 2009; Paez-Ribes et al. 2009). Although the exact molecular mechanisms which mediate resistance to anti-angiogenic therapies are not fully characterized, it is clear that testing therapeutic cocktails targeting different aspects of the tumor angiogenic responses may provide better therapeutic interventions.

9. Acknowledgements

Our studies on the role of growth factor signaling in cardiovascular diseases are supported by the Interdisziplinäres Zentrum für Klinische Forschung (IZKF) Münster and by the 'Innovative Medizinische Forschung' (IMF) programme of the Medical Faculty of the University of Münster.

REFERENCES

Aase, K., G. von Euler, X. Li, A. Ponten, P. Thoren, R. Cao, Y. Cao, B. Olofsson, S. Gebre-Medhin, M. Pekny, K. Alitalo, C. Betsholtz and U. Eriksson (2001). "Vascular endothelial growth factor-B-deficient mice display an atrial conduction defect." Circulation 104: 358-364.

Abu-Ghazaleh, R., J. Kabir, H. Jia, M. Lobo and I. Zachary (2001). "Src mediates stimulation by vascular endothelial growth factor of the phosphorylation of focal adhesion kinase at tyrosine 861, and migration and anti-apoptosis in endothelial cells." Biochem J 360: 255-264.

Achen, M. G. and S. A. Stacker (2008). "Molecular control of lymphatic metastasis." Ann NY Acad Sci 1131: 225-234.

Adams, J., P. Huang and D. Patrick (2002). "A strategy for the design of multiplex inhibitors for kinase-mediated signalling in angiogenesis." Curr Opin Chem Biol 6: 486-492.

Adams, R. H. and K. Alitalo (2007). "Molecular regulation of angiogenesis and lymphangiogenesis." Nat Rev Mol Cell Biol 8: 464-478.

Albuquerque, R. J., T. Hayashi, W. G. Cho, M. E. Kleinman, S. Dridi, A. Takeda, J. Z. Baffi, K. Yamada, H. Kaneko, M. G. Green, J. Chappell, J. Wilting, H. A. Weich, S. Yamagami, S. Amano, N. Mizuki, J. S. Alexander, M. L. Peterson, R. A. Brekken, M. Hirashima, S. Capoor, T. Usui, B. K. Ambati and J. Ambati (2009). "Alternatively spliced vascular endothelial growth factor receptor-2 is an essential endogenous inhibitor of lymphatic vessel growth." Nat Med 15: 1023-1030.

Autiero, M., J. Waltenberger, D. Communi, A. Kranz, L. Moons, D. Lambrechts, J. Kroll, S. Plaisance, M. De Mol, F. Bono, S. Kliche, G. Fellbrich, K. Ballmer-Hofer, D. Maglione, U. Mayr-Beyrle, M. Dewerchin, S. Dombrowski, D. Stanimirovic, P. Van Hummelen, C. Dehio, D. J. Hicklin, G. Persico, J. M. Herbert, M. Shibuya, D. Collen, E. M. Conway and P. Carmeliet (2003). "Role of PlGF in the intra- and intermolecular cross talk between the VEGF receptors Flt1 and Flk1." Nat Med 9: 936-943.

Bae, D. G., T. Kim, G. Li, W. H. Yoon and C. B. Chae (2005). "Anti-flt1 peptide, a vascular endothelial growth factor receptor 1-specific hexapeptide, inhibits tumor growth and metastasis." Clin Cancer Res 11: 2651-2661.

Bagley, R. G., L. Kurtzberg, W. Weber, T. H. Nguyen, S. Roth, R. Krumbholz, M. Yao, B. Richards, M. Zhang and P. Pechan (2011). "sFLT01: a novel fusion protein with antiangiogenic activity." Mol Cancer Ther 10: 404-415.

Bais, C., X. Wu, J. Yao, S. Yang, Y. Crawford, K. McCutcheon, C. Tan, G. Kolumam, J. M. Vernes and J. Eastham-Anderson (2010). "PlGF blockade does not inhibit angiogenesis during primary tumor growth." Cell 141: 166-177.

Baluk, P., H. Hashizume and D. M. McDonald (2005). "Cellular abnormalities of blood vessels as targets in cancer." Curr Opin Genet Dev 15: 102-111.

Banai, S., D. Shweiki, A. Pinson, M. Chandra, G. Lazarovici and E. Keshet (1994). "Upregulation of vascular endothelial growth factor expression induced by myocardial ischaemia: implications for coronary angiogenesis." Cardiovasc Res 28: 1176-1179.

Bates, D. O., T. G. Cui, J. M. Doughty, M. Winkler, M. Sugiono, J. D. Shields, D. Peat, D., Gillatt and S. J. Harper (2002). "VEGF165b, an inhibitory splice variant of vascular endothelial growth factor, is down-regulated in renal cell carcinoma." Cancer Res 62: 4123-4131.

Beck, B., G. Driessens, S. Goossens, K. K. Youssef, A. Kuchnio, A., Caauwe, P. A. Sotiropoulou, S. Loges, G. Lapouge, A. Candi, G. Mascre, B. Drogat, S. Dekoninck, J. J. Haigh, P. Carmeliet and C. Blanpain (2011). "A vascular niche and a VEGF-Nrp1 loop regulate the initiation and stemness of skin tumours." Nature 478: 399-403.

Beck, H., S. Raab, E. Copanaki, M. Heil, A. Scholz, M. Shibuya, T. Deller, M. Machein and K. H. Plate (2010). "VEGFR-1 signaling regulates the homing of bone marrow-derived cells in a mouse stroke model." J Neuropathol Exp Neurol 69: 168-175.

Becker, J., H. Pavlakovic, F. Ludewig, F. Wilting, H. A. Weich, R. Albuquerque, J. Ambati and J. Wilting (2010). "Neuroblastoma progression correlates with downregulation of the lymphangiogenesis inhibitor sVEGFR-2." Clin Cancer Res 16: 1431-1441.

Bellomo, D., J. P. Headrick, G. U. Silins, C. A. Paterson, P. S. Thomas, M. Gartside, A. Mould, M. M. Cahill, I. D. Tonks, S. M. Grimmond, S. Townson, C. Wells, M. Little, M. C. Cummings, N. K. Hayward and G. F. Kay (2000). "Mice lacking the vascular endothelial growth factor-B gene (Vegfb) have smaller hearts, dysfunctional coronary vasculature, and impaired recovery from cardiac ischemia." Circ Res 86: E29-35.

Ben-Av, P., L. J. Crofford, R. L Wilder and T. Hla (1995). "Induction of vascular endothelial growth factor expression in synovial fibroblasts by prostaglandin E and interleukin-1: a potential mechanism for inflammatory angiogenesis." FEBS Lett 372: 83-87.

Bergers, G. and D. Hanahan. 2008. "Modes of resistance to anti-angiogenic therapy." Nat Rev Cancer 8: 592-603.

Blumenschein, G. R. Jr., K. Reckamp, G. J. Stephenson, T. O'Rourke, G. Gladish, J. McGreivy, Y. N. Sun, Y. Ye, M. Parson and A. Sandler (2010). "Phase 1b study of motesanib, an oral angiogenesis inhibitor, in combination with carboplatin/paclitaxel and/or panitumumab for the treatment of advanced non-small cell lung cancer." Clin Cancer Res 16: 279-290.

Bosslet, K., R. Straub, R. Blumrich, J. Czech, M. Gerken, B. Sperker, H. K. Kroemer, J. P. Gesson, M. Koch and C. Monneret (1998). "Elucidation of the mechanism enabling tumor selective prodrug monotherapy." Cancer Res 58: 1195-1201.

Bruns, C. J., W. Liu, D. W. Davis, R. M. Shaheen, D. J. McConkey, M. R. Wilson, C. D. Bucana, D. J. Hicklin and L. M. Ellis (2000). "Vascular endothelial growth factor is an *in vivo* survival factor for tumor endothelium in a murine model of colorectal carcinoma liver metastases." Cancer 89: 488-499.

Burton, J. B., S. J. Priceman, J. L. Sung, E. Brakenhielm, D. S. An, B. Pytowski, K. Alitalo and L. Wu (2008). "Suppression of prostate cancer nodal and systemic metastasis by blockade of the lymphangiogenic axis." Cancer Res 68: 7828-7837.

Cai, J., S. Ahmad, W. G. Jiang, J. Huang, C. D. Kontos, M. Boulton and A. Ahmed (2003). "Activation of vascular endothelial growth factor receptor-1 sustains angiogenesis and Bcl-2 expression via the phosphatidylinositol 3-kinase pathway in endothelial cells." Diabetes 52: 2959-2968.

Carlevaro, M. F., S. Cermelli, R. Cancedda and F. Descalzi Cancedda (2000). "Vascular endothelial growth factor (VEGF) in cartilage neovascularization and chondrocyte differentiation: auto-paracrine role during endochondral bone formation." J Cell Sci 113: 56-59.

Carmeliet, P., F. De Smet, S. Loges and M. Mazzone (2009). "Branching morphogenesis and antiangiogenesis candidates: tip cells lead the way." Nat Rev Clin Oncol 6: 315-326.

Carmeliet, P., V. Ferreira, G. Breier, S. Pollefeyt, L. Kieckens, M. Gertsenstein, M. Fahrig, A. Vandenhoeck, K. Harpal, C. Eberhardt, C. Declercq, J. Pawling, L. Moons, D. Collen, W. Risau and A. Nagy (1996). "Abnormal blood vessel development and lethality in embryos lacking a single VEGF allele." Nature 380: 435-439.

Carmeliet, P. and R. K. Jain (2011a). "Molecular mechanisms and clinical applications of angiogenesis." Nature 473: 298-307.

Carmeliet, P. and R. K. Jain (2011b). "Principles and mechanisms of vessel normalization for cancer and other angiogenic diseases." Nat Rev Drug Discov 10: 417-427.

Carmeliet, P., L. Moons, A. Luttun, V. Vincenti, V. Compernolle, M. De Mol, Y. Wu, F. Bono, L. Devy, H. Beck, D. Scholz, T. Acker, T. DiPalma, M. Dewerchin, A. Noel, I. Stalmans, A. Barra, S. Blacher, T. Vandendriessche, A. Ponten, U. Eriksson, K. H. Plate, J. M. Foidart, W. Schaper, D. S. Charnock-Jones, D. J. Hicklin, J. M. Herbert, D. Collen and M. G. Persico (2001). "Synergism between vascular endothelial growth factor and placental growth factor contributes to angiogenesis and plasma extravasation in pathological conditions." Nat Med 7: 575-583.

Casanovas, O., D. J. Hicklin, G. Bergers and D. Hanahan (2005). "Drug resistance by evasion of antiangiogenic targeting of VEGF signaling in late-stage pancreatic islet tumors." Cancer Cell: 8, 299-309.

Caunt, M., J. Mak, W. C. Liang, S. Stawicki, Q. Pan, R. K. Tong, J. Kowalski, C. Ho, H. B. Reslan, J. Ross, L. Berry, I. Kasman, C. Zlot, Z. Cheng, J. Le Couter, E. H. Filvaroff, G. Plowman, F. Peale, D. French, R. Carano, A. W. Koch, Y. Wu, R. J. Watts, M. Tessier-Lavigne and A. Bagri (2008). "Blocking neuropilin-2 function inhibits tumor cell metastasis." Cancer Cell 13: 331-342.

Chung, E. S., S. K. Chauhan, Y. Jin, S. Nakao, A. Hafezi-Moghadam, N. van Rooijen, Q. Zhang, L. Chen and R. Dana (2009). "Contribution of macrophages to angiogenesis induced by vascular endothelial growth factor receptor-3-specific ligands." Am J Pathol 175: 1984-1992.

Clark, D. E., S. K. Smith, Y. He, K. A. Day, D. R. Licence, A. N. Corps, R. Lammoglia and D. S. Charnock-Jones (1998). "A vascular endothelial growth factor antagonist is produced by the human placenta and released into the maternal circulation." Biol Reprod 59: 1540-1548.

Conn, G., M. L. Bayne, D. D. Soderman, P. W. Kwok, K. A. Sullivan, T. M. Palisi, D. A. Hope and K. A. Thomas (1990). "Amino acid and cDNA sequences of a vascular endothelial cell mitogen that is homologous to platelet-derived growth factor." Proc Natl Acad Sci USA 87: 2628-2632.

Coultas, L., K. Chawengsaksophak and J. Rossant (2005). "Endothelial cells and VEGF in vascular development." Nature 438: 937-945.

Dawson, M. R., D. G. Duda, S. S. Chae, D. Fukumura and R. K. Jain (2009a). "VEGFR1 activity modulates myeloid cell infiltration in growing lung metastases but is not required for spontaneous metastasis formation." PLoS One 4: e6525.

Dawson, M. R., D. G. Duda, D. Fukumura and R. K. Jain (2009b). "VEGFR1-activity-independent metastasis formation." Nature 461: E4; discussion E5.

De Luca, A. and N. Normanno (2010). "Tivozanib, a pan-VEGFR tyrosine kinase inhibitor for the potential treatment of solid tumors." IDrugs 13: 636-645.

De Smet, F., I. Segura, K. De Bock, P. J. Hohensinner and P. Carmeliet (2009). "Mechanisms of vessel branching: filopodia on endothelial tip cells lead the way." Arterioscler Thromb Vasc Biol 29: 639-649.

Denekamp, J. (1990). "Vascular attack as a therapeutic strategy for cancer." Cancer Metastasis Reviews 9: 267-282.

Deveraux, Q. L., N. Roy, H. R. Stennicke, T. Van Arsdale, Q. Zhou, S. M. Srinivasula, E. S. Alnemri, G. S. Salvesen and J. C. Reed (1998). "IAPs block apoptotic events induced by caspase-8 and cytochrome c by direct inhibition of distinct caspases." EMBO J 17: 2215-2223.

Diaz-Padilla, I., and L. L. Siu (2011). "Brivanib alaninate for cancer." Expert Opin Investig Drugs 20: 577-586.

Dikov, M. M., J. E. Ohm, N. Ray, E. E. Tchekneva, J. Burlison, D. Moghanaki, S. Nadaf and D. P. Carbone (2005). "Differential roles of vascular endothelial growth factor receptors 1 and 2 in dendritic cell differentiation." J Immunol 174: 215-222.

DiSalvo, J., M. L. Bayne, G. Conn, P. W. Kwok, P. G. Trivedi, D. D. Soderman, T. M. Palisi, K. A. Sullivan and K. A. Thomas (1995). "Purification and characterization of a naturally occurring vascular endothelial growth factor · placenta growth factor." J Biol Chem 270: 7717-7723.

Dumont, D. J., L. Jussila, J. Taipale, A. Lymboussaki, T. Mustonen, K. Pajusola, M. Breitman and K.Alitalo (1998). "Cardiovascular failure in mouse embryos deficient in VEGF receptor-3." Science 282: 946-949.

Ebos, J. M., C. R. Lee and R. S. Kerbel (2009). "Tumor and host-mediated pathways of resistance and disease progression in response to antiangiogenic therapy." Clin Cancer Res 15: 5020-5025.

Eggert, A., N. Ikegaki, J. Kwiatkowski, H. Zhao, G. M. Brodeur and B. P Himelstein (2000). "High-level expression of angiogenic factors is associated with advanced tumor stage in human neuroblastomas." Clin Cancer Res 6: 1900-1908.

Eremina, V., M. Sood, J. Haigh, A. Nagy, G. Lajoie, N. Ferrara, H. P. Gerber, Y. Kikkawa, J. H. Miner and S. E. Quaggin (2003). "Glomerular-specific alterations of VEGF-A expression lead to distinct congenital and acquired renal diseases." The Journal of Clinical Investigation 111: 707-716.

Failla, C. M., T. Odorisio, F. Cianfarani, C. Schietroma, P. Puddu and G. Zambruno (2000). "Placenta growth factor is induced in human keratinocytes during wound healing." J Invest Dermatol 115: 388-395.

Favier, B., A. Alam, P. Barron, J. Bonnin, P. Laboudie, P. Fons, M. Mandron, J. P. Herault, G. Neufeld, P. Savi, J. M. Herbert and F. Bono (2006). "Neuropilin-2 interacts with VEGFR-2 and VEGFR-3 and promotes human endothelial cell survival and migration." Blood 108: 1243-1250.

Ferrara, N. and W. J. Henzel (1989). "Pituitary follicular cells secrete a novel heparin-binding growth factor specific for vascular endothelial cells." Biochem Biophys Res Commun 161: 851-858.

Ferrara, N., K. Carver-Moore, H. Chen, M. Dowd, L. Lu, K. S. O'Shea, L. Powell-Braxton, K. J. Hillan and M. W. Moore (1996). "Heterozygous embryonic lethality induced by targeted inactivation of the VEGF gene." Nature 380: 439-442.

Ferrara, N. (2010). "Binding to the extracellular matrix and proteolytic processing: two key mechanisms regulating vascular endothelial growth factor action." Mol Biol Cell 21: 687-690.

Fischer, C., B. Jonckx, M. Mazzone, S. Zacchigna, S. Loges, L. Pattarini, E. Chorianopoulos, L. Liesenborghs, M. Koch, M. De Mol, M. Autiero, S. Wyns, S. Plaisance, L. Moons, N. van Rooijen, M. Giacca, J.M. Stassen, M. Dewerchin, D. Collen and P. Carmeliet (2007). "Anti-PlGF inhibits growth of VEGF(R)-inhibitor-resistant tumors without affecting healthy vessels." Cell 131: 463-475.

Folkman, J. (1971). "Tumor angiogenesis: therapeutic implications." N Engl J Med 285: 1182-1186.

Fong, G. H., J. Rossant, M. Gertsenstein and M. L. Breitman (1995). "Role of the Flt-1 receptor tyrosine kinase in regulating the assembly of vascular endothelium." Nature 376: 66-70.

Frank, S., G. Hubner, G. Breier, M. T. Longaker, D. G. Greenhalgh and S. Werner (1995). "Regulation of vascular endothelial growth factor expression in cultured keratinocytes." Implications for normal and impaired wound healing. J Biol Chem 270: 12607-12613.

Fujisawa, H. and T. Kitsukawa (1998). "Receptors for collapsin/semaphorins." Curr Opin Neurobiol 8: 587-592.

Fukumura, D. and R. K. Jain (2007). "Tumor microenvironment abnormalities: causes, consequences, and strategies to normalize." J Cell Biochem 101: 937-949.

Gabrilovich, D. I., H. L. Chen, K. R. Girgis, H. T. Cunningham, G. M. Meny, S. Nadaf, D. Kavanaugh and D. P. Carbone (1996). "Production of vascular endothelial growth factor by human tumors inhibits the functional maturation of dendritic cells." Nat Med 2: 1096-1103.

Gerber, H. P., K. J. Hillan, A. M. Ryan, J. Kowalski, G. A. Keller, L. Rangell, B. D. Wright, F. Radtke, M. Aguet and N. Ferrara (1999a). "VEGF is required for growth and survival in neonatal mice." Development 126: 1149-1159.

Gerber, H. P., T. H. Vu, A. M. Ryan, J. Kowalski, Z. Werb and N. Ferrara (1999b). "VEGF couples hypertrophic cartilage remodeling, ossification and angiogenesis during endochondral bone formation." Nat Med 5: 623-628.

Gerhardt, H., M. Golding, M. Fruttiger, C. Ruhrberg, A. Lundkvist, A. Abramsson, M. Jeltsch, C. Mitchell, K. Alitalo, D. Shima and C. Betsholtz (2003). "VEGF guides angiogenic sprouting utilizing endothelial tip cell filopodia." J Cell Biol 161: 1163-1177.

Gerhardt, H., C. Ruhrberg, A. Abramsson, H. Fujisawa, D. Shima and C. Betsholtz (2004). "Neuropilin-1 is required for endothelial tip cell guidance in the developing central nervous system." Dev Dyn 231: 503-509.

Giordano, R. J., M. Cardo-Vila, A. Salameh, C. D. Anobom, B. D. Zeitlin, D. H. Hawke, A. P. Valente, F. C. Almeida, J. E. Nor, R. L. Sidman, R. Pasqualini and W. Arap (2010). "From combinatorial peptide selection to drug prototype (I): targeting the vascular endothelial growth factor receptor pathway." Proc Natl Acad Sci USA 107: 5112-5117.

Gotink, K. J. and H. M. Verheul (2010). "Anti-angiogenic tyrosine kinase inhibitors: what is their mechanism of action?" Angiogenesis 13: 1-14.

Green, C. J., P. Lichtlen, N. T. Huynh, M. Yanovsky, K. R. Laderoute, W. Schaffner and B. J. Murphy (2001). "Placenta growth factor gene expression is induced by hypoxia in fibroblasts: a central role for metal transcription factor-1." Cancer Res 61: 2696-2703.

Greenberg, J. I., D. J. Shields, S. G. Barillas, L. M. Acevedo, E. Murphy, J. Huang, L. Scheppke, C. Stockmann, R. S. Johnson, N. Angle and D. A. Cheresh (2008). "A role for VEGF as a negative regulator of pericyte function and vessel maturation." Nature 456: 809-813.

Gunningham, S. P., M. J. Currie, B. Han, B. A. Robinson, P. A. Scott, A. L. Harris and S. B. Fox (2001). "VEGF-B expression in human primary breast cancers is associated with lymph node metastasis but not angiogenesis." J Pathol 193: 325-332.

Hagberg, C. E., A. Falkevall, X. Wang, E. Larsson, J. Huusko, I. Nilsson, L. A. van Meeteren, E. Samen, L. Lu, M. Vanwildemeersch, J. Klar, G. Genove, K. Pietras,

S. Stone-Elander, L. Claesson-Welsh, S. Yla-Herttuala, P. Lindahl and U. Eriksson (2010). "Vascular endothelial growth factor B controls endothelial fatty acid uptake." Nature 464: 917-921.

Haiko, P., T. Makinen, S. Keskitalo, J. Taipale, M. J. Karkkainen, M. E. Baldwin, S. A. Stacker, M. G. Achen and K. Alitalo (2008). "Deletion of vascular endothelial growth factor C (VEGF-C) and VEGF-D is not equivalent to VEGF receptor 3 deletion in mouse embryos." Mol Cell Biol 28: 4843-4850.

Hamada, K., Y. Oike, N. Takakura, Y. Ito, L. Jussila, D.J. Dumont, K. Alitalo and T. Suda (2000). "VEGF-C signaling pathways through VEGFR-2 and VEGFR-3 in vasculoangiogenesis and hematopoiesis." Blood 96: 3793-3800.

Hattori, K., B. Heissig, Y. Wu, S. Dias, R. Tejada, B. Ferris, D. J. Hicklin, Z. Zhu, P. Bohlen, L. Witte, J. Hendrikx, N. R. Hackett, R. G. Crystal, M. A. Moore, Z. Werb, D. Lyden and S. Rafii (2002). "Placental growth factor reconstitutes hematopoiesis by recruiting VEGFR1(+) stem cells from bone-marrow microenvironment." Nat Med 8: 841-849.

He, Y., T. Karpanen and K. Alitalo (2004). "Role of lymphangiogenic factors in tumor metastasis." Biochim Biophys Acta 1654: 3-12.

He, Y., I. Rajantie, K. Pajusola, M. Jeltsch, T. Holopainen, S. Yla-Herttuala, T. Harding, K. Jooss, T. Takahashi and K. Alitalo (2005). "Vascular endothelial cell growth factor receptor 3-mediated activation of lymphatic endothelium is crucial for tumor cell entry and spread via lymphatic vessels." Cancer Res 65: 4739-4746.

Heckman, C. A., T. Holopainen, M. Wirzenius, S. Keskitalo, M. Jeltsch, S. Yla-Herttuala, S. R. Wedge, J. M. Jurgensmeier and K. Alitalo (2008). "The tyrosine kinase inhibitor cediranib blocks ligand-induced vascular endothelial growth factor receptor-3 activity and lymphangiogenesis." Cancer Res 68: 4754-4762.

Hicklin, D. J. and L. M. Ellis (2005). "Role of the vascular endothelial growth factor pathway in tumor growth and angiogenesis." J Clin Oncol 23: 1011-1027.

Hiratsuka, S., Y. Maru, A. Okada, M. Seiki, T. Noda and M. Shibuya (2001). "Involvement of Flt-1 tyrosine kinase (vascular endothelial growth factor receptor-1) in pathological angiogenesis." Cancer Res 61: 1207-1213.

Hiratsuka, S., K. Nakamura, S. Iwai, M. Murakami, T. Itoh, H. Kijima, J. M. Shipley, R. M. Senior and M. Shibuya (2002). "MMP9 induction by vascular endothelial growth factor receptor-1 is involved in lung-specific metastasis." Cancer Cell 2: 289-300.

Hiratsuka, S., Y. Kataoka, K. Nakao, K. Nakamura, S. Morikawa, S. Tanaka, M. Katsuki, Y. Maru and M. Shibuya (2005a). "Vascular endothelial growth factor A (VEGF-A) is involved in guidance of VEGF receptor-positive cells to the anterior portion of early embryos." Mol Cell Biol 25: 355-363.

Hiratsuka, S., K. Nakao, K. Nakamura, M. Katsuki, Y. Maru and M. Shibuya (2005b). "Membrane fixation of vascular endothelial growth factor receptor 1 ligand-binding domain is important for vasculogenesis and angiogenesis in mice." Mol Cell Biol 25: 346-354.

Holash, J., S. Davis, N. Papadopoulos, S. D. Croll, L. Ho, M. Russell, P. Boland, R. Leidich, D. Hylton, E. Burova, E. Ioffe, T. Huang, C. Radziejewski, K. Bailey, J. P. Fandl, T. Daly, S. J. Wiegand, G. D. Yancopoulos and J. S. Rudge (2002). "VEGF-Trap: a VEGF blocker with potent antitumor effects." Proc Natl Acad Sci USA 99: 11393-11398.

Houck, K. A., D. W. Leung, A. M. Rowland, J. Winer and N. Ferrara (1992). "Dual regulation of vascular endothelial growth factor bioavailability by genetic and proteolytic mechanisms." J Biol Chem 267: 26031-26037.

Igarashi, K., T. Isohara, T. Kato, K. Shigeta, T. Yamano and I. Uno (1998). "Tyrosine 1213 of Flt-1 is a major binding site of Nck and SHP-2." Biochem Biophys Res Commun 246: 95-99.

Ikeda, E., M. G. Achen, G. Breier and W. Risau (1995). "Hypoxia-induced transcriptional activation and increased mRNA stability of vascular endothelial growth factor in C6 glioma cells." J Biol Chem 270: 19761-19766.

Iruela-Arispe, M. L. and G. E. Davis (2009). "Cellular and molecular mechanisms of vascular lumen formation." Dev Cell 16: 222-231.

Ito, N., C. Wernstedt, U. Engstrom and L. Claesson-Welsh (1998). "Identification of vascular endothelial growth factor receptor-1 tyrosine phosphorylation sites and binding of SH2 domain-containing molecules." J Biol Chem 273: 23410-23418.

Jakeman, L. B., J. Winer, G. L. Bennett, C. A. Altar and N. Ferrara (1992). "Binding sites for vascular endothelial growth factor are localized on endothelial cells in adult rat tissues." J Clin Invest 89: 244-253.

Jingjing, L., Y. Xue, N. Agarwal and R. S. Roque (1999). "Human Muller cells express VEGF183, a novel spliced variant of vascular endothelial growth factor." Invest Ophthalmol Vis Sci 40: 752-759.

Kaipainen, A., J. Korhonen, T. Mustonen, V. W. van Hinsbergh, G. H. Fang, D. Dumont, M. Breitman and K. Alitalo (1995). "Expression of the fms-like tyrosine kinase 4 gene becomes restricted to lymphatic endothelium during development." Proc Natl Acad Sci USA 92: 3566-3570.

Kamba, T., B. Y. Tam, H. Hashizume, A. Haskell, B. Sennino, M. R. Mancuso, S. M. Norberg, S. M. O'Brien, R. B. Davis, L. C. Gowen, K. D. Anderson, G. Thurston, S. Joho, M. L. Springer, C. J. Kuo and D. M. McDonald (2006). "VEGF-dependent plasticity of fenestrated capillaries in the normal adult microvasculature." Am J Physiol Heart Circ Physiol 290: H560-576.

Kamei, M., W. B. Saunders, K. J. Bayless, L. Dye, G. E. Davis and B. M. Weinstein (2006). "Endothelial tubes assemble from intracellular vacuoles in vivo." Nature 442: 453-456.

Kami, J., K. Muranaka, Y. Yanagi, R. Obata, Y. Tamaki and M. Shibuya (2008). "Inhibition of choroidal neovascularization by blocking vascular endothelial growth factor receptor tyrosine kinase." Jpn J Ophthalmol 52: 91-98.

Kanda, M., S. Nomoto, Y. Nishikawa, H. Sugimoto, N. Kanazumi, S. Takeda and A. Nakao (2008). "Correlations of the expression of vascular endothelial growth factor B and its isoforms in hepatocellular carcinoma with clinico-pathological parameters." J Surg Oncol 98: 190-196.

Kaplan, R. N., R. D. Riba, S. Zacharoulis, A. H. Bramley, L. Vincent, C. Costa, D. D. MacDonald, D. K. Jin, K. Shido, S. A. Kerns, Z. Zhu, D. Hicklin, Y. Wu, J. L. Port, N. Altorki, E. R. Port, D. Ruggero, S. V. Shmelkov, K. K. Jensen, S. Rafii and D. Lyden (2005). "VEGFR1-positive haematopoietic bone marrow progenitors initiate the pre-metastatic niche." Nature 438: 820-827.

Kappas, N. C., G. Zeng, J. C. Chappell, J. B. Kearney, S. Hazarika, K. G. Kallianos, C. Patterson, B. H. Annex and V. L. Bautch (2008). "The VEGF receptor Flt-1 spatially modulates Flk-1 signaling and blood vessel branching." J Cell Biol 181: 847-858.

Kappert, K., K. G. Peters, F. D. Bohmer and A. Ostman (2005). "Tyrosine phosphatases in vessel wall signaling." Cardiovasc Res 65: 587-598.

Karadedou, C. T., A. R. Gomes, J. Chen, M. Petkovic, K. K. Ho, A. K. Zwolinska, A. Feltes, S. Y. Wong, K. Y. Chan, Y. N. Cheung, J. W. Tsang, J. J. Brosens,

U. S. Khoo and E. W. Lam (2011). "FOXO3a represses VEGF expression through FOXM1-dependent and -independent mechanisms in breast cancer." Oncogene 31: 1845-58.

Karkkainen, M. J., P. Haiko, K. Sainio, J. Partanen, J. Taipale, T. V. Petrova, M. Jeltsch, D. G. Jackson, M. Talikka, H. Rauvala, C. Betsholtz and K. Alitalo (2004). "Vascular endothelial growth factor C is required for sprouting of the first lymphatic vessels from embryonic veins." Nat Immunol 5: 74-80.

Karpanen, T., C. A. Heckman, S. Keskitalo, M. Jeltsch, H. Ollila, G. Neufeld, L. Tamagnone and K. Alitalo (2006). "Functional interaction of VEGF-C and VEGF-D with neuropilin receptors." FASEB J 20: 1462-1472.

Kawakami, M., T. Furuhata, Y. Kimura, K. Yamaguchi, F. Hata, K. Sasaki and K. Hirata (2003). "Expression analysis of vascular endothelial growth factors and their relationships to lymph node metastasis in human colorectal cancer." J Exp Clin Cancer Res 22: 229-237.

Kawamura, H., X. Li, , K. Goishi, L. A. van Meeteren, L. Jakobsson, S. Cebe-Suarez, A. Shimizu, D. Edholm, K. Ballmer-Hofer, L. Kjellen, M. Klagsbrun and L. Claesson-Welsh (2008). "Neuropilin-1 in regulation of VEGF-induced activation of p38MAPK and endothelial cell organization." Blood 112: 3638-3649.

Kawasaki, T., T. Kitsukawa, Y. Bekku, Y. Matsuda, M. Sanbo, T. Yagi and H. Fujisawa (1999). "A requirement for neuropilin-1 in embryonic vessel formation." Development 126: 4895-4902.

Keck, P. J., S. D. Hauser, G. Krivi, K. Sanzo, T. Warren, J. Feder and D. T. Connolly (1989). "Vascular permeability factor, an endothelial cell mitogen related to PDGF." Science 246: 1309-1312.

Kelly, R. J., C. Darnell and O. Rixe (2010). "Target inhibition in antiangiogenic therapy a wide spectrum of selectivity and specificity." Cancer J 16: 635-642.

Kendall, R. L., G. Wang, J. DiSalvo and K. A. Thomas (1994). "Specificity of vascular endothelial cell growth factor receptor ligand binding domains." Biochem Biophys Res Commun 201: 326-330.

Kerbel, R. S. (2000). "Tumor angiogenesis: past, present and the near future." Carcino-genesis 21: 505-515.

Kerber, M., Y. Reiss, A. Wickersheim, M, Jugold, F. Kiessling, M. Heil, V. Tchaikovski, J. Waltenberger, M. Shibuya, K. H. Plate and M. R. Machein (2008). "Flt-1 signaling in macrophages promotes glioma growth *in vivo*." Cancer Res 68: 7342-7351.

Keyt, B. A., L. T. Berleau, H. V. Nguyen, H. Chen, H. Heinsohn, R. Vandlen and N. Ferrara (1996). "The carboxyl-terminal domain (111-165) of vascular endothelial growth factor is critical for its mitogenic potency." J Biol Chem 271: 7788-7795.

Kieser, A., H. A. Weich, G. Brandner, D. Marme and W. Kolch (1994). "Mutant p53 potentiates protein kinase C induction of vascular endothelial growth factor expression." Oncogene 9: 963-969.

Kim, K. B., J. Chesney, D. Robinson, H. Gardner, M. M. Shi and J. M. Kirkwood (2011). "Phase I/II and pharmacodynamic study of dovitinib (TKI258), an inhibitor of fibroblast growth factor receptors, in patients with advanced melanoma." Clin Cancer Res 17: 7451-7461.

Kim, K. J., B. Li, J. Winer, M. Armanini, N. Gillett, H. S. Phillips and N. Ferrara (1993). "Inhibition of vascular endothelial growth factor-induced angiogenesis suppresses tumour growth *in vivo*." Nature 362: 841-844.

Kiselyov, A., K. V. Balakin and S. E. Tkachenko (2007). "VEGF/VEGFR signalling as a target for inhibiting angiogenesis." Expert Opin Investig Drugs 16: 83-107.

Klagsbrun, M. and A. Eichmann (2005). "A role for axon guidance receptors and ligands in blood vessel development and tumor angiogenesis." Cytokine Growth Factor Rev 16: 535-548.

Krueger, J., D. Liu, K. Scholz, A. Zimmer, Y. Shi, C. Klein, A. Siekmann, S. Schulte-Merker, M. Cudmore, A. Ahmed and F. le Noble (2011). "Flt1 acts as a negative regulator of tip cell formation and branching morphogenesis in the zebrafish embryo." Development 138: 2111-2120.

Krupitskaya, Y. and H. A. Wakelee (2009). "Ramucirumab, a fully human mAb to the transmembrane signaling tyrosine kinase VEGFR-2 for the potential treatment of cancer." Curr Opin Investig Drugs 10: 597-605.

Kukk, E., A. Lymboussaki, S. Taira, A. Kaipainen, M. Jeltsch, V. Joukov and K. Alitalo (1996). "VEGF-C receptor binding and pattern of expression with VEGFR-3 suggests a role in lymphatic vascular development." Development 122: 3829-3837.

Kumar, A., S. S. D'Souza, S. R. Nagaraj, S. L. Gaonkar, B. P. Salimath and K. M. Rai (2009). "Antiangiogenic and antiproliferative effects of substituted-1,3,4-oxadiazole derivatives is mediated by down regulation of VEGF and inhibition of translocation of HIF-1alpha in Ehrlich ascites tumor cells." Cancer Chemother Pharmacol 64: 1221-1233.

Laakkonen, P., M. Waltari, T. Holopainen, T. Takahashi, B. Pytowski, P. Steiner, D. Hicklin, K. Persaud, J. R. Tonra, L. Witte and K. Alitalo (2007). "Vascular endothelial growth factor receptor 3 is involved in tumor angiogenesis and growth." Cancer Res 67: 593-599.

Lamalice, L., F. Houle and J. Huot (2006). "Phosphorylation of Tyr1214 within VEGFR-2 triggers the recruitment of Nck and activation of Fyn leading to SAPK2/p38 activation and endothelial cell migration in response to VEGF." J Biol Chem 281: 34009-34020.

Lange, T., N. Guttmann-Raviv, L. Baruch, M. Machluf and G. Neufeld (2003). "VEGF162, a new heparin-binding vascular endothelial growth factor splice form that is expressed in transformed human cells." J Biol Chem 278, 17164-17169.

Laramee, M., C. Chabot, M. Cloutier, R. Stenne, M. Holgado-Madruga, A. J. Wong and I. Royal (2007). "The scaffolding adapter Gab1 mediates vascular endothelial growth factor signaling and is required for endothelial cell migration and capillary formation." J Biol Chem 282: 7758-7769.

Larcher, F., M. Franco, M. Bolontrade, M. Rodriguez-Puebla, L. Casanova, M. Navarro, G. Yancopoulos, J. L. Jorcano and C. J. Conti (2003). "Modulation of the angiogenesis response through Ha-ras control, placenta growth factor, and angiopoietin expression in mouse skin carcinogenesis." Mol Carcinog 37: 83-90.

Lee, J. H., S. Choi, Y. Lee, H. J. Lee, K. H. Kim, K. S. Ahn, H. Bae, E. O. Lee, S. Y. Ryu, J. Lu and S. H. Kim (2010). "Herbal compound farnesiferol C exerts antiangiogenic and antitumor activity and targets multiple aspects of VEGFR1 (Flt1) or VEGFR2 (Flk1) signaling cascades." Mol Cancer Ther 9: 389-399.

Lee, S., S. M. Jilani, G. V. Nikolova, D. Carpizo and M. L. Iruela-Arispe. (2005). "Processing of VEGF-A by matrix metalloproteinases regulates bioavailability and vascular patterning in tumors." J Cell Biol 169: 681-691.

Lee, S., T. T. Chen, C. L. Barber, M. C. Jordan, J. Murdock, S. Desai, N. Ferrara, A. Nagy, K. P. Roos and M. L. Iruela-Arispe (2007a). "Autocrine VEGF signaling is required for vascular homeostasis." Cell 130: 691-703.

Lee, T. H., S. Seng, M. Sekine, C. Hinton, Y. Fu, H. K. Avraham and S. Avraham (2007b). "Vascular endothelial growth factor mediates intracrine survival in human breast carcinoma cells through internally expressed VEGFR1/FLT1." PLoS Med 4: e186.

Leung, D. W., G. Cachianes, W. J. Kuang, D. V. Goeddel and N. Ferrara (1989). "Vascular endothelial growth factor is a secreted angiogenic mitogen." Science 246: 1306-1309.

Levine, A. M., A. Tulpule, D. I. Quinn, G. Gorospe 3rd, D. L. Smith, L. Hornor, W. D. Boswell, B. M. Espina, S. G. Groshen, R. Masood and P. S. Gill (2006). "Phase I study of antisense oligonucleotide against vascular endothelial growth factor: decrease in plasma vascular endothelial growth factor with potential clinical efficacy." J Clin Oncol 24: 1712-1719.

Li, B., E. E. Sharpe, A. B. Maupin, A. A. Teleron, A. L. Pyle, P. Carmeliet and P. P. Young (2006). "VEGF and PlGF promote adult vasculogenesis by enhancing EPC recruitment and vessel formation at the site of tumor neovascularization." FASEB J 20: 1495-1497.

Lin, J., A. S. Lalani, T. C. Harding, M. Gonzalez, W. W. Wu, B., Luan, G. H. Tu, K. Koprivnikar, M. J. VanRoey, Y. He, K. Alitalo and K. Jooss (2005). "Inhibition of lymphogenous metastasis using adeno-associated virus-mediated gene transfer of a soluble VEGFR-3 decoy receptor." Cancer Res 65: 6901-6909.

Lindsay, C. R., I. R. MacPherson and J. Cassidy (2009). "Current status of cediranib: the rapid development of a novel anti-angiogenic therapy." Future Oncol 5: 421-432.

Liu, Y., S. R. Cox, T. Morita and S. Kourembanas (1995). "Hypoxia regulates vascular endothelial growth factor gene expression in endothelial cells. Identification of a 5′ enhancer." Circ Res 77: 638-643.

Lorquet, S., S. Berndt, S. Blacher, E. Gengoux, E. Peulen, E. Maquoi, A. Noel, J. M. Foidart, C. Munaut and C. Pequeux (2010). "Soluble forms of VEGF receptor-1 and -2 promote vascular maturation via mural cell recruitment." FASEB J 24: 3782-3795.

Luo, J., Y. Xiong, X. Han and Y. Lu (2011). "VEGF non-angiogenic functions in adult organ homeostasis: therapeutic implications." J Mol Med (Berl) 89: 635-645.

Luttun, A., M. Tjwa and P. Carmeliet (2002). "Placental growth factor (PlGF) and its receptor Flt-1 (VEGFR-1): novel therapeutic targets for angiogenic disorders." Ann N Y Acad Sci 979: 80-93.

Mac Gabhann, F. and A. S. Popel (2007). "Dimerization of VEGF receptors and implications for signal transduction: a computational study." Biophys Chem 128: 125-139.

Mahasreshti, P. J., J. G. Navarro, M. Kataram, M. H. Wang, D. Carey, G. P. Siegal, M. N. Barnes, D. M. Nettelbeck, R. D. Alvarez, A. Hemminki and D. T. Curiel (2001). "Adenovirus-mediated soluble FLT-1 gene therapy for ovarian carcinoma." Clin Cancer Res 7: 2057-2066.

Matsumoto, T., S. Bohman, J. Dixelius, T. Berge, A. Dimberg, P. Magnusson, L. Wang, C. Wikner, J. H. Qi, C. Wernstedt, J. Wu, S. Bruheim, H. Mugishima, D. Mukhopadhyay, A. Spurkland and L. Claesson-Welsh (2005). "VEGF receptor-2 Y951 signaling and a role for the adapter molecule TSAd in tumor angiogenesis." EMBO J 24: 2342-2353.

Maynard, S. E., J. Y. Min, J. Merchan, K. H. Lim, J. Li, S. Mondal, T. A. Libermann, J. P. Morgan, F. W. Sellke, I. E. Stillman, F. H. Epstein, V. P. Sukhatme and S. A. Karumanchi (2003). "Excess placental soluble fms-like tyrosine kinase 1 (sFlt1) may contribute to endothelial dysfunction, hypertension, and proteinuria in preeclampsia." J Clin Invest 111: 649-658.

Meyer, R. D., M. Mohammadi and N. Rahimi (2006). "A single amino acid substitution in the activation loop defines the decoy characteristic of VEGFR-1/ FLT-1." J Biol Chem 281: 867-875.

Meyer, R. D., D. B. Sacks and N. Rahimi (2008). "IQGAP1-dependent signaling pathway regulates endothelial cell proliferation and angiogenesis." PLoS One 3: e3848.

Millauer, B., L. K. Shawver, K. H. Plate, W. Risau and A. Ullrich (1994). "Glioblastoma growth inhibited *in vivo* by a dominant-negative Flk-1 mutant." Nature 367: 576-579.

Millauer, B., S. Wizigmann-Voos, H. Schnurch, R. Martinez, N. P. Moller, W. Risau and A. Ullrich (1993). "High affinity VEGF binding and developmental expression suggest Flk-1 as a major regulator of vasculogenesis and angiogenesis." Cell 72: 835-846.

Minchenko, A., T. Bauer, S. Salceda and J. Caro (1994). "Hypoxic stimulation of vascular endothelial growth factor expression *in vitro* and *in vivo*." Lab Invest 71: 374-379.

Morabito, A., M. C. Piccirillo, F. Falasconi, G. De Feo, A. Del Giudice, J. Bryce, M. Di Maio, E. De Maio, N. Normanno and F. Perrone (2009). "Vandetanib (ZD6474), a dual inhibitor of vascular endothelial growth factor receptor (VEGFR) and epidermal growth factor receptor (EGFR) tyrosine kinases: current status and future directions." Oncologist 14: 378-390.

Muller, Y. A., B. Li, H. W. Christinger, J. A. Wells, B. C. Cunningham and A. M. de Vos (1997). "Vascular endothelial growth factor: crystal structure and functional mapping of the kinase domain receptor binding site." Proc Natl Acad Sci USA 94: 7192-7197.

Murakami, M., S. Iwai, S. Hiratsuka, M. Yamauchi, K. Nakamura, Y. Iwakura and M. Shibuya (2006). "Signaling of vascular endothelial growth factor receptor-1 tyrosine kinase promotes rheumatoid arthritis through activation of monocytes/ macrophages." Blood 108: 1849-1856.

Muramatsu, M., S. Yamamoto, T. Osawa and M. Shibuya (2010). "Vascular endothelial growth factor receptor-1 signaling promotes mobilization of macrophage lineage cells from bone marrow and stimulates solid tumor growth." Cancer Res 70: 8211-8221.

Mylona, E., P. Alexandrou, I. Giannopoulou, G. Liapis, M. Sofia, A. Keramopoulos and L. Nakopoulou (2007). "The prognostic value of vascular endothelial growth factors (VEGFs)-A and -B and their receptor, VEGFR-1, in invasive breast carcinoma." Gynecol Oncol 104: 557-563.

Neufeld, G., T. Cohen, H. Gitay-Goren, Z. Poltorak, S. Tessler, R. Sharon, S. Gengrinovitch and B. Z Levi (1996). "Similarities and differences between the vascular endothelial growth factor (VEGF) splice variants." Cancer Metastasis Rev 15: 153-158.

Neufeld, G., O. Kessler and Y. Herzog (2002). "The interaction of Neuropilin-1 and Neuropilin-2 with tyrosine-kinase receptors for VEGF." Adv Exp Med Biol 515: 81-90.

Nilsson, I., F. Bahram, X. Li, L. Gualandi, S. Koch, M. Jarvius, O. Soderberg, A. Anisimov, I. Kholova, B. Pytowski, M. Baldwin, S. Yla-Herttuala, K. Alitalo, J. Kreuger and L. Claesson-Welsh (2010). "VEGF receptor 2/-3 heterodimers detected *in situ* by proximity ligation on angiogenic sprouts." The EMBO J 29: 1377-1388.

None (2010). "Linifanib." Drugs R D 10: 111-122.

Oelrichs, R. B., H. H. Reid, O. Bernard, A. Ziemiecki and A. F. Wilks (1993). "NYK/ FLK-1: a putative receptor protein tyrosine kinase isolated from E10 embryonic neuroepithelium is expressed in endothelial cells of the developing embryo." Oncogene 8: 11-18.

Ohm, J. E., D. I. Gabrilovich, G. D. Sempowski, E. Kisseleva, K. S. Parman, S. Nadaf and D. P. Carbone (2003). "VEGF inhibits T-cell development and may contribute to tumor-induced immune suppression." Blood 101: 4878-4886.

Okada, F., J. W. Rak, B. S. Croix, B. Lieubeau, M. Kaya, L. Roncari, S. Shirasawa, T. Sasazuki and R. S. Kerbel (1998). "Impact of oncogenes in tumor angiogenesis: mutant K-ras up-regulation of vascular endothelial growth factor/vascular permeability factor is necessary, but not sufficient for tumorigenicity of human colorectal carcinoma cells." Proc Natl Acad Sci USA 95: 3609-3614.

Orlandini, M., A. Spreafico, M. Bardelli, M. Rocchigiani, A. Salameh, S. Nucciotti, C. Capperucci, B. Frediani and S. Oliviero (2006). "Vascular endothelial growth factor-D activates VEGFR-3 expressed in osteoblasts inducing their differentiation." J Biol Chem 281: 17961-17967.

Paavonen, K., P. Puolakkainen, L. Jussila, T. Jahkola and K. Alitalo (2000). "Vascular endothelial growth factor receptor-3 in lymphangiogenesis in wound healing." Am J Pathol 156: 1499-1504.

Paez-Ribes, M., E. Allen, J. Hudock, T. Takeda, H. Okuyama, F. Vinals, M. Inoue, G. Bergers, D. Hanahan and O. Casanovas (2009). "Antiangiogenic therapy elicits malignant progression of tumors to increased local invasion and distant metastasis." Cancer Cell 15: 220-231.

Pan, Q., Y. Chanthery, W. C. Liang, S. Stawicki, J. Mak, N. Rathore, R. K. Tong, J. Kowalski, S. F. Yee, G. Pacheco, S. Ross, Z. Cheng, J. Le Couter, G. Plowman, F. Peale, A. W. Koch, Y. Wu, A. Bagri, M. Tessier-Lavigne and R. J. Watts (2007a). "Blocking neuropilin-1 function has an additive effect with anti-VEGF to inhibit tumor growth." Cancer Cell 11: 53-67.

Pan, Q., Y. Chathery, Y. Wu, N. Rathore, R. K. Tong, F. Peale, A. Bagri, M. Tessier-Lavigne, A. W. Koch and R. J. Watts (2007b). "Neuropilin-1 binds to VEGF121 and regulates endothelial cell migration and sprouting." J Biol Chem 282: 24049-24056.

Park, J. E., G. A. Keller and N. Ferrara (1993). "The vascular endothelial growth factor (VEGF) isoforms: differential deposition into the subepithelial extracellular matrix and bioactivity of extracellular matrix-bound VEGF." Mol Biol Cell 4: 1317-1326.

Parr, C., G. Watkins, M. Boulton, J. Cai and W. G. Jiang (2005). "Placenta growth factor is over-expressed and has prognostic value in human breast cancer." Eur J Cancer 41: 2819-2827.

Pertovaara, L., A. Kaipainen, T. Mustonen, A. Orpana, N. Ferrara, O. Saksela and K. Alitalo (1994). "Vascular endothelial growth factor is induced in response to transforming growth factor-beta in fibroblastic and epithelial cells." J Biol Chem 269: 6271-6274.

Peters, K. G., C. De Vries and L. T. Williams (1993). "Vascular endothelial growth factor receptor expression during embryogenesis and tissue repair suggests a role in endothelial differentiation and blood vessel growth." Proc Natl Acad Sci USA 90: 8915-8919.

Pietras, K. and D. Hanahan (2005). "A multitargeted, metronomic, and maximum-tolerated dose "chemo-switch" regimen is antiangiogenic, producing objective responses and survival benefit in a mouse model of cancer." J Clin Oncol 23: 939-952.

Plate, K. H., G. Breier, B. Millauer, A. Ullrich and W. Risau (1993). "Up-regulation of vascular endothelial growth factor and its cognate receptors in a rat glioma model of tumor angiogenesis." Cancer Res 53: 5822-5827.

Podar, K. and K. C. Anderson (2005). "The pathophysiologic role of VEGF in hematologic malignancies: therapeutic implications." Blood 105: 1383-1395.

Prewett, M., J. Huber, Y. Li, A. Santiago, W. O'Connor, K. King, J. Overholser, A. Hooper, B. Pytowski, L. Witte, P. Bohlen and D. J. Hicklin (1999). "Antivascular endothelial growth factor receptor (fetal liver kinase 1) monoclonal antibody inhibits tumor angiogenesis and growth of several mouse and human tumors." Cancer Res 59: 5209-5218.

Quinn, T. P., K. G. Peters, C. De Vries, N. Ferrara and L. T. Williams (1993). "Fetal liver kinase 1 is a receptor for vascular endothelial growth factor and is selectively expressed in vascular endothelium." Proc Natl Acad Sci USA 90: 7533-7537.

Rafii, S., B. Heissig and K. Hattori (2002). "Efficient mobilization and recruitment of marrow-derived endothelial and hematopoietic stem cells by adenoviral vectors expressing angiogenic factors." Gene Ther 9: 631-641.

Rahimi, N., V. Dayanir and K. Lashkari (2000). "Receptor chimeras indicate that the vascular endothelial growth factor receptor-1 (VEGFR-1) modulates mitogenic activity of VEGFR-2 in endothelial cells." J Biol Chem 275: 16986-16992.

Ribatti, D. (2007). "The discovery of endothelial progenitor cells An historical review." Leuk Res 31: 439-444.

Risau, W. 1997. "Mechanisms of angiogenesis." Nature: 386: 671-674.

Roberts, N., B. Kloos, M. Cassella, S. Podgrabinska, K. Persaud, Y. Wu, B. Pytowski and M. Skobe (2006). "Inhibition of VEGFR-3 activation with the antagonistic antibody more potently suppresses lymph node and distant metastases than inactivation of VEGFR-2." Cancer Res 66: 2650-2657.

Rocha, S. F. and R. H. Adams (2009). "Molecular differentiation and specialization of vascular beds." Angiogenesis 12: 139-147.

Sawano, A., T. Takahashi, S. Yamaguchi and M. Shibuya (1997). "The phosphorylated 1169-tyrosine containing region of flt-1 kinase (VEGFR-1) is a major binding site for PLCgamma." Biochem Biophys Res Commun 238: 487-491.

Schmeisser, A., M. Christoph, A. Augstein, R. Marquetant, M. Kasper, R. C. Braun-Dullaeus and R. H. Strasser (2006). "Apoptosis of human macrophages by Flt-4 signaling: implications for atherosclerotic plaque pathology." Cardiovasc Res 71: 774-784.

Scott, E. N., G. Meinhardt, C. Jacques, D. Laurent and A. L. Thomas (2007). "Vatalanib: the clinical development of a tyrosine kinase inhibitor of angiogenesis in solid tumours." Expert Opin Investig Drugs 16: 367-379.

Senger, D. R., S. J. Galli, A. M. Dvorak, C. A. Perruzzi, V. S. Harvey and H. F. Dvorak (1983). "Tumor cells secrete a vascular permeability factor that promotes accumulation of ascites fluid." Science 219: 983-985.

Seto, T., M. Higashiyama, H. Funai, F. Imamura, K. Uematsu, N. Seki, K. Eguchi, T. Yamanaka and Y. Ichinose (2006). "Prognostic value of expression of vascular endothelial growth factor and its flt-1 and KDR receptors in stage I non-small-cell lung cancer." Lung Cancer 53: 91-96.

Shalaby, F., J. Rossant, T. P. Yamaguchi, M. Gertsenstein, X. F. Wu, M. L. Breitman and A. C. Schuh (1995). "Failure of blood-island formation and vasculogenesis in Flk-1-deficient mice." Nature 376: 62-66.

Shibata, M. A., J. Morimoto, E. Shibata and Y. Otsuki (2008). "Combination therapy with short interfering RNA vectors against VEGF-C and VEGF-A suppresses lymph node and lung metastasis in a mouse immunocompetent mammary cancer model." Cancer Gene Ther 15: 776-786.

Shifren, J. L., S. Mesiano, R. N. Taylor, N. Ferrara and R. B. Jaffe (1998). "Corticotropin regulates vascular endothelial growth factor expression in human fetal adrenal cortical cells." J Clin Endocrinol Metab 83: 1342-1347.

Shinkai, A., M. Ito, H. Anazawa, S. Yamaguchi, K. Shitara and M. Shibuya (1998). "Mapping of the sites involved in ligand association and dissociation at the extracellular domain of the kinase insert domain-containing receptor for vascular endothelial growth factor." J Biol Chem 273: 31283-31288.

Siekmann, A. F. and N. D. Lawson (2007). "Notch signalling and the regulation of angiogenesis." Cell Adh Migr 1: 104-106.

Smith, N. R., D. Baker, N. H. James, K. Ratcliffe, M. Jenkins, S. E. Ashton, G. Sproat, R. Swann, N. Gray, A. Ryan, J. M. Jurgensmeier and C. Womack (2010). "Vascular endothelial growth factor receptors VEGFR-2 and VEGFR-3 are localized primarily to the vasculature in human primary solid cancers." Clin Cancer Res 16: 3548-3561.

Soh, E. Y., S. A. Sobhi, M. G. Wong, Y. G. Meng, A. E. Siperstein, O. H. Clark and Q. Y. Duh (1996). "Thyroid-stimulating hormone promotes the secretion of vascular endothelial growth factor in thyroid cancer cell lines." Surgery 120: 944-947.

Soker, S., H. Fidder, G. Neufeld and M. Klagsbrun (1996). "Characterization of novel vascular endothelial growth factor (VEGF) receptors on tumor cells that bind VEGF165 via its exon 7-encoded domain." J Biol Chem 271 5761-5767.

Spratlin, J. (2011). "Ramucirumab (IMC-1121B): Monoclonal antibody inhibition of vascular endothelial growth factor receptor-2." Curr Oncol Rep 13: 97-102.

Stefater, J. A. 3rd, I. Lewkowich, S. Rao, G. Mariggi, A. C. Carpenter, A. R. Burr, J. Fan, R. Ajima, J. D. Molkentin, B. O. Williams, M. Wills-Karp, J. W. Pollard, T. Yamaguchi, N. Ferrara, H. Gerhardt and R. A. Lang (2011). "Regulation of angiogenesis by a non-canonical Wnt-Flt1 pathway in myeloid cells." Nature 474: 511-515.

Sternberg, C. N., I. D. Davis, J. Mardiak, C. Szczylik, E. Lee, J. Wagstaff, C. H. Barrios, P. Salman, O. A. Gladkov, A. Kavina, J. J. Zarba, M. Chen, L. McCann, L. Pandite, D. F. Roychowdhury and R. E. Hawkins (2010). "Pazopanib in locally advanced or metastatic renal cell carcinoma: results of a randomized phase III trial." J Clin Oncol 28: 1061-1068.

Stuttfeld, E. and K. Ballmer-Hofer (2009). "Structure and function of VEGF receptors." IUBMB Life 61: 915-922.

Sulkes, A. (2005). "The emerging role of the new aromatase inhibitors in the treatment of breast cancer." Isr Med Assoc J 7: 257-261.

Sulkes, A. (2010). Novel multitargeted anticancer oral therapies: sunitinib and sorafenib as a paradigm. Isr Med Assoc J 12: 628-632.

Takahashi, T., H. Ueno and M. Shibuya (1999). "VEGF activates protein kinase C-dependent, but Ras-independent Raf-MEK-MAP kinase pathway for DNA synthesis in primary endothelial cells." Oncogene 18: 2221-2230.

Tammela, T., G. Zarkada, E. Wallgard, A. Murtomaki, S. Suchting, M. Wirzenius, M. Waltari, M. Hellstrom, T. Schomber, R. Peltonen, C. Freitas, A. Duarte, H. Isoniemi, P. Laakkonen, G. Christofori, S. Yla-Herttuala, M. Shibuya, B. Pytowski, A. Eichmann, C. Betsholtz and K. Alitalo (2008). "Blocking VEGFR-3 suppresses angiogenic sprouting and vascular network formation." Nature 454: 656-660.

Tammela, T. and K. Alitalo (2010). "Lymphangiogenesis: Molecular mechanisms and future promise." Cell 140: 460-476.

Tammela, T., G. Zarkada, H. Nurmi, L. Jakobsson, K. Heinolainen, D. Tvorogov, W. Zheng, C. A. Franco, A. Murtomaki, E. Aranda, N. Miura, S. Yla-Herttuala,

M. Fruttiger, T. Makinen, A. Eichmann, J. W. Pollard, H. Gerhardt and K. Alitalo (2011). "VEGFR-3 controls tip to stalk conversion at vessel fusion sites by reinforcing Notch signalling." Nat Cell Biol 13: 1202-1213.

Tasaki, Y., R. Nishimura, M. Shibaya, H. Y. Lee, T. J. Acosta and K. Okuda (2010). "Expression of VEGF and its receptors in the bovine endometrium throughout the estrous cycle: effects of VEGF on prostaglandin production in endometrial cells." J Reprod Dev 56: 223-229.

Taylor, A. P. and D. M. Goldenberg (2007). "Role of placenta growth factor in malignancy and evidence that an antagonistic PlGF/Flt-1 peptide inhibits the growth and metastasis of human breast cancer xenografts." Mol Cancer Ther 6: 524-531.

Taylor, A. P., E. Leon and D. M. Goldenberg (2010). "Placental growth factor (PlGF) enhances breast cancer cell motility by mobilising ERK1/2 phosphorylation and cytoskeletal rearrangement." Br J Cancer 103: 82-89.

Tchaikovski, V., S. Olieslagers, F. D. Bohmer and J. Waltenberger (2009). "Diabetes mellitus activates signal transduction pathways resulting in vascular endothelial growth factor resistance of human monocytes." Circulation 120: 150-159.

Teng, L. S., K. T. Jin, K. F. He, J. Zhang, H. H. Wang and J. Cao (2010). "Clinical applications of VEGF-trap (aflibercept) in cancer treatment." J Chin Med Assoc 73: 449-456.

Toi, M., H. Bando, T. Ogawa, M. Muta, C. Hornig and H. A. Weich (2002). "Significance of vascular endothelial growth factor (VEGF)/soluble VEGF receptor-1 relationship in breast cancer." Int J Cancer 98: 14-18.

Tong, R. T., Y. Boucher, S. V. Kozin, F. Winkler, D. J. Hicklin and R. K. Jain (2004). "Vascular normalization by vascular endothelial growth factor receptor 2 blockade induces a pressure gradient across the vasculature and improves drug penetration in tumors." Cancer Res 64: 3731-3736.

Tvorogov, D., A. Anisimov, W. Zheng, V. M. Leppanen, T. Tammela, S. Laurinavicius, W. Holnthoner, H. Helotera, T. Holopainen, M. Jeltsch, N. Kalkkinen, H. Lankinen, P. M. Ojala and K. Alitalo (2010). "Effective suppression of vascular network formation by combination of antibodies blocking VEGFR ligand binding and receptor dimerization." Cancer Cell 18: 630-640.

Twombly, R. (2011). "Avastin's uncertain future in breast cancer treatment." J Natl Cancer Inst 103: 458-460.

Ushio-Fukai, M. and Y. Nakamura (2008). "Reactive oxygen species and angiogenesis: NADPH oxidase as target for cancer therapy." Cancer Lett 266: 37-52.

Van de Veire, S., I. Stalmans, F. Heindryckx, H. Oura, A. Tijeras-Raballand, T. Schmidt, S. Loges, I. Albrecht, B. Jonckx, S. Vinckier, C. Van Steenkiste, S. Tugues, C. Rolny, M. De Mol, D. Dettori, P. Hainaud, L. Coenegrachts, J. O. Contreres, T. Van Bergen, H. Cuervo, W. H. Xiao, C. Le Henaff, I. Buysschaert, B. Kharabi Masouleh, A. Geerts, T. Schomber, P. Bonnin, V. Lambert, J. Haustraete, S. Zacchigna, J. M. Rakic, W. Jimenez, A. Noel, M. Giacca, I. Colle, J. M. Foidart, G. Tobelem, M. Morales-Ruiz, J. Vilar, P. Maxwell, S. A. Vinores, G. Carmeliet, M. Dewerchin, L. Claesson-Welsh, E. Dupuy, H. Van Vlierberghe, G. Christofori, M. Mazzone, M. Detmar, D. Collen and P. Carmeliet (2010). "Further pharmacological and genetic evidence for the efficacy of PlGF inhibition in cancer and eye disease." Cell 141: 178-190.

Van Meter, M. E. and E. S. Kim (2010). "Bevacizumab: current updates in treatment." Curr Opin Oncol 22: 586-591.

Vander Kooi, C. W., M. A. Jusino, B. Perman, D. B. Neau, H. D. Bellamy and D. J. Leahy (2007). "Structural basis for ligand and heparin binding to neuropilin B domains." Proc Natl Acad Sci USA 104: 6152-6157.

Vincenti, V., C. Cassano, M. Rocchi and G. Persico. 1996. "Assignment of the vascular endothelial growth factor gene to human chromosome 6p21.3." Circulation 93: 1493-1495.

Waltenberger, J., L. Claesson-Welsh, A. Siegbahn, M. Shibuya and C. H. Heldin (1994). "Different signal transduction properties of KDR and Flt1, two receptors for vascular endothelial growth factor." J Biol Chem 269: 26988-26995.

Waltenberger, J., J. Lange and A. Kranz (2000). "Vascular endothelial growth factor-A-induced chemotaxis of monocytes is attenuated in patients with diabetes mellitus: A potential predictor for the individual capacity to develop collaterals." Circulation 102: 185-190.

Waltenberger, J. (2009). "VEGF resistance as a molecular basis to explain the angiogenesis paradox in diabetes mellitus." Biochem Soc Trans 37: 1167-1170.

Wang, F., M. Yamauchi, M. Muramatsu, T. Osawa, R. Tsuchida and M. Shibuya (2011). "RACK1 regulates VEGF/Flt1-mediated cell migration via activation of a PI3K/Akt pathway." J Biol Chem 286: 9097-9106.

Wang, F. Q., E. Barfield, S. Dutta, T. Pua and D. A. Fishman (2009). "VEGFR-2 silencing by small interference RNA (siRNA) suppresses LPA-induced epithelial ovarian cancer (EOC) invasion." Gynecol Oncol 115: 414-423.

Warner, A. J., J. Lopez-Dee, E. L. Knight, J. R. Feramisco and S. A. Prigent (2000). "The Shc-related adaptor protein, Sck, forms a complex with the vascular-endothelial-growth-factor receptor KDR in transfected cells." Biochem J 347: 501-509.

Warren, R. S., H. Yuan, M. R. Matli, N. Ferrara and D. B. Donner (1996). "Induction of vascular endothelial growth factor by insulin-like growth factor 1 in colorectal carcinoma." J Biol Chem 271: 29483-29488.

Wood, J. M., G. Bold, E. Buchdunger, R. Cozens, S. Ferrari, J. Frei, F. Hofmann, J. Mestan, H. Mett, T. O'Reilly, E. Persohn, J. Rosel, C. Schnell, D. Stover, A. Theuer, H. Towbin, F. Wenger, K. Woods-Cook, A. Menrad, G. Siemeister, M. Schirner, K. H. Thierauch, M. R. Schneider, J. Drevs, G. Martiny-Baron and F. Totzke (2000). "PTK787/ZK 222584, a novel and potent inhibitor of vascular endothelial growth factor receptor tyrosine kinases, impairs vascular endothelial growth factor-induced responses and tumor growth after oral administration." Cancer Res 60: 2178-2189.

Wu, Y., Z. Zhong, J. Huber, R. Bassi, B. Finnerty, E. Corcoran, H. Li, E. Navarro, P. Balderes, X. Jimenez, H. Koo, V. R. Mangalampalli, D. L. Ludwig, J. R. Tonra and D. J. Hicklin (2006). "Anti-vascular endothelial growth factor receptor-1 antagonist antibody as a therapeutic agent for cancer." Clin Cancer Res 12: 6573-6584.

Xu, L., D. M. Cochran, R. T. Tong, F. Winkler, S. Kashiwagi, R. K. Jain and D. Fukumura (2006). "Placenta growth factor overexpression inhibits tumor growth, angiogenesis, and metastasis by depleting vascular endothelial growth factor homodimers in orthotopic mouse models." Cancer Res 66: 3971-3977.

Yamagishi, S., S. Amano, Y. Inagaki, T. Okamoto, M. Takeuchi and H. Inoue (2003). "Pigment epithelium-derived factor inhibits leptin-induced angiogenesis by suppressing vascular endothelial growth factor gene expression through anti-oxidative properties." Microvasc Res 65: 186-190.

Yamaoka-Tojo, M., T. Tojo, H. W. Kim, L. Hilenski, N. A. Patrushev, L. Zhang, T. Fukai and M. Ushio-Fukai (2006). "IQGAP1 mediates VE-cadherin-based cell-

cell contacts and VEGF signaling at adherence junctions linked to angiogenesis." Arterioscler Thromb Vasc Biol 26: 1991-1997.

Yonekura, H., S. Sakurai, X. Liu, H. Migita, H. Wang, S. Yamagishi, M. Nomura, M. J. Abedin, H. Unoki, Y. Yamamoto and H. Yamamoto (1999). "Placenta growth factor and vascular endothelial growth factor B and C expression in microvascular endothelial cells and pericytes. Implication in autocrine and paracrine regulation of angiogenesis." J Biol Chem 274: 35172-35178.

Youssoufian, H., D. J. Hicklin and E. K. Rowinsky (2007). "Review: monoclonal antibodies to the vascular endothelial growth factor receptor-2 in cancer therapy." Clin Cancer Res 13: 5544s-5548s.

Yu, Y., J. D. Hulmes, M. T. Herley, R. G. Whitney, J. W. Crabb and J. D. Sato (2001). "Direct identification of a major autophosphorylation site on vascular endothelial growth factor receptor Flt-1 that mediates phosphatidylinositol 3'-kinase binding." Biochem J 358: 465-472.

Yu, Y., P. Lee, Y. Ke, Y. Zhang, Q. Yu, J. Lee, M. Li, J. Song, J. Chen, J. Dai, F. J. Do Couto, Z. An, W. Zhu and G. L. Yu (2010). "A humanized anti-VEGF rabbit monoclonal antibody inhibits angiogenesis and blocks tumor growth in xenograft models." PloS one 5: e9072.

Zhang, L., F. Zhou, W. Han, B. Shen, J. Luo, M. Shibuya and Y. He (2010). "VEGFR-3 ligand-binding and kinase activity are required for lymphangiogenesis but not for angiogenesis." Cell Res 20: 1319-1331.

Zhang, X., J. P. Gaspard and D. C. Chung (2001). "Regulation of vascular endothelial growth factor by the Wnt and K-ras pathways in colonic neoplasia." Cancer Res 61: 6050-6054.

Zhao, Q., K. Egashira, K. Hiasa, M. Ishibashi, S. Inoue, K. Ohtani, C. Tan, M. Shibuya, A. Takeshita and K. Sunagawa (2004). "Essential role of vascular endothelial growth factor and Flt-1 signals in neointimal formation after periadventitial injury." Arterioscler Thromb Vasc Biol 24: 2284-2289.

Ziche, M., D. Maglione, D. Ribatti, L. Morbidelli, C. T. Lago, M. Battisti, I. Paoletti, A. Barra, M. Tucci, G. Parise, V. Vincenti, H. J. Granger, G. Viglietto and M. G. Persico (1997). "Placenta growth factor-1 is chemotactic, mitogenic, and angiogenic." Lab Invest 76: 517-531.

Zimmermann, R. C., T. Hartman, P. Bohlen, M. V. Sauer and J. Kitajewski (2001a). "Preovulatory treatment of mice with anti-VEGF receptor 2 antibody inhibits angiogenesis in corpora lutea." Microvasc Res 62: 15-25.

Zimmermann, R. C., E. Xiao, N. Husami, M. V. Sauer, R. Lobo, J. Kitajewski and M. Ferin (2001b). "Short-term administration of antivascular endothelial growth factor antibody in the late follicular phase delays follicular development in the rhesus monkey." J Clin Endocrinol Metab 86: 768-772.

Chapter 2

Regulation of Angiogenesis via Notch Signaling in Human Malignancy

Shanchun Guo[1], Mingli Liu[2], Guangdi Wang[3], Lily Yang[4] and Ruben R. Gonzalez-Perez[5]*

[1]*Department of Microbiology, Biochemistry & Immunology, Morehouse School of Medicine, Atlanta, GA 30310, USA. Email: sguo@msm.edu.*
[2]*Department of Biochemistry & Immunology, Morehouse School of Medicine, Atlanta, GA 30310, USA. Email: mliu@msm.edu.*
[3]*Department of Chemistry, Xavier University of Louisiana, New Orleans, LA 70125, USA. Email: gwang@xula.edu.*
[4]*Department of Surgery, Emory University School of Medicine, Atlanta, GA 30322, USA. Email: lyang02@emory.edu.*
[5]*Department of Microbiology, Biochemistry & Immunology, Morehouse School of Medicine, Atlanta, GA 30310, USA. Email: rgonzalez@msm.edu*

CHAPTER OUTLINE

- ▶ Introduction
- ▶ Notch signaling in cancer–an overview
- ▶ Structure, activation and function of Notch receptors and ligands
- ▶ Notch signaling target gene in angiogenesis
- ▶ Notch signaling in vascular development
- ▶ Activation of Notch signaling in tumor angiogenesis
- ▶ Notch signaling, cancer stem cell and angiogenesis
- ▶ Crosstalk among Notch signaling and other oncogenic pathways in tumor angiogenesis
- ▶ Notch signaling as therapeutic drug target in human cancers
- ▶ Conclusion and perspectives
- ▶ Acknowledgements
- ▶ Glossary
- ▶ References

Correspondence Author Department of Microbiology, Biochemistry & Immunology, Morehouse School of Medicine, Atlanta, GA 30310, USA. Email: rgonzalez@msm.edu

ABSTRACT

The malignant cells undergo an angiogenic switch leading to secretion of angiogenic factors and proteolytic enzymes in response to hypoxia culminating in the activation of endothelial cell (EC) proliferation, migration and establishment of a robust capillary network. This irregular and non-well organized network is capable of providing the growing tumor mass with all the required metabolites. In addition, the tumor angiogenesis network also provides tumor cells with the opportunity to enter the circulation and to form distant metastases. Tumor angiogenesis is elicited and regulated by several factors. Among these factors, Notch signaling plays an important role. In light of several valuable reviews published on the role of Notch signaling in several types of cancer (Politi et al. 2004; Farnie and Clarke 2007; Wu et al. 2007; Wang et al. 2010), we wish to focus this chapter on the more recent advancements in our knowledge of aberrant Notch signaling contributing to tumor angiogenesis, as well as its crosstalk with other factors contributing to angiogenesis.

Key words: Notch, tumor angiogenesis, signaling crosstalk

1   Introduction

The formation of new blood vessels from existing ones (angiogenesis) is a crucial requirement for the growth, progression and metastatic spread of a tumor (Hanahan and Weinberg 2011). Low oxygen microenvironment triggers angiogenesis in normal and pathological conditions, i.e., tumor growth (Adams and Alitalo 2007). The tumor angiogenic switch is characterized by the actions of several factors on EC to form the vascular network needed by the tumor to growth and metastasis (Prager and Poettler 2011). Among these factors, Notch signaling plays an important role. Notch is essential for a variety of cell fate decisions and can regulate diverse cellular biological processes especially during embryogenesis. To signal, membrane-bound Notch receptors and ligands need to be co-expressed in adjacent cells. Notch receptors and ligands are expressed in tumor cells as well as in the stromal compartment and have been implicated in tumorigenesis (Bridges et al. 2011; Guo et al. 2011).

2   Notch Signaling in Cancer–An Overview

Notch genes encode transmembrane receptors that are highly conserved from invertebrates to mammals. Notch-mediated signals regulate cell-fate decisions in a large number of developmental systems (Simpson 1995; Lewis 1998). Such signals are mainly transmitted through direct contact between adjacent cells expressing Notch receptors and their ligands. Notch receptors activated in response to ligand expressed by adjacent cells have the potential to regulate cell fate specification, differentiation, proliferation, or survival (Borggrefe and Oswald 2009).

Notch signaling pathway is frequently deregulated in several human malignancies (Table 1). Over-expression of Notch receptors and their ligands have been found in cervical, colon, head and neck, lung, renal carcinoma, pancreatic cancer, breast cancer, acute myeloid, Hodgkin and Large-cell lymphomas (Miele 2006; Miele et al. 2006; Kopan and Ilagan 2009). Overall, it is well-established that Notch signaling plays an important role in tumor progression (Miele and Osborne 1999; Guo et al. 2011). Remarkably, several studies suggest that Notch signaling in tumorigenesis and angiogenesis can either be oncogenic or oncosuppressive and it is context dependent (Table 1). Signals exchanged between neighboring cells through the Notch pathway can amplify and consolidate molecular differences, which eventually dictate cell fates. Thus, Notch induced signaling influences how cells respond to intrinsic or extrinsic developmental cues that are necessary to unfold specific developmental programs (Artavanis-Tsakonas et al. 1999). Notch signaling has the potential to affect the path of differentiation, proliferation, and apoptosis in both normal developmental and abnormal cellular growth programs. Remarkably, the same signaling pathways within different contexts can trigger a variety of cellular activities. Therefore, cancer progression activities induced by Notch and its crosstalk with other signaling pathways are also context-dependent.

3 Structure, Activation and Function of Notch Receptors and Ligands

The Notch system in vertebrates comprises four receptors (Notch1- Notch4) and at least five ligands from the families Delta and JAG/Serrate (DSL): JAG1, JAG2, Delta-like (Dll)-1, Dll-3, and Dll-4 (Artavanis-Tsakonas et al. 1999; Miele 2006; Miele et al. 2006). Ligands of Notch receptors can be divided into several groups based on their domain composition. Canonical DSL ligands (JAG1, JAG2 and Dll-1) are type I cell surface proteins, consisting of the Delta/Serrate/LAG-2 (DSL), Delta and OSM-11-like proteins [DOS, which is specialized tandem EGF repeats] and EGF motifs. The other subtypes of DSL canonical ligands include Dll-3 and Dll-4 that lack the DOS motif (Cordle et al. 2008; D'Souza et al. 2008; Komatsu et al. 2008). Both the DSL and DOS domains are crucial for physical binding with Notch receptor (Kopan and Ilagan 2009). However some membrane-tethered and secreted non canonical ligands lacking DSL and DOS domains have also been documented to activate Notch signaling both *in vitro* and *in vivo* (Cui et al. 2004; Leask and Abraham 2006; Gupta et al. 2007; Albig et al. 2008; D'Souza et al. 2008; Heath et al. 2008; Lu et al. 2008), which may explain the diverse and frequent effects of Notch signaling with the small number of canonical DSL ligands and receptors in vertebrate genomes (D'Souza et al. 2008).

Notch receptors belong to a large single-pass type 1 transmembrane protein family; the extracellular domain consists of 29-36 tandem arrays of EGF (epidermal growth factor)-like repeats, followed by a conserved negative regulatory region (NRR or LNR) consisting of three cystein-rich Notch Lin12 repeats (N/Lin 12) and a heterodimerization (HD) domain (Milner and Bigas 1999). Notch family members differ in the number of EGF-like repeats, however they share many similarities in structure

Table 1.   Function of Notch signaling in human tumors and tumor cell lines

Tissue	Cancer types	Notch/ligand	Function	References
Bladder	Carcinoma	Dell4	Vascular differentiation	(Patel et al. 2006)
Brain	Medulloblastoma	Notch1	Tumor suppressor	
	Medulloblastoma	Notch2	Oncogenic	(Fan et al. 2004)
	Glioblastoma	Total Notch	Oncogenic	(Fan et al. 2010)
Breast	Cancer	Notch1	Oncogenic	(Dievart et al. 1999; Reedijk et al. 2005)
	Cancer	Notch2	Tumor suppressor	(Parr et al. 2004)
	MDA-MB-231 cell line	Notch2	Tumor suppressor	(O'Neill et al. 2007)
	Cell lines	Notch3	Oncogenic	(Yamaguchi et al. 2008)
	Cancer	Jagged1	Oncogenic	(Leong et al. 2007; Sethi et al. 2011)
Cervix	Cancer	Total Notch	Oncogenic	(Sun et al. 2009)
Colon	Carcinoma	Notch1	Oncogenic	(Meng et al. 2009)
	Carcinoma	Jagged ligands	Oncogenic	(Reedijk et al. 2008)
Esophagus	Adenocarcinoma	Hes1 / Jagged1	Oncogenic	(Mendelson et al. 2011)
Lung	Small cell	Notch1	Tumor suppressor	(Sriuranpong et al. 2002)
	Non-small cell	Notch3	Oncogenic	(Konishi et al. 2010)
Oral	Carcinoma cell	Notch1	Oncogenic	(Liao et al. 2011)
Ovary	Cancer cell line	Notch3	Oncogenic	(Park et al. 2006; Choi et al. 2008)
	Cancer cell line	Jagged1	Oncogenic	(Choi et al. 2008)
	Cancer cell line	Notch1	Oncogenic	(Rose et al. 2010; Wang et al. 2010c)
Pancreas	Cancer cell line	Notch1	Oncogenic	(Wang et al. 2006a; Wang et al. 2006b)
	Cancer cell line	Notch1	Angiogenesis inhibition	(Wang et al. 2007)
	Cancer cell line	Notch3	Oncogenic	(Yao and Qian 2010)
	Cancer cell line	DLL4	Angiogenesis inhibition	(Oishi et al. 2010)
Prostate	Cancer cell line	Jagged1	Oncogenic	(Zhang et al. 2006; Wang et al. 2010b)
	Cancer cell line	Notch1	Oncogenic	(Bin Hafeez et al. 2009; Wang et al. 2010b)
Skn	Cancer cell line	Notch1	Tumor suppressor	(Lefort et al. 2007)

Note: Notch signaling may act as a tumor suppressor or an oncogene depending on the cell type and cell context.

(Weng et al. 2004; Kopan and Ilagan 2009). EGF-like repeats mediate ligand binding, whereas NRR functions to prevent both ligand-dependent and -independent signaling (Weng et al. 2004). The cytoplasmic portion of Notch is composed of a DNA binding protein (RBP-Jk associated molecule or RAM) domain and six ankyrin (ANK) repeats, which are flanked by two nuclear localization signals (NLS), followed by a transactivation domain (TAD) and a domain rich in proline, glutamine, serine and threonine residues (PEST) that controls receptor half life (Okuyama et al. 2008; Kopan and Ilagan 2009; Tien et al. 2009) (Figure 1).

Membrane localization of Notch requires S1 cleavage of precursor of the Notch receptor. This event occurs in the Golgi network by the action of a furin-like convertase. Then, the two fragments are re-assembled as a non-covalently linked heterodimeric receptor at the cell surface (Lewis 1998). Mature Notch receptors are heterodimers made up of an extracellular subunit, a transmembrane subunit (N^{TM}) and a cytoplasmic subunit. Activation of Notch comprises two consecutive cleavages of the transmembrane receptor upon the binding of a Notch ligand. Binding of Notch heterodimer to ligand triggers S2 cleavage. This process takes places at the cell surface. N^{TM} subunit is cleaved by ADAM/Tumor necrosis factor-α-converting enzyme (TACE) metalloprotease family at Site 2 (located ~12 amino acids before the transmembrane domain). S2 cleavage releases the Notch extracelluar domain (NECD) from the heterodimer and creates a membrane-tethered Notch extracellular truncation (NEXT), which becomes a substrate for γ-secretase. S3 is cleaved by γ-secretase at Site 3 and 4 (Mumm et al. 2000). This last cleavage occurs on the plasma membrane and/or in endosome. The new mobile cytoplasmic subunit [Notch intracellular domain (NICD or N^{IC})] is translocated to the nucleus, where it interacts with members of the DNA-binding protein, recombination signal binding protein for immunoglobulin kappa J (RBP-Jk) or CBF1/Su(H)/Lag-1 (CSL) family of transcription factors (Borggrefe and Oswald 2009). Activated NICD-RBP-Jk complex displaces co-repressors and recruits coactivator (co-A) mediating the transcription of target genes such as Hes-1 (hairy enhancer of split), cyclin D, Hey-1 (hairy/enhancer-of-split related with YRPW motif) and others (Miele 2006; Miele et al. 2006). In the absence of NICD, CSL may interplay with the ubiquitous corepressor (Co-R) proteins and histone deacetylases (HDACs) to repress transcription of some target genes (Fortini and Artavanis-Tsakonas 1994; Fiuza and Arias 2007).

4   Notch Signaling Target Gene in Angiogenesis

The best-characterized Notch targets are transcriptional repressors of the Hes (Hes1-7) and Hey subfamilies (Hey1, Hey2, HeyL, HesL/HelT, Dec1/BHLHB2, Dec2/BHLHB3) (Iso et al. 2003; Schwanbeck et al. 2008; Zanotti and Canalis 2010). Both Hes and Hey proteins contain a basic domain which determines DNA binding specificity, and a helix-loop-helix domain which allows the proteins to form homo- or heterodimers. In contrast, Hes6 is a

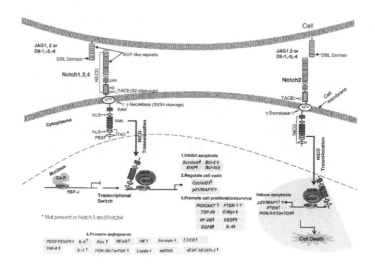

Figure 1   Notch signaling and its possible downstream targets in human cancers and angiogenesis (Reproduced from: Guo et al. 2011). Mammalian ligands of Notch are membrane-bound proteins containing an extracellular NH2-terminal Delta/Serrate/ LAG2 (DSL) motif followed by epidermal growth factor (EGF)-like repeats. Notch receptors are broadly expressed on the cell surface as heterodimers containing a Notch extracellular domain (NECD) composed by multiple extracellular EGF-like repeats and three Lin12/Notch repeats (LNR). Notch receptor cytoplasmatic region or Notch intracellular domain (NICD) contains one nuclear localization signals (NLS) linking RAM domain to six ankyrin (ANK) repeats (ANK domain) followed by an additional bipartite NLS, a loosely defined transactivation domain (TAD), and a conserved proline/glutamic acid/ser/threo-rich domain (PEST domain). In the absence of activated Notch signaling, the DNA binding protein RBP-Jk (CSL/CBF1/Su (H)/ Lag1, a transcription factor) forms a complex with corepressor molecules that represses transcription of target genes. Ligand binding to NECD triggers successive proteolytic cleavages of Notch cytoplasmatic region by ADAM and γ-secretase, proteases resulting in the release of NICD, which translocates into nucleus and removes corepressors from RBP-Jk. This allows RBP-Jk to recruit a coactivator complex composed of Mastermind (MAM) and several transcription factors to transcriptionally activate Notch target genes.Activation of Notch could impact on the following processes in human cancer: 1) inhibition of apoptosis through upregulation of Survivin (Lee et al. 2008a; Lee et al. 2008b) and Bcl-2 protein family (Sade et al. 2004); 2) activation of the cell cycle through upregulation of Cyclin D1 (Ling et al. 2010); 3) promotion of cell proliferation/survival through upregulation of PI-3K/Akt (Liu et al. 2006), TGF-β (Niimi et al. 2007), c-Myc (Efstratiadis et al. 2007), NF-κB (Pratt et al. 2009), EGFR (Dai et al. 2009) and IL-6 (Knupfer and Preiss 2007) pathways; 4) stimulation of angiogenesis and VEGF/VEGFR-2 autocrine/paracrine loop by upregulation of IL-1 system and VEGF/VEGFR-2 (Guo et al. 2010; Guo and Gonzalez-Perez 2011); 5) suppression of cancer growth in some cellular situations. For example, Notch2 signaling may function as a tumor suppressor through upregulation of PTEN or down-regulation of PI-3K/Akt/mTOR (Graziani et al. 2008). Coordinated actions of Notch affect cancer cell growth, migration, invasion, metastasis and angiogenesis, as well as cancer stem cell self-renewal.

Color image of this figure appears in the color plate section at the end of the book.

novel estrogen-regulated gene and a potential oncogene overexpressed in breast cancer, with tumor-promoting and proliferative functions (Hartman et al. 2009). Other Notch target genes include proteins and factors involved in the control of the cell cycle and survival processes such as p21^{WAF1}/^{Cip1}, a cyclin-dependent kinase inhibitor that acts as both a sensor and an effector of multiple anti-proliferative signals. Notch activation contributes to contact inhibition of ECs, in part through repression of p21^{Cip1} expression (Noseda et al. 2004). *Deltex* including *Deltex1, Deltex2, Deltex4* (Lehar and Bevan 2006) acts as a positive regulator of Notch signaling through interactions with the Notch ankyrin repeats (Matsuno et al. 1995). Deltex family also acts a transcriptional regulator downstream of the Notch receptor, but has no direct role in angiogenesis (Yamamoto et al. 2001). Nuclear factor-kappa B (NF-κB) (a transcriptional factor) was very early identified as a Notch target gene (Oswald et al. 1998). NF-κB was later found able to collaborate with Notch signaling in angiogenesis (Wang et al. 2007; Johnston et al. 2009). Deregulation of *cyclin D1* (a mitogenic sensor and allosteric activator of cyclin dependent kinase CDK4/6) (Ronchini and Capobianco 2001) and *c-myc* (an oncogene and cell cycle regulator) are considered the hallmarks for the development of many cancers (Artavanis-Tsakonas et al. 1999; Miele and Osborne 1999; Miele 2006; Miele et al. 2006; Efstratiadis et al. 2007). Both factors are involved in angiogenesis and identified as Notch target genes (Ronchini and Capobianco 2001; Weng et al. 2006).

Notch signaling can modulate apoptosis. NICD interacts and can inactivate p53 through phosphorylation (Kim et al. 2007). Recently, Survivin, a member of the inhibitor of apoptosis family of proteins (IAP) that induces cell proliferation, was identified as a novel Notch target gene (Altieri 2008; Ryan et al. 2009). Notch stimulation resulted in direct activation of Survivin gene transcription through at least one RPB-Jk site in the Survivin promoter (Bray 2006). Activation of Notch directly induced the transcriptional up-regulation of Survivin in ER-breast cancer cells (Lee et al. 2008a; Lee et al. 2008b). Accumulated evidence suggests that lack of Survivin in EC causes embryonic defects in angiogenesis (Zwerts et al. 2007). Moreover, the knockdown of Survivin in ECs could inhibit angiogenesis (Mesri et al. 2001; Coma et al. 2004). A recent study demonstrated that Survivin, through upregulation of VEGF and bFGF, plays an essential role in glioma angiogenesis (Wang et al. 2011).

5 Notch Signaling in Vascular Development

Gain- and loss-of-function mutations in humans, mice, and zebrafish have demonstrated the involvement of Notch signaling in multiple aspects of vascular development (Krebs et al. 2000; Duarte et al. 2004; Roca and Adams 2007; Phng and Gerhardt 2009). The vascular system comprises arteries, veins, and lymphatics which separated subdivisions into large vessels, small vessels and capillaries. These vessels primarily consist of ECs, supporting cells (smooth muscle cells, SMCs and pericytes), and surrounding matrix. Mounting evidence suggests that Notch ligands [Dll-4, Jagged-1 (JAG1), and JAG2], receptors (Notch1, 2, 3, and 4) and effectors (HERP1, 2, and 3) are involved in the development of the vascular system (Gridley 2007; Roca and Adams 2007; Gridley 2010). In this process Notch pathway induces the formation of distinct cell subpopulations from bipotential precursor cells, a process known as lateral specification or lateral inhibition (Egan et al. 1998; Greenwald 1998; Artavanis-Tsakonas et al. 1999). Homozygous mice for null mutations of Notch (including Notch1, plus Notch4, and JAG1), show embryonic lethality together with vascular remodeling defects (Xue et al. 1999; Krebs et al. 2000). Both Dll-1-deficient and homozygous Notch2 mutant mouse embryos showed hemorrhage, possibly resulting from poor development of vascular structures. However, Dll-1 and Notch2 were not detected in large vessels of mutant embryos (Hrabe de Angelis et al. 1997; McCright et al. 2001). EC-specific expression of Notch4 activated form in transgenic mice led to embryonic lethality with abnormal vessel structure and patterning. This phenotype was similar to that seen in Notch1– and Notch1/Notch4-deficient mice (Uyttendaele et al. 2001). The similarities between the vascular phenotypes observed in knockout mice (loss-of-function) and transgenic mice (gain-of-function) suggest that a window of appropriate Notch expression levels might be needed for proper development of the embryonic vasculature.

6 Activation of Notch Signaling in Tumor Angiogenesis

Roles of Notch1 and Notch4 in angiogenesis have been established using mutant mice. Notch1 and Notch1/Notch4 double mutant embryos displayed severe defects in angiogenic vascular remodeling (Krebs et al. 2000). Among Notch receptors with potential roles in carcinogenesis and angiogenesis, Notch1 (MW: 272kDa and NICD MW: 110-120kDa) is relatively the best studied (Dievart et al. 1999; Kiaris et al. 2004; Stylianou et al. 2006). It was found earlier that Notch1, a putative collaborator of *c-myc*, was mutated in high proportion (52%) in $CD4^+CD8^+$ T-cell tumors (Girard et al. 1996). These mutations led to high expression of truncated Notch1 proteins. The Notch1 gene was identified as a novel target for MMTV provirus insertional activation. MMTV insertion in the Notch1 gene induced the overexpression of 5′ truncated ~7 kb RNA (280 kDa mutant protein: Notch1 ectodomain) and truncated 3′ Notch1 transcripts (3.5-4.5 kb) and proteins (86-110 kDa) that can transform HC11 mouse mammary epithelial cells *in vitro* (Dievart et al. 1999). Notch1 was further found to be a mediator of oncogenic Ras (retrovirus-associated DNA sequence kinase, small cytoplasmic GTP-binding proteins) (Weijzen et al. 2002). Moreover, Notch1 and JAG1 were co-upregulated upon estrogen treatment not only in MCF-7 breast cancer cells, but also in ECs, suggesting a role of Notch1-JAG1 in angiogenesis (Soares et al. 2004).

The exact roles of Notch2 (MW: 205kDa) and NICD (MW: 110kDa) in angiogenesis have not been determined. $Notch2^{-/-}$ mice develop normally until E9.5, and then around E11.5 massive cell death occurs (Hamada et al. 1999). These results suggest that Notch2 plays an essential role in post implantation development in mice. The role of Notch 2 is dependent on its ankyrin repeats and probably linked to some aspects of cell specification and/or differentiation (Hamada et al. 1999). However, Notch2 (unlike Notch1) is not essential for generating hematopoietic stem cells from ECs (Kumano et al. 2003). The development of hematopoietic cells is closely related to angiogenesis, indicating the existence of hemangioblasts and hemogenic ECs. A recent report shows Notch2 has a role in EC and vascular dysfunction. Inflammatory cytokines can elicit a switch toward Notch2 expression over Notch4, which leads to the reduction of Notch activity and the promotion of apoptosis (Quillard et al. 2010).

Mutations in human Notch3 (MW:244 and NICD MW:86kDa) cause CADASIL (cerebral autosomal dominant arteriopathy with subcortical infarcts and leukoencephalopathy), a late-onset disorder causing stroke and dementia, which arises from slowly developing systemic vascular lesions ultimately resulting in the degeneration of vascular smooth muscle cell (Ruchoux et al. 1995; Joutel et al. 1996). Notch3 expression maintains a differentiated phenotype of mural cells (smooth muscle cells, pericytes, or fibroblasts) through an autoregulatory loop that requires endothelial-expressed JAG1 (Liu et al. 2009). Thus, Notch3 is considered to be critical for proper angiogenesis and mural cell investment (Liu et al. 2010). ECs and mural cell interactions support fully functional blood vessels and regulate vessel assembly and differentiation or maturation. Alterations in mural cell density and attachment to the endothelium are associated with several human diseases such as diabetic retinopathy, venous malformation, and hereditary stroke. In addition mural cells are implicated in regulating tumor growth and have thus been suggested as potential targets in tumor anti angiogenic therapy (Gaengel et al. 2009).

Notch4 (MW:230 and NICD MW:80kDa) actions in breast cancer are also cell-context dependent. Early studies indicated that Notch4 is an EC specific homologue of Notch and it may play a crucial role in vasculogenesis and angiogenesis (Shirayoshi et al. 1997). However, in contrast to Notch1, evidences suggest that constitutive Notch4 activation in ECs inhibits angiogenesis in part by promoting β1-integrin-mediated adhesion to the underlying matrix (Leong et al. 2002). In addition, Notch4 induced inhibition of angiogenesis requires the ankyrin repeats and appears to involve RBP-Jκ-dependent and -independent signaling (MacKenzie et al. 2004). On the other hand, mammary carcinogenesis was related to gain-of-function mutations of Notch4 leading to deregulated levels of the Notch4 NICD (Jhappan et al. 1992; Gallahan et al. 1996). Transgenic expression of the 1.8 Kb Notch4 RNA species in non-malignant human mammary epithelial cell line MCF-10A enabled these cells to grow in soft agar, suggesting Notch4 can transform MCF-10A cells (Imatani and Callahan 2000). Notch4 was also found to subvert normal epithelial morphogenesis and to promote invasion of the extracellular matrix. Moreover, Notch4 significantly increased the tumorigenic potential *in vitro* of mammary epithelial cells by changing the morphogenetic properties (Soriano et al. 2000; Callahan and Raafat 2001).

7   Notch Signaling, Cancer Stem Cell and Angiogenesis

A new theory about the initiation and progression of cancer is emerging from the idea that tumors, like normal adult tissues, contain stem cells (called cancer stem cells, CSC) and more importantly, could arise from them (Ischenko et al. 2008). Genetic mutations in genes encoding proteins involved in critical signaling pathways for stem cells such as BMP (bone morphogenetic protein), Notch, Hedgehog and Wnt would allow cells to undergo uncontrolled proliferation and form tumors. Notch receptors and/or ligands were demonstrated to correlate with CSCs in several cancer types (Table 2). In the case of breast cancer, Notch activity had increasingly been investigated in breast CSC (BCSC) subpopulation (Dontu et al. 2004; Sansone et al. 2007a; Harrison et al. 2010). Up-regulated Notch expression was found in BCSC and initiating cell populations identified by phenotypic markers $CD44^+/CD24^-$ (Farnie and Clarke 2007) and linked to tumor-initiating properties and CSC-like invasive features (Farnie and Clarke 2007). Notch1 NICD impairs mammary stem cell ($CD24^+CD29^{high}$) self-renewal and facilitates their transformation through a cyclin D1-dependent pathway (Ling et al. 2010). Moreover, Notch1 is related to BCSC self-renewal. The ErbB2 (HER2) promoter contains Notch-RBP-Jκ binding sequences (Chen et al. 1997) that can be activated by Notch1 signaling and increase HER2 transcription in both mammary stem/progenitor cells (Dontu et al. 2004; Politi et al. 2004) and BCSC. These Notch1 effects could impact the self-renewal properties of BCSC (Magnifico et al. 2009). Expression of erythropoietin receptor (EpoR) is found on the surface of BCSC (Phillips et al. 2007). It has been shown that Notch1 interacts with erythropoietin (Epo) to maintain the self-renewing capacity of BCSC. In addition, recombinant human Epo (rhEpo) increased the numbers of BCSC and self-renewing activity in a Notch-dependent manner through induction of JAG1 (Phillips et al. 2007).

Notch1 mRNA is primarily expressed in luminal cells of normal breast epithelium (Raouf et al. 2008). In contrast, Notch4 is mainly present in the basal cell population and in the BCSC-enriched population (Harrison et al. 2010). These data suggest that Notch1 and Notch4 may impact different cell subpopulations and have different roles in BCSC. Secretase inhibitors, DAPT and DBZ, which preferentially affect Notch1 activity, only partially abrogated mammosphere-forming units (MFUs) and tumor formation, whereas Notch4 knockdown caused a significantly greater inhibition in MFUs than Notch1 (Harrison et al. 2010). Therefore, it was suggested that Notch4 signaling regulates the route from BCSC into progenitor populations. In contrast, Notch1 activity regulates the progenitor proliferation and luminal differentiation. Then, the single activation of Notch1 receptor

Table 2,   Function of Notch signaling in cancer stem cells (CSC)

Cancer types	CSC marker	Notch/ligand	Function or changes	References
Bladder cancer	K5, p63, BMI-1, OCT-4	Notch1	Activation	(Tokar et al. 2010)
Brain medulloblastoma	CD133, CD15	DLL1	Cell proliferation	(de Antonellis et al. 2011)
Brain glioblastoma	CD133	Jagged1, DLL1, DLL4	Cancer stem-like cell self-renewal	(Zhu et al. 2011)
Brain glioblastoma	CD133	Notch1	Cell proliferation	(Wang et al. 2011)
Brain glioblastoma	CD133, NESTIN, BMI1	Notch2	Tumor growth	(Fan et al. 2010)
Brain medulloblastoma	CD133	HES1	Cell proliferation and anchorage	(Garzia et al. 2009)
Brain medulloblastoma	CD133	Notch pathway	Stem-like cells self-renewal	(Fan et al. 2006)
Breast cancer	CD44Hi/CD24Low	Notch1	CSC self-renewal	(Sharma et al. 2011)
Breast cancer	CD44Hi/CD24Low	Notch1	Brain metastases	(McGowan et al. 2011)
Breast cancer	CD24(-) CD44(+)	Notch1	Cell proliferation	(Du et al. 2010)
Breast cancer	ESA(+)/CD44(+)/CD24(low)	Notch4	Stem cell activity & tumor formation	(Harrison et al. 2010)
Colon cancer	CD133, CD44, ESA, ALDH1	Notch	Prevention of apoptosis	(Sikandar et al. 2010)
Lung cancer	ALDH(+)	Notch3	CSC maintenance	(Sullivan et al. 2010)
Liver cancer	Oct3/4, OV6, CD133, EpCAM	Notch	Self-renewal, extensive proliferation	(Cao et al. 2011)
Liver cancer	EpCAM	Jagged1	Tumorigenesis	(Nishina et al. 2011)
Ovary cancer	CD44 High/CD24 Low	Notch1	Proliferation/division and survival	(Mine et al. 2010)
Pancreatic cancer	CD44 and EpCAM	Notch1	Acquisition of EMT phenotype	(Bao et al. 2011)

gene might not be sufficient to generate mammary carcinogenesis in mice. Conversely, activation of the Notch4 receptor inhibited mammary epithelial cell differentiation and could be sufficient to generate mammary carcinogenesis in mice (Gallahan and Callahan 1987).

Notch2 and Notch3 have been also linked to BCSC. Recently, a single nucleotide polymorphism (SNP) rs11249433 in the 1p11.2 region has been identified as a novel risk factor for breast cancer and strongly associated with ER+ versus ER- cancer (Fu et al. 2010). Notch2 expression was particularly enhanced in carriers of the risk genotypes (AG/GG) of rs11249433, that may favor development of ER+ luminal tumors and affect tumor-initiating cells (Fu et al. 2010). Notch3 is a poor activator of Hairy/Enhancer of split 1 and 5 (Hes-1 and Hes-5), in contrast to that of Notch1 (Beatus et al. 1999). The Notch3 intracellular domain represses Notch1-mediated activation through Hes promoters. Notch3 is critical for the differentiation of human progenitor cells to luminal lineage *in vitro* (Raouf et al. 2008). Notch activation leads to the formation of dimers of Hes and/or Hey proteins that repress the transcription of a variety of genes by interacting with co-repressors or sequestering transcriptional activators. Moreover, activation of canonical Notch signaling induces the maintenance of stem or progenitor cells through the inhibition of normal cell differentiation (Artavanis-Tsakonas et al. 1999; Radtke and Raj 2003; Leong and Karsan 2006). In recent years, several oncogenes, such as HER2 (Korkaya et al. 2008; Korkaya and Wicha 2009), Akt (Korkaya et al. 2009) as well as transcriptional factors, such as STAT3 (Zhou et al. 2007), NF-κB (Pratt et al. 2009) were found to be associated with BCSC. Since Notch signaling crosstalks with these oncogenic pathways (will be discussed below), it could impact BCSC and breast cancer development through these signals.

CSCs also have critical roles in promoting tumor angiogenesis. Strong evidence comes from studies of correlation of CSCs and VEGF/VEGFR. The VEGF expression in CD133+ glioma CSCs was 10–20 fold up-regulated, combined with a dramatically increased vascular density identified by CD31 staining (Bao et al. 2006). In addition, VEGF neutralizing antibody (bevacizumab) can deplete glioma CSCs-induced vascular EC migration and tube formation (Bao et al. 2006). Recent studies further established that CXCL12 and its receptor CXCR4 may promote glioma CSCs growth and angiogenesis by stimulating VEGF production (Ping et al. 2011). Malignant melanoma-initiating cells (MMIC) were identified by an ATP-binding cassette (ABC) member ABCB5 (Zabierowski and Herlyn 2008; Gazzaniga et al. 2010). ABCB5 (+) melanoma cells have been shown to overexpress the vasculogenic differentiation markers CD144 (VE-cadherin) and TIE1 and are associated with CD31 (–) vasculogenic mimicry (VM), an established biomarker associated with tumor angiogenesis and increased patient mortality (Maniotis et al. 1999; Frank et al. 2011). Induced VEGF was found in ABCB5 (+) cells that constitutively expressed VEGFR-1 but not in ABCB5 (–) bulk populations that were predominantly VEGFR-1 (–). *In vivo*, melanoma-specific shRNA-mediated knockdown of VEGFR-1 inhibited the development of ABCB5 (+) VM morphology and ABCB5 (+) VM-associated production of the secreted melanoma mitogen laminin (Frank et al. 2011). These data support the notion that not only VEGF, but also VEGFR-1 in MMIC regulates VM and is associated with laminin production and tumor angiogenesis. Additional evidence was provided in a report in which tumors with larger CSC population recruited a substantially higher amount of endothelial progenitor cells (EPC), indicating that CSCs promote local angiogenesis and EPC mobilization via stimulating pro angiogenic factors such as VEGF and SDF-1 (Folkins et al. 2009).

To date, our understanding of the interplay between CSCs and angiogenesis is limited and continues to evolve with intense investigations. The readers are referred to Chapter 3 for more detailed discussions on this subject (Liu, et al. Endothelial and Cancer Stem Cell Regulation of Tumor Angiogenesis). Although we believe Notch signaling is critical for both CSCs and angiogenesis, other signaling pathways which may have direct or indirect interactions with Notch are also important as discussed below.

8 Crosstalk among Notch Signaling and Other Oncogenic Pathways in Tumor Angiogenesis

In human cancers, activation of Notch signaling can upregulate several factors that in turn transmit bidirectional signals among cancer cells expressing both ligands and receptors. Notch could also transmit signals among cancer, stroma and endothelium cells (Miele 2006; Rizzo et al. 2008a). Therefore, it is not surprising that Notch signaling crosstalks with many oncogenic signaling pathways, such as developmental signals, i.e., Wnt and Hegdehog signaling, growth factors, cytokines, oncogenic kinases as well as transcriptional factors.

8.1 Developmental Signaling

8.1.1 Hedgehog Signaling

Hedgehog is a developmental signaling pathway that plays key roles in a variety of processes, such as embryogenesis, maintenance of adult tissue homeostasis, tissue repair during chronic persistent inflammation, and carcinogenesis (Pasca di Magliano and Hebrok 2003; Lum and Beachy 2004; Hooper and Scott 2005). More recently, hedgehog signaling has also been implicated in angiogenesis. While hedgehog signaling in adult angiogenesis may constitute a simple recapitulation of that in embryonic development, it should be appreciated that Hedgehog signaling occurs in embryonic angiogenesis in different developmental contexts (Nagase et al. 2008). Hedgehog family ligands, Sonic hedgehog (Shh), Indian hedgehog (Ihh) and Desert hedgehog (Dhh), undergo autoprocessing and lipid modification to generate mature peptides (Marigo et al. 1995; Katoh and Katoh 2005a; Katoh and Katoh 2005b). Genetic evidence in mice as well as molecular biological studies in human cells clearly indicate that deregulated Hedgehog signaling can lead to mammary hyperplasia and tumor formation (Visbal and Lewis 2010). Notch ligand, JAG2 is induced by Hedgehog signaling during carcinogenesis (Katoh 2007).

Notch-Hedgehog crosstalk induces the expression of Hes3 and Shh through rapid activation of cytoplasmic signals, including Akt, STAT3 and the mammalian target of rapamycin (mTOR), promoting the survival of neural stem cells (Androutsellis-Theotokis et al. 2006). Hedgehog signals could induce Hes1 in both C3H/10T1/2 mesodermal and MNS70 neural cells (Ingram et al. 2008). In human breast cancer, deregulated Hedgehog, together with Notch and Wnt signals, could also regulate the self-renewal and differentiation ability of BCSC (Zhao et al. 2010).

8.1.2   Wnt Signaling

Deregulation of the Wnt pathway has been extensively studied in multiple diseases, including some angiogenic disorders. Wnt signaling activation is a major stimulator in pathological angiogenesis and thus, Wnt antagonists are considered to have a therapeutic role for neovascular disorders (Dejana 2010; Zhang and Ma 2010). Some Wnt antagonists have been identified directly from the anti-angiogenic factor family (Barcelos et al. 2009; Hu et al. 2009; Zhang et al. 2010). Wnt1 expression led to subsequent activation of Notch signaling in human mammary epithelial cells (HMEC) (Ayyanan et al. 2006). Concomitant up regulation of the Wnt target genes *Lef1* and *Axin2* along with Notch ligand Dll-3 and Dll-4 was found in breast carcinomas, suggesting that the same process takes place in other tumors (Ayyanan et al. 2006). The blockade of expression of Notch ligands abrogated HMEC transformation by Wnt1, demonstrating the requirement for Notch-Wnt crosstalk during mammary tumorigenesis (Ayyanan et al. 2006). Notch-regulated ankyrin repeat protein (Nrarp) acts as a molecular link between Notch- and Wnt signaling in ECs to control stability of new vessel connections in mouse and zebrafish (Phng et al. 2009). Dll4/Notch-induced expression of Nrarp limits Notch signaling and promotes Wnt signaling in endothelial stalk cells through interactions with Lymphoid enhancer-binding factor-1 (Lef1) (Phng et al. 2009). These results suggest that the balance between Notch and Wnt signaling determines whether to make or break new vessel connections.

8.2   Growth Factors

8.2.1   HER/ErbB

HER/ErbB genes (HER1/EGFR, HER3 and HER4) encode for receptor tyrosine kinase (RTK)-transmembrane proteins that, upon binding of several ligands (epidermal growth factor, EGF; amphiregulin; heregulin or *neu*-mouse and transforming growth factor alpha, TGF-α) induce signaling involved in the regulation of cell proliferation, differentiation and survival (Koutras et al. 2010; Tai et al. 2010). Deregulation of EGF receptor (EGFR) by over-expression or constitutive activation can promote tumor processes including angiogenesis and metastasis. Moreover, EGFR over-expression is associated with poor prognosis in many human malignancies (Jimeno and Hidalgo 2006; Flynn et al. 2009).

A number of studies have established that growth factors and their receptors play an essential role in regulating the proliferation of epithelial cells (Kumar and Yarmand-Bagheri 2001; Press and Lenz 2007). Over-expression of HER2 in human tumor cells is closely associated with increased angiogenesis and expression of VEGF-A. HER2 signaling may increase the rate of hypoxia-inducible factor 1 α (HIF-1α) synthesis which in turn mediates VEGF-A expression (Laughner et al. 2001). On the other hand, when the VEGF pathway is inhibited, tumor growth is suppressed. The anti-HER2 antibody trastuzumab has been shown to inhibit tumor cell growth and VEGF expression (Koukourakis et al. 2003; Kumar et al. 2003).

Notch signaling could regulate HER2 activity since the HER2 promoter contains Notch-binding sequences (Chen et al. 1997). Yamaguchi et al (Yamaguchi et al. 2008) observed that down-regulation of Notch3 significantly suppressed proliferation and promoted apoptosis of the ErbB2-negative tumor cell lines. Magnifico et al demonstrated that HER2-overexpressing cells display activated Notch1 signaling (Magnifico et al. 2009). Their results show that inhibition of Notch1 signaling by small interfering RNA or γ-secretase inhibitor resulted in down-regulation of HER2 expression and decrease of sphere formation (Magnifico et al. 2009).

In contrast to HER2, the long-standing relationships between the EGFR and Notch signaling pathways, and the opposing effects exerted by these signal transduction cascades, have been well documented in various developmental settings and organisms (Sundaram 2005; Hasson and Paroush 2006). Dai et al found that forced overexpression of Notch1 by transfection increased EGFR expression in human breast cancer cells (Dai et al. 2009). Moreover, overexpression of Notch1 reversed EGFR inhibitor-induced cell toxicity, suggesting that Notch and EGFR signaling may be positively cross-linked in human breast cancer. Dong et al (Dong et al. 2010) further observed that inhibition of either EGFR or Notch signaling alone was insufficient to suppress basal-like breast tumor cell survival and proliferation. However, simultaneous inhibition of EGFR and Notch signaling uncovered a lethal relationship between these two oncogenic pathways.

8.2.2 PDGF/PDGFR Signaling

Platelet-derived growth factor (PDGF) is a potent angiogenic family of molecules comprised four polypeptide chains encoded by different genes. PDGF-A and PDGF-B were identified earlier whereas PDGF-C and PDGF-D were discovered more recently (Bergsten et al. 2001; LaRochelle et al. 2001; Li and Eriksson 2003). The PDGF isoforms exert their cellular effects by specific binding to two structurally related tyrosine kinase receptors (α and β PDGFR). PDGF is a potent mitogen and chemoattractant for mesenchymal cells, neutrophils and monocytes (Li et al. 2007). Therefore, the expression of PDGF correlates with advanced tumor stages and unfavorable prognosis in human breast carcinomas (Seymour and Bezwoda 1994). PDGF produced in carcinomas is generally thought to act on the non-epithelial tumor stroma promoting angiogenesis (Pietras et al. 2003).

The growing body of literature strongly suggests that a crosstalk between PDGF-D and Notch signaling occurs in cancer (Wang et al. 2010a). Dr. Sarkar's group demonstrated that down-regulation of PDGF-D leads to the inactivation of Notch1 and NF-κB DNA-binding activity, as well as down-regulation of its target genes, such as VEGF and MMP-9 in pancreatic cancer cells (Wang et al. 2007). Therefore, the inactivation of PDGF-D-mediated cell invasion and angiogenesis could in part be attributable to inactivation of Notch1 (Wang et al. 2007). Additionally, down-regulation of PDGF-D also inhibited the Notch1 expression in breast cancer cells (Ahmad et al. 2010). Interestingly, mRNA and protein expressions of Notch1~4, Dll-1, Dll-3, Dll-4, JAG2 as well as Notch downstream targets, such as Hes and Hey were significantly higher in PC3 prostate cancer cells expressing PDGF-D, indicating PDGF-D was correlated to Notch signaling (Wang et al. 2010).

8.2.3   TGF-β Signaling

Genes encoding components of TGF-β signaling pathway, including ligands TGF-β1 and TGF-β2 and receptor TGFBRI are functionally polymorphic in humans (Cambien et al. 1996; Pasche et al. 1999; Beisner et al. 2006). TGF-β can regulate such diverse processes as cell proliferation, differentiation, motility, adhesion, organization, and programmed cell death. Both *in vitro* and *in vivo* experiments suggest that TGF-β can utilize these diverse programs to promote cancer metastasis through its effects on the tumor microenvironment, enhanced invasive properties, and inhibition of immune cell function (Barcellos-Hoff and Akhurst 2009; Padua and Massague 2009). In recent years, knockout studies of factors in TGF-β signaling have shown that this pathway is also indispensable for angiogenesis (Pardali and ten Dijke 2009; Pardali et al. 2010; van Meeteren et al. 2011).

TGF-β signaling is linked to Notch in many processes. First, TGF-β can upregulate Notch ligands. JAG1 has been shown to be a TGF-β target gene in multiple types of mammalian cells. JAG1 and Hey1 are critical for TGF-β-induced epithelial-mesenchymal transformation (EMT) in cells derived from several organs (Zavadil et al. 2004). In addition, JAG1 upregulation also contributes to TGF-β effects on cell cycle by stimulating p21 expression and cytostasis in epithelial cells (Niimi et al. 2007). Second, TGF-β and Notch can synergistically regulate common target genes in many cell types, for example, Smad3, a downstream transcription factor of TGF-β and Notch1 NICD can directly interact and form a complex with CSL that binds to specific DNA sequences as those found in the promoter of *Hes-1* (Blokzijl et al. 2003). Notch1 NICD not only interacts with activated Smad3 and facilitates its nuclear translocation (Asano et al. 2008), but also remains bound with pSmad3 in the nucleus where they jointly upregulate the transcription factor Forkhead box P3 (Foxp3) that is involved in immune processes (Samon et al. 2008).

8.2.4   VEGF/VEGFR-2 Signaling

Vascular endothelial growth factor (VEGF) is the major angiogenic factor in physiological and pathological angiogenesis (Ferrara 1999; Ferrara et al. 2003). The expression of the VEGF gene is enhanced in a variety of angiogenic tumors (Ferrara 1999; Ferrara et al. 2003). VEGFR-2, receptor type 2 (KDR or flk-1) is generally recognized to have a principal role in mediating VEGF-induced responses and is considered as the earliest marker for EC development (Guo et al. 2010). Moreover, VEGFR-2 directly regulates tumor angiogenesis (Ferrara et al. 2003; Shibuya and Claesson-Welsh 2006; Guo et al. 2010). In addition to its angiogenic actions in ECs, the VEGF/VEGFR-2 signaling paracrine-autocrine loop functions as an important survival process in cancer cells (Guo et al. 2010)   VEGF was first shown to act upstream of Notch in determining arterial cell fate in vascular development (Lawson et al. 2002). VEGF was demonstrated to increase Dll-4 and Notch expression, in turn leading to the activation of Notch signaling and arterial specification (expression of a set of arterial genes). Further studies in several systems established that VEGF regulates the expression of Notch signaling components (Patel et al. 2005; Hainaud et al. 2006; Ridgway et al. 2006). Blocking VEGF, by intravitreal injection of soluble VEGF receptors, results in decreased sprouting and reduced expression of Dll-4 in retinal vessels (Suchting et al. 2007). Similar interactions among VEGF signaling, growing vessels and Notch components expression were found in tumor vessels (Mailhos et al. 2001; Patel et al. 2005; Noguera-Troise et al. 2006; Sainson et al. 2008).

Providing a feedback mechanism, Notch signaling in turn can alter expression levels of all three VEGF receptors. For example, VEGFR-2 was down-regulated by either Notch1, 4 or Hey1 in ECs (Taylor et al. 2002). Reciprocally, VEGFR-2 expression increased in vessels of Dll-4 heterozygous mice or as a result of Dll-4 blockade (Suchting et al. 2007). Thus, Notch signaling can provide negative feedback to reduce the activity of the VEGF/VEGFR-2 axis in ECs.

The emerging picture is that VEGF pathway acts as a potent upstream activating stimulus for angiogenesis, whereas Notch pathway helps to shape that action appropriately (Siekmann et al. 2008; Thurston and Kitajewski 2008). Thus, an important feature of angiogenesis is the manifold ways in which the VEGF and Notch pathways interact (Thurston and Kitajewski 2008).

8.3   Inflammatory Cytokines

8.3.1   IL-6

IL-6, a multifunctional cytokine, produced by various types of cells, including macrophages and cancer cells, is an important factor for immune responses, cell survival, apoptosis, proliferation and angiogenesis (Kishimoto 2005; Neurath and Finotto 2011; Tawara et al. 2011). IL-6 signals via a heterodimeric IL-6R/gp130 receptor complex, whose engagement triggers the activation of Janus (JAK) kinases, and the downstream effectors STAT proteins (Kishimoto 2005). A number of studies implicated IL-6 and STAT3 as pro-tumorigenic and pro-angiogenic agents in many cancers (Nilsson et al. 2005; Shinriki et al. 2009; Liu et al. 2010; Anglesio et al. 2011; Wani et al. 2011).

Sasone et al first determined that Notch pathway was a critical downstream target of IL-6 (Sansone et al. 2007b). IL-6 treatment triggered Notch3-dependent upregulation of the Notch ligand JAG1 and promotion of primary human mammospheres and MCF-7-derived spheroid growth. Moreover, autocrine IL-6 signaling relied upon Notch3 activity to sustain the aggressive features of MCF-7-derived hypoxia-selected cells. Thus, these data support the hypothesis that IL-6 induces malignant features in Notch3-expressing stem/progenitor cells from human ductal breast carcinoma and normal mammary gland. These authors also pointed out that the hypoxia resistance gene carbonic anhydrase (CA-IX) is activated in breast cancer cells by IL-6/Notch/JAG action and provides survival advantages under hypoxic conditions. Very recently, within the IL-6 gene promoter region, the signature binding motif of CSL was identified and found to overlap with a consensus of NF-κB-binding site (Wongchana and Palaga 2011). These authors demonstrated that Notch1 positively regulates IL-6 expression via NF-κB in activated macrophages (Wongchana and Palaga 2011).

Lee et al established a HeLa/rtTAA/TRE-N1-IC cell line capable of doxycycline-induced expression of human Notch1 NICD (Lee et al. 2009). They found that the induction of Notch signaling activated HIF-1α and its target gene expression in the above cells. Interestingly, HIF-1α expression was required for Notch signaling enhanced STAT3 phosphorylation required under hypoxia conditions. Furthermore, Src (a proto-oncogenic tyrosine kinase) was also required for the enhanced STAT3 phosphorylation in response to Notch signaling. Notch signaling activated Src/STAT3 pathway was dependent on the Notch effector Hes1 transcription factor. However, the treatment of Trichostatin A (TSA) that interferes with Hes1 transcriptional regulation did not affect STAT3 phosphorylation, and dominant negative Hes1 failed to interfere with Hes1-dependent Src/STAT3 pathway and induction of HIF-1α. These observations indicate that Hes1-dependent activation of Src/STAT3 pathway is independent of Hes1 transcription regulation. Therefore, Hes1-dependent Src/STAT3 pathway provides a functional link between Notch signaling and hypoxia pathway.

8.3.2 IL-1 Signaling

IL-1 family belongs to pro-inflammatory/-angiogenesis cytokines that is represented by two ligands: IL-1α, IL-1β, an antagonist: interleukin-1 receptor antagonist (IL-1Ra) and two receptors: IL-1R tI (type I receptor) and IL-1R tII (type II receptor) (Dinarello 1998). IL-1 plays a key role in the onset and development of the host reaction to invasion, being an important factor in the initiation of the inflammatory response and immune functions. IL-1 is also abundant at tumor sites, where it may affect the process of carcinogenesis, tumor growth and invasiveness, the patterns of tumor-host interactions and tumor angiogenesis (Apte et al. 2006). There is also convincing evidence that IL-1 family and leptin (the major adipocytokine) crosstalk represents a major link among obesity, inflammation, angiogenesis and cancer progression (Perrier et al. 2009; Zhou et al. 2010; Guo and Gonzalez-Perez 2011; Zhou et al. 2011).

IL-1 activates Notch signaling pathway probably through NF-κB pathway (Cao et al. 1999; Verstrepen et al. 2008; Vallabhapurapu and Karin 2009). NF-κB is present as a latent, inactive, light polypeptide gene enhancer (I-κB, inhibitor of NF-κB)-bound complex in the cytoplasm in majority of cells. IL-1 activates NF-κB via IL-1 receptor-associated kinase (IRAK) and mitogen-activated protein kinase (MAPK) dependent inhibition of I-κB (Renard and Raes 1999; Yao et al. 2007). c-Rel (an NF-κB subunit) can trigger Notch1 signaling pathway by inducing expression of JAG1 (Bash et al. 1999; Osipo et al. 2008). Results from our laboratory suggest that leptin is an important inducer of IL-1 system in breast cancer cells (Zhou et al. 2010). Moreover, IL-1, Notch and leptin-induced upregulation of their gene components and NF-κB, HIF-1α and VEGF/VEGFR-2 are interconnected (Gonzalez-Perez et al. 2010; Guo et al. 2010).

8.3.3 Leptin Signaling

Leptin, a pluripotent cytokine secreted primarily by adipocytes but also by breast cancer cells, plays key roles in regulating energy intake and energy expenditure, including appetite and metabolism (Brennan and Mantzoros 2006). In the past decade, accumulating evidence indicates that leptin actions are related not only to energy metabolism, but also to reproduction, proliferation, inflammation and angiogenesis (Dardeno et al. 2010; Fernandez-Riejos et al. 2010). Breast cancer cells express higher levels of leptin and leptin receptor, OB-R, than normal mammary cells. Importantly, higher levels of leptin/OB-R levels correlated with metastasis and lower survival of breast cancer patients (Tessitore et al. 2000; Hu et al. 2002; Laud et al. 2002). *In vitro*, leptin was demonstrated to stimulate the proliferation of breast cancer cell lines (Dieudonne et al. 2002; Hu et al. 2002; Chen et al. 2006). *In vivo* studies clearly demonstrated a role for leptin in mammary tumor initiation and development as evidenced by the fact that mutant mice deficient in leptin ($Lep^{ob}Lep^{ob}$), or with non-functioning leptin receptors ($Lepr^{db}Lepr^{db}$) do not develop transgene-induced mammary tumors (Cleary et al. 2003; Cleary et al. 2004). In our previous reports, the disruption of leptin signaling using pegylated leptin peptide receptor antagonist (PEG-LPrA2) markedly reduced the growth of tumors in mouse models of syngeneic and human breast cancer xenografts (Gonzalez et al. 2006; Rene Gonzalez et al. 2009). These effects were accompanied by a significant decrease in VEGF/VEGFR-2, IL-1 R tI, cyclin D1 and PCNA levels. Moreover, tumor angiogenesis was also impaired (Gonzalez et al. 2006; Rene Gonzalez et al. 2009).

Leptin and IL-1 have been shown to be associated in several pathological situations (Kumar et al. 2003; Johnston et al. 2008), suggesting an interplay between them. Indeed, leptin regulates IL-1 family members in a diabetic context (Maedler et al. 2004) and in endometrial cancer cells (Carino et al. 2008). We also found that leptin increased protein and mRNA levels of all components of the IL-1 system in a mouse mammary cancer cell line. Leptin-induced canonical signaling pathways (JAK2/STAT3, MAPK/ ERK 1/2 and PI-3K/Akt1) were mainly involved in IL-1 up-regulation. In addition, leptin upregulation of IL-1α promoter involved the activation of SP1 and NF-κB transcription factors (Zhou et al. 2010).

Little information on leptin-Notch interactions is available. An earlier report shows that leptin regulates the expression of JAG1 and Notch4 in human cord blood CD34$^+$ cells and early differentiated ECs (HUVEC) and also promotes cell differentiation (Polus A. 2003). We and others recently observed that leptin was able to activate Notch signaling pathway in breast cancer cells (Guo and Gonzalez-Perez 2011; Knight et al. 2011). Moreover, leptin increased the expression of both Notch receptors and ligands (Guo and Gonzalez-Perez 2011). In these cells leptin also up-regulated Notch-target genes Hey2 and Survivin (Guo et al. 2010; Guo and Gonzalez-Perez 2011). Leptin-induced non-canonical signaling pathways (PKC, p38 and JNK) differentially impacted on CSL promoter activity and on the expression of IL-1 system in mouse 4T1 mammary cancer cell line (Zhou et al. 2010). Interestingly, effects of leptin upregulation on pro-angiogenic factors IL-1, VEGF/VEGFR-2 and Notch were significantly abrogated by a γ-secretase inhibitor, DAPT as well as siRNA against CSL in 4T1 cells (Guo and Gonzalez-Perez 2011).

We speculate that complex molecular mechanisms are involved in leptin interactions with Notch and IL-1 in breast cancer cells. Notch, IL-1 and leptin crosstalk outcome (NILCO) is required for leptin regulation of VEGF/VEGFR-2 (Guo and Gonzalez-Perez 2011). As discussed above, Notch could crosstalk with and be regulated by STAT3, NF-κB, HIF-1α and HER2. These factors are also regulated/activated by leptin and IL-1. Thus, NILCO could impact on Notch signaling regulation as follows: (1) STAT3 activation: leptin binding to OB-Rb (long and full-functional isoform) activates JAK2 that is recruited/autophosphorylated- and phosphorylates OB-R and STAT3 (Banks et al. 2000; Bates et al. 2003). Leptin–induced activation of STAT3 was required for the induction of CSL, IL-1 (unpublished data) and VEGF (Gonzalez-Perez et al. 2010). (2) Leptin activation of NF-κB is involved in the upregulation of IL-1 (Zhou et al. 2010). NF-κB and HIF-1α are essential factors for leptin regulation of VEGF in breast cancer cells (Gonzalez-Perez et al. 2010). Moreover, the blockade of leptin signaling markedly reduced the growth of tumors and the expression of VEGF in mouse models of syngeneic and human breast cancer xenografts (Gonzalez et al. 2006; Rene Gonzalez et al. 2009). Leptin increased VEGF/VEGFR-2 expression in endometrial cancer cells *in vitro* (Carino et al. 2008) and in breast cancer cells *in vitro* and *in vivo* (Gonzalez et al. 2006; Rene Gonzalez et al. 2009). (3) Leptin/HER2 crosstalk: Leptin treatment activates HER2 through transphosphorylation on Tyr-1248 in the HER2 intracytoplasmatic tail (Eisenberg et al. 2004). HER2 physically interacts with OB-R and, thus the leptin/OB-R signaling might contribute to enhanced HER2 activity in MCF-7 cells (Fiorio et al. 2008). (4) Leptin increases estrogen level and ERα activity: ERα and Notch have an active crosstalk (see section 8.5.1) (Soares et al. 2004; Rizzo et al. 2008b). Aromatase, a product of the CYP19 gene and a member of the cytochrome P450 (CYP) enzyme family, catalyzes a rate-limiting step for the conversion of androstenedione to estrone and testosterone to E2 (Ghosh et al. 2009). Leptin can enhance aromatase activity and expression either in stromal cells or breast cancer cell lines, thereby increasing estrogen production (Magoffin et al. 1999; Catalano et al. 2003). In addition, leptin transactivates ERα in MCF-7 cells (Catalano et al. 2004).

Obesity has become a pandemic particularly in western countries. Increased leptin signaling is associated with obesity and breast cancer incidence. Combinatory therapy targeting NILCO would enhance the disruption of specific signaling pathways involved in breast cancer progression. This might help to design new strategies aimed at controlling growth, angiogenesis and metastasis in breast cancer.

8.4 Oncogenic Kinases and Transcription Factors

8.4.1 Ras Signaling Pathway

Ras signaling plays an important role transmitting signaling from RTKs to ser/threo kinases. Among the effector molecules connected with the group of cell surface receptors, Ras plays essential roles in transducing extracellular signals to diverse intracellular events by controlling the activities of multiple signaling pathways (DeNicola and Tuveson 2009). The multifunctional signal transducer Ras is a proto-oncogene and frequently becomes mutated in a variety of human cancers, including angiosarcomas (Bajaj et al. 2010). Because Ras signaling impacts many cellular functions, including cell cycle regulation, apoptosis, cell survival, EC function and angiogenesis, it is a major target for the development of novel cancer treatments (Young et al. 2009; Bajaj et al. 2010). The signaling networks Ras regulates are very complex due to their multi-faceted functions and crosstalk (Zebisch et al. 2007).

Crosstalk between Ras and Notch pathways has been described in pancreatic ductal adenocarcinoma (Hanlon et al. 2010), colorectal tumors (Veenendaal et al. 2008), astrocytic gliomas (Xu et al. 2010), leukemia (Chiang et al. 2008; Kindler et al. 2008), as well as breast cancer (Gustafson et al. 2009; Mittal et al. 2009). In an early report (Weijzen et al. 2002), Weijzen et al demonstrated that oncogenic Ras activates Notch signaling. Notch1 was necessary to maintain the neoplastic phenotype in Ras-transformed human cells *in vitro* and *in vivo* (Weijzen et al. 2002). Ras increased the expression and activity of Notch1 NICD and upregulated Notch ligand Dll-1 and presenilin-1, a protein involved in Notch processing, through a p38-mediated pathway (Weijzen et al. 2002). These observations established that Notch signals were among the key downstream effectors of oncogenic Ras. Gustafson et al (Gustafson et al. 2009) observed that transformation of MCF-10A cells by Harvey-Ras (Ha-Ras) induced CCAAT/enhance binding protein beta (C/EBPβ), a transcriptional factor, and activated the Notch signaling pathway to block SIM2s (a transcriptional factor) gene expression. High expression level of Notch receptors, ligands and their cooperation with the Ras/MAPK pathway in several breast cancers and early precursors place Notch signaling as a key player in breast cancer pathogenesis. This offers combined inhibition of the two pathways as a new modality for breast cancer treatment (Mittal et al. 2009). Given the regulation of Ras is important for EC function, angiogenesis and activated Ras signaling is critical for vascular malformations and angiosarcoma, crosstalk between Ras and Notch pathways might occur in ECs. However, the exact role and mechanism of these two pathways in ECs need to be determined.

8.4.2 PI-3K/Akt Signaling Pathway

The phosphatidylinositol 3-kinase (PI-3K/Akt) pathway plays a central role in a variety of cellular processes including cell growth, proliferation, motility, survival and angiogenesis in tumor cells (Steelman et al. 2008; Dillon and Muller 2010; Wickenden and Watson 2010). The PI-3K/Akt pathway is also instrumental in EMT during carcinogenesis (Sabbah et al. 2008).

Notch has been shown to regulate the Akt (ser/threo) or Protein kinase B (PKB) pathway. Liu et al (Liu et al. 2006) reported that Notch1 activation enhanced melanoma cell survival and the effects of Notch signaling were mediated via activation of the Akt pathway. Palomero et al (Palomero et al. 2008) found that Notch1 induced up-regulation of the PI-3K/Akt pathway via Hes1, which negatively controlled the expression of phosphatase and tensin homolog on chromosome 10 (PTEN) in T-cell acute Lymphoblastic Leukemia (T-ALL). Additional reports also demonstrated that Notch1 crosstalks with Akt pathway in T-ALL, melanoma as well as breast epithelial cells (Gutierrez and Look 2007; Bedogni et al. 2008; Meurette et al. 2009). On one hand, activation of Akt was necessary for Notch-induced protection against apoptosis in MCF-10A. On the other hand, inhibiting Notch signaling in breast cancer cells induced a decrease in Akt activity and an increase in apoptotic sensitivity (Meurette et al. 2009). Down-regulation of Notch1 or JAG1 mediated the inhibition of cell growth, migration and invasion, and the induction of apoptosis in prostate cancer. These effects were in part due to inactivation of Akt, mTOR, and NF-κB signaling pathways (Wang et al. 2010). In an early report (Rangarajan et al. 2001), activated Notch1 synergizes with papillomavirus oncogenes in transformation of immortalized epithelial cells and leads to the generation of resistance to anoikis, an apoptotic response induced by matrix withdrawal. This resistance to anoikis by activated Notch1 is mediated through the activation of PKB/Akt. The cellular responsiveness to Notch signaling dependent PI-3K/Akt pathway also occurs in other types of cells, such as Chinese hamster ovary (CHO) cells, primary T-cells and hippocampal neurons (McKenzie et al. 2006).

PI3-K/Akt is a regulator of Notch signaling in melanoma development. Hyperactivated PI3-K/Akt signaling led to upregulation of Notch1 through NF-κB activity, while the low oxygen content normally found in skin increased mRNA and protein levels of Notch1 via stabilization of HIF-1α in melanoma development (Bedogni et al. 2008).

PI3-K /Akt pathway is also a regulator of Noch1 and DLL4 in ECs. VEGF can induce gene expression of Notch1 and DLL4 in human arterial ECs. Furthermore, the VEGF-induced specific signaling is mediated through VEGFR-1 & 2 and is transmitted via the PI3-K/Akt pathway (Liu et al. 2003). Other reports also confirmed that PI3-K/Akt pathway can regulate Notch signaling in ECs (Kikuchi et al. 2011; Zhang et al. 2011).

8.4.3   mTOR Signaling

mTOR, a key protein kinase, controls signal transduction from various growth factors and upstream proteins to the level of mRNA translation and ribosome biogenesis. mTOR is a ser/threo kinase that is often a downstream effector of PI-3K/Akt signaling pathway in breasts and many types of cancer cells. However, MAPK pathway was identified as the preferential upstream regulator of mTOR in the induction of inflammatory/pro-angiogenic molecules in endometrial cancer cells (Carino et al. 2008). mTOR can also phosphorylate Akt (Sparks and Guertin 2010). mTOR has been intensely studied for over a decade as a central regulator of cell growth, proliferation, differentiation, autophagy, angiogenesis and survival (Menon and Manning 2008; Ciuffreda et al. 2010; Jung et al. 2010). mTOR functions as two distinct multiprotein complexes, mTORC1 and mTORC2 (Sparks and Guertin 2010; Zhou and Huang 2010). mTORC1 phosphorylates p70 S6 kinase (S6K1), eukaryotic initiation factor 4E (eIF4E) binding protein 1 (4E-BP1) and integrates hormones, growth factors, nutrients, stressors and energy signals. In contrast, mTORC2 is insensitive to nutrients or energy conditions. However, in response to hormones or growth factors, mTORC2 phosphorylates Akt, and regulates actin cytoskeleton and cell survival (Sparks and Guertin 2010). Aberrant activation of mTOR pathway is found in many types of cancer and thus plays a major role in breast cancer cell proliferation and anti-cancer drug resistance (Hadad et al. 2008; Noh et al. 2008; Ghayad and Cohen 2010). mTOR signaling has been reported to crosstalk with the Notch signaling pathway in several malignant cell lines (Androutsellis-Theotokis et al. 2006; Mungamuri et al. 2006; Chan et al. 2007; Efferson et al. 2010).

Inhibition of p53 by Notch1 NICD mainly occurs through mTOR linked to PI-3K/Akt pathway. Moreover, rapamycin treatment abrogated NICD inhibition of p53 and reversed the chemoresistance (Mungamuri et al. 2006). Chemoresistant MCF-7 and MOLT4 (T-cell acute lymphoblastic leukemia) cells have aberrant Notch1 that can be reversed by using both PI-3K and mTOR inhibitors (Mungamuri et al. 2006). Efferson et al (Efferson et al. 2010) used an ERbB2-transgenic mouse model of breast cancer (neuT) to show that Notch signaling plays a critical role in tumor maintenance. Inhibition of the Notch pathway with a γ-secretase inhibitor (GSI) decreased both the Notch and mTOR/Akt pathways. Antitumor activity resulting from GSI treatment was associated with decreased cell proliferation (Mungamuri et al. 2006). Since mTOR is closely linked to PI-3K/Akt pathway in ECs (Minhajuddin et al. 2009; Rafiee et al. 2010; Zhang et al. 2011), it is reasonable to speculate that mTOR should also be a regulator of Notch pathway in ECs.

8.4.4 NF-κB Signaling Pathway

The family of NF-κB transcription factors is involved in expression of key genes for innate and adaptive immunity, cell proliferation and survival, and lymphoid organ development. NF-κB is activated in a variety of cancers (Hayden and Ghosh 2008; Prasad et al. 2010) linked to tumor angiogenesis (Gonzalez-Perez et al. 2010). NF-κB family, RelA (p65), RelB, c-Rel, p105/p50 and p100/p52 are evolutionarily conserved molecules that form hetero- or homodimers. The p65/p50 heterodimer, the most abundant form of NF-κB is regulated by the so-called canonical pathway (Hayden and Ghosh 2004; Hayden and Ghosh 2008).

Numerous reports have described the bidirectional regulation of Notch and NF-κB through different context-dependent mechanisms. First, Oswald et al. (Oswald et al. 1998) clearly demonstrated that Notch was able to transcriptionally regulate NF-κB members. RBP-Jk is a strong transcriptional repressor of p100/p52 whose effects can be overcome by activated Notch1, suggesting that p100/p52 is a Notch target gene. Cheng et al (Cheng et al. 2001) further observed that Notch1 upregulated the expression of p65, p50, RelB, and c-Rel subunits in hemopoietic progenitor cells using Notch1 antisense transgenic (Notch-AS-Tg) mice. Additionally, Notch1 regulated NF-κB in cervical cancer cells in part via cytoplasmic and nuclear IKK-mediated pathways (Song et al. 2008). Second, NF-κB subunits are also able to transcriptionally regulate Notch family members. This is supported by the findings of Bash et al (Bash et al. 1999) that demonstrated c-Rel can activate Notch signaling pathway by up-regulating JAG1 gene expression in lymphocytes. A role for JAG1 in B-cell activation, differentiation or function was also suggested (Bash et al. 1999). Lastly, members from Notch and NF-κB family could physically interact with each other. Wang et al. demonstrated that the N-terminal portion of Notch1 NICD interacted specifically with p50 subunit and inhibited p50 DNA binding in human NTera-2 embryonal carcinoma cells (Wang et al. 2001). In contrast, in T-cells Notch1 NICD was found to activate NF-κB by directly interacting with NF-κB and competing with IκBα. These processes lead to the retention of NF-κB in the nucleus. It seems that in T-cells there are two 'waves' of NF-κB activation: an initial, Notch-independent phase, and a later, sustained activation of NF-κB, which is Notch dependent (Shin et al. 2006). Two recent reports also confirmed that Notch activation was required for NF-κB activation in ECs (Johnston et al. 2009; Quillard et al. 2010).

8.4.5 HIF Signaling Pathway

A critical aspect of tumor biology is the sensation of oxygen in the microenvironment. In response to hypoxia, a hallmark of most solid tumors, cells try to adapt by regulating metabolism, erythropoiesis, and angiogenesis and by modulating pathways that result in survival or cell death. HIF is a key molecule unregulated in response to oxygen deficiency, as it acts as a master regulator of genes involved in tissue reoxygenation (Ke and Costa 2006). Additionally, HIF has been known to facilitate cancer progression by promoting tumor neoangiogenesis, cell motility, and invasion (Hirota and Semenza 2006). HIF is a heterodimer consisting of a constitutively expressed HIF-1β subunit and an oxygen-regulated, unstable HIF-1α subunit. HIF interactions with DNA are mediated through Hypoxia-Responsive Elements (HRE) (Ke and Costa 2006). Several studies have demonstrated that HIF-1 plays important roles in the development and progression of cancer through activation of various genes involved in crucial aspects of cancer biology, including energy metabolism, vasomotor function, erythropoiesis, cell survival and angiogenesis (Li and Ye 2010).

Gustafsson et al (Gustafsson et al. 2005) showed evidence that hypoxia promotes the undifferentiated cell state in various stem and precursor cell populations. In this process, hypoxia blocks neuronal and myogenic differentiation in a Notch-dependent manner. Upon Notch activation under hypoxic conditions, Notch1 NICD can interact with HIF-1α, and the complex is recruited to Notch1-responsive promoters. Sahlgren et al (Sahlgren et al. 2008) further demonstrated that a hypoxia/Notch/EMT axis exists in tumor cells, where Notch serves as a critical intermediate in conveying the hypoxic response into EMT. Hypoxia-induced increased motility and invasiveness of the tumor cells require Notch signaling, and activated Notch mimicked hypoxia in the induction of EMT. In this process, Notch signaling acts in synergy to control the expression of Snail-1, a zinc-finger transcriptional factor repressor of E-cadherin and a critical regulator of EMT. First, NICD could interact with the *Snail-1* promoter, and second, Notch potentiated HIF-1α recruitment to the lysyl oxidase (LOX; a copper-dependent amine oxidase) promoter and elevated the hypoxia-induced up-regulation of LOX, which stabilizes the Snail-1 protein (Sahlgren et al. 2008). Hypoxia increased Notch1 mRNA and protein level as well as Notch activity, measured as Hes1 and Hey1 expression and Hes1 promoter activity. This effect was dependent on HIF-1α (Bedogni et al. 2008). These results suggest that Notch1 is under the control of oncogenes and the tissue microenvironment. Therefore, HIF-1α and Notch signaling pathways play a critical role in the regulation of EMT and open up perspectives for pharmacological intervention within hypoxia-induced EMT, cell invasiveness and angiogenesis in tumors.

8.5 Other Crosstalk Signaling

8.5.1 ER Signaling

Estrogens, in particular 17beta-estradiol (E2), play a pivotal role in sexual development and reproduction and are also implicated in a large number of physiological processes, including the cardiovascular system. The recognized risk factors for breast cancer are ages at: menarche, first pregnancy, and menopause. This suggests endogenous ovarian steroids may profoundly affect initiation, promotion, and progression of carcinogenesis through a cascade of reactions initiated by activation of the ER (Bernstein and Ross 1993; Hankinson 2005). ERs are known to regulate a huge number of genes affecting cancer proliferation and vascular function (Mendelsohn and Karas 1999; Welboren et al. 2007).

Soares et al (Soares et al. 2004) first demonstrated that a crosstalk between estrogen and Notch signaling occurs in breast cancer and EC. The authors observed that E2 promoted 8-fold and 6-fold increase in Notch1 and JAG1 expression, respectively, in MCF-7 breast cancer cells. A similar up-regulation of both Notch1 receptor and JAG1 ligand was also found in EC. Notch gene expression was required for tubule-like structure formation in EC. Moreover, Notch gene expression, together with HIF-1α, was upregulated by E2. In another report, E2 and parathion (an organophosphate compound and potent insecticide) alone and in combination also led to the activation of Notch signaling in MCF-I0F, in the process of malignant transformation as indicated by anchorage independency and in vitro invasive capabilities (Calaf and Roy 2008). Notch and ERα crosstalk in breast cancer suggests that combinations of antiestrogens and Notch inhibitors may be more effective in treating ERα (+) breast cancers (Rizzo et al. 2008a). Overall, the crosstalk between Notch and estrogen signaling pathway has a significant role in human breast carcinogenesis and angiogenesis.

8.5.2 miRNA Actions

MicroRNAs (miRNAs) are short non-coding RNAs that bind to the 3' untranslated region (UTR) of cognate messenger RNAs (mRNAs) through fully complementary or imperfect base-pairing repressing the translation or decreasing the stability of the bound mRNAs (Denli et al. 2004). miRNAs are involved in biological and pathologic processes including cell differentiation, proliferation, apoptosis, metabolism and angiogenesis, and are emerging as highly tissue-specific biomarkers with potential clinical applicability for defining cancer types and origins (Rosenfeld et al. 2008; Wahid et al. 2010; Staszel et al. 2011; Yang et al. 2011). These RNAs can function as oncogenes or tumor suppressors depending upon the cell type in which they are expressed (Croce 2009; Inui et al. 2010).

There are several reports on the crosstalk between miRNA and Notch signaling pathways (Yoo and Greenwald 2005; Solomon et al. 2008; Li et al. 2009). Yoo and Greenwald first reported that Notch activation leads to miR-61 mediated down-regulation of Vav, a proto-oncogene in *Caenorhabditis elegans* (Yoo and Greenwald 2005). Interestingly, miR-61 could control the expression of oncogene orthologues Ras and Vav, indicating miRNA capacity to act as tumor suppressors (Jannot and Simard 2006). Therapeutic potential of let-7's in cancer (initially identified as a timing developmental regulator in *C. elegans*) was recently reviewed (Barh et al. 2010). In various human cell lines, Notch activation up-regulates miRNA let-7 (Solomon et al. 2008). Let-7 regulates self renewal and tumorigenicity of breast cancer cells (Yu et al. 2007), as well as ERα signaling in ER positive breast cancer (Zhao et al. 2010).

On the other hand, miRNAs can regulate Notch pathways. miR-34a down-regulated the expression of Notch1 and Notch2 proteins in glioma cells (Li et al. 2009). miR-34 down-regulated JAG1 and Notch1 in cervical carcinoma and choriocarcinoma cells (Pang et al. 2010). miR-34 was required for a normal cellular response to DNA damage *in vivo*. Therefore, a potential therapeutic use for anti-miR-34 as a radiosensitizing agent in p53-mutant breast cancer is predicted (Kato et al. 2009). In addition, altered miRNA signatures including miR-34 may be associated with breast carcinogenesis and metastasis (O'Day and Lal 2010). Loss of miR-8/200 has been commonly observed in advanced tumors (Valastyan and Weinberg 2009) and correlates with their invasion (Hurteau et al. 2007; Bracken et al. 2008; Gregory et al. 2008) and acquisition of stem-like properties (Shimono et al. 2009; Wellner et al. 2009). Recently, miR-8/200 was identified to have the ability to inhibit JAG1, thus attenuating Notch signaling and impeding proliferation of human metastatic prostate cancer cells (Vallejo et al. 2011).

9 Notch Signaling as Therapeutic Drug Target in Human Cancers

The prevailing new strategy for rationally targeted cancer treatment is aimed at the development of target-selective "smart" drugs on the basis of characterized mechanisms of action. The connection between Notch signaling, carcinogenesis and angiogenesis, as well as its crosstalk with many oncogenic signaling pathways suggest that Notch signaling may be such a candidate for multi-target drugs. The major therapeutic targets in the Notch pathway are the Notch receptors, in which GSIs prevent the generation of the oncogenic NICD and suppress the Notch activity (Shih Ie and Wang 2007; Imbimbo 2008).

Gamma-secretase is a large membrane-integral multisubunit protease complex, which is essential for Notch receptor activation (Bergmans and De Strooper 2010). Rasul et al (Rasul et al. 2009) tested the effects of three different GSIs in breast cancer cells. One inhibitor (GSI1) was lethal to breast cancer cell lines, but had a minimal effect on the non-malignant breast lines. GSI1 treatment resulted in a marked decrease in γ-secretase activity and down-regulation of the Notch signaling pathway with no effects on expression of the γ-secretase components or ligands. In a recent report (Efferson et al. 2010), the authors observed that inhibition of the Notch pathway with a GSI decreased both the Notch and mTOR/Akt pathways. Antitumor activity resulting from GSI treatment was associated with decreased cell proliferation as measured by Ki67 and decreased expression of glucose transporter Glut1 (Efferson et al. 2010). GSI effects are much higher in HER2/neu- positive cell lines where HER2 is amplified and/or over-expressed (ZR-75-1 and MDA-MB-453) compared with HER2-negative cells (MCF-7 and MDA-MB-231) that lack ERbB2 amplification and show low HER2 expression (Lee et al. 2008; Rizzo et al. 2008b; Rasul et al. 2009). Since HER2 can influence the activity of Notch (Chen et al. 1997) and inhibition of HER2 via trastuzumab can activate Notch signaling (Osipo et al. 2008), it will be important to consider GSI as a monotherapy or in combination with trastuzumab or lapatinib in HER2 breast cancer patients.

Although several GSIs have been developed into clinical trials (Shih Ie and Wang 2007), GSIs fail to distinguish individual Notch receptors. In addition, GSIs inhibit other signaling pathways (Beel and Sanders 2008) and cause intestinal toxicity (van Es et al. 2005), probably attributable to dual inhibition of Notch1 and Notch2 (Riccio et al. 2008). Very recently, Wu et al (Wu et al. 2010) utilized phage display technology to generate highly specialized antibodies that specifically antagonize each receptor paralogue, enabling the discrimination of Notch1 versus Notch2 function in human patients and rodent models. Their results showed that inhibition of either receptor alone reduces or avoids toxicity, demonstrating a clear advantage over pan-Notch inhibitors.

10   Conclusion and Perspectives

Notch signaling and its crosstalk with many signaling pathways play an important role in cancer cell growth, migration, invasion, metastasis and angiogenesis, as well as cancer stem cell self-renewal (see Fig. 1). Therefore, significant attention has been paid in recent years toward the development of clinically useful antagonists of Notch signaling. Better understanding of the structure, function and regulation of Notch intracellular signaling pathways, as well as its complex crosstalk with other oncogenic signals in cancer cells will be essential to ensure rational use of treatment and development of new combinatory therapeutic possibilities. Emerging novel opportunities arise from the discovery of Notch crosstalk with inflammatory and angiogenic cytokines (i.e., NILCO) and their links to obesity-related cancers. Combinatory treatments with drugs designed to prevent Notch oncogenic signal crosstalk may be advantageous over GSIs alone.

11   Acknowledgements

The authors' work cited in this chapter was funded by Grants from NIH/ NCI 1SC1CA138658-03, the Georgia Cancer Coalition Distinguished Cancer

Scholar Award to R.R.G-P., and facilities and support services at Morehouse School of Medicine (NIH RR03034 and 1C06 RR18386).

12 Glossary

4T1 cells: mouse mammary cancer cell line; ADAM: a disintegrin and metalloprotease; Akt: protein kinase B; ALDH: aldehyde dehydrogenase; ANK: ankyrin; Axin2: the Axin-related protein; Bcl-2: B-cell lymphoma 2; Bcl-xl: B-cell lymphoma-extra large; BCSC: breast cancer stem cells; BMP: bone morphogenetic protein; CBF1: Centromere-Binding Factor 1; CD4: cluster of differentiation 4; CD8: cluster of differentiation 8; c-myc: Myc proto-oncogene protein; Co-A: recruits coactivator; Co-R: co-repressor; CSC: cancer stem cell; CSL: CBF1/Su(H)/Lag-1; Cyclin D1: kinase and regulator of cell cycle D1; DAPT: N-[N-(3,5-Difluorophenacetyl)-L-alanyl]-S-phenylglycine t-butyl ester; DBZ: dibenzazipene; DLL-1: Delta-like 1; DOS: Delta and OSM-11-like proteins; DSL: Delta/Serrate/LAG-2; E2: 17β-estradiol; EC: endothelial cell; EGF: epidermal growth factor; EGFR: epidermal growth factor receptor; EMT: epithelial-mesenchymal transformation; EMT6: a mouse mammary adenocarcinoma cell line; ER: estrogen receptor; ERK 1/2: extracellular regulated kinase 1 and 2; GSI: a γ-secretase inhibitor; HD: heterodimerization; HDAC: histone deacetylases; HIF-1α: hypoxia regulated factor-1 α; HUVECs: human umbilical vein ECs; ICN: intracellular region of Notch; IGF-1: insulin like growth factor-1; IL-1: interleukin-1; IL-1R tI: interleukin-1 type I receptor; IL-6: interleukin-6; IL-6R: interleukin-6 receptor; JAK2: Janus kinase 2; MAPK: mitogen activated protein kinase; MCF-7: ER positive human breast cancer cell line; MDA-MB-231: ER negative human breast cancer cell line; MFE: Mammosphere-forming efficiency; miRNA: MicroRNA; mTOR: mammalian target of rapamycin; NECD: Notch extracellular domain; NEXT: Notch extracellular truncation; NF-κB: eukaryotic nuclear transcription factor kappa B; NICD: Notch intracellular domain; NRR: negative regulatory region OB-R: leptin receptor; PDGF: platelet-derived growth factor; PEST: proline, glutamine, serine and threonine residue; PI-3K: phosphoinositide 3-kinase; RhoC: Ras homolog gene family, member C; Src: a proto-oncogenic tyrosine kinase; STAT3: signal transducer and activator of transcription 3; TACE: tumor necrosis factor-α-converting enzyme; TAD: transactivation domain; TAM: tamoxifen; T-ALL: T-cell acute Lymphoblastic Leukemia; TGF-β: transforming growth factor beta; TNF-α: tumor necrosis factor alpha; TSA: Trichostatin A; VEGF: Vascular endothelial growth factor; VEGFR-2: Vascular endothelial growth factor receptor 2 or KDR or Flk-1.

REFERENCES

Adams, R. H. and K. Alitalo (2007). "Molecular regulation of angiogenesis and lymphangiogenesis." Nat Rev Mol Cell Biol 8(6): 464-478.

Ahmad, A., Z. Wang, D. Kong, R. Ali, S. Ali, S. Banerjee and F. H. Sarkar (2011). "Platelet-derived growth factor-D contributes to aggressiveness of breast cancer cells by up-regulating Notch and NF-kappaB signaling pathways." Breast Cancer Res Treat 126(1): 15-25.

Albig, A. R., D. J. Becenti, T. G. Roy and W. P. Schiemann (2008). "Microfibril-associate glycoprotein-2 (MAGP-2) promotes angiogenic cell sprouting by blocking notch signaling in endothelial cells." Microvasc Res 76(1): 7-14.

Altieri, D. C. (2008). "New wirings in the survivin networks." Oncogene 27(48): 6276-6284.

Androutsellis-Theotokis, A., R. R. Leker, F. Soldner, D. J. Hoeppner, R. Ravin, S. W. Poser, M. A. Rueger, S. K. Bae, R. Kittappa and R. D. McKay (2006). "Notch signalling regulates stem cell numbers in vitro and in vivo." Nature 442(7104): 823-826.

Anglesio, M. S., J. George, H. Kulbe, M. Friedlander, D. Rischin, C. Lemech, J. Power, J. Coward, P. A. Cowin, C. M. House, P. Chakravarty, K. L. Gorringe, I. G. Campbell, A. Okamoto, M. J. Birrer, D. G. Huntsman, A. de Fazio, S. E. Kalloger, F. Balkwill, C. B. Gilks and D. D. Bowtell (2011). "IL6-STAT3-HIF signaling and therapeutic response to the angiogenesis inhibitor sunitinib in ovarian clear cell cancer." Clin Cancer Res 17(8): 2538-2548.

Apte, R. N., S. Dotan, M. Elkabets, M. R. White, E. Reich, Y. Carmi, X. Song, T. Dvozkin, Y. Krelin and E. Voronov (2006). "The involvement of IL-1 in tumorigenesis, tumor invasiveness, metastasis and tumor-host interactions." Cancer Metastasis Rev 25(3): 387-408.

Artavanis-Tsakonas, S., M. D. Rand and R. J. Lake (1999). "Notch signaling: cell fate control and signal integration in development." Science 284(5415): 770-776.

Asano, N., T. Watanabe, A. Kitani, I. J. Fuss and W. Strober (2008). "Notch1 signaling and regulatory T cell function." J Immunol 180(5): 2796-2804.

Ayyanan, A., G. Civenni, L. Ciarloni, C. Morel, N. Mueller, K. Lefort, A. Mandinova, W. Raffoul, M. Fiche, G. P. Dotto and C. Brisken (2006). "Increased Wnt signaling triggers oncogenic conversion of human breast epithelial cells by a Notch-dependent mechanism." Proc Natl Acad Sci U S A 103(10): 3799-3804.

Bajaj, A., Q. Zheng, A. Adam, P. Vincent and K. Pumiglia (2010). "Activation of endothelial ras signaling bypasses senescence and causes abnormal vascular morphogenesis." Cancer Res 70(9): 3803-3812.

Banks, A. S., S. M. Davis, S. H. Bates and M. G. Myers, Jr. (2000). "Activation of downstream signals by the long form of the leptin receptor." J Biol Chem 275(19): 14563-14572.

Bao, B., Z. Wang, S. Ali, D. Kong, Y. Li, A. Ahmad, S. Banerjee, A. S. Azmi, L. Miele and F. H. Sarkar (2011). "Notch-1 induces epithelial-mesenchymal transition consistent with cancer stem cell phenotype in pancreatic cancer cells." Cancer Lett 307(1): 26-36.

Bao, S., Q. Wu, S. Sathornsumetee, Y. Hao, Z. Li, A. B. Hjelmeland, Q. Shi, R. E. McLendon, D. D. Bigner and J. N. Rich (2006). "Stem cell-like glioma cells promote tumor angiogenesis through vascular endothelial growth factor." Cancer Res 66(16): 7843-7848.

Barcellos-Hoff, M. H. and R. J. Akhurst (2009). "Transforming growth factor-beta in breast cancer: too much, too late." Breast Cancer Res 11(1): 202.

Barcelos, L. S., C. Duplaa, N. Krankel, G. Graiani, G. Invernici, R. Katare, M. Siragusa, M. Meloni, I. Campesi, M. Monica, A. Simm, P. Campagnolo,

G. Mangialardi, L. Stevanato, G. Alessandri, C. Emanueli and P. Madeddu (2009). "Human CD133+ progenitor cells promote the healing of diabetic ischemic ulcers by paracrine stimulation of angiogenesis and activation of Wnt signaling." Circ Res 104(9): 1095-1102.

Barh, D., R. Malhotra, B. Ravi and P. Sindhurani (2010). "Microrna let-7: an emerging next-generation cancer therapeutic." Curr Oncol 17(1): 70-80.

Bash, J., W. X. Zong, S. Banga, A. Rivera, D. W. Ballard, Y. Ron and C. Gelinas (1999). "Rel/NF-kappaB can trigger the Notch signaling pathway by inducing the expression of Jagged1, a ligand for Notch receptors." EMBO J 18(10): 2803-2811.

Bates, S. H., W. H. Stearns, T. A. Dundon, M. Schubert, A. W. Tso, Y. Wang, A. S. Banks, H. J. Lavery, A. K. Haq, E. Maratos-Flier, B. G. Neel, M. W. Schwartz and M. G. Myers, Jr. (2003). "STAT3 signalling is required for leptin regulation of energy balance but not reproduction." Nature 421(6925): 856-859.

Beatus, P., J. Lundkvist, C. Oberg and U. Lendahl (1999). "The notch 3 intracellular domain represses notch 1-mediated activation through Hairy/Enhancer of split (HES) promoters." Development 126(17): 3925-3935.

Bedogni, B., J. A. Warneke, B. J. Nickoloff, A. J. Giaccia and M. B. Powell (2008). "Notch1 is an effector of Akt and hypoxia in melanoma development." J Clin Invest 118(11): 3660-3670.

Beel, A. J. and C. R. Sanders (2008). "Substrate specificity of gamma-secretase and other intramembrane proteases." Cell Mol Life Sci 65(9): 1311-1334.

Beisner, J., M. B. Buck, P. Fritz, J. Dippon, M. Schwab, H. Brauch, G. Zugmaier, K. Pfizenmaier and C. Knabbe (2006). "A novel functional polymorphism in the transforming growth factor-beta2 gene promoter and tumor progression in breast cancer." Cancer Res 66(15): 7554-7561.

Bergmans, B. A. and B. De Strooper (2010). "gamma-secretases: from cell biology to therapeutic strategies." Lancet Neurol 9(2): 215-226.

Bergsten, E., M. Uutela, X. Li, K. Pietras, A. Ostman, C. H. Heldin, K. Alitalo and U. Eriksson (2001). "PDGF-D is a specific, protease-activated ligand for the PDGF beta-receptor." Nat Cell Biol 3(5): 512-516.

Bernstein, L. and R. K. Ross (1993). "Endogenous hormones and breast cancer risk." Epidemiol Rev 15(1): 48-65.

Bin Hafeez, B., V. M. Adhami, M. Asim, I. A. Siddiqui, K. M. Bhat, W. Zhong, M. Saleem, M. Din, V. Setaluri and H. Mukhtar (2009). "Targeted knockdown of Notch1 inhibits invasion of human prostate cancer cells concomitant with inhibition of matrix metalloproteinase-9 and urokinase plasminogen activator." Clin Cancer Res 15(2): 452-459.

Blokzijl, A., C. Dahlqvist, E. Reissmann, A. Falk, A. Moliner, U. Lendahl and C. F. Ibanez (2003). "Cross-talk between the Notch and TGF-beta signaling pathways mediated by interaction of the Notch intracellular domain with Smad3." J Cell Biol 163(4): 723-728.

Borggrefe, T. and F. Oswald (2009). "The Notch signaling pathway: transcriptional regulation at Notch target genes." Cell Mol Life Sci 66(10): 1631-1646.

Bracken, C. P., P. A. Gregory, N. Kolesnikoff, A. G. Bert, J. Wang, M. F. Shannon and G. J. Goodall (2008). "A double-negative feedback loop between ZEB1-SIP1 and the microRNA-200 family regulates epithelial-mesenchymal transition." Cancer Res 68(19): 7846-7854.

Bray, S. J. (2006). "Notch signalling: a simple pathway becomes complex." Nat Rev Mol Cell Biol 7(9): 678-689.

Brennan, A. M. and C. S. Mantzoros (2006). "Drug Insight: the role of leptin in human physiology and pathophysiology–emerging clinical applications." Nat Clin Pract Endocrinol Metab 2(6): 318-327.

Bridges, E., C. E. Oon and A. Harris (2011). "Notch regulation of tumor angiogenesis." Future Oncol 7(4): 569-588.

Calaf, G. M. and D. Roy (2008). "Cell adhesion proteins altered by 17beta estradiol and parathion in breast epithelial cells." Oncol Rep 19(1): 165-169.

Callahan, R. and A. Raafat (2001). "Notch signaling in mammary gland tumorigenesis." J Mammary Gland Biol Neoplasia 6(1): 23-36.

Cambien, F., S. Ricard, A. Troesch, C. Mallet, L. Generenaz, A. Evans, D. Arveiler, G. Luc, J. B. Ruidavets and O. Poirier (1996). "Polymorphisms of the transforming growth factor-beta 1 gene in relation to myocardial infarction and blood pressure. The Etude Cas-Temoin de l'Infarctus du Myocarde (ECTIM) Study." Hypertension 28(5): 881-887.

Cao, L., Y. Zhou, B. Zhai, J. Liao, W. Xu, R. Zhang, J. Li, Y. Zhang, L. Chen, H. Qian, M. Wu and Z. Yin (2011). "Sphere-forming cell subpopulations with cancer stem cell properties in human hepatoma cell lines." BMC Gastroenterol 11: 71.

Cao, Z., M. Tanaka, C. Regnier, M. Rothe, A. Yamit-hezi, J. D. Woronicz, M. E. Fuentes, M. H. Durnin, S. A. Dalrymple and D. V. Goeddel (1999). "NF-kappa B activation by tumor necrosis factor and interleukin-1." Cold Spring Harb Symp Quant Biol 64: 473-483.

Carino, C., A. B. Olawaiye, S. Cherfils, T. Serikawa, M. P. Lynch, B. R. Rueda and R. R. Gonzalez (2008). "Leptin regulation of proangiogenic molecules in benign and cancerous endometrial cells." Int J Cancer 123(12): 2782-2790.

Catalano, S., S. Marsico, C. Giordano, L. Mauro, P. Rizza, M. L. Panno and S. Ando (2003). "Leptin enhances, via AP-1, expression of aromatase in the MCF-7 cell line." J Biol Chem 278(31): 28668-28676.

Catalano, S., L. Mauro, S. Marsico, C. Giordano, P. Rizza, V. Rago, D. Montanaro, M. Maggiolini, M. L. Panno and S. Ando (2004). "Leptin induces, via ERK1/ERK2 signal, functional activation of estrogen receptor alpha in MCF-7 cells." J Biol Chem 279(19): 19908-19915.

Chan, S. M., A. P. Weng, R. Tibshirani, J. C. Aster and P. J. Utz (2007). "Notch signals positively regulate activity of the mTOR pathway in T-cell acute lymphoblastic leukemia." Blood 110(1): 278-286.

Chen, C., Y. C. Chang, C. L. Liu, K. J. Chang and I. C. Guo (2006). "Leptin-induced growth of human ZR-75-1 breast cancer cells is associated with up-regulation of cyclin D1 and c-Myc and down-regulation of tumor suppressor p53 and p21WAF1/CIP1." Breast Cancer Res Treat 98(2): 121-132.

Chen, Y., W. H. Fischer and G. N. Gill (1997). "Regulation of the ERBB-2 promoter by RBPJkappa and NOTCH." J Biol Chem 272(22): 14110-14114.

Cheng, P., A. Zlobin, V. Volgina, S. Gottipati, B. Osborne, E. J. Simel, L. Miele and D. I. Gabrilovich (2001). "Notch-1 regulates NF-kappaB activity in hemopoietic progenitor cells." J Immunol 167(8): 4458-4467.

Chiang, M. Y., L. Xu, O. Shestova, G. Histen, S. L'Heureux, C. Romany, M. E. Childs, P. A. Gimotty, J. C. Aster and W. S. Pear (2008). "Leukemia-associated NOTCH1 alleles are weak tumor initiators but accelerate K-ras-initiated leukemia." J Clin Invest 118(9): 3181-3194.

Choi, J. H., J. T. Park, B. Davidson, P. J. Morin, M. Shih Ie and T. L. Wang (2008). "Jagged-1 and Notch3 juxtacrine loop regulates ovarian tumor growth and adhesion." Cancer Res 68(14): 5716-5723.

Ciuffreda, L., C. Di Sanza, U. C. Incani and M. Milella (2010). "The mTOR pathway: a new target in cancer therapy." Curr Cancer Drug Targets 10(5): 484-495.

Cleary, M. P., S. C. Juneja, F. C. Phillips, X. Hu, J. P. Grande and N. J. Maihle (2004). "Leptin receptor-deficient MMTV-TGF-alpha/Lepr(db)Lepr(db) female mice do not develop oncogene-induced mammary tumors." Exp Biol Med (Maywood) 229(2): 182-193.

Cleary, M. P., F. C. Phillips, S. C. Getzin, T. L. Jacobson, M. K. Jacobson, T. A. Christensen, S. C. Juneja, J. P. Grande and N. J. Maihle (2003). "Genetically obese MMTV-TGF-alpha/Lep(ob)Lep(ob) female mice do not develop mammary tumors." Breast Cancer Res Treat 77(3): 205-215.

Coma, S., V. Noe, C. Lavarino, J. Adan, M. Rivas, M. Lopez-Matas, R. Pagan, F. Mitjans, S. Vilaro, J. Piulats and C. J. Ciudad (2004). "Use of siRNAs and antisense oligonucleotides against survivin RNA to inhibit steps leading to tumor angiogenesis." Oligonucleotides 14(2): 100-113.

Cordle, J., S. Johnson, J. Z. Tay, P. Roversi, M. B. Wilkin, B. H. de Madrid, H. Shimizu, S. Jensen, P. Whiteman, B. Jin, C. Redfield, M. Baron, S. M. Lea and P. A. Handford (2008). "A conserved face of the Jagged/Serrate DSL domain is involved in Notch trans-activation and cis-inhibition." Nat Struct Mol Biol 15(8): 849-857.

Croce, C. M. (2009). "Causes and consequences of microRNA dysregulation in cancer." Nat Rev Genet 10(10): 704-714.

Cui, X. Y., Q. D. Hu, M. Tekaya, Y. Shimoda, B. T. Ang, D. Y. Nie, L. Sun, W. P. Hu, M. Karsak, T. Duka, Y. Takeda, L. Y. Ou, G. S. Dawe, F. G. Yu, S. Ahmed, L. H. Jin, M. Schachner, K. Watanabe, Y. Arsenijevic and Z. C. Xiao (2004). "NB-3/ Notch1 pathway via Deltex1 promotes neural progenitor cell differentiation into oligodendrocytes." J Biol Chem 279(24): 25858-25865.

D'Souza, B., A. Miyamoto and G. Weinmaster (2008). "The many facets of Notch ligands." Oncogene 27(38): 5148-5167.

Dai, J., D. Ma, S. Zang, D. Guo, X. Qu, J. Ye and C. Ji (2009). "Cross-talk between Notch and EGFR signaling in human breast cancer cells." Cancer Invest 27(5): 533-540.

Dardeno, T. A., S. H. Chou, H. S. Moon, J. P. Chamberland, C. G. Fiorenza and C. S. Mantzoros (2010). "Leptin in human physiology and therapeutics." Front Neuroendocrinol 31(3): 377-393.

de Antonellis, P., C. Medaglia, E. Cusanelli, I. Andolfo, L. Liguori, G. De Vita, M. Carotenuto, A. Bello, F. Formiggini, A. Galeone, G. De Rosa, A. Virgilio, I. Scognamiglio, M. Sciro, G. Basso, J. H. Schulte, G. Cinalli, A. Iolascon and M. Zollo (2011). "MiR-34a targeting of Notch ligand delta-like 1 impairs CD15+/CD133+ tumor-propagating cells and supports neural differentiation in medulloblastoma." PLoS One 6(9): e24584.

Dejana, E. (2010). "The role of wnt signaling in physiological and pathological angiogenesis." Circ Res 107(8): 943-952.

DeNicola, G. M. and D. A. Tuveson (2009). "RAS in cellular transformation and senescence." Eur J Cancer 45 Suppl 1: 211-216.

Denli, A. M., B. B. Tops, R. H. Plasterk, R. F. Ketting and G. J. Hannon (2004). "Processing of primary microRNAs by the Microprocessor complex." Nature 432(7014): 231-235.

Dieudonne, M. N., F. Machinal-Quelin, V. Serazin-Leroy, M. C. Leneveu, R. Pecquery and Y. Giudicelli (2002). "Leptin mediates a proliferative response in human MCF7 breast cancer cells." Biochem Biophys Res Commun 293(1): 622-628.

Dievart, A., N. Beaulieu and P. Jolicoeur (1999). "Involvement of Notch1 in the development of mouse mammary tumors." Oncogene 18(44): 5973-5981.

Dillon, R. L. and W. J. Muller (2010). "Distinct biological roles for the akt family in mammary tumor progression." Cancer Res 70(11): 4260-4264.

Dinarello, C. A. (1998). "Interleukin-1, interleukin-1 receptors and interleukin-1 receptor antagonist." Int Rev Immunol 16(5-6): 457-499.

Dong, Y., A. Li, J. Wang, J. D. Weber and L. S. Michel (2010). "Synthetic lethality through combined Notch-epidermal growth factor receptor pathway inhibition in basal-like breast cancer." Cancer Res 70(13): 5465-5474.

Dontu, G., K. W. Jackson, E. McNicholas, M. J. Kawamura, W. M. Abdallah and M. S. Wicha (2004). "Role of Notch signaling in cell-fate determination of human mammary stem/progenitor cells." Breast Cancer Res 6(6): R605-615.

Du, Z., J. Li, L. Wang, C. Bian, Q. Wang, L. Liao, X. Dou, X. Bian and R. C. Zhao (2010). "Overexpression of DeltaNp63alpha induces a stem cell phenotype in MCF7 breast carcinoma cell line through the Notch pathway." Cancer Sci 101(11): 2417-2424.

Duarte, A., M. Hirashima, R. Benedito, A. Trindade, P. Diniz, E. Bekman, L. Costa, D. Henrique and J. Rossant (2004). "Dosage-sensitive requirement for mouse Dll4 in artery development." Genes Dev 18(20): 2474-2478.

Efferson, C. L., C. T. Winkelmann, C. Ware, T. Sullivan, S. Giampaoli, J. Tammam, S. Patel, G. Mesiti, J. F. Reilly, R. E. Gibson, C. Buser, T. Yeatman, D. Coppola, C. Winter, E. A. Clark, G. F. Draetta, P. R. Strack and P. K. Majumder (2010). "Downregulation of Notch pathway by a gamma-secretase inhibitor attenuates AKT/mammalian target of rapamycin signaling and glucose uptake in an ERBB2 transgenic breast cancer model." Cancer Res 70(6): 2476-2484.

Efstratiadis, A., M. Szabolcs and A. Klinakis (2007). "Notch, Myc and breast cancer." Cell Cycle 6(4): 418-429.

Egan, S. E., B. St-Pierre and C. C. Leow (1998). "Notch receptors, partners and regulators: from conserved domains to powerful functions." Curr Top Microbiol Immunol 228: 273-324.

Eisenberg, A., E. Biener, M. Charlier, R. V. Krishnan, J. Djiane, B. Herman and A. Gertler (2004). "Transactivation of erbB2 by short and long isoforms of leptin receptors." FEBS Lett 565(1-3): 139-142.

Fan, X., I. Mikolaenko, I. Elhassan, X. Ni, Y. Wang, D. Ball, D. J. Brat, A. Perry and C. G. Eberhart (2004). "Notch1 and Notch2 have opposite effects on embryonal brain tumor growth." Cancer Res 64(21): 7787-7793.

Fan, X., W. Matsui, L. Khaki, D. Stearns, J. Chun, Y. M. Li and C. G. Eberhart (2006). "Notch pathway inhibition depletes stem-like cells and blocks engraftment in embryonal brain tumors." Cancer Res 66(15): 7445-7452.

Fan, X., L. Khaki, T. S. Zhu, M. E. Soules, C. E. Talsma, N. Gul, C. Koh, J. Zhang, Y. M. Li, J. Maciaczyk, G. Nikkhah, F. Dimeco, S. Piccirillo, A. L. Vescovi and C. G. Eberhart (2010). "NOTCH pathway blockade depletes CD133-positive glioblastoma cells and inhibits growth of tumor neurospheres and xenografts." Stem Cells 28(1): 5-16.

Farnie, G. and R. B. Clarke (2007). "Mammary stem cells and breast cancer–role of Notch signalling." Stem Cell Rev 3(2): 169-175.

Fernandez-Riejos, P., S. Najib, J. Santos-Alvarez, C. Martin-Romero, A. Perez-Perez, C. Gonzalez-Yanes and V. Sanchez-Margalet (2010). "Role of leptin in the activation of immune cells." Mediators Inflamm 2010: 568343.

Ferrara, N. (1999). "Vascular endothelial growth factor: molecular and biological aspects." Curr Top Microbiol Immunol 237: 1-30.

Ferrara, N., H. P. Gerber and J. LeCouter (2003). "The biology of VEGF and its receptors." Nat Med 9(6): 669-676.

Fiorio, E., A. Mercanti, M. Terrasi, R. Micciolo, A. Remo, A. Auriemma, A. Molino, V. Parolin, B. Di Stefano, F. Bonetti, A. Giordano, G. L. Cetto and E. Surmacz (2008). "Leptin/HER2 crosstalk in breast cancer: in vitro study and preliminary in vivo analysis." BMC Cancer 8: 305.

Fiuza, U. M. and A. M. Arias (2007). "Cell and molecular biology of Notch." J Endocrinol 194(3): 459-474.

Flynn, J. F., C. Wong and J. M. Wu (2009). "Anti-EGFR Therapy: Mechanism and Advances in Clinical Efficacy in Breast Cancer." J Oncol 2009: 526963.

Folkins, C., Y. Shaked, S. Man, T. Tang, C. R. Lee, Z. Zhu, R. M. Hoffman and R. S. Kerbel (2009). "Glioma tumor stem-like cells promote tumor angiogenesis and vasculogenesis via vascular endothelial growth factor and stromal-derived factor 1." Cancer Res 69(18): 7243-7251.

Fortini, M. E. and S. Artavanis-Tsakonas (1994). "The suppressor of hairless protein participates in notch receptor signaling." Cell 79(2): 273-282.

Frank, N. Y., T. Schatton, S. Kim, Q. Zhan, B. J. Wilson, J. Ma, K. R. Saab, V. Osherov, H. R. Widlund, M. Gasser, A. M. Waaga-Gasser, T. S. Kupper, G. F. Murphy and M. H. Frank (2011). "VEGFR-1 expressed by malignant melanoma-initiating cells is required for tumor growth." Cancer Res 71(4): 1474-1485.

Fu, Y. P., H. Edvardsen, A. Kaushiva, J. P. Arhancet, T. M. Howe, I. Kohaar, P. Porter-Gill, A. Shah, H. Landmark-Hoyvik, S. D. Fossa, S. Ambs, B. Naume, A. L. Borresen-Dale, V. N. Kristensen and L. Prokunina-Olsson (2010). "NOTCH2 in breast cancer: association of SNP rs11249433 with gene expression in ER-positive breast tumors without TP53 mutations." Mol Cancer 9: 113.

Gaengel, K., G. Genove, A. Armulik and C. Betsholtz (2009). "Endothelial-mural cell signaling in vascular development and angiogenesis." Arterioscler Thromb Vasc Biol 29(5): 630-638.

Gallahan, D. and R. Callahan (1987). "Mammary tumorigenesis in feral mice: identification of a new int locus in mouse mammary tumor virus (Czech II)-induced mammary tumors." J Virol 61(1): 66-74.

Gallahan, D., C. Jhappan, G. Robinson, L. Hennighausen, R. Sharp, E. Kordon, R. Callahan, G. Merlino and G. H. Smith (1996). "Expression of a truncated Int3 gene in developing secretory mammary epithelium specifically retards lobular differentiation resulting in tumorigenesis." Cancer Res 56(8): 1775-1785.

Garzia, L., I. Andolfo, E. Cusanelli, N. Marino, G. Petrosino, D. De Martino, V. Esposito, A. Galeone, L. Navas, S. Esposito, S. Gargiulo, S. Fattet, V. Donofrio, G. Cinalli, A. Brunetti, L. D. Vecchio, P. A. Northcott, O. Delattre, M. D. Taylor, A. Iolascon and M. Zollo (2009). "MicroRNA-199b-5p impairs cancer stem cells through negative regulation of HES1 in medulloblastoma." PLoS One 4(3): e4998.

Gazzaniga, P., E. Cigna, V. Panasiti, V. Devirgiliis, U. Bottoni, B. Vincenzi, C. Nicolazzo, A. Petracca and A. Gradilone (2010). "CD133 and ABCB5 as stem cell markers on sentinel lymph node from melanoma patients." Eur J Surg Oncol 36(12): 1211-1214.

Ghayad, S. E. and P. A. Cohen (2010). "Inhibitors of the PI3K/Akt/mTOR pathway: new hope for breast cancer patients." Recent Pat Anticancer Drug Discov 5(1): 29-57.

Ghosh, D., J. Griswold, M. Erman and W. Pangborn (2009). "Structural basis for androgen specificity and oestrogen synthesis in human aromatase." Nature 457(7226): 219-223.

Girard, L., Z. Hanna, N. Beaulieu, C. D. Hoemann, C. Simard, C. A. Kozak and P. Jolicoeur (1996). "Frequent provirus insertional mutagenesis of Notch1 in thymomas of MMTVD/myc transgenic mice suggests a collaboration of c-myc and Notch1 for oncogenesis." Genes Dev 10(15): 1930-1944.

Gonzalez-Perez, R. R., Y. Xu, S. Guo, A. Watters, W. Zhou and S. J. Leibovich (2010). "Leptin upregulates VEGF in breast cancer via canonic and non-canonical signalling pathways and NFkappaB/HIF-1alpha activation." Cell Signal 22(9): 1350-1362.

Gonzalez, R. R., S. Cherfils, M. Escobar, J. H. Yoo, C. Carino, A. K. Styer, B. T. Sullivan, H. Sakamoto, A. Olawaiye, T. Serikawa, M. P. Lynch and B. R. Rueda (2006). "Leptin signaling promotes the growth of mammary tumors and increases the expression of vascular endothelial growth factor (VEGF) and its receptor type two (VEGF-R2)." J Biol Chem 281(36): 26320-26328.

Graziani, I., S. Eliasz, M. A. De Marco, Y. Chen, H. I. Pass, R. M. De May, P. R. Strack, L. Miele and M. Bocchetta (2008). "Opposite effects of Notch-1 and Notch-2 on mesothelioma cell survival under hypoxia are exerted through the Akt pathway." Cancer Res 68(23): 9678-9685.

Greenwald, I. (1998). "LIN-12/Notch signaling: lessons from worms and flies." Genes Dev 12(12): 1751-1762.

Gregory, P. A., A. G. Bert, E. L. Paterson, S. C. Barry, A. Tsykin, G. Farshid, M. A. Vadas, Y. Khew-Goodall and G. J. Goodall (2008). "The miR-200 family and miR-205 regulate epithelial to mesenchymal transition by targeting ZEB1 and SIP1." Nat Cell Biol 10(5): 593-601.

Gridley, T. (2007). "Notch signaling in vascular development and physiology." Development 134(15): 2709-2718.

Gridley, T. (2010). "Notch signaling in the vasculature." Curr Top Dev Biol 92: 277-309.

Guo, S., L. S. Colbert, M. Fuller, Y. Zhang and R. R. Gonzalez-Perez (2010). "Vascular endothelial growth factor receptor-2 in breast cancer." Biochim Biophys Acta 1806(1): 108-121.

Guo, S. and R. R. Gonzalez-Perez (2011). "Notch, IL-1 and leptin crosstalk outcome (NILCO) is critical for leptin-induced proliferation, migration and VEGF/VEGFR-2 expression in breast cancer." PLoS One 6(6): e21467(doi:10.1371/journal.pone.0021467).

Guo, S., M. Liu and R. R. Gonzalez-Perez (2011). "Role of Notch and its oncogenic signaling crosstalk in breast cancer." Biochim Biophys Acta 1815(2): 197-213.

Gupta, R., D. Hong, F. Iborra, S. Sarno and T. Enver (2007). "NOV (CCN3) functions as a regulator of human hematopoietic stem or progenitor cells." Science 316(5824): 590-593.

Gustafson, T. L., E. Wellberg, B. Laffin, L. Schilling, R. P. Metz, C. A. Zahnow and W. W. Porter (2009). "Ha-Ras transformation of MCF10A cells leads to repression of Singleminded-2s through NOTCH and C/EBPbeta." Oncogene 28(12): 1561-1568.

Gustafsson, M. V., X. Zheng, T. Pereira, K. Gradin, S. Jin, J. Lundkvist, J. L. Ruas, L. Poellinger, U. Lendahl and M. Bondesson (2005). "Hypoxia requires notch signaling to maintain the undifferentiated cell state." Dev Cell 9(5): 617-628.

Gutierrez, A. and A. T. Look (2007). "NOTCH and PI3K-AKT pathways intertwined." Cancer Cell 12(5): 411-413.

Hadad, S. M., S. Fleming and A. M. Thompson (2008). "Targeting AMPK: a new therapeutic opportunity in breast cancer." Crit Rev Oncol Hematol 67(1): 1-7.

Hainaud, P., J. O. Contreres, A. Villemain, L. X. Liu, J. Plouet, G. Tobelem and E. Dupuy (2006). "The role of the vascular endothelial growth factor-Delta-like 4 ligand/Notch4-ephrin B2 cascade in tumor vessel remodeling and endothelial cell functions." Cancer Res 66(17): 8501-8510.

Hamada, Y., Y. Kadokawa, M. Okabe, M. Ikawa, J. R. Coleman and Y. Tsujimoto (1999). "Mutation in ankyrin repeats of the mouse Notch2 gene induces early embryonic lethality." Development 126(15): 3415-3424.

Hanahan, D. and R. A. Weinberg (2011). "Hallmarks of cancer: the next generation." Cell 144(5): 646-674.

Hankinson, S. E. (2005). "Endogenous hormones and risk of breast cancer in postmenopausal women." Breast Dis 24: 3-15.

Hanlon, L., J. L. Avila, R. M. Demarest, S. Troutman, M. Allen, F. Ratti, A. K. Rustgi, B. Z. Stanger, F. Radtke, V. Adsay, F. Long, A. J. Capobianco and J. L. Kissil (2010). "Notch1 functions as a tumor suppressor in a model of K-ras-induced pancreatic ductal adenocarcinoma." Cancer Res 70(11): 4280-4286.

Harrison, H., G. Farnie, S. J. Howell, R. E. Rock, S. Stylianou, K. R. Brennan, N. J. Bundred and R. B. Clarke (2010). "Regulation of breast cancer stem cell activity by signaling through the Notch4 receptor." Cancer Res 70(2): 709-718.

Hartman, J., E. W. Lam, J. A. Gustafsson and A. Strom (2009). "Hes-6, an inhibitor of Hes-1, is regulated by 17beta-estradiol and promotes breast cancer cell proliferation." Breast Cancer Res 11(6): R79.

Hasson, P. and Z. Paroush (2006). "Crosstalk between the EGFR and other signalling pathways at the level of the global transcriptional corepressor Groucho/TLE." Br J Cancer 94(6): 771-775.

Hayden, M. S. and S. Ghosh (2004). "Signaling to NF-kappaB." Genes Dev 18(18): 2195-2224.

Hayden, M. S. and S. Ghosh (2008). "Shared principles in NF-kappaB signaling." Cell 132(3): 344-362.

Heath, E., D. Tahri, E. Andermarcher, P. Schofield, S. Fleming and C. A. Boulter (2008). "Abnormal skeletal and cardiac development, cardiomyopathy, muscle atrophy and cataracts in mice with a targeted disruption of the Nov (Ccn3) gene." BMC Dev Biol 8: 18.

Hirota, K. and G. L. Semenza (2006). "Regulation of angiogenesis by hypoxia-inducible factor 1." Crit Rev Oncol Hematol 59(1): 15-26.

Hooper, J. E. and M. P. Scott (2005). "Communicating with Hedgehogs." Nat Rev Mol Cell Biol 6(4): 306-317.

Hrabe de Angelis, M., J. McIntyre, 2nd and A. Gossler (1997). "Maintenance of somite borders in mice requires the Delta homologue Dll1." Nature 386(6626): 717-721.

Hu, J., A. Dong, V. Fernandez-Ruiz, J. Shan, M. Kawa, E. Martinez-Anso, J. Prieto and C. Qian (2009). "Blockade of Wnt signaling inhibits angiogenesis and tumor growth in hepatocellular carcinoma." Cancer Res 69(17): 6951-6959.

Hu, X., S. C. Juneja, N. J. Maihle and M. P. Cleary (2002). "Leptin–a growth factor in normal and malignant breast cells and for normal mammary gland development." J Natl Cancer Inst 94(22): 1704-1711.

Hurteau, G. J., J. A. Carlson, S. D. Spivack and G. J. Brock (2007). "Overexpression of the microRNA hsa-miR-200c leads to reduced expression of transcription factor 8 and increased expression of E-cadherin." Cancer Res 67(17): 7972-7976.

Imatani, A. and R. Callahan (2000). "Identification of a novel NOTCH-4/INT-3 RNA species encoding an activated gene product in certain human tumor cell lines." Oncogene 19(2): 223-231.

Imbimbo, B. P. (2008). "Therapeutic potential of gamma-secretase inhibitors and modulators." Curr Top Med Chem 8(1): 54-61.

Ingram, W. J., K. I. McCue, T. H. Tran, A. R. Hallahan and B. J. Wainwright (2008). "Sonic Hedgehog regulates Hes1 through a novel mechanism that is independent of canonical Notch pathway signalling." Oncogene 27(10): 1489-1500.

Inui, M., G. Martello and S. Piccolo (2010). "MicroRNA control of signal transduction." Nat Rev Mol Cell Biol 11(4): 252-263.

Ischenko, I., H. Seeliger, M. Schaffer, K. W. Jauch and C. J. Bruns (2008). "Cancer stem cells: how can we target them?" Curr Med Chem 15(30): 3171-3184.

Iso, T., L. Kedes and Y. Hamamori (2003). "HES and HERP families: multiple effectors of the Notch signaling pathway." J Cell Physiol 194(3): 237-255.

Jannot, G. and M. J. Simard (2006). "Tumour related microRNAs functions in Caenorhabditis elegans." Oncogene 25(46): 6197-6201.

Jhappan, C., D. Gallahan, C. Stahle, E. Chu, G. H. Smith, G. Merlino and R. Callahan (1992). "Expression of an activated Notch-related int-3 transgene interferes with cell differentiation and induces neoplastic transformation in mammary and salivary glands." Genes Dev 6(3): 345-355.

Jimeno, A. and M. Hidalgo (2006). "Pharmacogenomics of epidermal growth factor receptor (EGFR) tyrosine kinase inhibitors." Biochim Biophys Acta 1766(2): 217-229.

Johnston, A., S. Arnadottir, J. E. Gudjonsson, A. Aphale, A. A. Sigmarsdottir, S. I. Gunnarsson, J. T. Steinsson, J. T. Elder and H. Valdimarsson (2008). "Obesity in psoriasis: leptin and resistin as mediators of cutaneous inflammation." Br J Dermatol 159(2): 342-350.

Johnston, D. A., B. Dong and C. C. Hughes (2009). "TNF induction of jagged-1 in endothelial cells is NFkappaB-dependent." Gene 435(1-2): 36-44.

Joutel, A., C. Corpechot, A. Ducros, K. Vahedi, H. Chabriat, P. Mouton, S. Alamowitch, V. Domenga, M. Cecillion, E. Marechal, J. Maciazek, C. Vayssiere, C. Cruaud, E. A. Cabanis, M. M. Ruchoux, J. Weissenbach, J. F. Bach, M. G. Bousser and E. Tournier-Lasserve (1996). "Notch3 mutations in CADASIL, a hereditary adult-onset condition causing stroke and dementia." Nature 383(6602): 707-710.

Jung, C. H., S. H. Ro, J. Cao, N. M. Otto and D. H. Kim (2010). "mTOR regulation of autophagy." FEBS Lett 584(7): 1287-1295.

Kato, M., T. Paranjape, R. U. Muller, S. Nallur, E. Gillespie, K. Keane, A. Esquela-Kerscher, J. B. Weidhaas and F. J. Slack (2009). "The mir-34 microRNA is required for the DNA damage response in vivo in C. elegans and in vitro in human breast cancer cells." Oncogene 28(25): 2419-2424.

Katoh, M. (2007). "Networking of WNT, FGF, Notch, BMP, and Hedgehog signaling pathways during carcinogenesis." Stem Cell Rev 3(1): 30-38.

Katoh, Y. and M. Katoh (2005a). "Comparative genomics on Sonic hedgehog orthologs." Oncol Rep 14(4): 1087-1090.

Katoh, Y. and M. Katoh (2005b). "Identification and characterization of rat Desert hedgehog and Indian hedgehog genes in silico." Int J Oncol 26(2): 545-549.

Ke, Q. and M. Costa (2006). "Hypoxia-inducible factor-1 (HIF-1)." Mol Pharmacol 70(5): 1469-1480.

Kiaris, H., K. Politi, L. M. Grimm, M. Szabolcs, P. Fisher, A. Efstratiadis and S. Artavanis-Tsakonas (2004). "Modulation of notch signaling elicits signature tumors and inhibits hras1-induced oncogenesis in the mouse mammary epithelium." Am J Pathol 165(2): 695-705.

Kikuchi, R., K. Takeshita, Y. Uchida, M. Kondo, X. W. Cheng, T. Nakayama, K. Yamamoto, T. Matsushita, J. K. Liao and T. Murohara (2011). "Pitavastatin-induced angiogenesis and arteriogenesis is mediated by Notch1 in a murine hindlimb ischemia model without induction of VEGF." Lab Invest 91(5): 691-703.

Kim, S. B., G. W. Chae, J. Lee, J. Park, H. Tak, J. H. Chung, T. G. Park, J. K. Ahn and C. O. Joe (2007). "Activated Notch1 interacts with p53 to inhibit its phosphorylation and transactivation." Cell Death Differ 14(5): 982-991.

Kindler, T., M. G. Cornejo, C. Scholl, J. Liu, D. S. Leeman, J. E. Haydu, S. Frohling, B. H. Lee and D. G. Gilliland (2008). "K-RasG12D-induced T-cell lymphoblastic lymphoma/leukemias harbor Notch1 mutations and are sensitive to gamma-secretase inhibitors." Blood 112(8): 3373-3382.

Kishimoto, T. (2005). "Interleukin-6: from basic science to medicine—40 years in immunology." Annu Rev Immunol 23: 1-21.

Knight, B. B., G. M. Oprea-Ilies, A. Nagalingam, L. Yang, C. Cohen, N. K. Saxena and D. Sharma (2011). "Survivin upregulation, dependent on leptin-EGFR-Notch1 axis, is essential for leptin induced migration of breast carcinoma cells." Endocr Relat Cancer.

Knupfer, H. and R. Preiss (2007). "Significance of interleukin-6 (IL-6) in breast cancer (review)." Breast Cancer Res Treat 102(2): 129-135.

Komatsu, H., M. Y. Chao, J. Larkins-Ford, M. E. Corkins, G. A. Somers, T. Tucey, H. M. Dionne, J. Q. White, K. Wani, M. Boxem and A. C. Hart (2008). "OSM-11 facilitates LIN-12 Notch signaling during Caenorhabditis elegans vulval development." PLoS Biol 6(8): e196.

Konishi, J., F. Yi, X. Chen, H. Vo, D. P. Carbone and T. P. Dang (2010). "Notch3 cooperates with the EGFR pathway to modulate apoptosis through the induction of bim." Oncogene 29(4): 589-596.

Kopan, R. and M. X. Ilagan (2009). "The canonical Notch signaling pathway: unfolding the activation mechanism." Cell 137(2): 216-233.

Korkaya, H., A. Paulson, E. Charafe-Jauffret, C. Ginestier, M. Brown, J. Dutcher, S. G. Clouthier and M. S. Wicha (2009). "Regulation of mammary stem/progenitor cells by PTEN/Akt/beta-catenin signaling." PLoS Biol 7(6): e1000121.

Korkaya, H., A. Paulson, F. Iovino and M. S. Wicha (2008). "HER2 regulates the mammary stem/progenitor cell population driving tumorigenesis and invasion." Oncogene 27(47): 6120-6130.

Korkaya, H. and M. S. Wicha (2009). "HER-2, notch, and breast cancer stem cells: targeting an axis of evil." Clin Cancer Res 15(6): 1845-1847.

Koukourakis, M. I., C. Simopoulos, A. Polychronidis, S. Perente, S. Botaitis, A. Giatromanolaki and E. Sivridis (2003). "The effect of trastuzumab/docatexel combination on breast cancer angiogenesis: dichotomus effect predictable by the HIFI alpha/VEGF pre-treatment status?" Anticancer Res 23(2C): 1673-1680.

Koutras, A. K., G. Fountzilas, K. T. Kalogeras, I. Starakis, G. Iconomou and H. P. Kalofonos (2010). "The upgraded role of HER3 and HER4 receptors in breast cancer." Crit Rev Oncol Hematol 74(2): 73-78.

Krebs, L. T., Y. Xue, C. R. Norton, J. R. Shutter, M. Maguire, J. P. Sundberg, D. Gallahan, V. Closson, J. Kitajewski, R. Callahan, G. H. Smith, K. L. Stark and T. Gridley (2000). "Notch signaling is essential for vascular morphogenesis in mice." Genes Dev 14(11): 1343-1352.

Kumano, K., S. Chiba, A. Kunisato, M. Sata, T. Saito, E. Nakagami-Yamaguchi, T. Yamaguchi, S. Masuda, K. Shimizu, T. Takahashi, S. Ogawa, Y. Hamada and H. Hirai (2003). "Notch1 but not Notch2 is essential for generating hematopoietic stem cells from endothelial cells." Immunity 18(5): 699-711.

Kumar, R. and R. Yarmand-Bagheri (2001). "The role of HER2 in angiogenesis." Semin Oncol 28(5 Suppl 16): 27-32.

Kumar, S., H. Kishimoto, H. L. Chua, S. Badve, K. D. Miller, R. M. Bigsby and H. Nakshatri (2003). "Interleukin-1 alpha promotes tumor growth and cachexia in MCF-7 xenograft model of breast cancer." Am J Pathol 163(6): 2531-2541.

LaRochelle, W. J., M. Jeffers, W. F. McDonald, R. A. Chillakuru, N. A. Giese, N. A. Lokker, C. Sullivan, F. L. Boldog, M. Yang, C. Vernet, C. E. Burgess, E. Fernandes, L. L. Deegler, B. Rittman, J. Shimkets, R. A. Shimkets, J. M. Rothberg and H. S. Lichenstein (2001). "PDGF-D, a new protease-activated growth factor." Nat Cell Biol 3(5): 517-521.

Laud, K., I. Gourdou, L. Pessemesse, J. P. Peyrat and J. Djiane (2002). "Identification of leptin receptors in human breast cancer: functional activity in the T47-D breast cancer cell line." Mol Cell Endocrinol 188(1-2): 219-226.

Laughner, E., P. Taghavi, K. Chiles, P. C. Mahon and G. L. Semenza (2001). "HER2 (neu) signaling increases the rate of hypoxia-inducible factor 1alpha (HIF-1alpha) synthesis: novel mechanism for HIF-1-mediated vascular endothelial growth factor expression." Mol Cell Biol 21(12): 3995-4004.

Lawson, N. D., A. M. Vogel and B. M. Weinstein (2002). "sonic hedgehog and vascular endothelial growth factor act upstream of the Notch pathway during arterial endothelial differentiation." Dev Cell 3(1): 127-136.

Leask, A. and D. J. Abraham (2006). "All in the CCN family: essential matricellular signaling modulators emerge from the bunker." J Cell Sci 119(Pt 23): 4803-4810.

Lee, C. W., C. M. Raskett, I. Prudovsky and D. C. Altieri (2008a). "Molecular dependence of estrogen receptor-negative breast cancer on a notch-survivin signaling axis." Cancer Res 68(13): 5273-5281.

Lee, C. W., K. Simin, Q. Liu, J. Plescia, M. Guha, A. Khan, C. C. Hsieh and D. C. Altieri (2008b). "A functional Notch-survivin gene signature in basal breast cancer." Breast Cancer Res 10(6): R97.

Lee, J. H., J. Suk, J. Park, S. B. Kim, S. S. Kwak, J. W. Kim, C. H. Lee, B. Byun, J. K. Ahn and C. O. Joe (2009). "Notch signal activates hypoxia pathway through HES1-dependent SRC/signal transducers and activators of transcription 3 pathway." Mol Cancer Res 7(10): 1663-1671.

Lefort, K., A. Mandinova, P. Ostano, V. Kolev, V. Calpini, I. Kolfschoten, V. Devgan, J. Lieb, W. Raffoul, D. Hohl, V. Neel, J. Garlick, G. Chiorino and G. P. Dotto (2007). "Notch1 is a p53 target gene involved in human keratinocyte tumor suppression through negative regulation of ROCK1/2 and MRCKalpha kinases." Genes Dev 21(5): 562-577.

Lehar, S. M. and M. J. Bevan (2006). "T cells develop normally in the absence of both Deltex1 and Deltex2." Mol Cell Biol 26(20): 7358-7371.

Leong, K. G., X. Hu, L. Li, M. Noseda, B. Larrivee, C. Hull, L. Hood, F. Wong and A. Karsan (2002). "Activated Notch4 inhibits angiogenesis: role of beta 1-integrin activation." Mol Cell Biol 22(8): 2830-2841.

Leong, K. G. and A. Karsan (2006). "Recent insights into the role of Notch signaling in tumorigenesis." Blood 107(6): 2223-2233.

Leong, K. G., K. Niessen, I. Kulic, A. Raouf, C. Eaves, I. Pollet and A. Karsan (2007). "Jagged1-mediated Notch activation induces epithelial-to-mesenchymal transition through Slug-induced repression of E-cadherin." J Exp Med 204(12): 2935-2948.

Lewis, J. (1998). "Notch signalling and the control of cell fate choices in vertebrates." Semin Cell Dev Biol 9(6): 583-589.

Li, M., V. Jendrossek and C. Belka (2007). "The role of PDGF in radiation oncology." Radiat Oncol 2: 5.

Li, X. and U. Eriksson (2003). "Novel PDGF family members: PDGF-C and PDGF-D." Cytokine Growth Factor Rev 14(2): 91-98.

Li, Y., F. Guessous, Y. Zhang, C. Dipierro, B. Kefas, E. Johnson, L. Marcinkiewicz, J. Jiang, Y. Yang, T. D. Schmittgen, B. Lopes, D. Schiff, B. Purow and R. Abounader (2009). "MicroRNA-34a inhibits glioblastoma growth by targeting multiple oncogenes." Cancer Res 69(19): 7569-7576.

Li, Y. and D. Ye (2010). "Cancer Therapy By Targeting Hypoxia-Inducible Factor-1." Curr Cancer Drug Targets.

Liao, S., J. Xia, Z. Chen, S. Zhang, A. Ahmad, L. Miele, F. H. Sarkar and Z. Wang (2011). "Inhibitory effect of curcumin on oral carcinoma CAL-27 cells via suppression of Notch-1 and NF-kappaB signaling pathways." J Cell Biochem 112(4): 1055-1065.

Ling, H., J. R. Sylvestre and P. Jolicoeur (2010). "Notch1-induced mammary tumor development is cyclin D1-dependent and correlates with expansion of pre-malignant multipotent duct-limited progenitors." Oncogene 29(32): 4543-54.

Liu, H., S. Kennard and B. Lilly (2009). "NOTCH3 expression is induced in mural cells through an autoregulatory loop that requires endothelial-expressed JAGGED1." Circ Res 104(4): 466-475.

Liu, H., W. Zhang, S. Kennard, R. B. Caldwell and B. Lilly (2010). "Notch3 is critical for proper angiogenesis and mural cell investment." Circ Res 107(7): 860-870.

Liu, Q., G. Li, R. Li, J. Shen, Q. He, L. Deng, C. Zhang and J. Zhang (2010). "IL-6 promotion of glioblastoma cell invasion and angiogenesis in U251 and T98G cell lines." J Neurooncol 100(2): 165-176.

Liu, Z. J., T. Shirakawa, Y. Li, A. Soma, M. Oka, G. P. Dotto, R. M. Fairman, O. C. Velazquez and M. Herlyn (2003). "Regulation of Notch1 and Dll4 by vascular endothelial growth factor in arterial endothelial cells: implications for modulating arteriogenesis and angiogenesis." Mol Cell Biol 23(1): 14-25.

Liu, Z. J., M. Xiao, K. Balint, K. S. Smalley, P. Brafford, R. Qiu, C. C. Pinnix, X. Li and M. Herlyn (2006). "Notch1 signaling promotes primary melanoma progression by activating mitogen-activated protein kinase/phosphatidylinositol 3-kinase-Akt pathways and up-regulating N-cadherin expression." Cancer Res 66(8): 4182-4190.

Lu, L., X. Chen, C. W. Zhang, W. L. Yang, Y. J. Wu, L. Sun, L. M. Bai, X. S. Gu, S. Ahmed, G. S. Dawe and Z. C. Xiao (2008). "Morphological and functional characterization of predifferentiation of myelinating glia-like cells from human bone marrow stromal cells through activation of F3/Notch signaling in mouse retina." Stem Cells 26(2): 580-590.

Lum, L. and P. A. Beachy (2004). "The Hedgehog response network: sensors, switches, and routers." Science 304(5678): 1755-1759.

MacKenzie, F., P. Duriez, B. Larrivee, L. Chang, I. Pollet, F. Wong, C. Yip and A. Karsan (2004). "Notch4-induced inhibition of endothelial sprouting requires the ankyrin repeats and involves signaling through RBP-Jkappa." Blood 104(6): 1760-1768.

Maedler, K., P. Sergeev, J. A. Ehses, Z. Mathe, D. Bosco, T. Berney, J. M. Dayer, M. Reinecke, P. A. Halban and M. Y. Donath (2004). "Leptin modulates beta cell expression of IL-1 receptor antagonist and release of IL-1beta in human islets." Proc Natl Acad Sci USA 101(21): 8138-8143.

Magnifico, A., L. Albano, S. Campaner, D. Delia, F. Castiglioni, P. Gasparini, G. Sozzi, E. Fontanella, S. Menard and E. Tagliabue (2009). "Tumor-initiating cells of HER2-positive carcinoma cell lines express the highest oncoprotein levels and are sensitive to trastuzumab." Clin Cancer Res 15(6): 2010-2021.

Magoffin, D. A., S. R. Weitsman, S. K. Aagarwal and A. J. Jakimiuk (1999). "Leptin regulation of aromatase activity in adipose stromal cells from regularly cycling women." Ginekol Pol 70(1): 1-7.

Mailhos, C., U. Modlich, J. Lewis, A. Harris, R. Bicknell and D. Ish-Horowicz (2001). "Delta4, an endothelial specific notch ligand expressed at sites of physiological and tumor angiogenesis." Differentiation 69(2-3): 135-144.

Maniotis, A. J., R. Folberg, A. Hess, E. A. Seftor, L. M. Gardner, J. Pe'er, J. M. Trent, P. S. Meltzer and M. J. Hendrix (1999). "Vascular channel formation by human melanoma cells in vivo and in vitro: vasculogenic mimicry." Am J Pathol 155(3): 739-752.

Marigo, V., D. J. Roberts, S. M. Lee, O. Tsukurov, T. Levi, J. M. Gastier, D. J. Epstein, D. J. Gilbert, N. G. Copeland, C. E. Seidman and et al. (1995). "Cloning, expression, and chromosomal location of SHH and IHH: two human homologues of the Drosophila segment polarity gene hedgehog." Genomics 28(1): 44-51.

Matsuno, K., R. J. Diederich, M. J. Go, C. M. Blaumueller and S. Artavanis-Tsakonas (1995). "Deltex acts as a positive regulator of Notch signaling through interactions with the Notch ankyrin repeats." Development 121(8): 2633-2644.

McCright, B., X. Gao, L. Shen, J. Lozier, Y. Lan, M. Maguire, D. Herzlinger, G. Weinmaster, R. Jiang and T. Gridley (2001). "Defects in development of the kidney, heart and eye vasculature in mice homozygous for a hypomorphic Notch2 mutation." Development 128(4): 491-502.

McGowan, P. M., C. Simedrea, E. J. Ribot, P. J. Foster, D. Palmieri, P. S. Steeg, A. L. Allan and A. F. Chambers (2011). "Notch1 inhibition alters the CD44hi/CD24lo population and reduces the formation of brain metastases from breast cancer." Mol Cancer Res 9(7): 834-844.

McKenzie, G., G. Ward, Y. Stallwood, E. Briend, S. Papadia, A. Lennard, M. Turner, B. Champion and G. E. Hardingham (2006). "Cellular Notch responsiveness is defined by phosphoinositide 3-kinase-dependent signals." BMC Cell Biol 7: 10.

Mendelsohn, M. E. and R. H. Karas (1999). "The protective effects of estrogen on the cardiovascular system." N Engl J Med 340(23): 1801-1811.

Mendelson, J., S. Song, Y. Li, D. M. Maru, B. Mishra, M. Davila, W. L. Hofstetter and L. Mishra (2011). "Dysfunctional transforming growth factor-beta signaling with constitutively active Notch signaling in Barrett's esophageal adenocarcinoma." Cancer 117(16): 3691-3702.

Meng, R. D., C. C. Shelton, Y. M. Li, L. X. Qin, D. Notterman, P. B. Paty and G. K. Schwartz (2009). "gamma-Secretase inhibitors abrogate oxaliplatin-induced activation of the Notch-1 signaling pathway in colon cancer cells resulting in enhanced chemosensitivity." Cancer Res 69(2): 573-582.

Menon, S. and B. D. Manning (2008). "Common corruption of the mTOR signaling network in human tumors." Oncogene 27 Suppl 2: S43-51.

Mesri, M., M. Morales-Ruiz, E. J. Ackermann, C. F. Bennett, J. S. Pober, W. C. Sessa and D. C. Altieri (2001). "Suppression of vascular endothelial growth factor-mediated endothelial cell protection by survivin targeting." Am J Pathol 158(5): 1757-1765.

Meurette, O., S. Stylianou, R. Rock, G. M. Collu, A. P. Gilmore and K. Brennan (2009). "Notch activation induces Akt signaling via an autocrine loop to prevent apoptosis in breast epithelial cells." Cancer Res 69(12): 5015-5022.

Miele, L. (2006). "Notch signaling." Clin Cancer Res 12(4): 1074-1079.

Miele, L., H. Miao and B. J. Nickoloff (2006). "NOTCH signaling as a novel cancer therapeutic target." Curr Cancer Drug Targets 6(4): 313-323.

Miele, L. and B. Osborne (1999). "Arbiter of differentiation and death: Notch signaling meets apoptosis." J Cell Physiol 181(3): 393-409.

Milner, L. A. and A. Bigas (1999). "Notch as a mediator of cell fate determination in hematopoiesis: evidence and speculation." Blood 93(8): 2431-2448.

Mine, T., S. Matsueda, H. Gao, Y. Li, K. K. Wong, G. E. Peoples, S. Ferrone and C. G. Ioannides (2010). "Created Gli-1 duplex short-RNA (i-Gli-RNA) eliminates CD44 Hi progenitors of taxol-resistant ovarian cancer cells." Oncol Rep 23(6): 1537-1543.

Minhajuddin, M., K. M. Bijli, F. Fazal, A. Sassano, K. I. Nakayama, N. Hay, L. C. Platanias and A. Rahman (2009). "Protein kinase C-delta and phosphati-dylinositol 3-kinase/Akt activate mammalian target of rapamycin to modulate NF-kappaB activation and intercellular adhesion molecule-1 (ICAM-1) expression in endothelial cells." J Biol Chem 284(7): 4052-4061.

Mittal, S., D. Subramanyam, D. Dey, R. V. Kumar and A. Rangarajan (2009). "Cooperation of Notch and Ras/MAPK signaling pathways in human breast carcinogenesis." Mol Cancer 8: 128.

Mumm, J. S., E. H. Schroeter, M. T. Saxena, A. Griesemer, X. Tian, D. J. Pan, W. J. Ray and R. Kopan (2000). "A ligand-induced extracellular cleavage regulates gamma-secretase-like proteolytic activation of Notch1." Mol Cell 5(2): 197-206.

Mungamuri, S. K., X. Yang, A. D. Thor and K. Somasundaram (2006). "Survival signaling by Notch1: mammalian target of rapamycin (mTOR)-dependent inhibition of p53." Cancer Res 66(9): 4715-4724.

Nagase, T., M. Nagase, M. Machida and T. Fujita (2008). "Hedgehog signalling in vascular development." Angiogenesis 11(1): 71-77.

Neurath, M. F. and S. Finotto (2011). "IL-6 signaling in autoimmunity, chronic inflammation and inflammation-associated cancer." Cytokine Growth Factor Rev 22(2): 83-89.

Niimi, H., K. Pardali, M. Vanlandewijck, C. H. Heldin and A. Moustakas (2007). "Notch signaling is necessary for epithelial growth arrest by TGF-beta." J Cell Biol 176(5): 695-707.

Nilsson, M. B., R. R. Langley and I. J. Fidler (2005). "Interleukin-6, secreted by human ovarian carcinoma cells, is a potent proangiogenic cytokine." Cancer Res 65(23): 10794-10800.

Nishina, S., H. Shiraha, Y. Nakanishi, S. Tanaka, M. Matsubara, N. Takaoka, M. Uemura, S. Horiguchi, J. Kataoka, M. Iwamuro, T. Yagi and K. Yamamoto (2011). "Restored expression of the tumor suppressor gene RUNX3 reduces cancer stem cells in hepatocellular carcinoma by suppressing Jagged1-Notch signaling." Oncol Rep 26(3): 523-531.

Noguera-Troise, I., C. Daly, N. J. Papadopoulos, S. Coetzee, P. Boland, N. W. Gale, H. C. Lin, G. D. Yancopoulos and G. Thurston (2006). "Blockade of Dll4 inhibits tumour growth by promoting non-productive angiogenesis." Nature 444(7122): 1032-1037.

Noh, W. C., Y. H. Kim, M. S. Kim, J. S. Koh, H. A. Kim, N. M. Moon and N. S. Paik (2008). "Activation of the mTOR signaling pathway in breast cancer and its correlation with the clinicopathologic variables." Breast Cancer Res Treat 110(3): 477-483.

Noseda, M., L. Chang, G. McLean, J. E. Grim, B. E. Clurman, L. L. Smith and A. Karsan (2004). "Notch activation induces endothelial cell cycle arrest and participates in contact inhibition: role of p21Cip1 repression." Mol Cell Biol 24(20): 8813-8822.

O'Day, E. and A. Lal (2010). "MicroRNAs and their target gene networks in breast cancer." Breast Cancer Res 12(2): 201.

O'Neill, C. F., S. Urs, C. Cinelli, A. Lincoln, R. J. Nadeau, R. Leon, J. Toher, C. Mouta-Bellum, R. E. Friesel and L. Liaw (2007). "Notch2 signaling induces apoptosis and inhibits human MDA-MB-231 xenograft growth." Am J Pathol 171(3): 1023-1036.

Oishi, H., M. Sunamura, S. Egawa, F. Motoi, M. Unno, T. Furukawa, N. A. Habib and H. Yagita (2010). "Blockade of delta-like ligand 4 signaling inhibits both growth and angiogenesis of pancreatic cancer." Pancreas 39(6): 897-903.

Okuyama, R., H. Tagami and S. Aiba (2008). "Notch signaling: its role in epidermal homeostasis and in the pathogenesis of skin diseases." J Dermatol Sci 49(3): 187-194.

Osipo, C., T. E. Golde, B. A. Osborne and L. A. Miele (2008). "Off the beaten pathway: the complex cross talk between Notch and NF-kappaB." Lab Invest 88(1): 11-17.

Osipo, C., P. Patel, P. Rizzo, A. G. Clementz, L. Hao, T. E. Golde and L. Miele (2008). "ErbB-2 inhibition activates Notch-1 and sensitizes breast cancer cells to a gamma-secretase inhibitor." Oncogene 27(37): 5019-5032.

Oswald, F., S. Liptay, G. Adler and R. M. Schmid (1998). "NF-kappaB2 is a putative target gene of activated Notch-1 via RBP-Jkappa." Mol Cell Biol 18(4): 2077-2088.

Padua, D. and J. Massague (2009). "Roles of TGFbeta in metastasis." Cell Res 19(1): 89-102.

Palomero, T., M. Dominguez and A. A. Ferrando (2008). "The role of the PTEN/AKT Pathway in NOTCH1-induced leukemia." Cell Cycle 7(8): 965-970.

Pang, R. T., C. O. Leung, T. M. Ye, W. Liu, P. C. Chiu, K. K. Lam, K. F. Lee and W. S. Yeung (2010). "MicroRNA-34a suppresses invasion through downregulation of Notch1 and Jagged1 in cervical carcinoma and choriocarcinoma cells." Carcinogenesis 31(6): 1037-1044.

Pardali, E., M. J. Goumans and P. ten Dijke (2010). "Signaling by members of the TGF-beta family in vascular morphogenesis and disease." Trends Cell Biol 20(9): 556-567.

Pardali, E. and P. ten Dijke (2009). "Transforming growth factor-beta signaling and tumor angiogenesis." Front Biosci 14: 4848-4861.

Park, J. T., M. Li, K. Nakayama, T. L. Mao, B. Davidson, Z. Zhang, R. J. Kurman, C. G. Eberhart, M. Shih Ie and T. L. Wang (2006). "Notch3 gene amplification in ovarian cancer." Cancer Res 66(12): 6312-6318.

Parr, C., G. Watkins and W. G. Jiang (2004). "The possible correlation of Notch-1 and Notch-2 with clinical outcome and tumour clinicopathological parameters in human breast cancer." Int J Mol Med 14(5): 779-786.

Pasca di Magliano, M. and M. Hebrok (2003). "Hedgehog signalling in cancer formation and maintenance." Nat Rev Cancer 3(12): 903-911.

Pasche, B., P. Kolachana, K. Nafa, J. Satagopan, Y. G. Chen, R. S. Lo, D. Brener, D. Yang, L. Kirstein, C. Oddoux, H. Ostrer, P. Vineis, L. Varesco, S. Jhanwar, L. Luzzatto, J. Massague and K. Offit (1999). "TbetaR-I(6A) is a candidate tumor susceptibility allele." Cancer Res 59(22): 5678-5682.

Patel, N. S., M. S. Dobbie, M. Rochester, G. Steers, R. Poulsom, K. Le Monnier, D. W. Cranston, J. L. Li and A. L. Harris (2006). "Up-regulation of endothelial delta-like 4 expression correlates with vessel maturation in bladder cancer." Clin Cancer Res 12(16): 4836-4844.

Patel, N. S., J. L. Li, D. Generali, R. Poulsom, D. W. Cranston and A. L. Harris (2005). "Up-regulation of delta-like 4 ligand in human tumor vasculature and the role of basal expression in endothelial cell function." Cancer Res 65(19): 8690-8697.

Perrier, S., F. Caldefie-Chezet and M. P. Vasson (2009). "IL-1 family in breast cancer: potential interplay with leptin and other adipocytokines." FEBS Lett 583(2): 259-265.

Phillips, T. M., K. Kim, E. Vlashi, W. H. McBride and F. Pajonk (2007). "Effects of recombinant erythropoietin on breast cancer-initiating cells." Neoplasia 9(12): 1122-1129.

Phng, L. K. and H. Gerhardt (2009). "Angiogenesis: a team effort coordinated by notch." Dev Cell 16(2): 196-208.

Phng, L. K., M. Potente, J. D. Leslie, J. Babbage, D. Nyqvist, I. Lobov, J. K. Ondr, S. Rao, R. A. Lang, G. Thurston and H. Gerhardt (2009). "Nrarp coordinates endothelial Notch and Wnt signaling to control vessel density in angiogenesis." Dev Cell 16(1): 70-82.

Pietras, K., T. Sjoblom, K. Rubin, C. H. Heldin and A. Ostman (2003). "PDGF receptors as cancer drug targets." Cancer Cell 3(5): 439-443.

Ping, Y. F., X. H. Yao, J. Y. Jiang, L. T. Zhao, S. C. Yu, T. Jiang, M. C. Lin, J. H. Chen, B. Wang, R. Zhang, Y. H. Cui, C. Qian, J. Wang and X. W. Bian (2011). "The chemokine CXCL12 and its receptor CXCR4 promote glioma stem cell-mediated VEGF production and tumour angiogenesis via PI3K/AKT signalling." J Pathol 224(3): 344-354.

Politi, K., N. Feirt and J. Kitajewski (2004). "Notch in mammary gland development and breast cancer." Semin Cancer Biol 14(5): 341-347.

Polus, A., G. J., Piatkowska E., Dembinska-Kiec A. (2003). "Differences in leptin, VEGF, and bFGF-induced angiogenic differentiation of HUVEC and human umbilical blood CD34+ progenitor cells." European Journal of Biochemistry: Abstract number: P4.1-67.

Prager, G. W. and M. Poettler (2012). "Angiogenesis in cancer. Basic mechanisms and therapeutic advances." Hamostaseologie 32(2): 105-114.

Prasad, S., J. Ravindran and B. B. Aggarwal (2010). "NF-kappaB and cancer: how intimate is this relationship." Mol Cell Biochem 336(1-2): 25-37.

Pratt, M. A., E. Tibbo, S. J. Robertson, D. Jansson, K. Hurst, C. Perez-Iratxeta, R. Lau and M. Y. Niu (2009). "The canonical NF-kappaB pathway is required for formation of luminal mammary neoplasias and is activated in the mammary progenitor population." Oncogene 28(30): 2710-2722.

Press, M. F. and H. J. Lenz (2007). "EGFR, HER2 and VEGF pathways: validated targets for cancer treatment." Drugs 67(14): 2045-2075.

Quillard, T., J. Devalliere, S. Coupel and B. Charreau (2010). "Inflammation dysregulates Notch signaling in endothelial cells: implication of Notch2 and Notch4 to endothelial dysfunction." Biochem Pharmacol 80(12): 2032-2041.

Radtke, F. and K. Raj (2003). "The role of Notch in tumorigenesis: oncogene or tumour suppressor?" Nat Rev Cancer 3(10): 756-767.

Rafiee, P., D. G. Binion, M. Wellner, B. Behmaram, M. Floer, E. Mitton, L. Nie, Z. Zhang and M. F. Otterson (2010). "Modulatory effect of curcumin on survival of irradiated human intestinal microvascular endothelial cells: role of Akt/mTOR and NF-{kappa}B." Am J Physiol Gastrointest Liver Physiol 298(6): G865-877.

Rangarajan, A., R. Syal, S. Selvarajah, O. Chakrabarti, A. Sarin and S. Krishna (2001). "Activated Notch1 signaling cooperates with papillomavirus oncogenes in transformation and generates resistance to apoptosis on matrix withdrawal through PKB/Akt." Virology 286(1): 23-30.

Raouf, A., Y. Zhao, K. To, J. Stingl, A. Delaney, M. Barbara, N. Iscove, S. Jones, S. McKinney, J. Emerman, S. Aparicio, M. Marra and C. Eaves (2008). "Transcriptome analysis of the normal human mammary cell commitment and differentiation process." Cell Stem Cell 3(1): 109-118.

Rasul, S., R. Balasubramanian, A. Filipovic, M. J. Slade, E. Yague and R. C. Coombes (2009). "Inhibition of gamma-secretase induces G2/M arrest and triggers apoptosis in breast cancer cells." Br J Cancer 100(12): 1879-1888.

Reedijk, M., S. Odorcic, L. Chang, H. Zhang, N. Miller, D. R. McCready, G. Lockwood and S. E. Egan (2005). "High-level coexpression of JAG1 and NOTCH1 is observed in human breast cancer and is associated with poor overall survival." Cancer Res 65(18): 8530-8537.

Reedijk, M., S. Odorcic, H. Zhang, R. Chetty, C. Tennert, B. C. Dickson, G. Lockwood, S. Gallinger and S. E. Egan (2008). "Activation of Notch signaling in human colon adenocarcinoma." Int J Oncol 33(6): 1223-1229.

Renard, P. and M. Raes (1999). "The proinflammatory transcription factor NFkappaB: a potential target for novel therapeutical strategies." Cell Biol Toxicol 15(6): 341-344.

Rene Gonzalez, R., A. Watters, Y. Xu, U. P. Singh, D. R. Mann, B. R. Rueda and M. L. Penichet (2009). "Leptin-signaling inhibition results in efficient anti-tumor activity in estrogen receptor positive or negative breast cancer." Breast Cancer Res 11(3): R36.

Riccio, O., M. E. van Gijn, A. C. Bezdek, L. Pellegrinet, J. H. van Es, U. Zimber-Strobl, L. J. Strobl, T. Honjo, H. Clevers and F. Radtke (2008). "Loss of intestinal crypt progenitor cells owing to inactivation of both Notch1 and Notch2 is accompanied by derepression of CDK inhibitors p27Kip1 and p57Kip2." EMBO Rep 9(4): 377-383.

Ridgway, J., G. Zhang, Y. Wu, S. Stawicki, W. C. Liang, Y. Chanthery, J. Kowalski, R. J. Watts, C. Callahan, I. Kasman, M. Singh, M. Chien, C. Tan, J. A. Hongo, F. de Sauvage, G. Plowman and M. Yan (2006). "Inhibition of Dll4 signalling inhibits tumour growth by deregulating angiogenesis." Nature 444(7122): 1083-1087.

Rizzo, P., H. Miao, G. D'Souza, C. Osipo, L. L. Song, J. Yun, H. Zhao, J. Mascarenhas, D. Wyatt, G. Antico, L. Hao, K. Yao, P. Rajan, C. Hicks, K. Siziopikou, S. Selvaggi, A. Bashir, D. Bhandari, A. Marchese, U. Lendahl, J. Z. Qin, D. A. Tonetti, K. Albain, B. J. Nickoloff and L. Miele (2008a). "Cross-talk between notch and the estrogen receptor in breast cancer suggests novel therapeutic approaches." Cancer Res 68(13): 5226-5235.

Rizzo, P., C. Osipo, K. Foreman, T. Golde, B. Osborne and L. Miele (2008b). "Rational targeting of Notch signaling in cancer." Oncogene 27(38): 5124-5131.

Roca, C. and R. H. Adams (2007). "Regulation of vascular morphogenesis by Notch signaling." Genes Dev 21(20): 2511-2524.

Ronchini, C. and A. J. Capobianco (2001). "Induction of cyclin D1 transcription and CDK2 activity by Notch(ic): implication for cell cycle disruption in transformation by Notch(ic)." Mol Cell Biol 21(17): 5925-5934.

Rose, S. L., M. Kunnimalaiyaan, J. Drenzek and N. Seiler (2010). "Notch 1 signaling is active in ovarian cancer." Gynecol Oncol 117(1): 130-133.

Rosenfeld, N., R. Aharonov, E. Meiri, S. Rosenwald, Y. Spector, M. Zepeniuk, H. Benjamin, N. Shabes, S. Tabak, A. Levy, D. Lebanony, Y. Goren, E. Silberschein, N. Targan, A. Ben-Ari, S. Gilad, N. Sion-Vardy, A. Tobar, M. Feinmesser, O. Kharenko, O. Nativ, D. Nass, M. Perelman, A. Yosepovich, B. Shalmon, S. Polak-Charcon, E. Fridman, A. Avniel, I. Bentwich, Z. Bentwich, D. Cohen, A. Chajut and I. Barshack (2008). "MicroRNAs accurately identify cancer tissue origin." Nat Biotechnol 26(4): 462-469.

Ruchoux, M. M., D. Guerouaou, B. Vandenhaute, J. P. Pruvo, P. Vermersch and D. Leys (1995). "Systemic vascular smooth muscle cell impairment in cerebral autosomal dominant arteriopathy with subcortical infarcts and leukoencephalopathy." Acta Neuropathol 89(6): 500-512.

Ryan, B. M., N. O'Donovan and M. J. Duffy (2009). "Survivin: a new target for anti-cancer therapy." Cancer Treat Rev 35(7): 553-562.

Sabbah, M., S. Emami, G. Redeuilh, S. Julien, G. Prevost, A. Zimber, R. Ouelaa, M. Bracke, O. De Wever and C. Gespach (2008). "Molecular signature and therapeutic perspective of the epithelial-to-mesenchymal transitions in epithelial cancers." Drug Resist Updat 11(4-5): 123-151.

Sade, H., S. Krishna and A. Sarin (2004). "The anti-apoptotic effect of Notch-1 requires p56lck-dependent, Akt/PKB-mediated signaling in T cells." J Biol Chem 279(4): 2937-2944.

Sahlgren, C., M. V. Gustafsson, S. Jin, L. Poellinger and U. Lendahl (2008). "Notch signaling mediates hypoxia-induced tumor cell migration and invasion." Proc Natl Acad Sci U S A 105(17): 6392-6397.

Sainson, R. C., D. A. Johnston, H. C. Chu, M. T. Holderfield, M. N. Nakatsu, S. P. Crampton, J. Davis, E. Conn and C. C. Hughes (2008). "TNF primes endothelial cells for angiogenic sprouting by inducing a tip cell phenotype." Blood 111(10): 4997-5007.

Samon, J. B., A. Champhekar, L. M. Minter, J. C. Telfer, L. Miele, A. Fauq, P. Das, T. E. Golde and B. A. Osborne (2008). "Notch1 and TGFbeta1 cooperatively regulate Foxp3 expression and the maintenance of peripheral regulatory T cells." Blood 112(5): 1813-1821.

Sansone, P., G. Storci, C. Giovannini, S. Pandolfi, S. Pianetti, M. Taffurelli, D. Santini, C. Ceccarelli, P. Chieco and M. Bonafe (2007a). "p66Shc/Notch-3 interplay controls self-renewal and hypoxia survival in human stem/progenitor cells of the mammary gland expanded in vitro as mammospheres." Stem Cells 25(3): 807-815.

Sansone, P., G. Storci, S. Tavolari, T. Guarnieri, C. Giovannini, M. Taffurelli, C. Ceccarelli, D. Santini, P. Paterini, K. B. Marcu, P. Chieco and M. Bonafe (2007b). "IL-6 triggers malignant features in mammospheres from human ductal breast carcinoma and normal mammary gland." J Clin Invest 117(12): 3988-4002.

Schwanbeck, R., T. Schroeder, K. Henning, H. Kohlhof, N. Rieber, M. L. Erfurth and U. Just (2008). "Notch signaling in embryonic and adult myelopoiesis." Cells Tissues Organs 188(1-2): 91-102.

Sethi, N., X. Dai, C. G. Winter and Y. Kang (2011). "Tumor-derived JAGGED1 promotes osteolytic bone metastasis of breast cancer by engaging notch signaling in bone cells." Cancer Cell 19(2): 192-205.

Seymour, L. and W. R. Bezwoda (1994). "Positive immunostaining for platelet derived growth factor (PDGF) is an adverse prognostic factor in patients with advanced breast cancer." Breast Cancer Res Treat 32(2): 229-233.

Sharma, A., A. N. Paranjape, A. Rangarajan and R. R. Dighe (2012). "A Monoclonal Antibody against Human Notch1 Ligand Binding Domain Depletes Subpopulation of Breast Cancer Stem-like Cells." Mol Cancer Ther 11(1): 77-86.

Shibuya, M. and L. Claesson-Welsh (2006). "Signal transduction by VEGF receptors in regulation of angiogenesis and lymphangiogenesis." Exp Cell Res 312(5): 549-560.

Shih Ie, M. and T. L. Wang (2007). "Notch signaling, gamma-secretase inhibitors, and cancer therapy." Cancer Res 67(5): 1879-1882.

Shimono, Y., M. Zabala, R. W. Cho, N. Lobo, P. Dalerba, D. Qian, M. Diehn, H. Liu, S. P. Panula, E. Chiao, F. M. Dirbas, G. Somlo, R. A. Pera, K. Lao and M. F. Clarke (2009). "Downregulation of miRNA-200c links breast cancer stem cells with normal stem cells." Cell 138(3): 592-603.

Shin, H. M., L. M. Minter, O. H. Cho, S. Gottipati, A. H. Fauq, T. E. Golde, G. E. Sonenshein and B. A. Osborne (2006). "Notch1 augments NF-kappaB activity by facilitating its nuclear retention." EMBO J 25(1): 129-138.

Shinriki, S., H. Jono, K. Ota, M. Ueda, M. Kudo, T. Ota, Y. Oike, M. Endo, M. Ibusuki, A. Hiraki, H. Nakayama, Y. Yoshitake, M. Shinohara and Y. Ando (2009). "Humanized anti-interleukin-6 receptor antibody suppresses tumor angiogenesis and in vivo growth of human oral squamous cell carcinoma." Clin Cancer Res 15(17): 5426-5434.

Shirayoshi, Y., Y. Yuasa, T. Suzuki, K. Sugaya, E. Kawase, T. Ikemura and N. Nakatsuji (1997). "Proto-oncogene of int-3, a mouse Notch homologue, is expressed in endothelial cells during early embryogenesis." Genes Cells 2(3): 213-224.

Siekmann, A. F., L. Covassin and N. D. Lawson (2008). "Modulation of VEGF signalling output by the Notch pathway." Bioessays 30(4): 303-313.

Sikandar, S. S., K. T. Pate, S. Anderson, D. Dizon, R. A. Edwards, M. L. Waterman and S. M. Lipkin (2010). "NOTCH signaling is required for formation and self-renewal of tumor-initiating cells and for repression of secretory cell differentiation in colon cancer." Cancer Res 70(4): 1469-1478.

Simpson, P. (1995). "Developmental genetics. The Notch connection." Nature 375(6534): 736-737.

Soares, R., G. Balogh, S. Guo, F. Gartner, J. Russo and F. Schmitt (2004). "Evidence for the notch signaling pathway on the role of estrogen in angiogenesis." Mol Endocrinol 18(9): 2333-2343.

Solomon, A., Y. Mian, C. Ortega-Cava, V. W. Liu, C. B. Gurumurthy, M. Naramura, V. Band and H. Band (2008). "Upregulation of the let-7 microRNA with precocious development in lin-12/Notch hypermorphic Caenorhabditis elegans mutants." Dev Biol 316(2): 191-199.

Song, L. L., Y. Peng, J. Yun, P. Rizzo, V. Chaturvedi, S. Weijzen, W. M. Kast, P. J. Stone, L. Santos, A. Loredo, U. Lendahl, G. Sonenshein, B. Osborne, J. Z. Qin, A. Pannuti, B. J. Nickoloff and L. Miele (2008). "Notch-1 associates with IKKalpha and regulates IKK activity in cervical cancer cells." Oncogene 27(44): 5833-5844.

Soriano, J. V., H. Uyttendaele, J. Kitajewski and R. Montesano (2000). "Expression of an activated Notch4(int-3) oncoprotein disrupts morphogenesis and induces an invasive phenotype in mammary epithelial cells in vitro." Int J Cancer 86(5): 652-659.

Sparks, C. A. and D. A. Guertin (2010). "Targeting mTOR: prospects for mTOR complex 2 inhibitors in cancer therapy." Oncogene 29(26): 3733-3744.

Sriuranpong, V., M. W. Borges, C. L. Strock, E. K. Nakakura, D. N. Watkins, C. M. Blaumueller, B. D. Nelkin and D. W. Ball (2002). "Notch signaling induces rapid degradation of achaete-scute homolog 1." Mol Cell Biol 22(9): 3129-3139.

Staszel, T., B. Zapala, A. Polus, A. Sadakierska-Chudy, B. Kiec-Wilk, E. Stepien, I. Wybranska, M. Chojnacka and A. Dembinska-Kiec (2011). "Role of microRNAs in endothelial cell pathophysiology." Pol Arch Med Wewn 121(10): 361-367.

Steelman, L. S., K. M. Stadelman, W. H. Chappell, S. Horn, J. Basecke, M. Cervello, F. Nicoletti, M. Libra, F. Stivala, A. M. Martelli and J. A. McCubrey (2008). "Akt as a therapeutic target in cancer." Expert Opin Ther Targets 12(9): 1139-1165.

Stylianou, S., R. B. Clarke and K. Brennan (2006). "Aberrant activation of notch signaling in human breast cancer." Cancer Res 66(3): 1517-1525.

Suchting, S., C. Freitas, F. le Noble, R. Benedito, C. Breant, A. Duarte and A. Eichmann (2007). "The Notch ligand Delta-like 4 negatively regulates endothelial tip cell formation and vessel branching." Proc Natl Acad Sci USA 104(9): 3225-3230.

Sullivan, J. P., M. Spinola, M. Dodge, M. G. Raso, C. Behrens, B. Gao, K. Schuster, C. Shao, J. E. Larsen, L. A. Sullivan, S. Honorio, Y. Xie, P. P. Scaglioni, J. M. DiMaio, A. F. Gazdar, J. W. Shay, Wistuba, II and J. D. Minna (2010). "Aldehyde dehydrogenase activity selects for lung adenocarcinoma stem cells dependent on notch signaling." Cancer Res 70(23): 9937-9948.

Sun, X. M., H. W. Wen, C. L. Chen and Q. P. Liao (2009). "Expression of Notch intracellular domain in cervical cancer and effect of DAPT on cervical cancer cell." Zhonghua Fu Chan Ke Za Zhi 44(5): 369-373.

Sundaram, M. V. (2005). "The love-hate relationship between Ras and Notch." Genes Dev 19(16): 1825-1839.

Tai, W., R. Mahato and K. Cheng (2010). "The role of HER2 in cancer therapy and targeted drug delivery." J Control Release 146(3): 264-75.

Tawara, K., J. T. Oxford and C. L. Jorcyk (2011). "Clinical significance of interleukin (IL)-6 in cancer metastasis to bone: potential of anti-IL-6 therapies." Cancer Manag Res 3: 177-189.

Taylor, K. L., A. M. Henderson and C. C. Hughes (2002). "Notch activation during endothelial cell network formation in vitro targets the basic HLH transcription factor HESR-1 and downregulates VEGFR-2/KDR expression." Microvasc Res 64(3): 372-383.

Tessitore, L., B. Vizio, O. Jenkins, I. De Stefano, C. Ritossa, J. M. Argiles, C. Benedetto and A. Mussa (2000). "Leptin expression in colorectal and breast cancer patients." Int J Mol Med 5(4): 421-426.

Thurston, G. and J. Kitajewski (2008). "VEGF and Delta-Notch: interacting signalling pathways in tumour angiogenesis." Br J Cancer 99(8): 1204-1209.

Tien, A. C., A. Rajan and H. J. Bellen (2009). "A Notch updated." J Cell Biol 184(5): 621-629.

Tokar, E. J., B. A. Diwan and M. P. Waalkes (2010). "Arsenic exposure transforms human epithelial stem/progenitor cells into a cancer stem-like phenotype." Environ Health Perspect 118(1): 108-115.

Uyttendaele, H., J. Ho, J. Rossant and J. Kitajewski (2001). "Vascular patterning defects associated with expression of activated Notch4 in embryonic endothelium." Proc Natl Acad Sci USA 98(10): 5643-5648.

Valastyan, S. and R. A. Weinberg (2009). "MicroRNAs: Crucial multi-tasking components in the complex circuitry of tumor metastasis." Cell Cycle 8(21): 3506-3512.

Vallabhapurapu, S. and M. Karin (2009). "Regulation and function of NF-kappaB transcription factors in the immune system." Annu Rev Immunol 27: 693-733.

Vallejo, D. M., E. Caparros and M. Dominguez (2011). "Targeting Notch signalling by the conserved miR-8/200 microRNA family in development and cancer cells." EMBO J 30(4): 756-769.

van Es, J. H., M. E. van Gijn, O. Riccio, M. van den Born, M. Vooijs, H. Begthel, M. Cozijnsen, S. Robine, D. J. Winton, F. Radtke and H. Clevers (2005). "Notch/gamma-secretase inhibition turns proliferative cells in intestinal crypts and adenomas into goblet cells." Nature 435(7044): 959-963.

van Meeteren, L. A., M. J. Goumans and P. Ten Dijke (2011). "TGF-beta Receptor Signaling Pathways in Angiogenesis; Emerging Targets for Anti-Angiogenesis Therapy." Curr Pharm Biotechnol 12(12): 2108-20.

Veenendaal, L. M., O. Kranenburg, N. Smakman, A. Klomp, I. H. Borel Rinkes and P. J. van Diest (2008). "Differential Notch and TGFbeta signaling in primary colorectal tumors and their corresponding metastases." Cell Oncol 30(1): 1-11.

Verstrepen, L., T. Bekaert, T. L. Chau, J. Tavernier, A. Chariot and R. Beyaert (2008). "TLR-4, IL-1R and TNF-R signaling to NF-kappaB: variations on a common theme." Cell Mol Life Sci 65(19): 2964-2978.

Visbal, A. P. and M. T. Lewis (2010). "Hedgehog Signaling in the Normal and Neoplastic Mammary Gland." Curr Drug Targets 11(9): 1103-11.

Wahid, F., A. Shehzad, T. Khan and Y. Y. Kim (2010). "MicroRNAs: Synthesis, mechanism, function, and recent clinical trials." Biochim Biophys Acta 1803(11): 1231-43.

Wang, J., L. Shelly, L. Miele, R. Boykins, M. A. Norcross and E. Guan (2001). "Human Notch-1 inhibits NF-kappa B activity in the nucleus through a direct interaction involving a novel domain." J Immunol 167(1): 289-295.

Wang, J., C. Wang, Q. Meng, S. Li, X. Sun, Y. Bo and W. Yao (2011). "siRNA targeting Notch-1 decreases glioma stem cell proliferation and tumor growth." Mol Biol Rep 39(3): 2497-503.

Wang, M., L. Wu, L. Wang and X. Xin (2010). "Down-regulation of Notch1 by gamma-secretase inhibition contributes to cell growth inhibition and apoptosis in ovarian cancer cells A2780." Biochem Biophys Res Commun 393(1): 144-149.

Wang, P., H. Zhen, J. Zhang, W. Zhang, R. Zhang, X. Cheng, G. Guo, X. Mao, J. Wang and X. Zhang (2011). "Survivin promotes glioma angiogenesis through vascular endothelial growth factor and basic fibroblast growth factor in vitro and in vivo." Mol Carcinog 51(7): 586-95.

Wang, Z., A. Ahmad, Y. Li, D. Kong, A. S. Azmi, S. Banerjee and F. H. Sarkar (2010a). "Emerging roles of PDGF-D signaling pathway in tumor development and progression." Biochim Biophys Acta 1806(1): 122-130.

Wang, Z., S. Banerjee, Y. Li, K. M. Rahman, Y. Zhang and F. H. Sarkar (2006a). "Down-regulation of notch-1 inhibits invasion by inactivation of nuclear factor-kappaB, vascular endothelial growth factor, and matrix metalloproteinase-9 in pancreatic cancer cells." Cancer Res 66(5): 2778-2784.

Wang, Z., D. Kong, S. Banerjee, Y. Li, N. V. Adsay, J. Abbruzzese and F. H. Sarkar (2007). "Down-regulation of platelet-derived growth factor-D inhibits cell growth and angiogenesis through inactivation of Notch-1 and nuclear factor-kappaB signaling." Cancer Res 67(23): 11377-11385.

Wang, Z., Y. Li, S. Banerjee, D. Kong, A. Ahmad, V. Nogueira, N. Hay and F. H. Sarkar (2010b). "Down-regulation of Notch-1 and Jagged-1 inhibits prostate cancer cell growth, migration and invasion, and induces apoptosis via inactivation of Akt, mTOR, and NF-kappaB signaling pathways." J Cell Biochem 109(4): 726-736.

Wang, Z., Y. Li and F. H. Sarkar (2010c). "Notch Signaling Proteins: Legitimate Targets for Cancer Therapy." Curr Protein Pept Sci.

Wang, Z., Y. Zhang, Y. Li, S. Banerjee, J. Liao and F. H. Sarkar (2006b). "Down-regulation of Notch-1 contributes to cell growth inhibition and apoptosis in pancreatic cancer cells." Mol Cancer Ther 5(3): 483-493.

Wani, A. A., S. M. Jafarnejad, J. Zhou and G. Li (2011). "Integrin-linked kinase regulates melanoma angiogenesis by activating NF-kappaB/interleukin-6 signaling pathway." Oncogene 30(24): 2778-2788.

Weijzen, S., P. Rizzo, M. Braid, R. Vaishnav, S. M. Jonkheer, A. Zlobin, B. A. Osborne, S. Gottipati, J. C. Aster, W. C. Hahn, M. Rudolf, K. Siziopikou, W. M. Kast and L. Miele (2002). "Activation of Notch-1 signaling maintains the neoplastic phenotype in human Ras-transformed cells." Nat Med 8(9): 979-986.

Welboren, W. J., H. G. Stunnenberg, F. C. Sweep and P. N. Span (2007). "Identifying estrogen receptor target genes." Mol Oncol 1(2): 138-143.

Wellner, U., J. Schubert, U. C. Burk, O. Schmalhofer, F. Zhu, A. Sonntag, B. Waldvogel, C. Vannier, D. Darling, A. zur Hausen, V. G. Brunton, J. Morton, O. Sansom, J. Schuler, M. P. Stemmler, C. Herzberger, U. Hopt, T. Keck, S. Brabletz and T. Brabletz (2009). "The EMT-activator ZEB1 promotes tumorigenicity by repressing stemness-inhibiting microRNAs." Nat Cell Biol 11(12): 1487-1495.

Weng, A. P., A. A. Ferrando, W. Lee, J. P. t. Morris, L. B. Silverman, C. Sanchez-Irizarry, S. C. Blacklow, A. T. Look and J. C. Aster (2004). "Activating mutations of NOTCH1 in human T cell acute lymphoblastic leukemia." Science 306(5694): 269-271.

Weng, A. P., J. M. Millholland, Y. Yashiro-Ohtani, M. L. Arcangeli, A. Lau, C. Wai, C. Del Bianco, C. G. Rodriguez, H. Sai, J. Tobias, Y. Li, M. S. Wolfe, C. Shachaf, D. Felsher, S. C. Blacklow, W. S. Pear and J. C. Aster (2006). "c-Myc is an important direct target of Notch1 in T-cell acute lymphoblastic leukemia/lymphoma." Genes Dev 20(15): 2096-2109.

Wickenden, J. A. and C. J. Watson (2010). "Key signalling nodes in mammary gland development and cancer. Signalling downstream of PI3 kinase in mammary epithelium: a play in 3 Akts." Breast Cancer Res 12(2): 202.

Wongchana, W. and T. Palaga (2011). "Direct regulation of interleukin-6 expression by Notch signaling in macrophages." Cell Mol Immunol 9(2): 155-62.

Wu, F., A. Stutzman and Y. Y. Mo (2007). "Notch signaling and its role in breast cancer." Front Biosci 12: 4370-4383.

Wu, Y., C. Cain-Hom, L. Choy, T. J. Hagenbeek, G. P. de Leon, Y. Chen, D. Finkle, R. Venook, X. Wu, J. Ridgway, D. Schahin-Reed, G. J. Dow, A. Shelton, S. Stawicki, R. J. Watts, J. Zhang, R. Choy, P. Howard, L. Kadyk, M. Yan, J. Zha, C. A. Callahan, S. G. Hymowitz and C. W. Siebel (2010). "Therapeutic antibody targeting of individual Notch receptors." Nature 464(7291): 1052-1057.

Xu, P., M. Qiu, Z. Zhang, C. Kang, R. Jiang, Z. Jia, G. Wang, H. Jiang and P. Pu (2010). "The oncogenic roles of Notch1 in astrocytic gliomas in vitro and in vivo." J Neurooncol 97(1): 41-51.

Xue, Y., X. Gao, C. E. Lindsell, C. R. Norton, B. Chang, C. Hicks, M. Gendron-Maguire, E. B. Rand, G. Weinmaster and T. Gridley (1999). "Embryonic lethality and vascular defects in mice lacking the Notch ligand Jagged1." Hum Mol Genet 8(5): 723-730.

Yamaguchi, N., T. Oyama, E. Ito, H. Satoh, S. Azuma, M. Hayashi, K. Shimizu, R. Honma, Y. Yanagisawa, A. Nishikawa, M. Kawamura, J. Imai, S. Ohwada, K. Tatsuta, J. Inoue, K. Semba and S. Watanabe (2008). "NOTCH3 signaling pathway plays crucial roles in the proliferation of ErbB2-negative human breast cancer cells." Cancer Res 68(6): 1881-1888.

Yamamoto, N., S. Yamamoto, F. Inagaki, M. Kawaichi, A. Fukamizu, N. Kishi, K. Matsuno, K. Nakamura, G. Weinmaster, H. Okano and M. Nakafuku (2001). "Role of Deltex-1 as a transcriptional regulator downstream of the Notch receptor." J Biol Chem 276(48): 45031-45040.

Yang, W., D. Y. Lee and Y. Ben-David (2011). "The roles of microRNAs in tumorigenesis and angiogenesis." Int J Physiol Pathophysiol Pharmacol 3(2): 140-155.

Yao, J., T. W. Kim, J. Qin, Z. Jiang, Y. Qian, H. Xiao, Y. Lu, W. Qian, M. F. Gulen, N. Sizemore, J. DiDonato, S. Sato, S. Akira, B. Su and X. Li (2007). "Interleukin-1 (IL-1)-induced TAK1-dependent Versus MEKK3-dependent NFkappaB activation pathways bifurcate at IL-1 receptor-associated kinase modification." J Biol Chem 282(9): 6075-6089.

Yao, J. and C. Qian (2010). "Inhibition of Notch3 enhances sensitivity to gemcitabine in pancreatic cancer through an inactivation of PI3K/Akt-dependent pathway." Med Oncol 27(3): 1017-1022.

Yoo, A. S. and I. Greenwald (2005). "LIN-12/Notch activation leads to microRNA-mediated down-regulation of Vav in C. elegans." Science 310(5752): 1330-1333.

Young, A., J. Lyons, A. L. Miller, V. T. Phan, I. R. Alarcon and F. McCormick (2009). "Ras signaling and therapies." Adv Cancer Res 102: 1-17.

Yu, F., H. Yao, P. Zhu, X. Zhang, Q. Pan, C. Gong, Y. Huang, X. Hu, F. Su, J. Lieberman and E. Song (2007). "let-7 regulates self renewal and tumorigenicity of breast cancer cells." Cell 131(6): 1109-1123.

Zabierowski, S. E. and M. Herlyn (2008). "Learning the ABCs of melanoma-initiating cells." Cancer Cell 13(3): 185-187.

Zanotti, S. and E. Canalis (2010). "Notch and the skeleton." Mol Cell Biol 30(4): 886-896.

Zavadil, J., L. Cermak, N. Soto-Nieves and E. P. Bottinger (2004). "Integration of TGF-beta/Smad and Jagged1/Notch signalling in epithelial-to-mesenchymal transition." EMBO J 23(5): 1155-1165.

Zebisch, A., A. P. Czernilofsky, G. Keri, J. Smigelskaite, H. Sill and J. Troppmair (2007). "Signaling through RAS-RAF-MEK-ERK: from basics to bedside." Curr Med Chem 14(5): 601-623.

Zhang, B., J. G. Abreu, K. Zhou, Y. Chen, Y. Hu, T. Zhou, X. He and J. X. Ma (2010). "Blocking the Wnt pathway, a unifying mechanism for an angiogenic inhibitor in the serine proteinase inhibitor family." Proc Natl Acad Sci USA 107(15): 6900-6905.

Zhang, B. and J. X. Ma (2010). "Wnt pathway antagonists and angiogenesis." Protein Cell 1(10): 898-906.

Zhang, H., Y. Han, J. Tao, S. Liu, C. Yan and S. Li (2011). "Cellular repressor of E1A-stimulated genes regulates vascular endothelial cell migration by The ILK/AKT/mTOR/VEGF(165) signaling pathway." Exp Cell Res 317(20): 2904-2913.

Zhang, J., S. Fukuhara, K. Sako, T. Takenouchi, H. Kitani, T. Kume, G. Y. Koh and N. Mochizuki (2011). "Angiopoietin-1/Tie2 signal augments basal Notch signal controlling vascular quiescence by inducing delta-like 4 expression through AKT-mediated activation of beta-catenin." J Biol Chem 286(10): 8055-8066.

Zhang, Y., Z. Wang, F. Ahmed, S. Banerjee, Y. Li and F. H. Sarkar (2006). "Down-regulation of Jagged-1 induces cell growth inhibition and S phase arrest in prostate cancer cells." Int J Cancer 119(9): 2071-2077.

Zhao, X., G. K. Malhotra, S. M. Lele, M. S. Lele, W. W. West, J. D. Eudy, H. Band and V. Band (2010). "Telomerase-immortalized human mammary stem/progenitor cells with ability to self-renew and differentiate." Proc Natl Acad Sci USA 107(32): 14146-14151.

Zhao, Y., C. Deng, J. Wang, J. Xiao, Z. Gatalica, R. R. Recker and G. G. Xiao (2010). "Let-7 family miRNAs regulate estrogen receptor alpha signaling in estrogen receptor positive breast cancer." Breast Cancer Res Treat 127(1): 69-80.

Zhou, H. and S. Huang (2010). "The Complexes of Mammalian Target of Rapamycin." Curr Protein Pept Sci 11(6): 409-24.

Zhou, J., J. Wulfkuhle, H. Zhang, P. Gu, Y. Yang, J. Deng, J. B. Margolick, L. A. Liotta, E. Petricoin, 3rd and Y. Zhang (2007). "Activation of the PTEN/mTOR/STAT3 pathway in breast cancer stem-like cells is required for viability and maintenance." Proc Natl Acad Sci USA 104(41): 16158-16163.

Zhou, W., S. Guo and R. R. Gonzalez-Perez (2011). "Leptin pro-angiogenic signature in breast cancer is linked to IL-1 signalling." Br J Cancer 104(1): 128-37.

Zhu, T. S., M. A. Costello, C. E. Talsma, C. G. Flack, J. G. Crowley, L. L. Hamm, X. He, S. L. Hervey-Jumper, J. A. Heth, K. M. Muraszko, F. DiMeco, A. L. Vescovi and X. Fan (2011). "Endothelial cells create a stem cell niche in glioblastoma by providing NOTCH ligands that nurture self-renewal of cancer stem-like cells." Cancer Res 71(18): 6061-6072.

Zwerts, F., F. Lupu, A. De Vriese, S. Pollefeyt, L. Moons, R. A. Altura, Y. Jiang, P. H. Maxwell, P. Hill, H. Oh, C. Rieker, D. Collen, S. J. Conway and E. M. Conway (2007). "Lack of endothelial cell survivin causes embryonic defects in angiogenesis, cardiogenesis, and neural tube closure." Blood 109(11): 4742-4752.

Endothelial and Cancer Stem Cell Regulation of Tumor Angiogenesis

Mingli Liu[*1], **Lily Yang**[2], **Shanchun Guo**[3] and
Ruben R. Gonzalez-Perez[*4]

[1]*Microbiology, Biochemistry & Immunology, Morehouse School of Medicine
HG 352,720 Westview Dr. SW, Atlanta, GA 30310. Email: mliu@msm.edu*
[2]*Department of Surgery, Emory University School of Medicine, 1365-C Clifton Road
NE, Atlanta, Georgia 30322. Email: lyang02@emory.edu*
[3]*Microbiology, Biochemistry & Immunology, Morehouse School of Medicine
HG 330,720 Westview Dr. SW, Atlanta, GA 30310. Email: sguo@msm.edu*
[4]*Microbiology, Biochemistry & Immunology, Morehouse School of Medicine
HG 332,720 Westview Dr. SW, Atlanta, GA 30310. Email: rgonzalez@msm.edu*

CHAPTER OUTLINE

- ▶ Introduction
- ▶ CSCs and tumor angiogenesis
- ▶ EPCs and tumor angiogenesis
- ▶ Conclusions and perspectives
- ▶ Acknowledgements
- ▶ Glossary
- ▶ References

ABSTRACT

Cancer stem cells (CSCs) and endothelial progenitor cells (EPCs) play important roles in tumor angiogenesis. CSCs contribute to the tumor vasculature and vasculogenic mimicry through interactions with the microenvironment. EPCs, originated in the bone marrow, contribute to tumor growth and angiogenesis-induced neovascularization via vasculogenesis and angiogenesis. EPCs are derived

Correspondence Author   Department of Microbiology, Biochemistry & Immunology, Morehouse School of Medicine, 720 Westview Dr. SW, Atlanta, GA 30310. Email: rgonzalez@msm.edu

from hematopoietic stem cells (HSC) which also produces the hematopoietic (HPCs) and myeloid progenitor cells. This chapter will discuss the functions of the above cells associated with angiogenesis and vasculogenic mimicry (VM) in tumorigenesis. The sources, isolation, cellular markers, role of specific signaling players and their crosstalk effects on VM formation will be discussed. Additionally, the clinical relevance of CSC, EPCs and other cells in tumor growth and angiogenesis will also be discussed. Specific targeting of the tumor pro-angiogenic actions of CSCs and EPCs might provide promising choices of cancer anti-angiogenic therapy that might prove to be advantageous over current therapeutic interventions.

Key words: endothelial stem cells, cancer stem cells, tumor angiogenesis, vasculogenic mimicry

1   Introduction

Tumor angiogenesis is shaped through the actions of several factors. These factors include classical and non-classical angiogenic molecules secreted by cancer cells and supportive stroma. In addition, other cell types can also influence the development of the vasculature required by tumors to growth. Two cell types: cancer stem cells (CSCs) and endothelial progenitor cells (EPCs) play important roles in tumor angiogenesis and other scenarios. CSCs can contribute to the tumor vasculature through interactions with the microenvironment inducing the formation of non-endothelial vascular structures via vasculogenic mimicry (tumor cells mimicking endothelial cell functions). EPCs, coming from the bone marrow can also contribute to the neovascularization (vasculogenesis) and angiogenic processes within tumors, thereby promoting tumor growth.

2   CSCs and Tumor Angiogenesis

CSCs and non-tumor forming progenitor cells (endothelial progenitor cells, EPCs) are two populations of stem/progenitor cells involved in angiogenic programs. CSCs are a subpopulation of tumor cells that display stem cell-like characteristics; possess the capacity to self-renewal and differentiation and indefinite proliferation (Bjerkvig et al. 2009; Frank et al. 2010). There is emerging evidence showing that CSCs are able to trigger angiogenesis (Hillen and Griffioen 2007), the strategies include secretion of factors or directly influence on stromal and endothelial cells inducing angiogenesis (sprouting of existing blood vessels close to the tumor) (Folkman 1995; Risau 1997), vasculogenesis (*de novo* formation of vessels from pluripotent endothelial stem cells), vascular mimicry (tumor cells mimicking endothelial cell functions), vascular co-option (tumor cell growth along existing vessels) and intussusceptive angiogenesis (cleavage of vessels by septal invagination leading to the creation of new blood vessel via the splitting of an existing blood vessel in two) (Hillen and Griffioen 2007). All these processes could regulate the path and extension of tumor angiogenesis, thereby, influencing tumor growth and response to therapies.

2.1 Role of CSCs in Angiogenesis (Paracrine Factors and direct Cell-Cell Communication)

CSCs regulate angiogenesis by mutually modulately their microenvironment through direct cell-cell communication (Frank et al. 2010) or paracrine factors secretion. First, CSCs positive for Nestin+ [a type VI intermediate filament (IF) protein] and CD133+ (that mark neural stem and progenitor cells) reside within a specific perivascular niches in various brain tumors (oligodendrogliomas, gliobalstomas, medulloblastomas and ependymomas), where they interact tightly with endothelial cells. It is postulated that through this interaction the endothelial cells maintain Nestin+/CD133+ CSCs in a self-renewing and undifferentiated state (Calabrese et al. 2007). Second, paracrine factors released from CSCs could strength the process of angiogenesis (Bao et al. 2006). For instance, the CD133+ glioma-initiating cells in human glioma cell-derived xenografts secreted high levels of VEGF, which specifically promoted tumor angiogenesis and therefore tumor xenograft growth (Bao et al. 2006). Inhibition of either ERBB2 or VEGF signaling significantly reduced the CD133+ CSCs abundance and suppressed the growth of tumor xenograft growth (Bao et al. 2006; Calabrese et al. 2007). Glioma CSCs also preferentially expressed HIF2α and many HIF-regulated genes, which was directly linked to stem cell-like tumor cells (Bleau et al. 2009) and is critical for self-renewal of hematopoietic stem cells (Miyamoto et al. 2007; Li et al. 2009). Third, cancer stem cells such as glioma stem/progenitor cells (GSPCs) could participate in angiogenesis by trans-differentiating themselves into vascular endothelial cells (VEC). It has been suggested that GSPCs transcribe or copy functions and also can express VEC characteristic molecular markers (Zhao et al. 2010).

2.2 CSC-associated Angiogenesis and Vasculogenic Mimicry (VM)

Tumor cell-driven vasculogenic mimicry (VM) was first described in 1999 by Maniotis et al (Maniotis et al. 1999) in human melanomas. The term VM was used to differentiate the blood-like capillary from those blood vessels formed and lined by typical endothelial cells. VM is characterized by expression of vasculogenic markers such as tyrosine kinase with Ig-like, EGF-like domains 1 (TIE-1), CD144 (VE-cadherin; a calcium-dependent cell-cell adhesion glycoprotein), bone morphogenetic protein receptor, type IA (BMPR1A) (Folberg et al. 2000; Schatton et al. 2008), and other VM-associated proteins including VIII-associated antigen (FVIII associate antigen; also known as von Willebrand factor or vWF) and laminin 5γ2 chain domain III fragment (Petty et al. 2007). CSCs have been considered to be responsible for VM (Maniotis et al. 1999). This process seems to be independent of angiogenesis (Le Bourhis et al. 2010). Therefore, it is possible that the failure of anti-angiogenic drugs to completely eradicate tumor growth could be partially due to the actions of blood-like capillary formed by CSC (Frank et al. 2010).

2.2.1   Signaling Pathways Involved in VM

In addition to melanomas, VM has also been identified in breast cancer (Shirakawa et al. 2001), ovarian cancer (Sood et al. 2001; Sood et al. 2002), prostate carcinomas (Sharma et al. 2002), lung (Passalidou et al. 2002), clear cell renal cell carcinoma (Vartanian et al. 2009), Ewing sarcoma (van der Schaft et al. 2005), soft tissue sarcomas, rhabdomyosarcoma, osteosarcoma, and pheochromocytoma (Folberg and Maniotis 2004). CSCs-driven vasculogenic mimicry has been linked to the actions of bone morphogenetic proteins (BMPs) (Rothhammer et al. 2007), TGF-β superfamily (Liu et al. 2004) and VEGF (Frank et al. 2010), which play critical roles in vascular development. Diverse adhesion molecules and kinases are involved in the development of CSC-induced VM.

In melanoma, EphA2 (an isoform of EphA belonging to the tyrosine kinase receptor family) and VE-cadherin (CD144) co-localize on the surface of aggressive tumor cells, resulting in the phosphorylation of EphA2 (Hess et al. 2006). It has been suggested that activated EphA2 can phosphorylate FAK, which in turn results in the activation of ERK1/2 followed by the subsequent activation of PI-3K. Nevertheless, VE-cadherin/EphA2 signaling can also activate PI-3K in a way independent of FAK (Hendrix et al. 2003; Hess et al. 2006; Hess et al. 2007). Increased PI-3K and ERK1/2 phosphorylation can regulate the conversion of proteases, i.e., pro-MT1-MMP to active MT1-MMP, which subsequently activates pro-MMP2 (Hess et al. 2006; Paulis et al. 2010) or MMP9 (Zhang et al. 2007; Yao et al. 2011). Then, enzymatically active MT1-MMP and MMP-2/MMP-9 may cleave the laminin 5γ chain into pro-migratory fragments, laminin 5γ2′ or laminin 5γ2x, which eventually cause the formation of VM (vasculogenic mimicry) networks (Hess et al. 2006; Paulis et al. 2010; Yao et al. 2011).

Other reports suggest that Mig-7 (a novel human gene and promising cancer cell marker for diagnosis and disease progression) can co-localize with von vWF and VE-cadherin inducing VM (Petty et al. 2007; Robertson 2007). Additionally, c-Met gene (MET or hepatocyte growth factor receptor, HGFR proto-oncogene) product MET is essential for embryonic development and wound healing (Tang et al. 2009). Abnormal MET activation triggers tumor growth, angiogenesis and metastasis and correlates with poor prognosis in cancer. Mitochondrial reactive oxygen species (ROS) promote MET (proto-oncogene)-dependent VM via HIF-1α stabilization (Tang et al. 2009; Comito et al. 2011) in melanomas and ovarian cancer. Moreover, Twist1, an epithelial-mesenchymal transition (EMT) regulator, binds to VE-cadherin promoter and increases its transcriptional activity, therefore enhancing the formation of VM networks (Sun et al. 2010).

The proposed signaling molecules and pathways involved in the formation of VM are summarized in Fig. 1. Galectin-3 (Gal-3), which is known to have oncogenic and angiogenic potential, has recently been found to contribute to the formation of VM networks (see Fig. 1). Gal-3 positively influences VM by repression of the EGR-1 transcription factor, which represses transcription of VE-cadherin and IL-8 (Paulis et al. 2010).

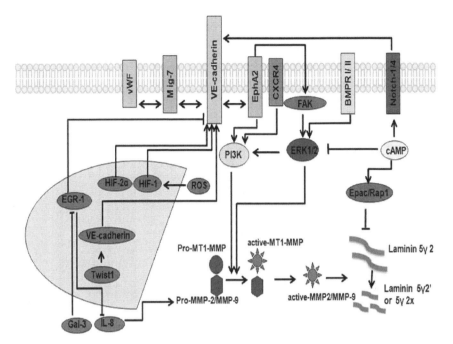

Figure 1 Proposed cellular pathways in the regulation of Vasculogenic Mimicry. The major signaling pathways activated are through: 1) bone morphogenetic proteins (BMPs) (Rothhammer et al. 2007); 2) EphA2 and VE-cadherin: co-localization of EphA2 and VE-cadherin results in the sequentially phosphorylation of EphA2 (Hess et al. 2006) and FAK, which in turn activates ERK1/2 and subsequent PI3K (Hendrix et al. 2003; Hess et al. 2006; Hess et al. 2007); 3) Gal-3 and EGR-1. These signaling cascades increase the expression of VE-cadherin and activated MMP-2 (Paulis et al. 2010); 4) cAMP can affect the formation of the VM network in either positive or negative ways by activation of Notch1/4 (Mitsuda et al. 2001) or Epac/Rap1 (Bos 2006; Roscioni et al. 2008) and/or inhibition of ERK1/2 signaling 2 (Lissitzky et al. 2009), respectively; 5) Mig-7 promotes the crosstalk with von Willebrand factor (vWF) and VE-cadherin (Petty et al. 2007; Robertson 2007), thereby inducing VM formation.

Color image of this figure appears in the color plate section at the end of the book.

High expression levels of Gal-3 were found in highly metastatic melanoma and breast carcinoma cells (Paulis et al. 2010). Furthermore, cAMP can also affect VM network formation in either positive (by activation Notch1/4 signaling (Mitsuda et al. 2001) or negative ways by activation of Epac/ Rap1 (Ras-proximate-1 or Ras-related protein 1; a small cytosolic GTPase) (Bos 2006; Roscioni et al. 2008) or inhibition of ERK1/2 (Lissitzky et al. 2009). However, the stimulatory effect of cAMP on Notch-1 and Notch-4 may increase the expression of Nodal (a member of the TGF-β superfamily), which subsequently upregulates VE-cadherin expression. By this latter effect cAMP may facilitate VM network formation (Paulis et al. 2010). In addition, impaired activity of BMPs had significant reduced effects on VM in melanoma (Rothhammer et al. 2007) (see Fig 1).

3 EPCs and Tumor Angiogenesis

EPCs are defined as bone marrow-derived cells that move into the circulatory system in response to cytokines such as VEGF (Li et al. 2006). As mentioned above, EPC contributes to vasculogenesis and tumor angiogenesis. Interestingly, significantly higher EPCs levels have been found in the peripheral circulation of breast (Naik et al. 2008), ovarian (Su et al. 2010) and non-small cell lung cancer (NSCLC) patients (Dome et al. 2006) and those suffering from glioblastoma (Greenfield et al. 2009), chronic lymphocytic (Gora-Tybor et al. 2009), acute myeloid leukemia (Wierzbowska et al. 2008) and lymphoma (Igreja et al. 2007).

3.1 Potential Sources of EPCs

EPCs are a type of lineage-restricted progenitor cell, which have limited self-renewal and differentiation abilities (Melero-Martin and Dudley 2011). EPC are derived from hematopoietic stem cells (HSCs) (Melero-Martin and Dudley 2011) and usually identified and enumerated by two primary methodologies, *in vitro* cell culture or flow cytometry (see Table 1) (Prater et al. 2007).

3.1.1 Bone Marrow (BM)-Derived EPCs Source

3.1.1.1 Hematopoietic Stem Cells (HSC)

The HSC produces the hematopoietic progenitor cells (HPCs), which in turn produce the myeloid progenitor cells, the lymphoid progenitor cells and the EPCs (Li Calzi et al. 2010). EPCs, HSCs, HPCs and mesenchymal stem cells (MSCs) share several surface markers (see Table 1) such as CD45, CD34, CD133 (CD117 for mouse), CD146, CD31, CD105, CD144, vascular endothelial growth factor receptor 2 (VEGFR2) and von Willebrand factor (vWF) (Shi et al. 1998; Gehling et al. 2000; Peichev et al. 2000; Monestiroli et al. 2001; Capillo et al. 2003; Bertolini et al. 2006; Woywodt et al. 2006; Widemann et al. 2008). Table 1 shows common markers used to identify EPCs, HSCs, HPCs and MPCs (mesenchymal progenitor stem cells) (Le Bourhis et al. 2010; Patenaude et al. 2010; Melero-Martin and Dudley 2011). It is important to mention that the difficulty in isolation of EPCs is due to lack of specific markers. Therefore, any data based on immunological phenotype strategies need to be interpreted with caution. CD133 (CD117 as a surrogate marker of CD133 in mouse (Bertolini et al. 2006)) is usually used to differentiate EPCs from circulating endothelial cell (CEC) and myelomonocytic cells (Rehman et al. 2003). HSCs can be distinguished from EPCs by CD45 expression. Endothelial markers, for example CD31, CD146 and VEGFR2 can be used to isolate EPCs from a CD45-negative pool (Patenaude et al. 2010).

3.1.1.2 Myeloid Cells

There is increasing evidence showing that myeloid-derived cells are indirectly (Murdoch et al. 2008) or directly involved in tumor neovascularization

(Kuwana et al. 2003; Zhao et al. 2003; Bailey et al. 2006; Murdoch et al. 2008; Li et al. 2009). Myeloid progenitor derived EPCs result from circulating CD14+/CD45+/CD34+ population. These cells display some characteristics of mesenchymal cells, but also express endothelial cell markers and form tubules *in vitro* in Matrigel assays (Kuwana et al. 2006).

Table 1 Common markers used to identify EPCs, circulating endothelial cell, mature endothelial cell, HSCs, HPCs and MPCs

Markers	EPCs	Circulating endothelial cell	HSCs	HPCs	MPCs
CD45	−	−	+	+	−
CD34	+	+/−	+	N/A	N/A
CD133	+	−	+	−	N/A
CD117	+	−	+	N/A	N/A
CD146	+/−	+	−	N/A	N/A
CD31	+/−	+	+/−	+	-
CD105	+/−	+	+/−	N/A	+
CD144	+/−	+	+/−	N/A	N/A
VEGFR2	+	+	+/−	+	N/A
VEGFR1	+	N/A	N/A	+	N/A
vWF	+/−	+	−	N/A	N/A
VE-cadherin	+	N/A	N/A	N/A	−
eNOS	+	N/A	N/A	N/A	N/A
UEA-1 lectin	+	N/A	N/A	+	N/A
Ac-LDL	+	N/A	N/A	+	N/A
CD14	−	N/A	N/A	+	−
CD11b	−	N/A	N/A	+	−
CD90	−	N/A	N/A	N/A	+
CD44	N/A	N/A	N/A	N/A	+
CD29	N/A	N/A	N/A	N/A	+
NG2	N/A	N/A	N/A	N/A	+
PDGF-Rβ	N/A	N/A	N/A	N/A	+
CD3	N/A	N/A	N/A	+	N/A
CD4	N/A	N/A	N/A	+	N/A
CD8	N/A	N/A	N/A	+	N/A
Glycophorin-A	N/A	N/A	N/A	+	N/A
Ter-119	N/A	N/A	N/A	+	N/A
CD41	N/A	N/A	N/A	+	N/A
CD42	N/A	N/A	N/A	+	N/A

N/A: not available

3.1.1.3 Mesenchymal Stem Cells (MSCs) and Mesenchymal Progenitor Cells (MPCs)

MSCs have the ability to differentiate into multilineage cells, such as the osteogenic, chodrogenic, adipogenic, fibroblastic, epithelial and neuronal lineages (Prockop 1997; Liu et al. 2009). MPCs also give rise to pericytes (Song et al. 2005; Beckermann et al. 2008) or an endothelial phenotype (Silva et al. 2005) and are a major source of VEGF secretion. MPCs and pericytes share CD146, NG2, PDGF-Rβ as well as CD44, CD73, CD90 and CD105 (surface molecules commonly used as MSC markers), which supports a perivascular origin of MSCs (Crisan et al. 2008). Wang et al found that both mRNA and protein levels of angiogenic factors such as VEGF, bFGF and HGF were elevated in the conditioned medium of MSCs from multiple myeloma, which significantly promoted the proliferation and capillary formation (Wang et al. 2011). Together, MPCs in cancers are major structural and paracrine cellular components of new blood vessels (Crisan et al. 2008).

3.1.2 Non-BM Sources

Whether non-BM putative EPCs contribute to tumor vasculogenesis remains under evaluated. An elegant experiment conducted by Ahn et al (Ahn and Brown 2008) with MMP9-deficient mouse model supports the possibility that CD11b+myelomonocytic cells expressing MMP9 would be important for vasculogenesis indirectly by triggering the generation and development of non-BM derived EPCs (Ahn and Brown 2008). Although the origin is unknown, it was suggested that putative sources for non bone marrow-derived progenitors may derive from organs of small intestine and liver (Aicher et al. 2007). The blood vessel wall has been found as a source for EPCs (Ingram et al. 2005). Using single cell assays Yoder's group identified that EPCs reside in vessel walls by testing the proliferative and clonogenic potential of single human umbilical vein endothelial cells (HUVECs) and human aortic endothelial cells (HAECs) (Ingram et al. 2005). The high proliferative potential-endothelial colony forming cell (HPP-ECFC) and the low proliferative potential-endothelial colony forming cell LPP-ECFCs exist in HUVEC and HAEC. In fact, the progeny derived from HPP-ECFCs contains higher telomerase activity compared with those derived from LPP-ECFCs, which provides a selective growth advantage for HPP-ECFCs. This feature makes HPP-ECFCs to grow and divide continuously (Ingram et al. 2005).

3.2 EPCs Identification and Isolation

EPCs can be isolated from peripheral blood (PB), umbilical cord blood and bone marrow (Li Calzi et al. 2010). Umbilical cord blood cells represent an important alternative and rich source of EPCs with practical and ethical advantages (Phuc et al. 2011). Traditional methods to isolate and enumerate EPCs include flow cytometry-based cell sorting (FACS), magnetic bead-based cell sorting (MACS), and colony-forming unit (CFU) assay (Thomas et al. 2009; Phuc et al. 2011; Yang et al. 2011). Recently some novel sorting techniques have been developed, such as EPC capture chip (Hansmann et al. 2011), and nano iron particle-labeling based magnetic resonance imaging (MRI) (Chen et al. 2011). Different enumeration techniques for EPCs are summarized in Table 2.

Table 2 Typical EPCs enumeration techniques

	Samples	Pre-treat	Length of time	Enumeration technique	Advantages	Disadvantages
FACS	BM, PB,UBC	MNC isolation	2.5 h	Gating based on FSC/SS/ markers	1. High efficiency and purity of isolation	1. Expensive equipment and experienced workers
	0.1-2 ml				2. Quantitative measurement	2. Gates discrimination
MACS	BM, PB, UBC	MNC isolation	0.5-2 h	1. Manual counting 2. CFU assay 3. Flow Cytometry	High purity isolation	1.Efficiency variable
	0.5-50 ml					2.Difficulty in removing beads
CFU assay	BM, PB, UBC	MNC isolation	5-7 days	Manual counting Via phase contrast microscope	Enumerate myeloid progenitor cell activity	1.High probability of non-target cells
	5-10 ml					2.Cannot distinguish colonies by naked eyes
EPC capture chip	BM, PB, UBC	None	1 h	Manual enumeration by immuno-fluorescence	1. High purity isolation	Expensive technique
	0.2 ml				2. Easy quantification 3. High reproducibility	
Nano iron-MRI	BM	Isolation of EPC	7-10 days	MR imaging	Monitor temporal and spatial migration of EPCs	Expensive equipment
	0.2 ml					

3.3 The Role of EPCs in Tumor Angiogenesis

3.3.1 General Introduction

The contribution of EPCs to tumor vasculature is under extensive investigations. The initial studies have yielded findings that were inconsistent. The contribution of EPCs to tumor endothelium varied from 35% to 100% (Garcia-Barros et al. 2003; Hilbe et al. 2004). Some evidence in mouse models supports the observation that EPCs structurally incorporate into tumor blood vessels. Indeed, BM-derived EPCs in the initial steps of mouse tumor development have been shown to differentiate into mature BM-derived endothelial cells (ECs) and incorporate into a subset of sprouting tumor neovessel lumina. This process dilutes the EC with non-BM-derived vessels from the periphery (Larrivee et al. 2005; Nolan et al.

2007). On the contrary, some studies suggest lower or even undetectable integration of EPCs in the vasculature of tumors. Alternatively, Patenaude et al pointed out that EPCs contribute significantly to tumor angiogenesis via paracrine mechanisms rather than forming tumor endothelium (Patenaude et al. 2010). Several reasons can explain the observed variations of EPCs integration in tumor vasculature. As already mentioned, lack of specific biomarkers for EPCs, cell fusion between stem cells and different cell types including endothelial cells, cell and microparticle engulfment (cell particles retain membrane antigens from the cells that they came from) could be various possible explanations (Patenaude et al. 2010) (Alvarez-Dolado et al. 2003; Dan and Yeoh 2008).

3.3.2 Role of EPC in Angiogenesis of Various Tumors

Many chimeric mouse models have been developed to investigate the contribution of EPCs to vascularization in breast cancer. By using the model systems of transplanting LacZ+ or GFP+ tagged bone marrow-derived cells into a lethally irradiated mouse tissue (to prevent angiogenesis) followed by tracing the migration of GFP or LacZ-tagged endothelial progenitor cells to tumor tissues (Dwenger et al. 2004; Duda et al. 2006; Nolan et al. 2007; Ahn and Brown 2008; Gao et al. 2008; Suriano et al. 2008), several research groups found that bone marrow-derived EPCs play a crucial role in neovascularization in early stages of breast cancer (Le Bourhis et al. 2010). In addition, the angiogenic cytokines released from EPCs such as VEGF, platelet-derived growth factor (PDGF), fibroblast growth factor (FGF) and estradiol also improve neovascularization in a paracrine manner (Urbich et al. 2005; Gao et al. 2008). The contribution to angiogenesis via EPCs was conducted by stimulating EPC mobilization using via proangiogenic cytokines. VEGF activates EPCs from a quiescent to a proliferative state (Heissig et al. 2003). VEGF also upregulated the CXC chemokine stromal-derived factor (SDF-1/CXCL12) and its receptor CXCR4, which control the release and homing of EPCs (Kryczek et al. 2005; Schatteman et al. 2007). In addition, other growth factors and cytokines such as CCL2, CCL5, the neurotrophin brain-derived neurotrophic factor (BDNF) (Kermani et al. 2005) and nerve growth factor (NGF) (Adriaenssens et al. 2008; Lagadec et al. 2009) released by breast tumor also attracted the EPCs from the circulation and are involved in EPC mobilization.

Peripheral blood levels of bone marrow-derived EPCs (CD34+, CD133+, and VEGFR2+) are significantly increased in patients with non-small-cell lung cancer (NSCLC) and correlate with response to chemotherapy. EPCs may offer a possible biomarker for efficiency of treatment and prognosis for NSCLC (Fleitas et al. 2010; Morita et al. 2011). This might indicate an association linking the number of circulating EPC and tumor stage and progression (Nowak et al. 2010). Wickersheim et al. pointed out that during the progression of Lewis lung carcinoma, vascularization occurs primarily via classical tumor angiogenesis (sprouting of pre-existing ECs), whereas

BM-EPCs do not incorporate into the vessel wall to any significant extent (Wickersheim et al. 2009).

The key roles of EPCs migration and adhesion in tumor metastasis were conducted by Su et al. in ovarian cancer. They found that DNA-binding protein inhibitor (Id1) induced EPCs mobilization and recruitment is mediated by the PI-3K/Akt signaling pathway and is related to activation of integrin α4 (Su et al. 2010). EPCs activity can be enhanced by tumor cells. One example is described in glioma by Zhang et al. They found that glioma cells enhance EPCs angiogenesis through VEGFR2 (not VEGFR1) via the MMP-9, Akt and ERK signaling pathways (Zhang et al. 2008). Sun et al. (Sun et al. 2005) demonstrated that not only EPCs but differentiated endothelial cells from tissue peripheral contribute to melanoma angiogenesis. Also, a progeny of human MSCs have similar capacities of participating in angiogenesis.

3.4 Clinical Relevance of EPCs in Tumor Vascularization

3.4.1 Circulating EPCs as Biomarkers in Clinical Oncology

High EPC levels were found to correlate with poor prognosis and/or survival of cancer patients (Dome et al. 2006; Wierzbowska et al. 2008; Greenfield et al. 2009; Su et al. 2010). The number of post treatment EPCs (CD31+, CD34+, CD133+, and VEGFR2+) in peripheral blood from NSCLC tumor patients were significantly lower in nonresponding patients (Dome et al. 2006). In acute myeloid leukemia (AML) patients, the numbers of circulating EPCs (CEPCs) and circulating endothelial cells (CEC) were significantly higher than in healthy subjects and correlated with response to treatment (Wierzbowska et al. 2008). Greenfield et al (Greenfield et al. 2009) observed a statistically significant relationship between the number of EPCs (CD133+ VEGFR2+) in the peripheral blood at the time of initial glioblastoma multiforme (GBM) resection and survival. These data suggest that peripheral blood circulating EPCs from glioma patients can be used as a surrogate biomarker to measure tumor angiogenicity. The quantity of circulating EPCs (CD34+, VEGFR2+) detected in human ovarian cancer pre- and post-treatment suggests that these cells are significantly decreased after treatment (Su et al. 2010). Circulating EPCs are correlated with tumor stage and residual tumor size, thus they may serve as a biomarker to monitoring tumor progression and angiogenesis (Su et al. 2010).

3.4.2 EPCs as Therapeutic Tools

As mentioned EPCs may serve as a biomarker for cancer progression and predict the response to cancer therapy (a target for therapy). These cells hold great potential as a diagnostic and/or therapeutic marker of anti-tumor therapies (Janic and Arbab 2010). High levels of circulating EPCs were found in both a breast cancer xenograft model (Shirakawa et al. 2002; Shaked et al. 2005) and in breast cancer patients (Kim et al.

2003; Mancuso et al. 2009), where EPCs number mirrored clinical severity. Therefore, interventions designed to block EPC mobilization (e.g. the use of an antibody against CXCR4, which binds SDF-1) improved the antitumor effect of chemotherapy (Hassan et al. 2009).

However, accelerated metastasis after short-term treatment with a potent inhibitor of tumor angiogenesis (VEGFR/PDGFR) such as sunitinib/SU11248 was reported by different groups (Ebos et al. 2009; Paez-Ribes et al. 2009). One possible explanation for this unexpected phenomenon could be the contrasting effects on tumor growth induced by the inhibition of VEGF in myeloid and tumor cells. Specifically targeting the myeloid-cell derived VFGF resulted in reduced phosphorylation of VEGFR2 but induced tumor progression (Greenberg et al. 2008; Stockmann et al. 2008), whereas targeting of both myeloid cell- and tumor-derived VEGF causes inhibition of tumor growth. It was suggested that angiogenic drug combinations might be more effective than cancer monotherapies (Seaman et al. 2007).

4   Conclusions and Perspectives

Accumulated evidence suggests that EPCs play important roles in tumor angiogenesis and vasculogenesis. These cells are potential biomarkers for both cancer diagnosis and treatment. However, many questions related to EPCs biology and functions remain to be addressed. Future studies need to focus on the identification of EPC-specific biomarkers so that true lineage of EPCs can be traced. Furthermore, there are discrepant data that need to be clarified. The definition of EPCs roles in tumor angiogenesis via neovasculogenesis or angiogenesis or both need to be further elucidated. The extent of EPC contribution to blood vessel maintenance is another important issue that requires additional investigations.

5   Acknowledgements

This work was supported by NIH/NCI 5SC1CA138658-03; NIH/ARRA/3SC1CA138658-02S1 and the Georgia Cancer Coalition Distinguished Cancer Scholar Award. Other funds were from facilities and support services at Morehouse School of Medicine (NIH RR03034 and 1C06 RR18386).

6   Glossary

BDNF: the neurotrophin brain-derived neurotrophic factor; BM: bone marrow; BMPR1A: bone morphogenetic protein receptor, type IA; CEC: circulating endothelial cells; CEPCs: circulating EPCs; CFU: colony-forming unit assay; CSCs: Cancer stem cells; CXCR-4: C-X-C chemokine receptor type 4; EPCs: endothelial progenitors cells; ERBB2: Human Epidermal growth factor Receptor 2; FACS: flow cytometry-based cell sorting; GBM:

glioblastoma multiforme; GSPCs: glioma stem/progenitor cells; HPCs: hematopoietic progenitor cells; HSC: hematopoietic stem cells; Id1: DNA-binding protein inhibitor; MACS: magnetic bead-based cell sorting; MRI: magnetic resonance imaging; MSCs: mesenchymal stem cells; MPCs: mesenchymal progenitor cells; NGF: nerve growth factor; NSCLC: non-small cell lung cancer; PB: peripheral blood; SDF-1/CXCL12: CXC chemokine stromal-derived factor; TIE-1: tyrosine kinase with Ig-like and EGF-like domains 1; UCB: umbilical cord blood; VM: vasculogenic mimicry; WF: von Willebrand factor

REFERENCES

Adriaenssens, E., E. Vanhecke, P. Saule, A. Mougel, A. Page, R. Romon, V. Nurcombe, X. Le Bourhis and H. Hondermarck (2008). "Nerve growth factor is a potential therapeutic target in breast cancer." Cancer Res 68(2): 346-351.

Ahn, G. O. and J. M. Brown (2008). "Matrix metalloproteinase-9 is required for tumor vasculogenesis but not for angiogenesis: role of bone marrow-derived myelomonocytic cells." Cancer Cell 13(3): 193-205.

Aicher, A., M. Rentsch, K. Sasaki, J. W. Ellwart, F. Fandrich, R. Siebert, J. P. Cooke, S. Dimmeler and C. Heeschen (2007). "Nonbone marrow-derived circulating progenitor cells contribute to postnatal neovascularization following tissue ischemia." Circ Res 100(4): 581-589.

Alvarez-Dolado, M., R. Pardal, J. M. Garcia-Verdugo, J. R. Fike, H. O. Lee, K. Pfeffer, C. Lois, S. J. Morrison and A. Alvarez-Buylla (2003). "Fusion of bone-marrow-derived cells with Purkinje neurons, cardiomyocytes and hepatocytes." Nature 425(6961): 968-973.

Bailey, A. S., H. Willenbring, S. Jiang, D. A. Anderson, D. A. Schroeder, M. H. Wong, M. Grompe and W. H. Fleming (2006). "Myeloid lineage progenitors give rise to vascular endothelium." Proc Natl Acad Sci U S A 103(35): 13156-13161.

Bao, S., Q. Wu, S. Sathornsumetee, Y. Hao, Z. Li, A. B. Hjelmeland, Q. Shi, R. E. McLendon, D. D. Bigner and J. N. Rich (2006). "Stem cell-like glioma cells promote tumor angiogenesis through vascular endothelial growth factor." Cancer Res 66(16): 7843-7848.

Beckermann, B. M., G. Kallifatidis, A. Groth, D. Frommhold, A. Apel, J. Mattern, A. V. Salnikov, G. Moldenhauer, W. Wagner, A. Diehlmann, R. Saffrich, M. Schubert, A. D. Ho, N. Giese, M. W. Buchler, H. Friess, P. Buchler and I. Herr (2008). "VEGF expression by mesenchymal stem cells contributes to angiogenesis in pancreatic carcinoma." Br J Cancer 99(4): 622-631.

Bertolini, F., Y. Shaked, P. Mancuso and R. S. Kerbel (2006). "The multifaceted circulating endothelial cell in cancer: towards marker and target identification." Nat Rev Cancer 6(11): 835-845.

Bjerkvig, R., M. Johansson, H. Miletic and S. P. Niclou (2009). "Cancer stem cells and angiogenesis." Semin Cancer Biol 19(5): 279-284.

Bleau, A. M., D. Hambardzumyan, T. Ozawa, E. I. Fomchenko, J. T. Huse, C. W. Brennan and E. C. Holland (2009). "PTEN/PI3K/Akt pathway regulates the side population phenotype and ABCG2 activity in glioma tumor stem-like cells." Cell Stem Cell 4(3): 226-235.

Bos, J. L. (2006). "Epac proteins: multi-purpose cAMP targets." Trends Biochem Sci 31(12): 680-686.

Calabrese, C., H. Poppleton, M. Kocak, T. L. Hogg, C. Fuller, B. Hamner, E. Y. Oh, M. W. Gaber, D. Finklestein, M. Allen, A. Frank, I. T. Bayazitov, S. S. Zakharenko, A. Gajjar, A. Davidoff and R. J. Gilbertson (2007). "A perivascular niche for brain tumor stem cells." Cancer Cell 11(1): 69-82.

Capillo, M., P. Mancuso, A. Gobbi, S. Monestiroli, G. Pruneri, C. Dell'Agnola, G. Martinelli, L. Shultz and F. Bertolini (2003). "Continuous infusion of endostatin inhibits differentiation, mobilization, and clonogenic potential of endothelial cell progenitors." Clin Cancer Res 9(1): 377-382.

Chen, R., H. Yu, Z. Y. Jia, Q. L. Yao and G. J. Teng (2011). "Efficient nano iron particle-labeling and noninvasive MR imaging of mouse bone marrow-derived endothelial progenitor cells." Int J Nanomedicine 6: 511-519.

Comito, G., M. Calvani, E. Giannoni, F. Bianchini, L. Calorini, E. Torre, C. Migliore, S. Giordano and P. Chiarugi (2011). "HIF-1alpha stabilization by mitochondrial ROS promotes Met-dependent invasive growth and vasculogenic mimicry in melanoma cells." Free Radic Biol Med 51(4): 893-904.

Crisan, M., S. Yap, L. Casteilla, C. W. Chen, M. Corselli, T. S. Park, G. Andriolo, B. Sun, B. Zheng, L. Zhang, C. Norotte, P. N. Teng, J. Traas, R. Schugar, B. M. Deasy, S. Badylak, H. J. Buhring, J. P. Giacobino, L. Lazzari, J. Huard and B. Peault (2008). "A perivascular origin for mesenchymal stem cells in multiple human organs." Cell Stem Cell 3(3): 301-313.

Dan, Y. Y. and G. C. Yeoh (2008). "Liver stem cells: a scientific and clinical perspective." J Gastroenterol Hepatol 23(5): 687-698.

Dome, B., J. Timar, J. Dobos, L. Meszaros, E. Raso, S. Paku, I. Kenessey, G. Ostoros, M. Magyar, A. Ladanyi, K. Bogos and J. Tovari (2006). "Identification and clinical significance of circulating endothelial progenitor cells in human non-small cell lung cancer." Cancer Res 66(14): 7341-7347.

Duda, D. G., K. S. Cohen, S. V. Kozin, J. Y. Perentes, D. Fukumura, D. T. Scadden and R. K. Jain (2006). "Evidence for incorporation of bone marrow-derived endothelial cells into perfused blood vessels in tumors." Blood 107(7): 2774-2776.

Dwenger, A., F. Rosenthal, M. Machein, C. Waller and A. Spyridonidis (2004). "Transplanted bone marrow cells preferentially home to the vessels of in situ generated murine tumors rather than of normal organs." Stem Cells 22(1): 86-92.

Ebos, J. M., C. R. Lee, W. Cruz-Munoz, G. A. Bjarnason, J. G. Christensen and R. S. Kerbel (2009). "Accelerated metastasis after short-term treatment with a potent inhibitor of tumor angiogenesis." Cancer Cell 15(3): 232-239.

Fleitas, T., V. Martinez-Sales, J. Gomez-Codina, M. Martin and G. Reynes (2010). "Circulating endothelial and endothelial progenitor cells in non-small-cell lung cancer." Clin Transl Oncol 12(8): 521-525.

Folberg, R., M. J. Hendrix and A. J. Maniotis (2000). "Vasculogenic mimicry and tumor angiogenesis." Am J Pathol 156(2): 361-381.

Folberg, R. and A. J. Maniotis (2004). "Vasculogenic mimicry." Apmis 112(7-8): 508-525.

Folkman, J. (1995). "Seminars in Medicine of the Beth Israel Hospital, Boston. Clinical applications of research on angiogenesis." N Engl J Med 333(26): 1757-1763.

Frank, N. Y., T. Schatton and M. H. Frank (2010). "The therapeutic promise of the cancer stem cell concept." J Clin Invest 120(1): 41-50.

Gao, D., D. J. Nolan, A. S. Mellick, K. Bambino, K. McDonnell and V. Mittal (2008). "Endothelial progenitor cells control the angiogenic switch in mouse lung metastasis." Science 319(5860): 195-198.

Garcia-Barros, M., F. Paris, C. Cordon-Cardo, D. Lyden, S. Rafii, A. Haimovitz-Friedman, Z. Fuks and R. Kolesnick (2003). "Tumor response to radiotherapy regulated by endothelial cell apoptosis." Science 300(5622): 1155-1159.

Gehling, U. M., S. Ergun, U. Schumacher, C. Wagener, K. Pantel, M. Otte, G. Schuch, P. Schafhausen, T. Mende, N. Kilic, K. Kluge, B. Schafer, D. K. Hossfeld and W. Fiedler (2000). "In vitro differentiation of endothelial cells from AC133-positive progenitor cells." Blood 95(10): 3106-3112.

Gora-Tybor, J., K. Jamroziak, A. Szmigielska-Kaplon, A. Krawczynska, E. Lech-Maranda, A. Wierzbowska, D. Jesionek-Kupnicka, J. Z. Blonski and T. Robak (2009). "Evaluation of circulating endothelial cells as noninvasive marker of angiogenesis in patients with chronic lymphocytic leukemia." Leuk Lymphoma 50(1): 62-67.

Greenberg, J. I., D. J. Shields, S. G. Barillas, L. M. Acevedo, E. Murphy, J. Huang, L. Scheppke, C. Stockmann, R. S. Johnson, N. Angle and D. A. Cheresh (2008). "A role for VEGF as a negative regulator of pericyte function and vessel maturation." Nature 456(7223): 809-813.

Greenfield, J. P., D. K. Jin, L. M. Young, P. J. Christos, L. Abrey, S. Rafii and P. H. Gutin (2009). "Surrogate markers predict angiogenic potential and survival in patients with glioblastoma multiforme." Neurosurgery 64(5): 819-826; discussion 826-817.

Hansmann, G., B. D. Plouffe, A. Hatch, A. von Gise, H. Sallmon, R. T. Zamanian and S. K. Murthy (2011). "Design and validation of an endothelial progenitor cell capture chip and its application in patients with pulmonary arterial hypertension." J Mol Med (Berl) 89(10): 971-983.

Hassan, S., C. Ferrario, U. Saragovi, L. Quenneville, L. Gaboury, A. Baccarelli, O. Salvucci and M. Basik (2009). "The influence of tumor-host interactions in the stromal cell-derived factor-1/CXCR4 ligand/receptor axis in determining metastatic risk in breast cancer." Am J Pathol 175(1): 66-73.

Heissig, B., Z. Werb, S. Rafii and K. Hattori (2003). "Role of c-kit/Kit ligand signaling in regulating vasculogenesis." Thromb Haemost 90(4): 570-576.

Hendrix, M. J., E. A. Seftor, A. R. Hess and R. E. Seftor (2003). "Molecular plasticity of human melanoma cells." Oncogene 22(20): 3070-3075.

Hess, A. R., N. V. Margaryan, E. A. Seftor and M. J. Hendrix (2007). "Deciphering the signaling events that promote melanoma tumor cell vasculogenic mimicry and their link to embryonic vasculogenesis: role of the Eph receptors." Dev Dyn 236(12): 3283-3296.

Hess, A. R., E. A. Seftor, L. M. Gruman, M. S. Kinch, R. E. Seftor and M. J. Hendrix (2006). "VE-cadherin regulates EphA2 in aggressive melanoma cells through a novel signaling pathway: implications for vasculogenic mimicry." Cancer Biol Ther 5(2): 228-233.

Hilbe, W., S. Dirnhofer, F. Oberwasserlechner, T. Schmid, E. Gunsilius, G. Hilbe, E. Woll and C. M. Kahler (2004). "CD133 positive endothelial progenitor cells contribute to the tumour vasculature in non-small cell lung cancer." J Clin Pathol 57(9): 965-969.

Hillen, F. and A. W. Griffioen (2007). "Tumour vascularization: sprouting angiogenesis and beyond." Cancer Metastasis Rev 26(3-4): 489-502.

Igreja, C., M. Courinha, A. S. Cachaco, T. Pereira, J. Cabecadas, M. G. da Silva and S. Dias (2007). "Characterization and clinical relevance of circulating and biopsy-derived endothelial progenitor cells in lymphoma patients." Haematologica 92(4): 469-477.

Ingram, D. A., L. E. Mead, D. B. Moore, W. Woodard, A. Fenoglio and M. C. Yoder (2005). "Vessel wall-derived endothelial cells rapidly proliferate because they contain a complete hierarchy of endothelial progenitor cells." Blood 105(7): 2783-2786.

Janic, B. and A. S. Arbab (2010). "The role and therapeutic potential of endothelial progenitor cells in tumor neovascularization." ScientificWorldJournal 10: 1088-1099.

Kermani, P., D. Rafii, D. K. Jin, P. Whitlock, W. Schaffer, A. Chiang, L. Vincent, M. Friedrich, K. Shido, N. R. Hackett, R. G. Crystal, S. Rafii and B. L. Hempstead (2005). "Neurotrophins promote revascularization by local recruitment of TrkB+ endothelial cells and systemic mobilization of hematopoietic progenitors." J Clin Invest 115(3): 653-663.

Kim, H. K., K. S. Song, H. O. Kim, J. H. Chung, K. R. Lee, Y. J. Lee, D. H. Lee, E. S. Lee, K. W. Ryu and J. M. Bae (2003). "Circulating numbers of endothelial progenitor cells in patients with gastric and breast cancer." Cancer Lett 198(1): 83-88.

Kryczek, I., A. Lange, P. Mottram, X. Alvarez, P. Cheng, M. Hogan, L. Moons, S. Wei, L. Zou, V. Machelon, D. Emilie, M. Terrassa, A. Lackner, T. J. Curiel, P. Carmeliet and W. Zou (2005). "CXCL12 and vascular endothelial growth factor synergistically induce neoangiogenesis in human ovarian cancers." Cancer Res 65(2): 465-472.

Kuwana, M., Y. Okazaki, H. Kodama, K. Izumi, H. Yasuoka, Y. Ogawa, Y. Kawakami and Y. Ikeda (2003). "Human circulating CD14+ monocytes as a source of progenitors that exhibit mesenchymal cell differentiation." J Leukoc Biol 74(5): 833-845.

Kuwana, M., Y. Okazaki, H. Kodama, T. Satoh, Y. Kawakami and Y. Ikeda (2006). "Endothelial differentiation potential of human monocyte-derived multipotential cells." Stem Cells 24(12): 2733-2743.

Lagadec, C., S. Meignan, E. Adriaenssens, B. Foveau, E. Vanhecke, R. Romon, R. A. Toillon, B. Oxombre, H. Hondermarck and X. Le Bourhis (2009). "TrkA overexpression enhances growth and metastasis of breast cancer cells." Oncogene 28(18): 1960-1970.

Larrivee, B., K. Niessen, I. Pollet, S. Y. Corbel, M. Long, F. M. Rossi, P. L. Olive and A. Karsan (2005). "Minimal contribution of marrow-derived endothelial precursors to tumor vasculature." J Immunol 175(5): 2890-2899.

Le Bourhis, X., R. Romon and H. Hondermarck (2010). "Role of endothelial progenitor cells in breast cancer angiogenesis: from fundamental research to clinical ramifications." Breast Cancer Res Treat 120(1): 17-24.

Li, B., E. E. Sharpe, A. B. Maupin, A. A. Teleron, A. L. Pyle, P. Carmeliet and P. P. Young (2006). "VEGF and PlGF promote adult vasculogenesis by enhancing EPC recruitment and vessel formation at the site of tumor neovascularization." Faseb J 20(9): 1495-1497.

Li, B., A. Vincent, J. Cates, D. M. Brantley-Sieders, D. B. Polk and P. P. Young (2009). "Low levels of tumor necrosis factor alpha increase tumor growth by inducing an endothelial phenotype of monocytes recruited to the tumor site." Cancer Res 69(1): 338-348.

Li Calzi, S., M. B. Neu, L. C. Shaw, J. L. Kielczewski, N. I. Moldovan and M. B. Grant (2010). "EPCs and pathological angiogenesis: when good cells go bad." Microvasc Res 79(3): 207-216.

Li, Z., S. Bao, Q. Wu, H. Wang, C. Eyler, S. Sathornsumetee, Q. Shi, Y. Cao, J. Lathia, R. E. McLendon, A. B. Hjelmeland and J. N. Rich (2009). "Hypoxia-inducible factors regulate tumorigenic capacity of glioma stem cells." Cancer Cell 15(6): 501-513.

Lissitzky, J. C., D. Parriaux, E. Ristorcelli, A. Verine, D. Lombardo and P. Verrando (2009). "Cyclic AMP signaling as a mediator of vasculogenic mimicry in aggressive human melanoma cells in vitro." Cancer Res 69(3): 802-809.

Liu, W., J. Selever, D. Wang, M. F. Lu, K. A. Moses, R. J. Schwartz and J. F. Martin (2004). "Bmp4 signaling is required for outflow-tract septation and branchial-arch artery remodeling." Proc Natl Acad Sci U S A 101(13): 4489-4494.

Liu, Z. J., Y. Zhuge and O. C. Velazquez (2009). "Trafficking and differentiation of mesenchymal stem cells." J Cell Biochem 106(6): 984-991.

Mancuso, P., P. Antoniotti, J. Quarna, A. Calleri, C. Rabascio, C. Tacchetti, P. Braidotti, H. K. Wu, A. J. Zurita, L. Saronni, J. B. Cheng, D. R. Shalinsky, J. V. Heymach and F. Bertolini (2009). "Validation of a standardized method for enumerating circulating endothelial cells and progenitors: flow cytometry and molecular and ultrastructural analyses." Clin Cancer Res 15(1): 267-273.

Maniotis, A. J., R. Folberg, A. Hess, E. A. Seftor, L. M. Gardner, J. Pe'er, J. M. Trent, P. S. Meltzer and M. J. Hendrix (1999). "Vascular channel formation by human melanoma cells in vivo and in vitro: vasculogenic mimicry." Am J Pathol 155(3): 739-752.

Melero-Martin, J. M. and A. C. Dudley (2011). "Concise review: Vascular stem cells and tumor angiogenesis." Stem Cells 29(2): 163-168.

Mitsuda, N., N. Ohkubo, M. Tamatani, Y. D. Lee, M. Taniguchi, K. Namikawa, H. Kiyama, A. Yamaguchi, N. Sato, K. Sakata, T. Ogihara, M. P. Vitek and M. Tohyama (2001). "Activated cAMP-response element-binding protein regulates neuronal expression of presenilin-1." J Biol Chem 276(13): 9688-9698.

Miyamoto, K., K. Y. Araki, K. Naka, F. Arai, K. Takubo, S. Yamazaki, S. Matsuoka, T. Miyamoto, K. Ito, M. Ohmura, C. Chen, K. Hosokawa, H. Nakauchi, K. Nakayama, K. I. Nakayama, M. Harada, N. Motoyama, T. Suda and A. Hirao (2007). "Foxo3a is essential for maintenance of the hematopoietic stem cell pool." Cell Stem Cell 1(1): 101-112.

Monestiroli, S., P. Mancuso, A. Burlini, G. Pruneri, C. Dell'Agnola, A. Gobbi, G. Martinelli and F. Bertolini (2001). "Kinetics and viability of circulating endothelial cells as surrogate angiogenesis marker in an animal model of human lymphoma." Cancer Res 61(11): 4341-4344.

Morita, R., K. Sato, M. Nakano, H. Miura, H. Odaka, K. Nobori, T. Kosaka, M. Sano, H. Watanabe, T. Shioya and H. Ito (2011). "Endothelial progenitor cells are associated with response to chemotherapy in human non-small-cell lung cancer." J Cancer Res Clin Oncol 137(12): 1849-1857.

Murdoch, C., M. Muthana, S. B. Coffelt and C. E. Lewis (2008). "The role of myeloid cells in the promotion of tumour angiogenesis." Nat Rev Cancer 8(8): 618-631.

Naik, R. P., D. Jin, E. Chuang, E. G. Gold, E. A. Tousimis, A. L. Moore, P. J. Christos, T. de Dalmas, D. Donovan, S. Rafii and L. T. Vahdat (2008). "Circulating endothelial progenitor cells correlate to stage in patients with invasive breast cancer." Breast Cancer Res Treat 107(1): 133-138.

Nolan, D. J., A. Ciarrocchi, A. S. Mellick, J. S. Jaggi, K. Bambino, S. Gupta, E. Heikamp, M. R. McDevitt, D. A. Scheinberg, R. Benezra and V. Mittal (2007). "Bone marrow-derived endothelial progenitor cells are a major determinant of nascent tumor neovascularization." Genes Dev 21(12): 1546-1558.

Nowak, K., N. Rafat, S. Belle, C. Weiss, C. Hanusch, P. Hohenberger and G. Beck (2010). "Circulating endothelial progenitor cells are increased in human lung cancer and correlate with stage of disease." Eur J Cardiothorac Surg 37(4): 758-763.

Paez-Ribes, M., E. Allen, J. Hudock, T. Takeda, H. Okuyama, F. Vinals, M. Inoue, G. Bergers, D. Hanahan and O. Casanovas (2009). "Antiangiogenic therapy elicits malignant progression of tumors to increased local invasion and distant metastasis." Cancer Cell 15(3): 220-231.

Passalidou, E., M. Trivella, N. Singh, M. Ferguson, J. Hu, A. Cesario, P. Granone, A. G. Nicholson, P. Goldstraw, C. Ratcliffe, M. Tetlow, I. Leigh, A. L. Harris, K. C. Gatter and F. Pezzella (2002). "Vascular phenotype in angiogenic and non-angiogenic lung non-small cell carcinomas." Br J Cancer 86(2): 244-249.

Patenaude, A., J. Parker and A. Karsan (2010). "Involvement of endothelial progenitor cells in tumor vascularization." Microvasc Res 79(3): 217-223.

Paulis, Y. W., P. M. Soetekouw, H. M. Verheul, V. C. Tjan-Heijnen and A. W. Griffioen (2010). "Signalling pathways in vasculogenic mimicry." Biochim Biophys Acta 1806(1): 18-28.

Peichev, M., A. J. Naiyer, D. Pereira, Z. Zhu, W. J. Lane, M. Williams, M. C. Oz, D. J. Hicklin, L. Witte, M. A. Moore and S. Rafii (2000). "Expression of VEGFR-2 and AC133 by circulating human CD34(+) cells identifies a population of functional endothelial precursors." Blood 95(3): 952-958.

Petty, A. P., K. L. Garman, V. D. Winn, C. M. Spidel and J. S. Lindsey (2007). "Overexpression of carcinoma and embryonic cytotrophoblast cell-specific Mig-7 induces invasion and vessel-like structure formation." Am J Pathol 170(5): 1763-1780.

Phuc, P. V., V. B. Ngoc, D. H. Lam, N. T. Tam, P. Q. Viet and P. K. Ngoc (2011). "Isolation of three important types of stem cells from the same samples of banked umbilical cord blood." Cell Tissue Bank 8: 8.

Prater, D. N., J. Case, D. A. Ingram and M. C. Yoder (2007). "Working hypothesis to redefine endothelial progenitor cells." Leukemia 21(6): 1141-1149.

Prockop, D. J. (1997). "Marrow stromal cells as stem cells for nonhematopoietic tissues." Science 276(5309): 71-74.

Rehman, J., J. Li, C. M. Orschell and K. L. March (2003). "Peripheral blood "endothelial progenitor cells" are derived from monocyte/macrophages and secrete angiogenic growth factors." Circulation 107(8): 1164-1169.

Risau, W. (1997). "Mechanisms of angiogenesis." Nature 386(6626): 671-674.

Robertson, G. P. (2007). "Mig-7 linked to vasculogenic mimicry." Am J Pathol 170(5): 1454-1456.

Roscioni, S. S., C. R. Elzinga and M. Schmidt (2008). "Epac: effectors and biological functions." Naunyn Schmiedebergs Arch Pharmacol 377(4-6): 345-357.

Rothhammer, T., F. Bataille, T. Spruss, G. Eissner and A. K. Bosserhoff (2007). "Functional implication of BMP4 expression on angiogenesis in malignant melanoma." Oncogene 26(28): 4158-4170.

Schatteman, G. C., M. Dunnwald and C. Jiao (2007). "Biology of bone marrow-derived endothelial cell precursors." Am J Physiol Heart Circ Physiol 292(1): H1-18.

Schatton, T., G. F. Murphy, N. Y. Frank, K. Yamaura, A. M. Waaga-Gasser, M. Gasser, Q. Zhan, S. Jordan, L. M. Duncan, C. Weishaupt, R. C. Fuhlbrigge, T. S. Kupper, M. H. Sayegh and M. H. Frank (2008). "Identification of cells initiating human melanomas." Nature 451(7176): 345-349.

Seaman, S., J. Stevens, M. Y. Yang, D. Logsdon, C. Graff-Cherry and B. St Croix (2007). "Genes that distinguish physiological and pathological angiogenesis." Cancer Cell 11(6): 539-554.

Shaked, Y., F. Bertolini, S. Man, M. S. Rogers, D. Cervi, T. Foutz, K. Rawn, D. Voskas, D. J. Dumont, Y. Ben-David, J. Lawler, J. Henkin, J. Huber, D. J. Hicklin, R. J. D'Amato and R. S. Kerbel (2005). "Genetic heterogeneity of the vasculogenic phenotype parallels angiogenesis; Implications for cellular surrogate marker analysis of antiangiogenesis." Cancer Cell 7(1): 101-111.

Sharma, N., R. E. Seftor, E. A. Seftor, L. M. Gruman, P. M. Heidger, Jr., M. B. Cohen, D. M. Lubaroff and M. J. Hendrix (2002). "Prostatic tumor cell plasticity involves cooperative interactions of distinct phenotypic subpopulations: role in vasculogenic mimicry." Prostate 50(3): 189-201.

Shi, Q., S. Rafii, M. H. Wu, E. S. Wijelath, C. Yu, A. Ishida, Y. Fujita, S. Kothari, R. Mohle, L. R. Sauvage, M. A. Moore, R. F. Storb and W. P. Hammond (1998). "Evidence for circulating bone marrow-derived endothelial cells." Blood 92(2): 362-367.

Shirakawa, K., S. Furuhata, I. Watanabe, H. Hayase, A. Shimizu, Y. Ikarashi, T. Yoshida, M. Terada, D. Hashimoto and H. Wakasugi (2002). "Induction of vasculogenesis in breast cancer models." Br J Cancer 87(12): 1454-1461.

Shirakawa, K., H. Tsuda, Y. Heike, K. Kato, R. Asada, M. Inomata, H. Sasaki, F. Kasumi, M. Yoshimoto, T. Iwanaga, F. Konishi, M. Terada and H. Wakasugi (2001). "Absence of endothelial cells, central necrosis, and fibrosis are associated with aggressive inflammatory breast cancer." Cancer Res 61(2): 445-451.

Silva, G. V., S. Litovsky, J. A. Assad, A. L. Sousa, B. J. Martin, D. Vela, S. C. Coulter, J. Lin, J. Ober, W. K. Vaughn, R. V. Branco, E. M. Oliveira, R. He, Y. J. Geng, J. T. Willerson and E. C. Perin (2005). "Mesenchymal stem cells differentiate into an endothelial phenotype, enhance vascular density, and improve heart function in a canine chronic ischemia model." Circulation 111(2): 150-156.

Song, S., A. J. Ewald, W. Stallcup, Z. Werb and G. Bergers (2005). "PDGFRbeta+ perivascular progenitor cells in tumours regulate pericyte differentiation and vascular survival." Nat Cell Biol 7(9): 870-879.

Sood, A. K., M. S. Fletcher, C. M. Zahn, L. M. Gruman, J. E. Coffin, E. A. Seftor and M. J. Hendrix (2002). "The clinical significance of tumor cell-lined vasculature in ovarian carcinoma: implications for anti-vasculogenic therapy." Cancer Biol Ther 1(6): 661-664.

Sood, A. K., E. A. Seftor, M. S. Fletcher, L. M. Gardner, P. M. Heidger, R. E. Buller, R. E. Seftor and M. J. Hendrix (2001). "Molecular determinants of ovarian cancer plasticity." Am J Pathol 158(4): 1279-1288.

Stockmann, C., A. Doedens, A. Weidemann, N. Zhang, N. Takeda, J. I. Greenberg, D. A. Cheresh and R. S. Johnson (2008). "Deletion of vascular endothelial growth factor in myeloid cells accelerates tumorigenesis." Nature 456(7223): 814-818.

Su, Y., L. Zheng, Q. Wang, J. Bao, Z. Cai and A. Liu (2010). "The PI3K/Akt pathway upregulates Id1 and integrin alpha4 to enhance recruitment of human ovarian cancer endothelial progenitor cells." BMC Cancer 10(459): 459.

Su, Y., L. Zheng, Q. Wang, W. Li, Z. Cai, S. Xiong and J. Bao (2010). "Quantity and clinical relevance of circulating endothelial progenitor cells in human ovarian cancer." J Exp Clin Cancer Res 29(27): 27.

Sun, B., S. Zhang, C. Ni, D. Zhang, Y. Liu, W. Zhang, X. Zhao, C. Zhao and M. Shi (2005). "Correlation between melanoma angiogenesis and the mesenchymal stem cells and endothelial progenitor cells derived from bone marrow." Stem Cells Dev 14(3): 292-298.

Sun, T., N. Zhao, X. L. Zhao, Q. Gu, S. W. Zhang, N. Che, X. H. Wang, J. Du, Y. X. Liu and B. C. Sun (2010). "Expression and functional significance of Twist1 in hepatocellular carcinoma: its role in vasculogenic mimicry." Hepatology 51(2): 545-556.

Suriano, R., D. Chaudhuri, R. S. Johnson, E. Lambers, B. T. Ashok, R. Kishore and R. K. Tiwari (2008). "17Beta-estradiol mobilizes bone marrow-derived endothelial progenitor cells to tumors." Cancer Res 68(15): 6038-6042.

Tang, H. S., Y. J. Feng and L. Q. Yao (2009). "Angiogenesis, vasculogenesis, and vasculogenic mimicry in ovarian cancer." Int J Gynecol Cancer 19(4): 605-610.

Thomas, R. A., D. C. Pietrzak, M. S. Scicchitano, H. C. Thomas, D. C. McFarland and K. S. Frazier (2009). "Detection and characterization of circulating endothelial progenitor cells in normal rat blood." J Pharmacol Toxicol Methods 60(3): 263-274.

Urbich, C., A. Aicher, C. Heeschen, E. Dernbach, W. K. Hofmann, A. M. Zeiher and S. Dimmeler (2005). "Soluble factors released by endothelial progenitor cells promote migration of endothelial cells and cardiac resident progenitor cells." J Mol Cell Cardiol 39(5): 733-742.

van der Schaft, D. W., F. Hillen, P. Pauwels, D. A. Kirschmann, K. Castermans, M. G. Egbrink, M. G. Tran, R. Sciot, E. Hauben, P. C. Hogendoorn, O. Delattre, P. H. Maxwell, M. J. Hendrix and A. W. Griffioen (2005). "Tumor cell plasticity in Ewing sarcoma, an alternative circulatory system stimulated by hypoxia." Cancer Res 65(24): 11520-11528.

Vartanian, A. A., E. V. Stepanova, S. L. Gutorov, E. Solomko, I. N. Grigorieva, I. N. Sokolova, A. Y. Baryshnikov and M. R. Lichinitser (2009). "Prognostic significance of periodic acid-Schiff-positive patterns in clear cell renal cell carcinoma." Can J Urol 16(4): 4726-4732.

Wang, X., Z. Zhang and C. Yao (2011). "Angiogenic activity of mesenchymal stem cells in multiple myeloma." Cancer Invest 29(1): 37-41.

Wickersheim, A., M. Kerber, L. S. de Miguel, K. H. Plate and M. R. Machein (2009). "Endothelial progenitor cells do not contribute to tumor endothelium in primary and metastatic tumors." Int J Cancer 125(8): 1771-1777.

Widemann, A., F. Sabatier, L. Arnaud, L. Bonello, G. Al-Massarani, F. Paganelli, P. Poncelet and F. Dignat-George (2008). "CD146-based immunomagnetic enrichment followed by multiparameter flow cytometry: a new approach to counting circulating endothelial cells." J Thromb Haemost 6(5): 869-876.

Wierzbowska, A., T. Robak, A. Krawczynska, A. Pluta, A. Wrzesien-Kus, B. Cebula, E. Robak and P. Smolewski (2008). "Kinetics and apoptotic profile of circulating endothelial cells as prognostic factors for induction treatment failure in newly diagnosed acute myeloid leukemia patients." Ann Hematol 87(2): 97-106.

Woywodt, A., A. D. Blann, T. Kirsch, U. Erdbruegger, N. Banzet, M. Haubitz and F. Dignat-George (2006). "Isolation and enumeration of circulating endothelial cells by immunomagnetic isolation: proposal of a definition and a consensus protocol." J Thromb Haemost 4(3): 671-677.

Yang, J., M. Ii, N. Kamei, C. Alev, S. M. Kwon, A. Kawamoto, H. Akimaru, H. Masuda, Y. Sawa and T. Asahara (2011). "CD34+ cells represent highly functional endothelial progenitor cells in murine bone marrow." PLoS One 6(5): e20219.

Yao, X. H., Y. F. Ping and X. W. Bian (2011). "Contribution of cancer stem cells to tumor vasculogenic mimicry." Protein Cell 2(4): 266-272.

Zhang, J., P. Zhao, Z. Fu, X. Chen, N. Liu, A. Lu, R. Li, L. Shi, P. Pu, C. Kang and Y. You (2008). "Glioma cells enhance endothelial progenitor cell angiogenesis via VEGFR-2, not VEGFR-1." Oncol Rep 20(6): 1457-1463.

Zhang, S., D. Zhang and B. Sun (2007). "Vasculogenic mimicry: current status and future prospects." Cancer Lett 254(2): 157-164.

Zhao, Y., J. Dong, Q. Huang, M. Lou, A. Wang and Q. Lan (2010). "Endothelial cell transdifferentiation of human glioma stem progenitor cells in vitro." Brain Res Bull 82(5-6): 308-312.

Zhao, Y., D. Glesne and E. Huberman (2003). "A human peripheral blood monocyte-derived subset acts as pluripotent stem cells." Proc Natl Acad Sci USA 100(5): 2426-2431.

Differential Activation of Macrophages in Tumors: Roles in the Regulation of Tumor Development and Angiogenesis

Shalini P. Outram and S. Joseph Leibovich*

Department of Cell Biology & Molecular Medicine & The Cardiovascular Research Institute
New Jersey Medical School, UMDNJ, 185 South Orange Avenue
Newark, NJ 07103, U.S.A.

CHAPTER OUTLINE

ABSTRACT

Cancer is a progressive disease that can manipulate both tumor and invading stromal cells to orchestrate tumor growth and angiogenesis. Inflammatory cells such as macrophages are recruited to the tumor microenvironment and can perform functions as tumor-activated macrophages to aid in the progression of cancer, in

Correspondence Author Department of Cell Biology & Molecular Medicine & The Cardio-vascular Research Institute, New Jersey Medical School, UMDNJ, 185 South Orange Avenue, Newark, NJ 07103, USA. Email: leibovic@umdnj.edu

a manner analogous to their roles in the process of wound repair. Macrophages are heterogeneous in nature and respond to environmental cues such as growth factors, cytokines, hypoxia, ischemia, and factors produced by tumors, to express pro- or anti-tumoral functions. Activation of macrophages has generally been described as following two major activation schemes, "classical" and "alternative activation". Classical activation involves stimulation of macrophages by factors such as bacterial endotoxin (lipopolysaccharide, LPS) through surface receptors such as Toll-like Receptor-4 (TLR4) to induce an inflammatory response and expression of the M1 phenotype. Alternative activation involves stimulation of macrophages by IL4 and IL13 through the IL4Rα to induce M2 macrophages, which exhibit an anti-inflammatory and angiogenic phenotype. The tumor micro-environment is generally anti-inflammatory and angiogenic, and tumor-associated macrophages generally adopt an M2-like alternatively activated phenotype. The tumor microenvironment is also predominantly hypoxic and ischemic, and provides a microenvironment that favors induction of an angiogenic phenotype by inducing increased expression of factors such as VEGF. Both low oxygen tension and adenosine generated by breakdown of ATP stimulate VEGF expression via regulation of Hypoxia-Inducible Factors (HIFs). We have described an M2-like macrophage phenotype that is induced by adenosine signaling through adenosine receptors in conjunction with TLR agonists, generating an "angiogenic switch" of M1 to M2-like macrophages (which we have termed "M2d") through down-regulation of genes such as TNFα and IL12 and up-regulation of genes such as VEGF and IL10. This angiogenic switch requires only TLR agonists and adenosine, and does not depend on specific cytokines as part of an adaptive immune response. Since both adenosine and VEGF play an important role in the regulation of tumor development and angiogenesis, analyzing the role of this angiogenic switch of macrophages should provide important insight towards identifying novel molecular targets for therapeutic intervention of cancer progression. In this chapter, the evidence for the roles of macrophages and their expression of adenosine receptors in the regulation of tumor development will be considered.

Key words: Macrophage activation; Tumor Development; Angiogenesis; Adenosine Receptors, TLRs

1   Introduction

Cancer is a complex disease that harnesses both malignant and non-malignant cells in the tumor microenvironment to drive tumor progression and metastasis. The progression of tumor growth has much in common with the normal host response to injury, in particular with the process of wound repair. In response to injury and infection, the host initiates an inflammatory reaction that is geared to destroy foreign organisms and clear the injury site of dead and damaged cells and tissues. This response is characterized by the influx of myeloid cells, including neutrophils, monocytes/macrophages and lymphocytes. Macrophages recruited as part of the inflammatory response differentiate at the site of injury, and respond to microenvironmental cues by adopting a variety of activation phenotypes that participate either as inflammatory cells or as angiogenic, wound healing cells that promote tissue repair. For successful repair, migration and proliferation of epithelial cells, mesenchymal cells and microvascular blood vessels are stimulated in response to factors produced at the site of injury, resulting in the formation of granulation tissue. In healing wounds, the signals mediating the development of granulation tissue resolve, and proliferation of epithelium, mesenchyme and blood vessels ceases, followed by the formation of a connective tissue scar. Similar events occur in the development of solid tumors. In response to signals that follow cellular transformation, the host mounts an inflammatory response that recruits neutrophils, macrophages and lymphocytes to the incipient tumor site. These cells perform as they do in a healing wound, and the inflammatory response ideally mediates destruction of the transformed cells. However, in many cases, developing transformed cells are able to evade the destructive effects of host immune cells. Recruited inflammatory cells, in particular macrophages, adopt an angiogenic, wound healing phenotype that promotes angiogenesis and cell proliferation. While the wound healing process leads to granulation tissue formation, which is essential for normal tissue repair, in tumors this leads to angiogenesis and the promotion of tumor cell growth. Tumors thus hijack the normal wound healing response to establish their own growth. As long ago as 1863, Rudolph Virchow, a German pathologist, noted that cancers often spring from sites that are chronically inflamed (Balkwill and Mantovani 2001). The ability of tumors to stimulate angiogenesis and perpetuate their growth was noted by Dvorak in 1986, who referred to tumors as wounds that fail to heal (Dvorak 1986). We propose that tumors utilize and hijack the normal host responses of inflammation and wound healing to induce angiogenesis and to establish their own growth, thus providing a key link between the host response to injury and tumorigenesis (Leibovich 1996).

The communication between tumor cells, their surrounding stromal cells (fibroblasts, endothelial cells, pericytes, and mesenchymal cells) and immune cells (macrophages, neutrophils, mast cells, myeloid-derived suppressor cells, dendritic cells, natural killer cells and T and B lymphocytes) by direct contact or via cytokine and chemokine production can regulate and control tumor progression (Grivennikov et al. 2010). Tumor-associated macrophages (TAMs) represent a major inflammatory component of the tumor stroma and are involved in both pro- and anti-tumoral functions. Pro-tumoral functions include direct stimulation of tumor cells, promotion of angiogenesis, stimulation of matrix synthesis and remodeling, and suppression of adaptive immunity (Lee et al. 2006; Shih et al. 2006; Coffelt et al. 2009). Anti-tumoral functions include direct and indirect cytotoxicity and regulation of acquired immunity by antigen presentation to lymphocytes. Macrophages are well-known for their ability to mediate innate immune responses against microbes by recognizing pathogen-associated molecular patterns (PAMPs) such as lipopolysaccharide (LPS) through Pattern Recognition Receptors (PRRs), and damage-associated molecular patterns (DAMPs) such as mitochondrial DNA from apoptotic cells, High Mobility Group type B (HMGB) proteins from cell nuclei and HSP70, among others (Bjorkbacka et al. 2004; Taylor et al. 2005; Rubartelli and Lotze 2007).

2 Role of Macrophages in Cancer

Macrophages are mononuclear phagocytes that are derived from precursor stem cells in the bone marrow, and are recruited from the circulation to sites of injury, where they play key roles in both inflammation and tissue repair. As inflammatory cells, macrophages recognize and destroy micro-organisms, either directly or by presenting antigens to lymphocytes, thus activating acquired immunity. As cells mediating tissue repair, macrophages produce chemokines and growth factors that regulate the migration, proliferation and biosynthetic activity of connective tissue cells, as well as angiogenic factors that mediate neovascularization. To mediate these diverse roles, macrophages exhibit exquisite plasticity in terms of their gene expression profile, and adopt markedly different activation phenotypes, depending upon the nature of the micro-environment in which they reside. Classically-activated macrophages are induced by interferon gamma (IFNγ) and Toll-like receptor (TLR) agonists to express inflammatory cytokines, including tumor necrosis factor-α (TNFα), IL-1, IL-6 and IL-12, as well activated oxygen radicals and nitric oxide (NO) (Nathan 1991; Stein et al. 1992; Ehrt et al. 2001; Domachowske et al. 2002; Ma et al. 2003). This phenotype has been termed "M1", and M1 macrophages act as strong promoters of the type 1 helper T (Th1) immune response. The persistent presence of M1 macrophages can be detrimental to the host, by causing DNA damage that might contribute to cancer development. In contrast, in response to a different set of microenvironmental signals, macrophages can be activated through distinct pathways to acquire anti-inflammatory, wound healing, and angiogenic phenotypes. These phenotypes have been generically termed "M2", and M2 macrophages express the anti-inflammatory cytokine IL-10, and the angiogenic growth factor VEGF rather than inflammatory cytokines (Goerdt et al. 1999; Mantovani 2006; Mosser and Edwards 2008; Classen et al. 2009; Martinez et al. 2009; Gordon and Martinez 2010). In an effort to promote survival, tumor cells can evade innate and adaptive immune responses by hijacking macrophages and inducing them to adopt an M2 phenotype. These M2 macrophages then play an anti-inflammatory, immunosuppressive and angiogenic role, thus promoting a pro-tumoral environment.

The M2 designation of macrophages is further subdivided into M2a (activated by IL-4 or IL-13 through the IL-4Rα sub-unit), M2b (induced by immune complexes and IL-1β or TLR stimulation), and M2c (stimulated by glucocorticoids, IL-10 or TGF-β) (Martinez et al. 2009). In addition, we have found that production of inflammatory cytokines and VEGF by macrophages is strongly regulated by adenosine, a ubiquitous metabolite which is induced by conditions of ischemia and hypoxia (Hasko et al. 1996; Hasko et al. 2000; Pinhal-Enfield et al. 2003; Hasko and Cronstein 2004). Adenosine synergizes with TLR agonists such as endotoxin (LPS) to strongly down-regulate the expression of TNFα and other inflammatory chemokines and cytokines, and to up-regulate the expression of IL-10,

VEGF and other anti-inflammatory and angiogenic factors (Leibovich et al. 2002). The synergy between TLR and adenosine receptor signaling in these alternatively activated macrophages results in an "angiogenic switch" that converts macrophages from an inflammatory to an angiogenic phenotype (Leibovich 2007). We have termed macrophages activated in this manner "M2d" (Enfield and Leibovich 2011) (Fig. 1).

Several studies have shown that TAMs often exhibit an M2-like phenotype, with reduced antimicrobial and anti-tumoral activity and increased production of mediators such as IL-10, TGFβ and VEGF, that suppress immunity and promote angiogenesis (Mantovani and Sica 2010). Since tumors exhibit regions of hypoxia and ischemia and thus generate high levels of adenosine, we propose that tumor cells utilize the adenosine receptor-TLR synergy to switch macrophages from an inflammatory to an angiogenic state. This chapter focuses on reviewing the importance of adenosine receptor and TLR signaling in macrophages and its role in regulating angiogenesis and tumor progression. Understanding the mechanisms involved in the angiogenic switch of macrophages may provide insight leading to the development of novel therapeutics for inflammation, wound healing and cancer.

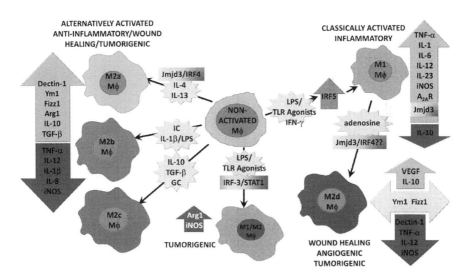

Figure 1 **Macrophage activation phenotypes.** Classically activated (M1), inflammatory macrophages are characterized by an increased production of pro-inflammatory cytokines and decreased production of anti-inflammatory cytokines. Alternatively activated macrophages are involved in wound healing, angiogenesis and tumorigenesis. These macrophages are characterized as M2 or M2-like depending on the activation stimuli, and exhibit an increased production of anti-inflammatory cytokines and angiogenic growth factors, and decreased production of pro-inflammatory cytokines.

Color image of this figure appears in the color plate section at the end of the book.

3   Role of Macrophages in Tumor Progression

Macrophages infiltrate developing tumors in response to growth factors such as macrophage colony stimulating factor (M-CSF) and chemokines (CCL2, CCL5, CCL7, CCL8, and CXCL12) produced by tumors. Within the tumor, due to their remarkable functional plasticity, they respond to factors present in the local tumor microenvironment and change their phenotype to express either anti- or pro-tumoral functions (Balkwill 2004; Pollard 2004; Loberg et al. 2007). The presence of PAMPs and DAMPs, local ischemia and hypoxia, and cross-talk between cancer cells and macrophages, play important roles in modulating the phenotype of macrophages. PAMPs and DAMPs generally induce an M1 (anti-tumoral and inflammatory) phenotype, while hypoxia/ ischemia and cross-talk between tumor cells and macrophages tend to activate a polarization pathway to an M2 (pro-tumoral and angiogenic) phenotype (Mantovani et al. 2002; Biswas et al. 2008). TAMs represent a distinct M2-polarized macrophage population and play important roles in all stages of tumor progression, including proliferation, migration, metastasis, invasion, angiogenesis, survival under hypoxic conditions and immunosuppression (Mantovani et al. 2004; Lee et al. 2006; Lewis and Pollard 2006; Shih et al. 2006; Sica et al. 2008; Siveen and Kuttan 2009) (Fig. 2). TAMs express a number of factors that stimulate tumor growth, proliferation and migration, including epidermal growth factor (EGF), platelet-derived growth factor (PDGF), TGF-β, fibroblast growth factor (FGF), VEGF and an assortment

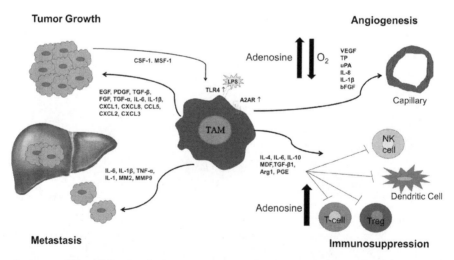

Figure 2   The role of TAMs in tumor development. TAMs generally exhibit an M2-like phenotype characterized by increased production of growth factors, cytokines, chemokines and enzymes involved in tumor growth, angiogenesis, metastasis and immuno-suppression. TLR stimulation upregulates expression of A2ARs, which in the presence of high concentrations of adenosine, switch macrophages into an M2-like (M2d) phenotype.

Color image of this figure appears in the color plate section at the end of the book.

of chemokines and cytokines (Table 1). Polverini and Leibovich showed a number of years ago that depletion of macrophages in hamster buccal pouch carcinoma using hydrocortisone acetate and intratumoral injections of anti-macrophage serum resulted in a reduction in endothelial and tumor cell DNA synthesis, suggesting that the presence of TAMs is essential for tumor growth and neovascularization (Polverini and Leibovich 1987). Recent studies from the laboratory of Pollard et al have supported this observation. Qian and colleagues depleted macrophages by three independent means. The first, used a genetic model with CSF-1 knockout mice; the second

Table 1 Factors secreted by tumor-associated-macrophages (TAMs)

Secreted factors	Role in tumor promotion	Reference
Growth Factors		
EGF	Enhanced growth	(Ono et al. 1999; Goswami et al. 2005)
PDGF	Enhanced growth, angiogenesis, invasion and migration	(Uutela et al. 2004)
TGF-β	Enhanced growth, angiogenesis, invasion and migration	(Mantovani et al. 2002)
VEGF	Enhanced angiogenesis and immunosuppression	(Balkwill and Mantovani 2001; Crowther et al. 2001)
FGF	Enhanced growth, angiogenesis and invasion	(Crowther et al. 2001)
TGF-α	Enhanced growth	(Ono et al. 1999)
Cytokines		
IL-6	Enhanced growth, angiogenesis and metastasis	(Balkwill and Mantovani 2001; Balkwill and Mantovani 2010)
IL-1β	Enhanced growth, angiogenesis and metastasis	(Saijo et al. 2002)
TNF-α	Enhanced metastasis	(Crowther et al. 2001)
IL-1	Enhanced metastasis	(Giavazzi et al. 1990)
IL-8	Enhanced angiogenesis	(Varney et al. 2002; Chen et al. 2003)
IL-10	Enhanced immunosuppression	(Sharma et al. 1999)
M-CSF	Enhanced tumor cell migration and angiogenesis	(Eubank et al. 2003; Goswami et al. 2005)
Chemokines		
CCL2	Enhanced angiogenesis	(Ueno et al. 2000; Mantovani et al. 2002)
CXCL12	Enhanced angiogenesis	(Azenshtein et al. 2002)
CXCL8	Enhanced growth and angiogenesis	(Crowther et al. 2001; Azenshtein et al. 2002)
CXCL1	Enhanced growth and angiogenesis	(Azenshtein et al. 2002)
CXCL13	Enhanced growth	(Azenshtein et al. 2002)
CCL5	Enhanced growth	(Azenshtein et al. 2002)
CXCL2	Enhanced growth and angiogenesis	(Haghnegahdar et al. 2000)
CXCL3	Enhanced growth and angiogenesis	(Haghnegahdar et al. 2000)
Thymidine phosphorylase	Enhanced angiogenesis and migration	(Hotchkiss et al. 2003)
MMP2	Enhanced invasion and metastasis	(Egeblad and Werb 2002)
MMP9	Enhanced invasion and metastasis	(Egeblad and Werb 2002; Hiratsuka et al. 2002)

used the classic macrophage depletion model of treatment with liposome encapsulated clodronate; the third used a suicide gene approach to ablate the CD11b+ macrophage population in mice expressing the human EGFR under the control of the CD11b promoter, using diphtheria toxin. All three depletion strategies resulted in inhibition of breast cancer tumor growth, inhibition of metastatic seeding and survival, and inhibition of subsequent growth after established breast cancer metastasis (Qian et al. 2009). Lin et al demonstrated, also using macrophage depletion strategies, that TAM recruitment is a requirement for the angiogenic switch and metastatic growth in a mouse model of spontaneous mammary carcinoma (Lin et al. 2006). Increased numbers of TAMs in primary tumors correlate with early formation of metastases and regulate the establishment of secondary tumors at distant sites (Leck et al. 1996; Hagemann and Lawrence 2009; Zhou et al. 2010). Macrophages can produce enzymes and inhibitors which aide in the breakdown of the extracellular matrix leading to tumor invasion. TAMs produce matrix metalloproteases (MMP-2, MMP-9) which are key players in the digestion of tissue barriers to invasion (Coussens et al. 2000; Hagemann et al. 2004). TAMs also produce the "secreted protein acidic and rich in cysteine" (SPARC), which modulates collagen density, leukocyte infiltration and blood vessel growth (Sangaletti et al. 2003). Table 1 summarizes factors produced by TAMs and their potential roles in tumor development.

TAMs have been shown to accumulate in poorly vascularized and hypoxic areas of several tumors, including human endometrial, breast, prostate and ovarian carcinoma. In response to hypoxia, TAMs produce various pro-angiogenic factors such as VEGF, thymidine phosphorylase, urokinase plasminogen activator (uPA), IL-8 and bFGF. Expression of these factors by TAMs correlates with increased angiogenesis, tumor progression and ultimately poor patient prognosis. Lin et al demonstrated that depletion of TAMs in Polyoma Middle T oncoprotein (PyMT)-induced mammary tumors in mice prevented the angiogenic switch and delayed progression to malignancy. They further discovered that transgenic expression of VEGF stimulated angiogenesis and tumor progression in macrophage depleted mice, suggesting that TAMs and TAM-derived VEGF regulate the angiogenic switch in this mouse model of breast cancer (Lin et al. 2006; Lin et al. 2007; Lin and Pollard 2007). TAMs respond to low oxygen levels by expressing elevated levels of hypoxia-inducible factor (HIF)-1α, which binds to HIF-1β to increase transcription of genes involved in angiogenesis and cell proliferation (Burke et al. 2002). TAMs express high levels of VEGF in areas of hypoxia and necrosis in human tumors, thus acting to replenish oxygen supply by inducing neovascularization (Lewis et al. 2000; Burke et al. 2002). As discussed in more detail below, we have reported that adenosine 2A receptor (A2AR) agonists and the TLR4 agonist (LPS) synergistically augment VEGF expression by macrophages under normoxic conditions, and this increase in VEGF is regulated transcriptionally by the binding of HIF-1α to the hypoxia response element (HRE) of the VEGF promoter (Leibovich et al. 2002; Ramanathan et al. 2007). This pathway provides an additional mechanism for VEGF induction by macrophages in response to adenosine produced under ischemic conditions.

Tumor-derived products, such as cytokines, chemokines, growth factors, and proteases influence TAMs in the tumor environment to dampen their cytotoxic capability and enhance immunosuppression. These factors include IL-4, IL-6, IL-10, Macrophage Deactivating Factor (MDF), TGF-β1, arginase I and PGE2. Many of these factors promote the polarization of TAMs to an M2 phenotype characterized by the reduced ability to present antigens, to stimulate T cell or natural killer (NK) cells and to produce pro-inflammatory cytokines (Elgert et al. 1998; Ben-Baruch 2006; Jeannin et al. 2011). In this way, tumor cells create an environment where they evade immune surveillance and manipulate macrophages to promote angiogenesis, tumor growth, survival and metastasis.

Several reports indicate that activation of transcription factors such as nuclear factor (NF)-κB and signal transducer and activator of transcription (STAT)-1, -3 and -6 molecules are up-regulated in TAMs through activation by IL-10. TAMs lacking STAT3 or STAT6 retain their M1 functions to present antigens, display tumoricidal activities and subsequently delay tumor growth (Kortylewski et al. 2005; Sinha et al. 2005). STAT1 activity contributes to the immunosuppressive effects of TAMs by up-regulating arginase I and inducible nitric oxide synthase (Kusmartsev and Gabrilovich 2005). Another transcription factor that contributes to macrophage polarization is interferon-regulatory factor (IRF)-5. High expression of IRF5 in macrophages results in an M1 phenotype by activating transcription of IL-12p40, IL-12p35 and IL-23p19 cytokines and repressing the transcription of IL-10 (Krausgruber et al. 2011). Therefore, inducing the expression of IRF5 in TAMs may restore anti-tumor responses in the tumor-microenvironment. On the opposite side of the spectrum, Jumonji domain containing-3 (Jmjd3), a histone 3 Lys27 demethylase, is critical for canonical (IL-4Rα-dependent) M2 polarization in response to helminth infection and chitin by targeting the *Irf4* gene (Satoh et al. 2010). Thus, the targeted inhibition of Jmjd3 and *Irf4* in macrophages may also play a role in regulating macrophage phenotype in tumors and may be important to reverse tumor immunosuppression. Standiford and colleagues showed that TAMs express high levels of Interleukin Receptor-Associated Kinase (IRAK)-M which is a negative regulator of TLR signaling and anti-tumoral responses (Standiford et al. 2011). IRAK-M is induced by TGF-β secreted by tumors into their microenvironment. In IRAK-M deficient mice implanted with Lewis lung carcinoma, a five-fold reduction in tumor growth was demonstrated as compared to the wild-type mice. Furthermore, TAMs isolated from IRAK-M deficient mice exhibited an M1 rather than M2 phenotype characterized by increased expression of IL-12 and IFNγ. IRAK-M expression thus represents a means by which tumor cells might evade anti-tumoral responses by immune cells such as macrophages. Restoring immune surveillance by macrophages by targeting reversal of tumor-induced immunosuppression could prove to be a promising strategy for cancer therapy.

4   Macrophage Activation Pathways

As mentioned earlier, macrophages polarize into different phenotypes (inflammatory and anti-inflammatory) by responding to their local microenvironment. In response to injury, macrophages are recruited to sites of injury by responding to chemo-attractant signals produced during inflammation. While macrophages exhibit considerable heterogeneity in their activation profiles, two major activation schemes, "classical activation" and "alternative activation" have been characterized in detail in recent years. The classical pathway involves recognition of PAMPs such as LPS by PRRs and stimulation by IFNγ produced by T helper 1 (T_H1) or CD8+ cells or natural killer (NK) cells leading to a T_H1-type inflammatory response (Ma et al. 2003; Taylor et al. 2005; Rubartelli and Lotze 2007; Classen et al. 2009; Martinez et al. 2009). The alternatively activated macrophage pathway involves stimulation of macrophages by IL4 and/or IL-13, inducing cells such as eosinophils, basophils and naive T lymphocytes to generate a T_H2-type anti-inflammatory response (Ma et al. 2003; Taylor et al. 2005; Mantovani 2006; Rubartelli and Lotze 2007; Martinez et al. 2009). The nomenclature of classically activated macrophages "M1" and alternatively activated macrophages "M2" was first designated by analogy to the T_H1 and T_H2 immune responses (Rubartelli and Lotze 2007).

In addition to signaling through IFNγ, antigen presenting cells are stimulated through TLRs to secrete TNFα, a pro-inflammatory cytokine which synergizes with IFNγ to sustain inflammation (Mosser and Edwards 2008). This M1 inflammatory macrophage is geared towards phagocytosis, recognition and destruction of foreign organisms, pathogens and tumor cells. M1 macrophages in turn secrete pro- inflammatory cytokines including TNFα, IL-1, IL-6 and IL-12, up-regulate expression of reactive oxygen radicals, iNOS, MMP9 and MHC class II receptors on the cell surface, and release nitric oxide (NO) which aids in their microbicidal and tumoricidal capability. Agonists of several TLRs, such as LPS (TLR4 agonist) and lipotechoic acid (TLR2 agonist), bind to TLRs on the cell surface in conjunction with other cell surface co-receptors. These include CD14, CD11b/CD18, and MD2 for TLR4 and CD36 for TLR2. In conjunction with these co-receptors, TLRs trigger intracellular signaling pathways that lead to gene induction through the action of several transcription factors, including NF-κB, STATs and interferon-regulatory factors (IRFs) that are involved in M1 macrophage polarization.

Alternatively activated macrophages (M2) differ from classically activated macrophages by their activation patterns, expression of cytokines and growth factors and their involvement in anti-inflammatory, wound healing and T_H2 responses (Gordon 2003; Gordon and Taylor 2005). M2 macrophages generally express high levels of IL-10, TGF-β, arginase-1 (*Arg1*), receptors such as the SR-A/scavenger receptor (CD204), MRCI/mannose receptor (CD206), β-glucan receptor (dectin-1) and cell surface markers such as Found in Inflammatory Zone-1 (Fizz1) and chitinase 3-like 3/chitinase 3-like

4 (Ym1/Ym2) (Enfield and Leibovich 2011). They also express low levels of TNFα, IL-12, IL-1β, IL-6, IL-8, IL-23 and iNOS (Enfield and Leibovich 2011). Exceptions exist, however, where distinct forms of M2 cells have been described to express low IL-12 and high IL-10 phenotype with the capacity to produce TNF, IL-1 and IL-6 (Mantovani et al. 2004). This highlights the complex ability of macrophages to change their phenotype to serve their required functions in different microenvironments. As mentioned previously, recent studies have categorized alternatively activated M2 macrophages into subsets (M2a, M2b, M2c and M2d) depending on their responses to different cytokines and growth factors (Martinez et al. 2009). M2a macrophages are prototypically induced by IL-4 and IL-13 through the common IL-4Rα subunit of the receptors for these cytokines (Martinez et al. 2009). However, in mouse wounds lacking IL-4 and IL-13, or lacking the IL-4Rα subunit, macrophages having M1 and M2-like characteristics were still found, clearly indicating that IL-4 and IL-13 are not essential for the induction of macrophages with an M2-like phenotype (Daley et al. 2010). M2b macrophages are induced by immune complex and IL-1β or TLR agonists. M2c macrophages are induced by glucocorticoids, IL-10 and TGFβ (Martinez et al. 2009). M2d macrophages are induced by TLR agonists and adenosine in an IL-4Rα-independent manner (Enfield and Leibovich 2011) (Fig. 1). A more detailed description of the M2d macrophage activation program in the context of wound healing and angiogenesis will be presented in the next section.

Several reports indicate that TAMs have diverse phenotypes, having both M1 and M2 markers, but are generally associated with M2 functions that support tumor progression and angiogenesis (Van Ginderachter et al. 2006). Mantovani and colleagues present TAMs as polarized M2 mononuclear phagocytes producing high IL-10 and low IL-12 (Mantovani et al. 2004; Mantovani and Sica 2010). Biswas et al report that TAMs from murine fibrosarcoma express a mixed phenotype with M1 and M2 properties (high dectin-1, high IL-10 and low IL-12), coexisting with M1 expression of IFN-inducible cytokines such as CXCL10. They further demonstrated that upon treatment with LPS, NF-κB activation was inhibited, but IFNγ-inducible chemokines were expressed through enhanced activation of IRF-3 and STAT1 (Biswas et al. 2006). Additionally, they reported that arginase-1, which is considered to be an M2 marker, is regulated in a complex manner in TAMs from murine fibrosarcoma. They suggest that both arginase-1 and iNOS could be induced by the same IRF-3/STAT1 pathway, which could explain the mixed M1/M2 gene expression profile in TAMs. Thus, macrophages can take on different phenotypes (M1, M2, M2-like) to achieve the goal of tumor promotion in the tumor microenviroment.

5 Macrophage Activation by Adenosine and Toll-like Receptors (TLRs)

Adenosine triphosphate (ATP) and the purine nucleoside adenosine were first discovered to function in an intracellular manner in the heart and blood vessels by Drury and Szent-Györgyi in 1929 (Drury and Szent-Györgyi 1929). They were later shown to be released from cells as neurotransmitters by Burnstock in the 1970s (Burnstock et al. 1970; Burnstock 1972). Sattin and Rall in 1970 showed that adenosine performs its physiological functions by interacting with cell surface receptors (Sattin and Rall 1970). To date, there are at least 4 known subtypes of adenosine receptors (A1, A2A, A2B and A3) which have been cloned and identified as members of the seven-transmembrane spanning G protein coupled receptor family (Ralevic and Burnstock 1998; Fredholm et al. 2001; Linden 2001; Hasko and Cronstein 2004; Fredholm et al. 2005). In addition, a large family of receptors for purine nucleotides called purinergic receptors or purinoceptors has been characterized (Burnstock 1980). Adenosine receptors are referred to as P1 receptors and receptors for ATP and other nucleotides are referred to as P2 receptors (Burnstock 1980). Although early studies demonstrated the role of purinergic receptors in neurotransmission, it is now known that adenosine is also an important regulator of inflammatory responses (Hasko and Cronstein 2004; Novitskiy et al. 2008) and general biological roles such as cell survival, proliferation, differentiation and motility, in diseases such as cancer. Adenosine receptors are well documented to be expressed on monocytes/macrophages and tumor cells, where they function to regulate immunosuppression and angiogenesis (Fig 2) (Hasko et al. 2002; Auchampach 2007; Hasko et al. 2007; Leibovich 2007; Feoktistov et al. 2009; Kumar and Sharma 2009; Stagg and Smyth 2010).

Adenosine functions to stabilize tissue function by balancing energy intake to meet metabolic demand (Newby et al. 1985; Fredholm 2007). Adenosine is ubiquitous and can be released by virtually all cells. Under metabolically unfavorable conditions such as ischemia, hypoxia and inflammation, adenosine concentration is augmented by the breakdown of ATP inside the cell, and is then released into the extracellular space through nucleoside transporters, thus contributing to high extracellular adenosine levels (Zimmermann 2000; Hyde et al. 2001; Ritzel et al. 2001; Eltzschig et al. 2004). In addition, ATP is released directly from stressed cells, at least in part through a vesicle-mediated transport mechanism, and membrane-bound ecto-nucleotidases such as nucleoside triphosphate disphosphohydrolase (CD39) hydrolyze/dephosphorylate ATP to ADP and AMP (Kaczmarek et al. 1996). Further dephosphorylation to adenosine by 5'-nucleotidase (CD73) represents the limiting step for adenosine formation in the extracellular space (Resta et al. 1998; Deussen 2000; Zimmermann 2000). Along with dephosphorylation of ATP, the activity of the salvage enzyme adenosine kinase is suppressed under hypoxic conditions and rephosphorylation of adenosine is prevented, contributing to high extracellular adenosine concentrations (Deussen 2000). Adenosine concentration is kept at equilibrium by cellular reuptake (Pastor-Anglada et al. 2001), phosphorylation to AMP by adenosine kinase or degradation to inosine by adenosine deaminase (Cristalli et al. 2001; Hasko et al. 2004). Several cell types produce extracellular adenosine in response to metabolic stress, including neutrophils, endothelial cells, nerve terminals and activated macrophages (Dowdall 1978; Burnstock 1980; James and Richardson 1993; Cronstein 1994; Sperlagh et al. 1998).

The level of adenosine dictates which adenosine receptor subtypes are engaged to induce downstream signaling pathways. A1, A2A and A3 receptors are activated in response to physiological concentrations of adenosine (30-300 nM), whereas A2BRs require higher levels of adenosine generated in response to metabolic stress or pathological conditions (Traut 1994). A1 and A3 receptors are generally coupled to Gi/o proteins to inhibit adenylyl cyclase and protein kinase A (PKA) activity to decrease cyclic AMP levels (Burnstock et al. 1970). In contrast, A2A and A2B receptors are generally coupled to Gs proteins, which increase cAMP levels through adenylyl cyclase and PKA activation (van Calker et al. 1979; Londos et al. 1980). A2BRs can couple to Gq proteins to signal through phospholipase-Cβ (PLCβ) (Linden et al. 1999; Aherne et al. 2010). Adenosine receptors are expressed on most immune cells, including macrophages (Hasko and Cronstein 2004). In macrophages, adenosine signaling is generally immunosuppressive and angiogenic. Adenosine interferes with the actions of PPRs such as TLRs resulting in suppression of TNFα and IL-12 production and augmentation of IL-10 and VEGF expression in monocytes/macrophages (Hasko et al. 2007). Adenosine receptor signaling has been reviewed for its potential therapeutic context in several disease processes, such as Parkinson's disease, schizophrenia, ischemia and cancer (von Lubitz 1999; Wardas 2008; Jenner et al. 2009; Gessi et al. 2011). We will focus on the role of adenosine receptors in cancer in the latter sections of this chapter.

The recognition of PAMPs by PRRs such as TLRs is the foundation of innate immunity, and also allows immune cells such as macrophages to elicit an efficient immune response against pathogens. In addition to PAMPs, TLRs also recognize DAMPs which are endogenous molecules that are released as danger signals during tissue damage and necrosis (Piccinini and Midwood 2010). Thus, TLRs regulate inflammatory responses during microbial infection, wound healing and tumorigenesis (Kluwe et al. 2009). TLRs are trans-membrane receptors that are highly conserved in both vertebrates and invertebrates. To date, there are 11 known murine TLRs and 10 known human TLRs. The existence of a large number of TLRs enables the immune system to respond to several PAMPs to launch specific defense mechanisms. TLR4 was the first Toll-like receptor described in humans and is the surface receptor for LPS from Gram negative bacteria (Beutler 2000). TLR2 recognizes cell membrane components and lipotechoic acid from Gram-positive bacteria. TLR3, TLR7/8 and TLR9 are located on endosomes and recognize double stranded RNA, single stranded RNA or non-methylated CpG DNA respectively. TLR5 recognizes the major protein component of bacterial flagellins. In addition, TLRs such as TLR2 and TLR4 recognize DAMPs, including HMGB1, heat shock protein 60, S100A8 and serum amyloid A, which play a role in disease processes such as cancer (Hiratsuka et al. 2008; Yang et al. 2009; Sims et al. 2010).

TLRs signal through two major pathways, myeloid differentiation primary response protein (MyD88)-dependent and MyD88-independent. All TLRs, with the exception of TLR3, signal through the MyD88-dependent pathway. Briefly, upon activation with their specific ligands, TLRs recruit and bind to the MyD88 adapter protein through the Toll-IL-1 receptor (TIR) domain on the cytoplasmic domain of TLRs to activate downstream IL-1R-associated kinase (IRAK)-4 and -1 proteins, with subsequent activation of targets such as NF-κB, mitogen associated protein (MAP) kinase and interferon regulatory factors (IRFs), which are involved in tumorigenic responses (So and Ouchi 2010). TLR3 recruits TIR-domain-containing-adapter-inducing interferon-β (TRIF) on endosomes, targeting either IRF3 leading to antiviral immune responses, or TNF receptor-associated factor-6 (TRAF-6) leading to MAP kinase activation and increased survival or proliferation (So and Ouchi 2010). TLR4 signals through both MyD88-dependent and MyD88-independent pathways. In macrophages, TLR signaling induces the expression of several cytokines such as TNFα and IL-12 that participate in inflammation and wound healing.

As mentioned earlier, we have described a distinct pathway of macrophage activation involving the synergy between TLRs and A2ARs (Leibovich et al. 2002). The ligation of TLR4 by its agonist LPS and of A2ARs by either specific A2AR agonist 2'-[p-92-carboxyethyl)-phenylethyl amino]-NECA (CGS21680) or non-specific adenosine receptor agonist 5'-N-ethyl-carboxamido-adenosine (NECA), switches macrophage activation from an inflammatory M1 to an anti-inflammatory and angiogenic M2-like (M2d) phenotype. This switch is characterized by a decrease in expression of the inflammatory cytokine TNFα and an increase in the angiogenic cytokine VEGF. In addition to TLR4, we have also shown in mice that agonists of TLR2, TLR7 and TLR9, but not TLR3 and TLR5 synergize with A2AR agonists and adenosine to up-regulate VEGF and simultaneously strongly down-regulate TNFα expression (Pinhal-Enfield et al. 2003). We reported that this "angiogenic switch" of macrophages requires A2ARs, since mice lacking A2ARs did not display the characteristic increase in VEGF production and decrease in TNFα production after treatment with LPS and NECA. Furthermore, co-stimulation with CGS21680 and LPS resulted in an increase in VEGF production to a similar extent as NECA (Leibovich et al. 2002). In addition, we found that the up-regulation of VEGF by ligation of TLRs and A2ARs is transcriptionally regulated by the hypoxia response element (HRE) in the VEGF promoter (Ramanathan et al. 2007). Recently, we have shown that LPS induces HIF-1α at the transcriptional level in an NF-κB-dependent manner. A key event in the angiogenic switch involves the LPS-dependent upregulation of the expression of A2ARs. Resting macrophages express low levels of A2ARs, but in response to LPS they strongly express A2ARs in a sustained manner (Murphree et al. 2005; Ramanathan et al. 2007). This induction of A2ARs is NF-κB-dependent and is upregulated by TLR agonists. Co-stimulation with LPS and A2AR agonists (NECA or CGS21680) triggers a pathway in murine peritoneal macrophages that leads to the stabilization of HIF-1α at the mRNA and protein levels, thus providing a mechanism for the synergistic effects of these two pathways on macrophage VEGF expression (Ramanathan et al. 2009). We have also reported that the "angiogenic switch" involving TLRs and A2ARs critically requires MyD88, IRAK4 and TRAF6 signaling (Macedo et al. 2007). Macrophages from mice lacking MyD88 or IRAK4 display impaired wound healing and do not respond to TLR and A2AR agonist-mediated VEGF production. Additionally, suppression of TRAF6 by siRNA in the murine macrophage cell line RAW264.7 also resulted in inhibition of the synergistic increase in VEGF and decrease in TNF production characteristic of the angiogenic switch. Furthermore, phosopholipase C-β2 (PLCβ2) is downregulated by LPS treatment in macrophages at the post-transcriptional level. This downregulation is critically associated with the upregulation of VEGF induced by co-stimulation with LPS and A2AR agonists (Grinberg et al. 2009).

The processes of wound healing and tumorigenesis involve comparable cellular process, such as cell proliferation, migration and survival that are controlled by both inflammation and angiogenesis. We propose therefore that targeting the synergy between TLRs and adenosine receptors, and thus modulating the inflammatory-angiogenic axis of macrophages, could be an important therapeutic approach to cancer therapy.

6   Role of Adenosine and Adenosine Receptors in Tumor Growth

One of the first indications that adenosine may play a role in tumor promotion was based on the well-defined role of adenosine in angiogenesis, immunosuppression, cytoprotection and growth promotion, all of which are essential tumor-promoting mechanisms (Spychala 2000). On further investigation into the tumor microenvironment, adenosine levels were found to be greatly increased in solid tumors, especially within hypoxic regions. Adenosine exerts immunosuppressive effects within hypoxic areas of tumors to inhibit anti-tumor responses by immune cells such as T-cells, macrophages and NK cells. Adenosine receptors are expressed on both cancer and immune cells in the tumor microenvironment. In this section, we will describe tumor-promoting roles of adenosine and adenosine receptors in the tumor stroma and possible therapeutic strategies using adenosine receptor ligands. Table 2 summarizes data that addresses adenosine receptor signaling in cancer.

The counter-intuitive notion that cancer cells and immune cells co-exist in the tumor microenvironment in patients and mouse models has been referred to as the "Hellstrom Paradox" (Hellstrom et al. 1968). In the presence of adenosine, immune cells acquire a functionally different phenotype as opposed to immune cells under normal conditions. Recent studies implicate adenosine and adenosine receptors in a tumor-protective role, which provides clues to the mechanism by which cancer cells can evade the cytotoxic effects of immune cells. Ohta and colleagues demonstrated that genetic deletion of A2ARs in mice bearing immunogenic tumors from both CL8-1 melanoma and RMA T lymphoma resulted in a 60% survival rate compared to 0% survival of their wild-type (A2AR +/+) counterparts (Ohta et al. 2006). Additionally, antagonists of A2ARs enhanced CD8+ T-cell mediated inhibition of tumor growth. These results implicate A2ARs as a negative regulator of T-cell anti-tumor activity. The authors suggest that the mechanism by which the anti-tumor activity of T-cells is suppressed by adenosine is through the A2AR-mediated increase in intracellular cAMP levels, which diminishes IFN-γ production in T-cells and hinders cytotoxic effects (Ohta et al. 2006). Another example of the immunosuppressive and anti-inflammatory effects of adenosine is presented in the study by Nowak et al who showed that all four adenosine receptors are expressed on mouse invariant NKT (iNKT) cells also found in the tumor microenvironment, but only A2ARs are required for the secretion of IL-4 and IL-10 by these cells (Nowak et al. 2010). As mentioned in early sections, IL-4 and IL-10 are anti-inflammatory cytokines and the enhanced secretion of these cytokines by ligation of A2ARs supports the role of A2ARs in tumor protection against attack by immune cells.

Table 2   Adenosine receptor signaling in cancer

Adenosine receptor type	Cancer type	Role	Reference
A2AR	Melanoma	Protects tumors from cytotoxic immune responses.	(Ohta et al. 2006)
A3R	Hepatocellular Carcinoma	CF102, a selective A3R agonist induces apoptosis and inhibits tumor growth.	(Bar-Yehuda et al. 2008)
$A_{2B}R$	Lewis Lung Carcinoma	Promotes tumor growth by regulating VEGF levels.	(Ryzhov et al. 2008)
$A_{2B}R$ and A3R	Human Melanoma	$A_{2B}R$ regulates IL-8 protein levels and A3R regulates VEGF protein levels in response to chemotherapeutic drugs etoposide and VP-16.	(Merighi et al. 2009)
$A_{2B}R$	Colon Carcinoma	Promotes tumor cell growth and proliferation under hypoxic conditions	(Ma et al. 2010)
A3R	Melanoma	A3R agonists inhibit tumor growth through Wnt signaling pathway	(Fishman et al. 2002)

As tumors develop, they outgrow their blood supply, leaving areas of the tumor that are oxygen-deprived or hypoxic. Adenosine accumulates in these hypoxic regions and triggers mechanisms that support tumor promotion, such as angiogenesis. This phenomenon has been documented in various cancer types (Table 2). In a model proposed by Sitkovsky et al, the "hypoxia-adenosinergic" suppression of antitumor T-cells is mediated by elevated adenosine levels during hypoxia which bind to A2ARs and A2BRs on T-cells, triggering intracellular cAMP accumulation. Elevated cAMP suppresses T-Cell Receptor (TCR) signaling to inhibit IFN-γ production. T-cells also exhibit a concomitant increase in HIF-1α in response to hypoxia which also results in suppression of TCR-signaling and IFN-γ production (Sitkovsky et al. 2008). This model suggests that tumor hypoxia results in increased levels of adenosine and HIF-1α which act to change the function of T-cells from cytotoxic to tumor protective. Other groups have demonstrated the role of the A2BR in hypoxia-mediated immunosuppression. Kong et al demonstrated that A2BRs expressed on human microvascular endothelial cells (HMEC-1) contain a functional binding site for HIF-1α within the A2BR promoter suggesting a mechanism by which A2BRs are induced during hypoxia (Kong et al. 2006). Ma et al using a panel of human colorectal tissues and human colon carcinoma cell lines found that A2BR mRNA was expressed and induced during hypoxic conditions. Furthermore, colon carcinoma cell proliferation was inhibited using an A2BR –specific antagonist, suggesting that A2BRs play an important role in colon carcinoma cell proliferation during hypoxia (Ma et al. 2010).

Recent work has focused on the role of CD73 and CD39 cell surface enzymes involved in extracellular adenosine production in tumor progression and metastases. Significant tumor growth suppression of MC38 colon cancer, EG7 lymphoma, AT-3 mammary tumors and B16F10 melanoma was found in CD73-deficient mice. This suppression involved increased anti-tumor immunity by increased expression of CD8+ T cells in tumors and peripheral blood and increased IFN-γ production. CD73-deficient mice were also resistant to metastasis of B16F10 melanoma after intravenous injection. Additionally, the authors unexpectedly found that CD73 expression on non-hematopoietic cells such as endothelial cells was essential for promoting lung metastasis, which was independent of the immunosuppressive effects of CD73. Therefore CD73 may be targeted at different levels: on tumor cells, T cells and nonhematopoietic cells to induce anti-tumor effects (Stagg et al. 2011). Other groups have found that suppression of CD73 activity using inhibitors to prevent adenosine accumulation in the host can render cancer cells more susceptible to anti-tumor immune responses and inhibit tumor progression (Yegutkin et al. 2011). Clayton et al studied the role of exosomes, which are small vesicles secreted by cancer cells, in extracellular adenosine production. They found that exosomes from several diverse cancer cell types express both CD39 and CD73 enzymes which contribute to 20% of the total ATP-hydrolytic activity. This study suggests that exosomes may contribute to the immunosuppressive effects in the tumor microenvironment through CD39 and CD73 production of adenosine (Clayton et al. 2011).

One of the major contributors to angiogenesis and metastasis is the elevated expression and production of VEGF by cancer cells and by infiltrating immune cells. Several studies have linked adenosine receptors with VEGF modulation in cancer. Using an isograft model of Lewis lung carcinoma in A2BR knockout mice, Ryzhov et al showed that A2BR knockout mice exhibited a decrease in tumor growth and increase in survival rate compared to wild-type control mice. They attributed this difference to the decrease in VEGF production from tumor-infiltrating CD45+ immune cells in the A2BR knockout mice, suggesting that tumor cells sustain their growth and survival through A2BR-mediated VEGF production in host immune cells (Ryzhov et al. 2008). Adenosine, through the A3 receptor (A3R), was shown to increase HIF-1α and VEGF expression in human glioblastoma cells through the activation of p44/p42 and p38 mitogen-activated protein kinase signaling pathways (Merighi et al. 2006). Furthermore, in human melanoma cells treated with DNA damaging agents such as etoposide and doxorubicin under hypoxic conditions, A2AR antagonists impaired IL-8 production and A3R antagonists decreased VEGF secretion. Since hypoxic cancers are resistant to conventional chemotherapeutic agents, this study suggests a possible role of A2AR and A3R blockade for improving human melanoma cell response to chemotherapeutic agents through inhibition of tumor-promoting growth factors and cytokines such as VEGF and IL-8 (Merighi et al. 2009). Adenosine in human macrophages was shown to upregulate VEGF expression (Ernens et al. 2010), and as in murine macrophages, this effect was synergistically amplified by LPS (Leibovich et al. 2002). In mammary tumors depleted of macrophages, transgenic overexpression of VEGF in macrophages restored tumor progression through stimulation of tumor angiogenesis, leukocyte infiltration and tumor cell invasion (Lin et al. 2007). These studies highlight the importance of VEGF production by immune cells such as macrophages in tumor promotion.

Adenosine receptor ligands have been tested in clinical trials of several diseases such as cardiovascular disease, Parkinson's, arthritis and cancer (Jacobson and Gao 2006). Of all the adenosine receptor ligands, only the highly selective A3R agonist Cl-IB-MECA (2-Chloro-N6-(3-iodobenzyl) adenosine-5′-N-methyluronamide) or CF102 is currently being used in a phase I-II clinical trial for patients with advanced hepatocellular carcinoma (Can-Fite BioPharma, Clinicaltrials.gov identifier NCT00790218). This trial is based on the work by Bar-Yehuda and colleagues who found that treatment with CF102 in rats with hepatocellular carcinoma resulted in a dose-responsive inhibition in tumor growth via induction of apoptosis by de-regulation of NF-κB and Wnt signaling pathways (Bar-Yehuda et al. 2008). The canonical Wnt signaling pathway is active during tumorigenesis and embryogenesis, and is involved in cell proliferation and cell cycle regulation (Fishman et al. 2002; Lustig and Behrens 2003). In three different tumor models: 1) B16-F10 melanoma in C57Bl/J mice 2) xenograft model of HCT-116 human colon carcinoma in nude mice and 3) xenograft model of PC-3 prostate carcinoma in nude mice, CI-IB-MECA and another A3R agonist, IB-MECA (1-deoxy-1-[6-[[(3-iodophenyl)methyl]amino]-9H-purine-9-yl]-N-methyl-β-$_D$-ribofuranuronamide) when administered orally at low dosages inhibited tumor growth. Furthermore, a synergistic effect was seen when given in combination with conventional chemotherapy and a myeloprotective effect on normal cells was determined through the induction of G-CSF production (Fishman et al. 2002). These results emphasize the possible role of A3R agonists in the treatment of several cancer models by inhibition of tumor growth at low doses by itself, and by enhancing anti-cancer effects in combination with conventional chemotherapeutic agents and myeloprotection.

7 Genetic and Pharmacological Modification of Macrophage Activation in Cancer Therapy

As mentioned previously, macrophages are critical components of the tumor microenvironment and play an important role in all aspects of tumor promotion including proliferation, angiogenesis and metastasis. Several research groups have implemented genetic and pharmacological approaches to evaluate the role of macrophage activation/inactivation as a potential target for cancer therapy. Genetic studies that will be discussed in this section include macrophage depletion studies in the context of cancer therapy, and pharmacological studies will include the use of VEGF-targeted inhibitors and antibodies that potentiate antitumor T-cell responses.

Genetic studies using macrophage depletion strategies in murine cancer models have led to an increased understanding of the role of macrophages in tumor progression. Recent studies by Pollard et al elucidated a fundamental relationship between macrophages and poor prognosis. Genetic depletion of macrophages in a mouse PyMT model of human breast cancer resulted in a delay in tumor progression and inhibition of metastasis involving VEGF and EGF (Lin et al. 2001; Lin et al. 2006; Lin et al. 2007). In a macrophage depletion model, using CSF-1 null mice, the authors found that these mice did not develop high density vessel formation in their primary tumors caused by expression of Polyoma middle T oncoprotein. These mice displayed delayed tumor progression and reduced metastasis compared to their CSF-1 positive counterparts. Furthermore, restoration of macrophage infiltration by expression of CSF-1 in CSF-1 null mice increased vessel density, indicating that macrophages recruited to tumors play an important role in angiogenesis to enhance tumor cell progression and migration. Additionally, through the use of fluorescently labeled cells in tumors, Pollard and collaborators developed a model of crosstalk between macrophages and tumor cells through CSF-1 and EGF signaling respectively, and this signaling pathway recruits tumors cells to regions of vessels where macrophages are abundant, to promote tumor cell intravasation into the circulation (Wyckoff et al. 2007).

Other groups have used macrophage depletion strategies to evaluate the role of macrophages in tumorigenesis. Zeisberger et al studied the effects of macrophage depletion by clondronate encapsulated in liposomes (Clodrolip) in two mouse tumor models: murine F9 teratocarcinoma and human A673 rhabdomyosarcoma. Treatment with Clodrolip resulted in a decrease in blood vessel density in the tumor tissue and inhibition of tumor growth. These workers also tested a combination therapy with Clodrolip and a VEGF-neutralizing antibody, and found a stronger effect on tumor growth inhibition than with Clodrolip alone. The combination therapy also depleted both TAMs and tumor-associated dendritic cell (TADC) populations (Zeisberger et al. 2006). This study suggests a role for both TAMs and TADCs in tumor promotion and the use of both Clodrolip and angiogenesis inhibitors as effective cancer therapeutics. Zhang et al tested the effect of macrophage depletion and Sorafenib, an angiogenesis inhibitor that is a VEGFR tyrosine kinase inhibitor currently approved by the FDA for treatment in patients with renal cell carcinoma and hepatocellular carcinoma (Zhang et al. 2010). In this study, two human hepatocellular carcinoma xenograft mouse models and two methods of macrophage depletion (Clodrolip and zolendronic acid) were used. Sorafenib inhibited tumor growth and lung metastasis, but induced an increase in recruitment of F4/80+ and CD11b+ cells (characteristic of macrophages). Sorafenib also increased CSF-1 and VEGF in the tumor and mouse VEGF in the peripheral blood, as well as stromal-derived factor-1α and plasma colony-stimulating factor-1, suggesting an influx in macrophages and host-derived growth factors. Depletion of macrophages by Clodrolip and zolendronic acid in combination with Sorafenib inhibited tumor progression, angiogenesis and lung metastasis compared to the single treatment with Sorafenib. This study suggests that macrophage depletion enhances the therapeutic effect of Sorafenib in metastatic liver cancer models.

There are several mechanisms for VEGF-targeted therapy, including anti-angiogenic effects, antitumor effects, and immune modulation (Ellis and Hicklin 2008). Bevacizumab (Avastin, Genentech Inc, San Francisco, CA) is an anti-VEGF antibody currently approved by the FDA for patients with metastatic colorectal cancer and non-small cell lung cancer in combination with chemotherapy. Bevacizumab binds and neutralizes VEGFA to prevent its binding to VEGFR-1, 2 and associated receptors neuropilin (NP)1 and NP2. In a phase III clinical trial, Bevacizumab in combination with fluorouracil-based chemotherapy improved survival rate by 20.3 months in patients with metastatic colorectal cancer (Hurwitz et al. 2004). Although the mechanism for increased patient survival by Bevacizumab and chemotherapy is not well-defined, the authors suggest that the primary mechanism is through inhibition of tumor growth. Since VEGFRs are present on tumor cells, VEGF-targeted therapy could act directly on the tumor cells to produce anti-tumor effects. In vitro studies in human colon cancer cells expressing VEGFR1 treatment with VEGFR1 monoclonal antibody blocked tumor cell migration and invasion (Fan et al. 2005). Tumor-derived VEGF contributes to immune tolerance in the tumor microenvironment by blocking dendritic cell and T-cell differentiation and maturation (Gabrilovich et al. 1996; Ohm et al. 2003). Fricke et al tested the effect of VEGF-trap (Aventis Pharmaceuticals, Inc, Bridgewater, NJ) on dendritic cell function in a phase I clinical trial in patients with refractory solid tumors (Fricke et al. 2007). VEGF-trap is a fusion protein consisting of the extracellular domains of human VEGFR1 and VEGFR2 coupled with the Fc region of human IgG1. This fusion protein binds to all isoforms of VEGFA and placental growth factors 1 and 2 to inhibit binding to their corresponding receptors. VEGF-trap treatment did not affect the total population of dendritic cells, myeloid subsets, myeloid-derived suppressor cells or regulatory T-cells, but increased the proportion of mature dendritic cells. This increase was not associated with improved immune function, suggesting the effect of VEGF-trap to enhance dendritic cell maturation was not sufficient to restore antitumor immune responses.

8 Conclusion and Perspectives

Macrophages play critical roles in processes such as inflammation, wound healing and tumorigenesis. The versatile nature of macrophages allows them to undergo major phenotypic changes in response to their environment. Within developing tumor microenvironment, macrophages respond to metabolic factors, including those induced by hypoxia and ischemia, as well as to factors derived from tumor cells. These factors drive macrophages towards an angiogenic M2 phenotype that is pro-angiogenic and pro-tumorigenic. As described in this review, the elevated expression of VEGF by tumor cells and infiltrating macrophages is critical in augmenting tumor angiogenesis and metastasis. While induction of the canonical "alternatively activated" M2 phenotype requires stimulation through the IL-4Rα, M2-like phenotypes can be induced by a variety of independent pathways. We have found that there is an M2-like phenotype that we have designated as "M2d" that is induced by signaling through TLRs and adenosine receptors, in a synergistic manner. Agonists of TLRs include both PAMPs and DAMPs, and macrophages respond to TLR agonists by up-regulating their expression of adenosine receptors, in particular A2ARs and A2BRs. Adenosine is a retaliatory metabolite produced in response to ischemia. M2d macrophages produce high levels of VEGF and IL-10, and express a wound healing, angiogenic phenotype. Tumors thus hijack normal macrophages to establish their own growth by inducing the M2d phenotype. Further studies on the signaling pathways involved in the induction of the M2d phenotype should provide insights into potential therapeutic approaches to modulating macrophage responses to enable them to combat rather than promote tumor growth.

9 Acknowledgements

This work is supported by grants from the US Public Health Services National Institutes of Health (NIH) (RO1-GM068636 and NIH training grant T32-HL069752).

REFERENCES

Aherne, C. M., E. M. Kewley and H. K. Eltzschig (2010). "The resurgence of A2B adenosine receptor signaling." Biochim et Biophysi Acta 1808(5): 1329-1339.

Auchampach, J. A. (2007). "Adenosine receptors and angiogenesis." Circulation Res 101(11): 1075-1077.

Azenshtein, E., G. Luboshits, S. Shina, E. Neumark, D. Shahbazian, M. Weil, N. Wigler, I. Keydar and A. Ben-Baruch (2002). "The CC chemokine RANTES in breast carcinoma progression: regulation of expression and potential mechanisms of promalignant activity." Cancer Res 62(4): 1093-1102.

Balkwill, F. (2004). "Cancer and the chemokine network." Nat. Rev. Cancer 4(7): 540-550.

Balkwill, F. and A. Mantovani (2001). "Inflammation and cancer: back to Virchow?" Lancet 357(9255): 539-545.

Balkwill, F. and A. Mantovani (2010). "Cancer and inflammation: implications for pharmacology and therapeutics." Clin Pharmacology and Therapeutics 87(4): 401-406.

Bar-Yehuda, S., S. M. Stemmer, L. Madi, D. Castel, A. Ochaion, S. Cohen, F. Barer, A. Zabutti, G. Perez-Liz, L. Del Valle and P. Fishman (2008). "The A3 adenosine receptor agonist CF102 induces apoptosis of hepatocellular carcinoma via de-regulation of the Wnt and NF-kappaB signal transduction pathways." Int J of Oncol 33(2): 287-295.

Ben-Baruch, A. (2006). "Inflammation-associated immune suppression in cancer: the roles played by cytokines, chemokines and additional mediators." Seminars in Cancer Biol 16(1): 38-52.

Beutler, B. (2000). "Tlr4: central component of the sole mammalian LPS sensor." Curr Op in Immunol 12(1): 20-26.

Biswas, S. K., L. Gangi, S. Paul, T. Schioppa, A. Saccani, M. Sironi, B. Bottazzi, A. Doni, B. Vincenzo, F. Pasqualini, L. Vago, M. Nebuloni, A. Mantovani and A. Sica (2006). "A distinct and unique transcriptional program expressed by tumor-associated macrophages (defective NF-kappaB and enhanced IRF-3/STAT1 activation)." Blood 107(5): 2112-2122.

Biswas, S. K., A. Sica and C. E. Lewis (2008). "Plasticity of macrophage function during tumor progression: regulation by distinct molecular mechanisms." J Immunol 180(4): 2011-2017.

Bjorkbacka, H., K. A. Fitzgerald, F. Huet, X. Li, J. A. Gregory, M. A. Lee, C. M. Ordija, N. E. Dowley, D. T. Golenbock and M. W. Freeman (2004). "The induction of macrophage gene expression by LPS predominantly utilizes Myd88-independent signaling cascades." Physiol Genomics 19(3): 319-330.

Burke, B., N. Tang, K. P. Corke, D. Tazzyman, K. Ameri, M. Wells and C. E. Lewis (2002). "Expression of HIF-1alpha by human macrophages: implications for the use of macrophages in hypoxia-regulated cancer gene therapy." J Pathol 196(2): 204-212.

Burnstock, G. (1972). "Purinergic nerves." Pharmacological Reviews 24(3): 509-581.

Burnstock, G. (1980). "Purinergic nerves and receptors." Progress Biochem Pharmacol 16: 141-154.

Burnstock, G., G. Campbell, D. Satchell and A. Smythe (1970). "Evidence that adenosine triphosphate or a related nucleotide is the transmitter substance released by non-adrenergic inhibitory nerves in the gut." British J Pharmacol 40(4): 668-688.

Chen, J. J., P. L. Yao, A. Yuan, T. M. Hong, C. T. Shun, M. L. Kuo, Y. C. Lee and P. C. Yang (2003). "Up-regulation of tumor interleukin-8 expression by infiltrating macrophages: its correlation with tumor angiogenesis and patient survival in non-small cell lung cancer." Clin Cancer Res 9(2): 729-737.

Classen, A., J. Lloberas and A. Celada (2009). "Macrophage activation: classical versus alternative." Methods Mol Biol 531: 29-43.

Clayton, A., S. Al-Taei, J. Webber, M. D. Mason and Z. Tabi (2011). "Cancer exosomes express CD39 and CD73, which suppress T cells through adenosine production." J Immunol 187(2): 676-683.

Coffelt, S. B., R. Hughes and C. E. Lewis (2009). "Tumor-associated macrophages: effectors of angiogenesis and tumor progression." Bioch Biophy Acta 1796(1): 11-18.

Coussens, L. M., C. L. Tinkle, D. Hanahan and Z. Werb (2000). "MMP-9 supplied by bone marrow-derived cells contributes to skin carcinogenesis." Cell 103(3): 481-490.

Cristalli, G., S. Costanzi, C. Lambertucci, G. Lupidi, S. Vittori, R. Volpini and E. Camaioni (2001). "Adenosine deaminase: functional implications and different classes of inhibitors." Medicinal Res Rev 21(2): 105-128.

Cronstein, B. N. (1994). "Adenosine, an endogenous anti-inflammatory agent." J Applied Physiol 76(1): 5-13.

Crowther, M., N. J. Brown, E. T. Bishop and C. E. Lewis (2001). "Microenvironmental influence on macrophage regulation of angiogenesis in wounds and malignant tumors." J Leukocyte Biol 70(4): 478-490.

Daley, J. M., S. K. Brancato, A. A. Thomay, J. S. Reichner and J. E. Albina (2010). "The phenotype of murine wound macrophages." J Leukocyte Biol 87(1): 59-67.

Deussen, A. (2000). "Metabolic flux rates of adenosine in the heart." Naunyn Schmiedebergs Arch Pharmacol 362(4-5): 351-363.

Domachowske, J. B., C. A. Bonville, A. J. Easton and H. F. Rosenberg (2002). "Differential expression of proinflammatory cytokine genes in vivo in response to pathogenic and nonpathogenic pneumovirus infections." J Infectious Dis 186(1): 8-14.

Dowdall, M. J. (1978). "Adenine nucleotides in cholinergic transmission: presynaptic aspects." J Physiol 74(5): 497-501.

Drury, A. N. and A. Szent-Gyorgyi (1929). "The physiological activity of adenine compounds with especial reference to their action upon the mammalian heart." J Physiol 68(3): 213-237.

Dvorak, H. F. (1986). "Tumors: wounds that do not heal. Similarities between tumor stroma generation and wound healing." New England J Med 315(26): 1650-1659.

Egeblad, M. and Z. Werb (2002). "New functions for the matrix metalloproteinases in cancer progression." Nat Rev Cancer 2(3): 161-174.

Ehrt, S., D. Schnappinger, S. Bekiranov, J. Drenkow, S. Shi, T. R. Gingeras, T. Gaasterland, G. Schoolnik and C. Nathan (2001). "Reprogramming of the macrophage transcriptome in response to interferon-gamma and Mycobacterium tuberculosis: signaling roles of nitric oxide synthase-2 and phagocyte oxidase." J Exp Med 194(8): 1123-1140.

Elgert, K. D., D. G. Alleva and D. W. Mullins (1998). "Tumor-induced immune dysfunction: the macrophage connection." J Leukocyte Biol 64(3): 275-290.

Ellis, L. M. and D. J. Hicklin (2008). "VEGF-targeted therapy: mechanisms of anti-tumour activity." Nat Rev Cancer 8(8): 579-591.

Eltzschig, H. K., L. F. Thompson, J. Karhausen, R. J. Cotta, J. C. Ibla, S. C. Robson and S. P. Colgan (2004). "Endogenous adenosine produced during hypoxia attenuates neutrophil accumulation: coordination by extracellular nucleotide metabolism." Blood 104(13): 3986-3992.

Enfield, G. P. and S. J. Leibovich (2011). Macrophage Heterogeneity and Wound Healing. Advances in Wound Care, Mary Ann Liebert, Inc. 2: 89-95.

Ernens, I., F. Leonard, M. Vausort, M. Rolland-Turner, Y. Devaux and D. R. Wagner (2010). "Adenosine up-regulates vascular endothelial growth factor in human macrophages." Biochem Biophys Res Comm 392(3): 351-356.

Eubank, T. D., M. Galloway, C. M. Montague, W. J. Waldman and C. B. Marsh (2003). "M-CSF induces vascular endothelial growth factor production and angiogenic activity from human monocytes." J Immunol 171(5): 2637-2643.

Fan, F., J. S. Wey, M. F. McCarty, A. Belcheva, W. Liu, T. W. Bauer, R. J. Somcio, Y. Wu, A. Hooper, D. J. Hicklin and L. M. Ellis (2005). "Expression and function of vascular endothelial growth factor receptor-1 on human colorectal cancer cells." Oncogene 24(16): 2647-2653.

Feoktistov, I., I. Biaggioni and B. N. Cronstein (2009). "Adenosine receptors in wound healing, fibrosis and angiogenesis." Handb Exp Pharmacol(193): 383-397.

Fishman, P., S. Bar-Yehuda, L. Madi and I. Cohn (2002). "A3 adenosine receptor as a target for cancer therapy." Anti-Cancer Drugs 13(5): 437-443.

Fishman, P., L. Madi, S. Bar-Yehuda, F. Barer, L. Del Valle and K. Khalili (2002). "Evidence for involvement of Wnt signaling pathway in IB-MECA mediated suppression of melanoma cells." Oncogene 21(25): 4060-4064.

Fredholm, B. B. (2007). "Adenosine, an endogenous distress signal, modulates tissue damage and repair." Cell Death Differentiation 14(7): 1315-1323.

Fredholm, B. B., I. J. AP, K. A. Jacobson, K. N. Klotz and J. Linden (2001). "International Union of Pharmacology. XXV. Nomenclature and classification of adenosine receptors." Pharmacol Rev 53(4): 527-552.

Fredholm, B. B., J. F. Chen, S. A. Masino and J. M. Vaugeois (2005). "Actions of adenosine at its receptors in the CNS: insights from knockouts and drugs." Ann Rev Pharmacol Toxicol 45: 385-412.

Fricke, I., N. Mirza, J. Dupont, C. Lockhart, A. Jackson, J. H. Lee, J. A. Sosman and D. I. Gabrilovich (2007). "Vascular endothelial growth factor-trap overcomes defects in dendritic cell differentiation but does not improve antigen-specific immune responses." Clin Cancer Res 13(16): 4840-4848.

Gabrilovich, D. I., H. L. Chen, K. R. Girgis, H. T. Cunningham, G. M. Meny, S. Nadaf, D. Kavanaugh and D. P. Carbone (1996). "Production of vascular endothelial growth factor by human tumors inhibits the functional maturation of dendritic cells." Nature Med 2(10): 1096-1103.

Gessi, S., S. Merighi, V. Sacchetto, C. Simioni and P. A. Borea (2011). "Adenosine receptors and cancer." Biochim Biophy Acta 1808(5): 1400-1412.

Giavazzi, R., A. Garofalo, M. R. Bani, M. Abbate, P. Ghezzi, D. Boraschi, A. Mantovani and E. Dejana (1990). "Interleukin 1-induced augmentation of experimental metastases from a human melanoma in nude mice." Cancer Res 50(15): 4771-4775.

Goerdt, S., O. Politz, K. Schledzewski, R. Birk, A. Gratchev, P. Guillot, N. Hakiy, C. D. Klemke, E. Dippel, V. Kodelja and C. E. Orfanos (1999). "Alternative versus classical activation of macrophages." Pathobiol 67(5-6): 222-226.

Gordon, S. (2003). "Alternative activation of macrophages." Nat Rev Immunol 3(1): 23-35.

Gordon, S. and F. O. Martinez (2010). "Alternative activation of macrophages: mechanism and functions." Immunity 32(5): 593-604.

Gordon, S. and P. R. Taylor (2005). "Monocyte and macrophage heterogeneity." Nat Rev Immunol 5(12): 953-964.

Goswami, S., E. Sahai, J. B. Wyckoff, M. Cammer, D. Cox, F. J. Pixley, E. R. Stanley, J. E. Segall and J. S. Condeelis (2005). "Macrophages promote the invasion of breast carcinoma cells via a colony-stimulating factor-1/epidermal growth factor paracrine loop." Cancer Res 65(12): 5278-5283.

Grinberg, S., G. Hasko, D. Wu and S. J. Leibovich (2009). "Suppression of PLCbeta2 by endotoxin plays a role in the adenosine A(2A) receptor-mediated switch of macrophages from an inflammatory to an angiogenic phenotype." Am J Pathol 175(6): 2439-2453.

Grivennikov, S. I., F. R. Greten and M. Karin (2010). "Immunity, inflammation, and cancer." Cell 140(6): 883-899.

Hagemann, T. and T. Lawrence (2009). "Investigating macrophage and malignant cell interactions in vitro." Methods Mol Biol 512: 325-332.

Hagemann, T., S. C. Robinson, M. Schulz, L. Trumper, F. R. Balkwill and C. Binder (2004). "Enhanced invasiveness of breast cancer cell lines upon co-cultivation with macrophages is due to TNF-alpha dependent up-regulation of matrix metalloproteases." Carcinogenesis 25(8): 1543-1549.

Haghnegahdar, H., J. Du, D. Wang, R. M. Strieter, M. D. Burdick, L. B. Nanney, N. Cardwell, J. Luan, R. Shattuck-Brandt and A. Richmond (2000). "The tumorigenic and angiogenic effects of MGSA/GRO proteins in melanoma." J Leukocyte Biol 67(1): 53-62.

Hasko, G. and B. N. Cronstein (2004). "Adenosine: an endogenous regulator of innate immunity." Trends Immunol 25(1): 33-39.

Hasko, G., E. A. Deitch, C. Szabo, Z. H. Nemeth and E. S. Vizi (2002). "Adenosine: a potential mediator of immunosuppression in multiple organ failure." Curr Opin Pharmacol 2(4): 440-444.

Hasko, G., D. G. Kuhel, J. F. Chen, M. A. Schwarzschild, E. A. Deitch, J. G. Mabley, A. Marton and C. Szabo (2000). "Adenosine inhibits IL-12 and TNF-[alpha] production via adenosine A2a receptor-dependent and independent mechanisms." FASEB J 14(13): 2065-2074.

Hasko, G., P. Pacher, E. A. Deitch and E. S. Vizi (2007). "Shaping of monocyte and macrophage function by adenosine receptors." Pharmacol Therapeut 113(2): 264-275.

Hasko, G., M. V. Sitkovsky and C. Szabo (2004). "Immunomodulatory and neuroprotective effects of inosine." Trends Pharmacol Sci 25(3): 152-157.

Hasko, G., C. Szabo, Z. H. Nemeth, V. Kvetan, S. M. Pastores and E. S. Vizi (1996). "Adenosine receptor agonists differentially regulate IL-10, TNF-alpha, and nitric oxide production in RAW 264.7 macrophages and in endotoxemic mice." J Immunol 157(10): 4634-4640.

Hellstrom, I., K. E. Hellstrom and G. E. Pierce (1968). "In vitro studies of immune reactions against autochthonous and syngeneic mouse tumors induced by methylcholanthrene and plastic discs." Int J Cancer 3(4): 467-482.

Hiratsuka, S., K. Nakamura, S. Iwai, M. Murakami, T. Itoh, H. Kijima, J. M. Shipley, R. M. Senior and M. Shibuya (2002). "MMP9 induction by vascular endothelial growth factor receptor-1 is involved in lung-specific metastasis." Cancer Cell 2(4): 289-300.

Hiratsuka, S., A. Watanabe, Y. Sakurai, S. Akashi-Takamura, S. Ishibashi, K. Miyake, M. Shibuya, S. Akira, H. Aburatani and Y. Maru (2008). "The S100A8-serum amyloid A3-TLR4 paracrine cascade establishes a pre-metastatic phase." Nat Cell Biol 10(11): 1349-1355.

Hotchkiss, K. A., A. W. Ashton, R. S. Klein, M. L. Lenzi, G. H. Zhu and E. L. Schwartz (2003). "Mechanisms by which tumor cells and monocytes expressing the angiogenic factor thymidine phosphorylase mediate human endothelial cell migration." Cancer Res 63(2): 527-533.

Hurwitz, H., L. Fehrenbacher, W. Novotny, T. Cartwright, J. Hainsworth, W. Heim, J. Berlin, A. Baron, S. Griffing, E. Holmgren, N. Ferrara, G. Fyfe, B. Rogers, R. Ross and F. Kabbinavar (2004). "Bevacizumab plus irinotecan, fluorouracil, and leucovorin for metastatic colorectal cancer." New England J Med 350(23): 2335-2342.

Hyde, R. J., C. E. Cass, J. D. Young and S. A. Baldwin (2001). "The ENT family of eukaryote nucleoside and nucleobase transporters: recent advances in the investigation of structure/function relationships and the identification of novel isoforms." Mol Membrane Biol 18(1): 53-63.

Jacobson, K. A. and Z. G. Gao (2006). "Adenosine receptors as therapeutic targets." Nat Rev Drug Discov 5(3): 247-264.

James, S. and P. J. Richardson (1993). "Production of adenosine from extracellular ATP at the striatal cholinergic synapse." J Neurochem 60(1): 219-227.

Jeannin, P., D. Duluc and Y. Delneste (2011). "IL-6 and leukemia-inhibitory factor are involved in the generation of tumor-associated macrophage: regulation by IFN-gamma." Immunotherapy 3(4 Suppl): 23-26.

Jenner, P., A. Mori, R. Hauser, M. Morelli, B. B. Fredholm and J. F. Chen (2009). "Adenosine, adenosine A 2A antagonists, and Parkinson's disease." Parkinsonism Relat Disord 15(6): 406-413.

Kaczmarek, E., K. Koziak, J. Sevigny, J. B. Siegel, J. Anrather, A. R. Beaudoin, F. H. Bach and S. C. Robson (1996). "Identification and characterization of CD39/vascular ATP diphosphohydrolase." J Biol Chem 271(51): 33116-33122.

Kluwe, J., A. Mencin and R. F. Schwabe (2009). "Toll-like receptors, wound healing, and carcinogenesis." J Mol Med (Berl) 87(2): 125-138.

Kong, T., K. A. Westerman, M. Faigle, H. K. Eltzschig and S. P. Colgan (2006). "HIF-dependent induction of adenosine A2B receptor in hypoxia." FASEB J 20(13): 2242-2250.

Kortylewski, M., M. Kujawski, T. Wang, S. Wei, S. Zhang, S. Pilon-Thomas, G. Niu, H. Kay, J. Mule, W. G. Kerr, R. Jove, D. Pardoll and H. Yu (2005). "Inhibiting Stat3 signaling in the hematopoietic system elicits multicomponent antitumor immunity." Nat Med 11(12): 1314-1321.

Krausgruber, T., K. Blazek, T. Smallie, S. Alzabin, H. Lockstone, N. Sahgal, T. Hussell, M. Feldmann and I. A. Udalova (2011). "IRF5 promotes inflammatory macrophage polarization and T_H1-T_H17 responses." Nat Immunol 12(3): 231-238.

Kumar, V. and A. Sharma (2009). "Adenosine: an endogenous modulator of innate immune system with therapeutic potential." Eur J Pharmacol 616(1-3): 7-15.

Kusmartsev, S. and D. I. Gabrilovich (2005). "STAT1 signaling regulates tumor-associated macrophage-mediated T cell deletion." J Immunol 174(8): 4880-4891.

Lee, C.-C., K.-J. Liu and T.-S. Huang (2006). "Tumor-Associated Macrophage: Its Role in Tumor Angiogenesis." J Cancer Mol 2(4): 135-140.

Leek, R. D., C. E. Lewis, R. Whitehouse, M. Greenall, J. Clarke and A. L. Harris (1996). "Association of macrophage infiltration with angiogenesis and prognosis in invasive breast carcinoma." Cancer Res 56(20): 4625-4629.

Leibovich, S. J. (1996). The Role of Cytokines and Growth Factors in Tumor Angiogenesis. Human Cytokines: Their Role in Disease and Therapy. B. Aggarwal, Puri, R, Blackwell Scientific: 539-564.

Leibovich, S. J. (2007). Regulation of Macrophage-Dependent Angiogenesis by Adenosine and Toll-Like Receptors. Adenosine Receptors: Therapeutic Aspects for Inflammatory and Immune Diseases. G. Haskó, Cronstein, B N, Szabó, C. Boca Raton, CRC Press Taylor & Francis Group: 325-346.

Leibovich, S. J., J. F. Chen, G. Pinhal-Enfield, P. C. Belem, G. Elson, A. Rosania, M. Ramanathan, C. Montesinos, M. Jacobson, M. A. Schwarzschild, J. S. Fink and B. Cronstein (2002). "Synergistic up-regulation of vascular endothelial growth factor expression in murine macrophages by adenosine A(2A) receptor agonists and endotoxin." Am J Pathol 160(6): 2231-2244.

Lewis, C. E. and J. W. Pollard (2006). "Distinct role of macrophages in different tumor microenvironments." Cancer Res 66(2): 605-612.

Lewis, J. S., R. J. Landers, J. C. Underwood, A. L. Harris and C. E. Lewis (2000). "Expression of vascular endothelial growth factor by macrophages is up-regulated in poorly vascularized areas of breast carcinomas." J Pathol 192(2): 150-158.

Lin, E. Y., J. F. Li, G. Bricard, W. Wang, Y. Deng, R. Sellers, S. A. Porcelli and J. W. Pollard (2007). "Vascular endothelial growth factor restores delayed tumor progression in tumors depleted of macrophages." Mol Oncol 1(3): 288-302.

Lin, E. Y., J. F. Li, L. Gnatovskiy, Y. Deng, L. Zhu, D. A. Grzesik, H. Qian, X. N. Xue and J. W. Pollard (2006). "Macrophages regulate the angiogenic switch in a mouse model of breast cancer." Cancer Res 66(23): 11238-11246.

Lin, E. Y., A. V. Nguyen, R. G. Russell and J. W. Pollard (2001). "Colony-stimulating factor 1 promotes progression of mammary tumors to malignancy." J Exp Med 193(6): 727-740.

Lin, E. Y. and J. W. Pollard (2007). "Tumor-associated macrophages press the angiogenic switch in breast cancer." Cancer Res 67(11): 5064-5066.

Linden, J. (2001). "Molecular approach to adenosine receptors: receptor-mediated mechanisms of tissue protection." Ann Rev Pharmacol Toxicol 41: 775-787.

Linden, J., T. Thai, H. Figler, X. Jin and A. S. Robeva (1999). "Characterization of human A(2B) adenosine receptors: radioligand binding, western blotting, and coupling to G(q) in human embryonic kidney 293 cells and HMC-1 mast cells." Mol Pharmacol 56(4): 705-713.

Loberg, R. D., C. Ying, M. Craig, L. Yan, L. A. Snyder and K. J. Pienta (2007). "CCL2 as an important mediator of prostate cancer growth in vivo through the regulation of macrophage infiltration." Neoplasia 9(7): 556-562.

Londos, C., D. M. Cooper and J. Wolff (1980). "Subclasses of external adenosine receptors." Proc Nat Acad Sci USA 77(5): 2551-2554.

Lustig, B. and J. Behrens (2003). "The Wnt signaling pathway and its role in tumor development." J Cancer Res Clin Oncol 129(4): 199-221.

Ma, D. F., T. Kondo, T. Nakazawa, D. F. Niu, K. Mochizuki, T. Kawasaki, T. Yamane and R. Katoh (2010). "Hypoxia-inducible adenosine A2B receptor modulates proliferation of colon carcinoma cells." Human Pathology 41(11): 1550-1557.

Ma, J., T. Chen, J. Mandelin, A. Ceponis, N. E. Miller, M. Hukkanen, G. F. Ma and Y. T. Konttinen (2003). "Regulation of macrophage activation." Cell Mol Life Sci 60(11): 2334-2346.

Macedo, L., G. Pinhal-Enfield, V. Alshits, G. Elson, B. N. Cronstein and S. J. Leibovich (2007). "Wound healing is impaired in MyD88-deficient mice: a role for MyD88 in the regulation of wound healing by adenosine A2A receptors." Am J Pathol 171(6): 1774-1788.

Mantovani, A. (2006). "Macrophage diversity and polarization: in vivo veritas." Blood 108(2): 408-409.

Mantovani, A., P. Allavena and A. Sica (2004). "Tumour-associated macrophages as a prototypic type II polarised phagocyte population: role in tumour progression." Eur J Cancer 40(11): 1660-1667.

Mantovani, A. and A. Sica (2010). "Macrophages, innate immunity and cancer: balance, tolerance, and diversity." Curr Op Immunol 22(2): 231-237.

Mantovani, A., A. Sica, S. Sozzani, P. Allavena, A. Vecchi and M. Locati (2004). "The chemokine system in diverse forms of macrophage activation and polarization." Trends Immunol 25(12): 677-686.

Mantovani, A., S. Sozzani, M. Locati, P. Allavena and A. Sica (2002). "Macrophage polarization: tumor-associated macrophages as a paradigm for polarized M2 mononuclear phagocytes." Trends Immunol 23(11): 549-555.

Martinez, F. O., L. Helming and S. Gordon (2009). "Alternative activation of macrophages: an immunologic functional perspective." Ann Rev Immunol 27: 451-483.

Merighi, S., A. Benini, P. Mirandola, S. Gessi, K. Varani, E. Leung, S. Maclennan and P. A. Borea (2006). "Adenosine modulates vascular endothelial growth factor expression via hypoxia-inducible factor-1 in human glioblastoma cells." Biochemical Pharmacology 72(1): 19-31.

Merighi, S., C. Simioni, S. Gessi, K. Varani, P. Mirandola, M. A. Tabrizi, P. G. Baraldi and P. A. Borea (2009). "A(2B) and A(3) adenosine receptors modulate vascular endothelial growth factor and interleukin-8 expression in human melanoma cells treated with etoposide and doxorubicin." Neoplasia 11(10): 1064-1073.

Mosser, D. M. and J. P. Edwards (2008). "Exploring the full spectrum of macrophage activation." Nat Rev Immunol 8(12): 958-969.

Murphree, L. J., G. W. Sullivan, M. A. Marshall and J. Linden (2005). "Lipopolysaccharide rapidly modifies adenosine receptor transcripts in murine and human macrophages: role of NF-kappaB in A(2A) adenosine receptor induction." Bioch J 391(Pt 3): 575-580.

Nathan, C. (1991). "Mechanisms and modulation of macrophage activation." Behring Institute Mitteilungen (88): 200-207.

Newby, A. C., Y. Worku and C. A. Holmquist (1985). "Adenosine formation. Evidence for a direct biochemical link with energy metabolism." Ad Myocardiol 6: 273-284.

Novitskiy, S. V., S. Ryzhov, R. Zaynagetdinov, A. E. Goldstein, Y. Huang, O. Y. Tikhomirov, M. R. Blackburn, I. Biaggioni, D. P. Carbone, I. Feoktistov and M. M. Dikov (2008). "Adenosine receptors in regulation of dendritic cell differentiation and function." Blood 112(5): 1822-1831.

Nowak, M., L. Lynch, S. Yue, A. Ohta, M. Sitkovsky, S. P. Balk and M. A. Exley (2010). "The A2aR adenosine receptor controls cytokine production in iNKT cells." Eur J Immunol 40(3): 682-687.

Ohm, J. E., D. I. Gabrilovich, G. D. Sempowski, E. Kisseleva, K. S. Parman, S. Nadaf and D. P. Carbone (2003). "VEGF inhibits T-cell development and may contribute to tumor-induced immune suppression." Blood 101(12): 4878-4886.

Ohta, A., E. Gorelik, S. J. Prasad, F. Ronchese, D. Lukashev, M. K. Wong, X. Huang, S. Caldwell, K. Liu, P. Smith, J. F. Chen, E. K. Jackson, S. Apasov, S. Abrams and M. Sitkovsky (2006). "A2A adenosine receptor protects tumors from antitumor T cells." Proc Nat Acad Sci USA 103(35): 13132-13137.

Ono, M., H. Torisu, J. Fukushi, A. Nishie and M. Kuwano (1999). "Biological implications of macrophage infiltration in human tumor angiogenesis." Cancer Chemother Pharmacol 43 Suppl: S69-71.

Pastor-Anglada, M., F. J. Casado, R. Valdes, J. Mata, J. Garcia-Manteiga and M. Molina (2001). "Complex regulation of nucleoside transporter expression in epithelial and immune system cells." Mol Membrane Biol 18(1): 81-85.

Piccinini, A. M. and K. S. Midwood (2010). "DAMPening inflammation by modulating TLR signalling." Mediators of Inflammation. 2010: 1-21.

Pinhal-Enfield, G., M. Ramanathan, G. Hasko, S. N. Vogel, A. L. Salzman, G. J. Boons and S. J. Leibovich (2003). "An angiogenic switch in macrophages involving synergy between Toll-like receptors 2, 4, 7, and 9 and adenosine A(2A) receptors." Am J Pathol 163(2): 711-721.

Pollard, J. W. (2004). "Tumour-educated macrophages promote tumour progression and metastasis." Nat Rev Cancer 4(1): 71-78.

Polverini, P. J. and S. J. Leibovich (1987). "Effect of macrophage depletion on growth and neovascularization of hamster buccal pouch carcinomas." J Oral Pathol 16(9): 436-441.

Qian, B., Y. Deng, J. H. Im, R. J. Muschel, Y. Zou, J. Li, R. A. Lang and J. W. Pollard (2009). "A distinct macrophage population mediates metastatic breast cancer cell extravasation, establishment and growth." PLoS ONE 4(8): e6562.

Ralevic, V. and G. Burnstock (1998). "Receptors for purines and pyrimidines." Pharmacol Rev 50(3): 413-492.

Ramanathan, M., W. Luo, B. Csoka, G. Hasko, D. Lukashev, M. V. Sitkovsky and S. J. Leibovich (2009). "Differential regulation of HIF-1alpha isoforms in murine macrophages by TLR4 and adenosine A(2A) receptor agonists." J Leukocyte Biol 86(3): 681-689.

Ramanathan, M., G. Pinhal-Enfield, I. Hao and S. J. Leibovich (2007). "Synergistic up-regulation of vascular endothelial growth factor (VEGF) expression in macrophages by adenosine A2A receptor agonists and endotoxin involves transcriptional regulation via the hypoxia response element in the VEGF promoter." Mol Biol Cell 18(1): 14-23.

Resta, R., Y. Yamashita and L. F. Thompson (1998). "Ecto-enzyme and signaling functions of lymphocyte CD73." Immunol Rev 161: 95-109.

Ritzel, M. W., A. M. Ng, S. Y. Yao, K. Graham, S. K. Loewen, K. M. Smith, R. J. Hyde, E. Karpinski, C. E. Cass, S. A. Baldwin and J. D. Young (2001). "Recent molecular advances in studies of the concentrative Na+-dependent nucleoside transporter

(CNT) family: identification and characterization of novel human and mouse proteins (hCNT3 and mCNT3) broadly selective for purine and pyrimidine nucleosides (system cib)." Mol Membrane Biol 18(1): 65-72.

Rubartelli, A. and M. T. Lotze (2007). "Inside, outside, upside down: damage-associated molecular-pattern molecules (DAMPs) and redox." Trends Immunol 28(10): 429-436.

Ryzhov, S., S. V. Novitskiy, R. Zaynagetdinov, A. E. Goldstein, D. P. Carbone, I. Biaggioni, M. M. Dikov and I. Feoktistov (2008). "Host A(2B) adenosine receptors promote carcinoma growth." Neoplasia 10(9): 987-995.

Saijo, Y., M. Tanaka, M. Miki, K. Usui, T. Suzuki, M. Maemondo, X. Hong, R. Tazawa, T. Kikuchi, K. Matsushima and T. Nukiwa (2002). "Proinflammatory cytokine IL-1 beta promotes tumor growth of Lewis lung carcinoma by induction of angiogenic factors: in vivo analysis of tumor-stromal interaction." J Immunol 169(1): 469-475.

Sangaletti, S., A. Stoppacciaro, C. Guiducci, M. R. Torrisi and M. P. Colombo (2003). "Leukocyte, rather than tumor-produced SPARC, determines stroma and collagen type IV deposition in mammary carcinoma." J Ex Med 198(10): 1475-1485.

Satoh, T., O. Takeuchi, A. Vandenbon, K. Yasuda, Y. Tanaka, Y. Kumagai, T. Miyake, K. Matsushita, T. Okazaki, T. Saitoh, K. Honma, T. Matsuyama, K. Yui, T. Tsujimura, D. M. Standley, K. Nakanishi, K. Nakai and S. Akira (2010). "The Jmjd3-Irf4 axis regulates M2 macrophage polarization and host responses against helminth infection." Nat Immunol 11(10): 936-944.

Sattin, A. and T. W. Rall (1970). "The effect of adenosine and adenine nucleotides on the cyclic adenosine 3',5'-phosphate content of guinea pig cerebral cortex slices." Mol Pharmacol 6(1): 13-23.

Sharma, S., M. Stolina, Y. Lin, B. Gardner, P. W. Miller, M. Kronenberg and S. M. Dubinett (1999). "T cell-derived IL-10 promotes lung cancer growth by suppressing both T cell and APC function." J Immunol 163(9): 5020-5028.

Shih, J.-Y., A. Yuan, J. J.-W. Chen and P.-C. Yang (2006). "Tumor-Associated Macrophage: Its Role in Cancer Invasion and Metastasis." J Cancer Mol 2(3): 101-106.

Sica, A., P. Allavena and A. Mantovani (2008). "Cancer related inflammation: the macrophage connection." Cancer Lett 267(2): 204-215.

Sims, G. P., D. C. Rowe, S. T. Rietdijk, R. Herbst and A. J. Coyle (2010). "HMGB1 and RAGE in inflammation and cancer." Ann Rev Immunol 28: 367-388.

Sinha, P., V. K. Clements and S. Ostrand-Rosenberg (2005). "Reduction of myeloid-derived suppressor cells and induction of M1 macrophages facilitate the rejection of established metastatic disease." J Immunol 174(2): 636-645.

Sitkovsky, M. V., J. Kjaergaard, D. Lukashev and A. Ohta (2008). "Hypoxia-adenosinergic immunosuppression: tumor protection by T regulatory cells and cancerous tissue hypoxia." Clin Cancer Res 14(19): 5947-5952.

Siveen, K. S. and G. Kuttan (2009). "Role of macrophages in tumour progression." Immunol Lett 123(2): 97-102.

So, E. Y. and T. Ouchi (2010). "The application of Toll like receptors for cancer therapy." Int J Biol Sci 6(7): 675-681.

Sperlagh, B., G. Hasko, Z. Nemeth and E. S. Vizi (1998). "ATP released by LPS increases nitric oxide production in raw 264.7 macrophage cell line via P2Z/P2X7 receptors." Neurochem Int 33(3): 209-215.

Spychala, J. (2000). "Tumor-promoting functions of adenosine." Pharmacology and Therapeut 87(2-3): 161-173.

Stagg, J., U. Divisekera, H. Duret, T. Sparwasser, M. W. Teng, P. K. Darcy and M. J. Smyth (2011). "CD73-deficient mice have increased antitumor immunity and are resistant to experimental metastasis." Cancer Res 71(8): 2892-2900.

Stagg, J. and M. J. Smyth (2010). "Extracellular adenosine triphosphate and adenosine in cancer." Oncogene 29(39): 5346-5358.

Standiford, T. J., R. Kuick, U. Bhan, J. Chen, M. Newstead and V. G. Keshamouni (2011). "TGF-beta-induced IRAK-M expression in tumor-associated macrophages regulates lung tumor growth." Oncogene 30(21): 2475-2484.

Stein, M., S. Keshav, N. Harris and S. Gordon (1992). "Interleukin 4 potently enhances murine macrophage mannose receptor activity: a marker of alternative immunologic macrophage activation." J Ex Med 176(1): 287-292.

Taylor, P. R., L. Martinez-Pomares, M. Stacey, H. H. Lin, G. D. Brown and S. Gordon (2005). "Macrophage receptors and immune recognition." Ann Rev Immunol 23: 901-944.

Traut, T. W. (1994). "Physiological concentrations of purines and pyrimidines." Mol Cell Biochem 140(1): 1-22.

Ueno, T., M. Toi, H. Saji, M. Muta, H. Bando, K. Kuroi, M. Koike, H. Inadera and K. Matsushima (2000). "Significance of macrophage chemoattractant protein-1 in macrophage recruitment, angiogenesis, and survival in human breast cancer." Clin Cancer Res 6(8): 3282-3289.

Uutela, M., M. Wirzenius, K. Paavonen, I. Rajantie, Y. He, T. Karpanen, M. Lohela, H. Wiig, P. Salven, K. Pajusola, U. Eriksson and K. Alitalo (2004). "PDGF-D induces macrophage recruitment, increased interstitial pressure, and blood vessel maturation during angiogenesis." Blood 104(10): 3198-3204.

van Calker, D., M. Muller and B. Hamprecht (1979). "Adenosine regulates via two different types of receptors, the accumulation of cyclic AMP in cultured brain cells." J Neurochem 33(5): 999-1005.

Van Ginderachter, J. A., K. Movahedi, G. Hassanzadeh Ghassabeh, S. Meerschaut, A. Beschin, G. Raes and P. De Baetselier (2006). "Classical and alternative activation of mononuclear phagocytes: picking the best of both worlds for tumor promotion." Immunobiol 211(6-8): 487-501.

Varney, M. L., K. J. Olsen, R. L. Mosley, C. D. Bucana, J. E. Talmadge and R. K. Singh (2002). "Monocyte/macrophage recruitment, activation and differentiation modulate interleukin-8 production: a paracrine role of tumor-associated macrophages in tumor angiogenesis." In Vivo 16(6): 471-477.

von Lubitz, D. K. (1999). "Adenosine and cerebral ischemia: therapeutic future or death of a brave concept?" Eur J Pharmacol 371(1): 85-102.

Wardas, J. (2008). "Potential role of adenosine A2A receptors in the treatment of schizophrenia." Frontiers in Biosci 13: 4071-4096.

Wyckoff, J. B., Y. Wang, E. Y. Lin, J. F. Li, S. Goswami, E. R. Stanley, J. E. Segall, J. W. Pollard and J. Condeelis (2007). "Direct visualization of macrophage-assisted tumor cell intravasation in mammary tumors." Cancer Res 67(6): 2649-2656.

Yang, H. Z., B. Cui, H. Z. Liu, S. Mi, J. Yan, H. M. Yan, F. Hua, H. Lin, W. F. Cai, W. J. Xie, X. X. Lv, X. X. Wang, B. M. Xin, Q. M. Zhan and Z. W. Hu (2009). "Blocking TLR2 activity attenuates pulmonary metastases of tumor." PLoS One 4(8): e6520.

Yegutkin, G. G., F. Marttila-Ichihara, M. Karikoski, J. Niemela, J. P. Laurila, K. Elima, S. Jalkanen and M. Salmi (2011). "Altered purinergic signaling in CD73-deficient mice inhibits tumor progression." Eur J Immunol 41(5): 1231-1241.

Zeisberger, S. M., B. Odermatt, C. Marty, A. H. Zehnder-Fjallman, K. Ballmer-Hofer and R. A. Schwendener (2006). "Clodronate-liposome-mediated depletion of tumour-associated macrophages: a new and highly effective antiangiogenic therapy approach." Br J Cancer 95(3): 272-281.

Zhang, W., X. D. Zhu, H. C. Sun, Y. Q. Xiong, P. Y. Zhuang, H. X. Xu, L. Q. Kong, L. Wang, W. Z. Wu and Z. Y. Tang (2010). "Depletion of tumor-associated macrophages enhances the effect of sorafenib in metastatic liver cancer models by antimetastatic and antiangiogenic effects." Clin Cancer Res 16(13): 3420-3430.

Zhou, Q., R. Q. Peng, X. J. Wu, Q. Xia, J. H. Hou, Y. Ding, Q. M. Zhou, X. Zhang, Z. Z. Pang, D. S. Wan, Y. X. Zeng and X. S. Zhang (2010). "The density of macrophages in the invasive front is inversely correlated to liver metastasis in colon cancer." J Transl Med 8: 13.

Zimmermann, H. (2000). "Extracellular metabolism of ATP and other nucleotides." Naunyn-Schmiedebergs Arch Pharmacol 362(4-5): 299-309.

Paracrine and Autocrine Control of Endothelial Cell Function by Inflammatory and Bioactive Lipid Mediators:
Relevance for Tumor Angiogenesis

Charlotte Farrar, Elena Garonna, Alexandra McSloy and
Caroline PD Wheeler-Jones*

Comparative Biomedical Sciences, Royal Veterinary College
Royal College Street, London NW1 0TU, U.K.

CHAPTER OUTLINE

▶ Introduction
▶ Pathological angiogenesis and inflammation
▶ Tumor endothelial cells
▶ Inflammatory mediators influence the angiogenic functions of endothelial cells
▶ Biologically active lipids and EC function in angiogenesis
▶ Molecular regulation of pro-inflammatory gene expression in endothelial cells
▶ Anti-inflammatory drugs as regulators of tumor angiogenesis: focus on NSAIDs
▶ Conclusions and perspectives
▶ References

ABSTRACT

Angiogenesis is a term used to describe the sprouting of new capillaries from the endothelial cells of post-capillary venules. Angiogenesis in the post-natal animal is important for tissue repair, wound healing and tumor growth. Under normal conditions the microvascular endothelium is maintained in a quiescent state by the balanced expression and production of a range of pro- and anti-angiogenic

Correspondence Author: Comparative Biomedical Sciences, Royal Veterinary College, University of London, Royal College Street, London NW1 0TU. Email: cwheeler@rvc.ac.uk

factors. During physiological and pathological angiogenesis endothelial cells (ECs) respond to diverse paracrine inflammatory signals that support their increased proliferation and migration, and trigger formation of a capillary network. Increasing evidence indicates that ECs concomitantly up-regulate their own production of pro-angiogenic (angiocrine) mediators, which facilitates their angiogenic functions through autocrine mechanisms. In the tumor environment, enhanced bi-directional cross-talk between tumor cells and ECs results in persistent EC activation which contributes to excessive and deregulated angiogenesis, driving enhanced tumor growth and metastases. Here, we provide a general overview of the association between inflammation and angiogenesis and critically review the evidence supporting a pivotal role for bioactive lipids in regulating the pro-angiogenic functions of ECs through paracrine and autocrine signaling.

Key words: endothelial cells, cyclo-oxygenase, NSAIDs, angiogenesis, inflammation, eicosanoids, prostaglandins

1   Introduction

Inflammation is a normal physiological response to injury or infection, which results in changes in pro-inflammatory gene expression and consequent production of inflammatory mediators. Chronic, unresolved inflammation is a prominent feature of the tumor microenvironment. Cytokines, chemokines, growth factors, eicosanoids and numerous other biologically active molecules produced by the cellular components of the tumor act through autocrine and paracrine mechanisms to influence tumor cell-stromal cell interactions. The products generated as a result of these interactions ultimately create an environment which facilitates tumor cell and endothelial cell (EC) proliferation and survival, enhances angiogenesis, and influences the potential for metastases. In this article we provide a general overview of the involvement of inflammatory mediators in EC functions relevant for angiogenesis and where possible compare their documented roles in normal versus tumor endothelial cells (TECs). We focus particularly on the paracrine and autocrine actions of prostaglandins (PG) and eicosanoids generated through cyclo-oxygenase (COX) activities and also highlight the EC-directed actions of non-steroidal anti-inflammatory drugs (NSAIDs) that suppress the pathways driving persistent EC activation.

2   Pathological Angiogenesis and Inflammation

Angiogenesis is the term used to describe the formation of new blood vessels from pre-existing vasculature and is a sequence of finely controlled events that characterises a range of physiological and pathological conditions. Physiological angiogenesis is required for embryonic development, wound healing and ocular maturation, whereas excessive pathological angiogenesis is associated with chronic inflammatory conditions, including tumor maintenance (reviewed in (Carmeliet and Jain 2000; Carmeliet 2003; Ferrara and Kerbel 2005; Chung et al. 2010).

The major players in sprouting angiogenesis are endothelial cells (ECs) that become activated by a diverse range of local stimuli to adopt a pro-angiogenic phenotype and undergo a program of extensive reorganisation culminating in the formation of new, stable blood vessels (see Chung et al. 2010). The response of ECs to an angiogenic stimulus includes several components. In the initial stages activated ECs retract, loosen contact with adjacent cells and produce matrix metalloproteinases (MMPs), which assist with digesting the extracellular matrix (ECM)-rich basal lamina. Specialized polarized and motile ECs (tip cells) sprout filopodia and migrate towards the angiogenic stimulus, thus breaching the surrounding tissue and initiating the angiogenic process. Proliferation then occurs, temporary tubes are formed and vacuoles within adjoining ECs coalesce to generate the vessel lumen. Ultimately, the vessels are matured and stabilized through recruitment of perivascular cells (e.g. pericytes), forming a functional microvascular network (Jain and Booth 2003; Chung et al. 2010). Pathological angiogenesis relies on many of the same processes as those involved in physiological angiogenesis, with ischemia and hypoxic conditions initiating a cascade of coordinated cellular functions, resulting in neo-angiogenesis. Thus, the ability of ECs to proliferate, migrate and form capillary-like tubes are key components of the angiogenic cascade and all of these pro-angiogenic properties are commonly exploited in *in vitro* assays designed to evaluate EC angiogenic potential (Garonna et al. 2011).

There is now evidence that angiogenesis is coupled with inflammation under physiological and pathological conditions (Arroyo and Iruela-Arispe 2010). During wound healing, for instance, inflammation results in the generation of cytokines and growth factors that directly trigger the angiogenic program to facilitate repair processes, but homeostasis is restored when the vessels regress during the resolution phase of inflammation. However, because the tumor experiences persistent, unresolved inflammation tumor-associated ECs (TECs) are permanently exposed to a cocktail of inflammatory, pro-angiogenic mediators and the vascular networks generated are distinct from their normal counterparts, exhibiting disorganized branching, incomplete pericyte coverage, abnormal blood flow and enhanced permeability (see Weis 2008). In either setting the extent of angiogenesis depends upon the balance between production of pro-angiogenic versus anti-angiogenic factors and the stability of this so-called 'angiogenic switch' determines the initiation of angiogenesis. Perhaps not surprisingly, deregulation of the angiogenic switch has been shown in the early stages of tumor development, with tumors displaying a distinct shift towards a pro-angiogenic mediator profile early on in the disease (Raica et al. 2009).

3 Tumor Endothelial Cells

It is widely accepted that the molecular regulation of angiogenesis, whether physiological or pathological, is highly complex but it is increasingly acknowledged that tissue environments directly control resident EC phenotype and function, accounting for the marked heterogeneity between ECs isolated from different sources. The tumor microenvironment represents an extreme example of this relationship where TEC characteristics and behaviors depend upon paracrine and autocrine signals emanating from both tumor and stromal cell compartments. The relative expression of angiogenic factors is likely to be highly tumor-dependent (see Hanahan and Weinberg 2011) and thus, the characteristics of the tumor vasculature will vary. However, the tumor microenvironment directly influences TEC function, hence contributing to the deregulated angiogenic processes occurring in this setting. In tumors, excessive production of pro-angiogenic factors leads to EC proliferation, migration and tube formation, a chronically activated deregulated angiogenic cascade, and a change in EC phenotype. There are obvious difficulties associated with isolating pure populations of EC from tumors but it is now known that tumor-derived ECs display genetic instability (Li and Harris 2007) and differentially express a range of TEC-specific markers in addition to the pan endothelial markers also expressed by normal EC populations (NECs) (Nagy et al. 2009). In keeping with their constant exposure to angiogenic factors it is also evident that the gene expression signature of TECs resembles that of normal angiogenic ECs rather than quiescent cells (Olsen et al. 2004); (van Beijnum et al. 2006); (Bussolati et al. 2010; Ruoslahti et al. 2010). From a functional perspective, it has been shown that TECs exhibit higher proliferation and migration rates and greater resistance to serum-starvation induced apoptosis when compared to standard models (human umbilical vein ECs (HUVEC) and microvascular ECs) (Bussolati et al. 2003; Du et al. 2008). However, unlike NECs they formed capillary-like structures *in vitro* that persisted for up to 7 days in the absence of serum, demonstrating their overt angiogenic phenotype (Bussolati et al. 2003).

Despite these significant advances it still remains a challenge to isolate and characterise TECs. As a result, it must be pointed out that the vast majority of studies addressing questions relating to the regulation of angiogenesis, as well as to tumor-associated angiogenesis, have been carried out using normal ECs. The following discussions will examine the involvement of bioactive lipid mediators in the angiogenic functions of ECs, and where possible will highlight studies that have attempted to delineate pathways and functional responses in TECs.

4 Inflammatory Mediators Influence the Angiogenic Functions of Endothelial Cells

Vascularization of tumors is essential for their growth and viability and was originally thought to be directly controlled by tumor cells. However, it is now evident that several inflammatory cell types and ECs themselves also play pivotal roles in orchestrating tumor angiogenesis through multi-directional paracrine and autocrine cross-talk mechanisms involving cytokines, growth factors and eicosanoids. Tumor cells produce numerous mediators that attract leukocytes (macrophages, neutrophils, dendritic cells, eosinophils, mast cells, lymphocytes) to the tumor site, and there is now ample evidence to suggest that inflammatory cell recruitment can also be modulated by EC-derived factors. To date, the inflammatory cell attributed the most importance in influencing tumor angiogenesis is the macrophage, whose secretome is up-regulated in environments which actively encourage neo-angiogenesis (e.g hypoxia).

The involvement of TAMs in regulating tumor vascularization is undisputed and is well documented (see Lewis and Pollard 2006; Ono 2008). These cells are recruited from a pool of circulating monocytes by chemokines (including MCP-1, CSF-1 and VEGF) released by tumor and stromal cells as well as by ECs, and are converted to a pro-angiogenic M2 phenotype by factors present within the tumor. These cells subsequently synthesize and release an array of mediators that affect all stages of the angiogenic process (see Table 1). Importantly, these M2-polarized TAMs not only express a range of pro-inflammatory and pro-angiogenic cytokines that directly influence EC function but also exhibit high levels of cyclo-oxygenase-2 (COX-2) enzyme expression coupled to synthesis of prostaglandin E2 (PGE_2), a major regulator of the pro-angiogenic phenotype of ECs.

The chronically inflamed tumor microenvironment and the heightened opportunity for cell-cell communication strongly suggest that the angiogenic functions of TECs are likely to be regulated by inflammatory signals operating through both autocrine and paracrine mechanisms. As mentioned previously there is a paucity of information relating specifically to TEC function so the majority of available evidence documenting mediator production by and influence on ECs is derived from studies employing normal ECs. NECs are capable of synthesizing and/or responding to numerous mediators including cytokines (e.g. TNFα, IL-1α/β, TGFβ, IL-8), growth factors (e.g. VEGF, PDGF), extracellular proteases (which release pro-angiogenic factors sequestered within the ECM) and metabolites of the polyunsaturated fatty acid (PUFA) arachidonate (AA) (e.g. eicosanoids); all of these mediators have been implicated in the direct or indirect regulation of tumor angiogenesis.

TAMs and other stromal cells produce a range of cytokines and chemokines that directly influence EC behavior and angiogenesis (see Table 1). TNFα and IL-1α/β, for example, are both capable of promoting pro-angiogenic mediator synthesis (e.g. VEGF, bFGF, IL-8) in cancer cells, as well as ECs themselves, suggesting involvement of autocrine and paracrine actions (Torisu et al. 2000; Nakao et al. 2005). Notably, there is still debate concerning the anti-/pro-cancer actions of TNFα since low doses are considered to support neo-vascularization whereas high doses are cytotoxic to cancer cells, and presumably TECs (see Bertazza and Mocellin 2010). Nevertheless, antibody neutralization of TNFα, and IL-1, has been reported to abrogate TAM-induced angiogenesis (Torisu et

Table 1 Inflammatory mediators: actions on NECs versus TECs

Inflammatory mediator	Secreted by macrophages	Secreted by NECs	Effect on NECs	Effect on TECs	Tumor angiogenesis
Chemokines and Cytokines					
IL-1	✓	✓	Proliferation Migration Differentiation (Matsuo et al. 2009)	Undefined	Promotes (Voronov et al. 2003)
IL-6	✓	✓	Proliferation (Zhu et al. 2011) Migration (Liu et al. 2010) Differentiation (Zhu et al. 2011)	Undefined	Promotes (Saidi et al. 2009)
IL-10	✓	✓	Undefined	Undefined	Promotes (Martinez et al. 2008) Reduces prostate (van Moorselaar and Voest 2002)
TGFβ	✓	✓	Undefined (Holderfield and Hughes 2008)	Undefined	Promotes (Niu and Xia 2009)
TNFα	✓	✓	Proliferation (Bouchentouf et al. 2010) Migration (Valacchi et al. 2011) Differentiation (Aoki et al. 2010)	Undefined	Promotes (Sasi et al. 2011)
IL-8	✓	✓	Proliferation (Li et al. 2003) Migration (Salcedo 2003) Differentiation (Li et al. 2003)	Undefined	Promotes (Brat et al. 2005)
Growth Factors					
VEGF	✓	✓	Proliferation Migration Differentiation (Garonna et al. 2011)	Proliferation (Yakes et al. 2011) Migration (Pupo et al. 2011) Differentiation (Grau et al. 2011)	Promotes (Taeger et al. 2011; Yadav et al. 2011)
bFGF	✓	✓	Proliferation Migration Differentiation (Boosani et al. 2010)	Undefined	Promotes (Dempke and Zippel 2010; Taeger et al. 2011)
PDGF	✓	✓	Undefined	Undefined	Promotes (Guo et al. 2003; Taeger et al. 2011)

Prostanoids					
PGE_2	✓	Proliferation Migration (Finetti et al. 2008; Finetti et al. 2009) Differentiation (Zhang and Daaka 2011)	✓	Undefined	Promotes (Ghosh et al. 2010)
PGF_{2a}	✓	Proliferation (Keightley et al. 2010)	✓	Undefined	Promotion through increasing VEGF production in tumor cells (Sales et al. 2005)
PGD_2	✓	Undefined	✓	Undefined	Reduces (Murata et al. 2012)
15-deoxy-$\Delta(12,14)$-PGJ_2	✓	Inihibits proliferation (Xin et al. 1999) Apoptosis (Bishop-Bailey and Hla 1999)	✓	Undefined	Reduces (Wang and DuBois 2010)
PGI_2	✓	Undefined	✓	Undefined	Reduces (Pradono et al. 2002)
TxA_2	✓	Migration Differentiation (Nie et al. 2000)	✓	Undefined	Promotes (Pradono et al. 2002)

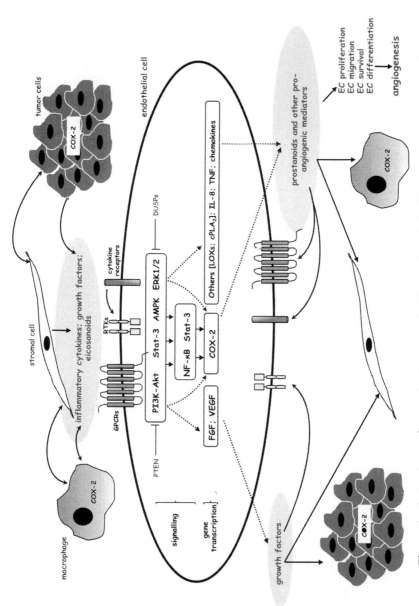

Figure 1 Autocrine and paracrine loops controlling endothelial cell function in angiogenesis. *Color image of this figure appears in the color plate section at the end of the book.*

al. 2000) and TNFα also up-regulates IL-6 and MMPs in ECs (Mawatari et al. 1989; Okamura et al. 1991), both of which would facilitate autocrine angiogenic responses. Interestingly, Ligresti and colleagues have recently demonstrated in an aortic model of angiogenesis that TNFα exerts pro-angiogenic actions primarily through macrophage VEGF production but that TNFα also directly influences angiogenesis through mechanisms that do not involve macrophage-derived mediators (Ligresti et al. 2011). IL-1α is also directly pro-angiogenic in NECs *in vitro* (HUVEC) and colon cancer cell-derived IL-1α up-regulates EC migration and tubulogenesis (Matsuo et al. 2009). In contrast, other studies using an aortic ring sprouting assay have suggested that macrophage-derived IL-1β regulates EC migration and proliferation indirectly through VEGF production and has no ability to directly modify angiogenic responses in ECs (Carmi et al. 2009). Thus, while cellular cross-talk mechanisms are likely to be engaged differentially by different cytokines, current evidence suggests that TNFα and IL-1α/β are both capable of influencing the angiogenic behaviors of ECs and hence contributing to pathological and physiological angiogenesis. Importantly, these cytokines can also enhance COX-2 expression in cancer cells and ECs, a property which is shared by numerous other factors enriched within the tumor microenvironment (e.g. VEGF; IL-8; PGE$_2$).

5 Biologically Active Lipids and EC Function in Angiogenesis

Unraveling how bioactive lipid mediators derived from AA and other fatty acids influence the angiogenic functions of ECs is a major current research focus that has relevance for understanding tumorigenesis. This has been driven largely by the recognition that a vast number of tumors are characterized by over-expression of COX-2 and heightened levels of COX-2-derived mediators and other inflammatory factors.

5.1 COX Enzymes and Prostanoid Synthesis

COX enzymes, in conjunction with terminal prostanoid synthases, catalyse the production of prostanoids from AA (Hata and Breyer 2004; Wheeler-Jones et al. 2009). This process tends to be regulated in a cell-specific manner, with differentiated cells producing a predominant prostanoid (Smith et al. 1991; Wheeler-Jones et al. 2009). COX-1 is constitutively expressed in most cells whereas COX-2 is an early response gene (Wu et al. 2003) located on a separate chromosome to COX-1 and is an inducible COX enzyme in many tissues, including the vasculature, thus accounting for its role in inflammation and associated pathologies. The constitutive expression of COX-1 versus the inducible nature of COX-2 expression implies distinct roles in prostanoid production in different phases of cell activation. Accordingly, studies have shown that in un-stimulated

HUVECs, which express COX-1 and low basal levels of COX-2, prostacyclin (PGI_2) is the predominant prostanoid present whereas following COX-2 induction, a sustained increase in prostanoid production is evident with COX-2 preferentially metabolizing AA to PGE_2 and PGI_2 in ECs (Brock et al. 1999; Caughey et al. 2001). Thus, COX-2 activity effectively maintains prostanoid production under inflammatory conditions and has a similar role in ECs exposed to pro-angiogenic mediators, many of which strongly induce COX-2 expression (Syeda et al. 2006; Clarkin et al. 2008; Garonna et al. 2011).

Prostanoids produced downstream of COX/terminal synthases activities bind and activate cell surface seven transmembrane domain G-protein coupled receptors (GPCRs) (Hata and Breyer 2004). ECs express the appropriate receptor repertoire that permits activation by PGI_2 and PGE_2 and other products, suggesting the potential for these mediators to modify EC signaling through autocrine mechanisms (Wheeler-Jones et al. 2009). In support of this hypothesis, exogenous PGE_2 and PGI_2 activate a range of signaling pathways and transcription factors to promote angiogenic and pro-survival effects in ECs and enhance COX-2 expression, thus providing an autocrine feed-forward mechanism for maintaining angiogenic function (Hata and Breyer 2004; Clarkin et al. 2008a; Clarkin et al. 2008b; Wheeler-Jones et al. 2009). In addition, PGI_2 is known to act intracellularly through peroxisome proliferator-activated receptors (PPARs), a nuclear receptor family of ligand-inducible transcription factors (Youssef and Badr 2011) suggesting that multiple mechanisms are likely to mediate the angiogenic functions of these prostanoids.

5.2   COX-2 Activity and Angiogenesis

Genetic, biochemical and clinical evidence now supports an important role for COX-2 in tumor angiogenesis. COX-2 is highly expressed in many cancers, cancer development in $COX-2^{-/-}$ mice is reduced, and clinical blockade of COX-2 activity can be therapeutically effective in cancer treatment. COX-2 over-expression also increases the production of numerous pro-angiogenic factors, including VEGF (Kaidi et al. 2006). Thus, prominent involvement of COX-2 in tumor initiation, progression and promotion has been established in many studies (Liu et al. 2001; Müller-Decker et al. 2002; Ghosh et al. 2010). However, although the level of COX-2 in tumors is generally elevated, the expression of this enzyme is heterogeneous among malignant, stromal and inflammatory cell types. In tumor ECs, on the other hand, relatively homogeneous expression of COX-2 has been demonstrated (Muraki et al. 2012). COX-2 expression in tumors suppresses apoptosis and promotes angiogenesis, effects mediated through the actions of its major downstream products, PGE_2, PGI_2 and thromboxane A_2. Accordingly, there is evidence that these mediators regulate VEGF production, enhance EC survival, and promote vascular sprouting, migration and tubulogenesis (Ghosh et al. 2010).

5.2.1 PGE$_2$

PGE$_2$ is produced by several cell types within the tumor microenvironment and consequently is considered to be one of the most important mediators in cancer promotion (Greenhough et al. 2009; Ghosh et al. 2010). PGE$_2$ is now known to exert a range of effects on NECs in culture, enhancing the activities of pro-angiogenic signaling pathways and promoting changes in proliferation, motility and differentiation. These effects have invariably been ascribed to activation of four prostanoid receptors (EP1-4) belonging to the GPCR family (see Wheeler-Jones et al. 2009). It is also conceivable that PGE$_2$ could also indirectly activate PPARs since exogenous PGE$_2$ can promote endothelial PGI$_2$ synthesis (Clarkin et al. 2008a). PGE$_2$ is thought to promote tumor progression through multiple mechanisms, including inhibition of cancer cell apoptosis, adhesion, migration, MMP production and stimulation of angiogenesis (Ghosh et al. 2010). Although it is likely that PGE$_2$ plays a role in the majority of cancers, thus implicating EPs, the EP receptor(s) responsible differ between tumors (Wang and DuBois 2010). With respect to ECs, however, recent data derived from studies in NECs have clearly demonstrated the overall importance of EP4 in the angiogenic behaviors of microvascular ECs; in these studies EP4 agonism induced tubulogenesis to the same extent as PGE$_2$, whereas EP1, EP2 and EP3 agonists were ineffective (Zhang and Daaka 2011). EP4 has also been shown to mediate PGE$_2$-induced synthesis of the pro-angiogenic cytokine, IL-8, by human ECs (Aso et al. 2012). However, despite a clear role for EP receptors in cancer progression and evidence of involvement in angiogenic responses in NECs, their specific roles in TECs and thus in tumor angiogenesis have not been elucidated.

In addition to the importance of COX-2 and EP receptors in defining the functional effects of PGE$_2$ in tumor angiogenesis, expression of the inducible downstream terminal synthase, membrane PGE synthase-1 (mPGES-1), is also critical and, accordingly, this enzyme is up-regulated in many cancers (reviewed in Murakami and Kudo 2006). In this respect, significant suppression of intestinal cancer growth has been reported in mPGES-1$^{-/-}$ mice, with tumors displaying a disorganized vasculature, strongly suggesting a role for mPGES-1, and thus PGE$_2$, in tumor angiogenesis (Nakanishi et al. 2008). Interestingly, recent *in vitro* studies have shown that formation of PGE$_2$ by NECs is PGES-independent, and that COX-derived PGH$_2$ is non-enzymatically degraded to PGE$_2$ in these cells (HUVEC). These studies also provided evidence that EC-derived PGH$_2$ can be metabolized to PGE$_2$ by co-cultured tumor cells in which mPGES-1 upregulation is evident (Salvado et al. 2009). This apparent inability of ECs to regulate PGE$_2$ production through increased PGES activity may be relevant for understanding the link between COX-2 over-expression and PGE$_2$ upregulation in cancers. It should also be noted that increased levels of PGE$_2$ within the tumor microenvironment could be facilitated by reduced expression of prostaglandin metabolizing enzymes, including 15-prostaglandin dehydrogenase (PGDH). COX-2 and PGDH are often reciprocally regulated in cancer, thus permitting unopposed expression of tumorigenic and pro-angiogenic prostanoids (Smartt et al. 2012). The functional roles, if any, of PG metabolizing enzymes in NECs or TECs await clarification.

5.2.2 PGI_2

As described above, PGI_2 acts as an agonist at the cell surface IP receptor but also has the ability to bind intracellularly to PPARs, of which there are 3 defined isoforms (α, β/δ and γ). In contrast to enhanced tumor expression of mPGES-1 and PGE_2, current evidence suggests that PGIS, PGI_2 and IP expression are generally low in tumors, potentially suggesting that re-instating this axis may be anti-tumorigenic. In keeping with this, tumors over-expressing PGIS have been shown to exhibit reduced vascularization which was reversed by administration of a PGIS inhibitor (Pradono et al. 2002). Lung-specific over-expression of PGIS decreased tumor incidence in smoke-exposed mice and these animals had significantly elevated PGI_2 levels (Keith et al. 2004). However, mice over-expressing PGIS, but lacking the IP receptor are still protected from tumor development. This effect was shown to be mediated through PPARγ activation (Nemenoff et al. 2008). Interestingly, Tennis et al have recently reported that the ability of iloprost to inhibit cancer cell proliferation through PPARγ activation requires Frizzled-9, a GPCR for Wnt-7a (Tennis et al. 2010). Thus, the PPAR-dependent mechanisms accounting for the anti-tumorigenic effects of iloprost are likely to be complex.

There is *in vitro* and *in vivo* evidence that the stable PGI_2 analogue iloprost exerts pro-angiogenic actions (Biscetti et al. 2009). The apparent anti-carcinogenic effects of PGI_2 analogues seem to be at odds with the ability of these compounds to promote angiogenesis in *in vivo* models and to increase the pro-angiogenic potential of NECs *in vitro* (Biscetti et al. 2009). However, it is conceivable that the latter relate more specifically to a role for PGI_2's actions in the processes of normal tissue repair, and are not relevant in the context of the tumor microenvironment which is dominated by preferential coupling of COX-2 to PGE_2 synthesis. In this regard, pro-repair functions of endothelial progenitor cells (proliferation and tubulogenesis) have been shown to depend upon a COX-1-PGI_2-PPARβ/δ pathway (He et al. 2008).

5.2.3   PPARβ/δ

The precise role of PPARβ/δ in the context of tumor progression and associated angiogenesis has not been fully defined. Muller-Brusselbach and co-workers reported impaired lung tumor growth in PPARβ/δ$^{-/-}$ mice and showed that this was due, in part, to diminished blood flow and abnormal microvascular structures formed with immature hyperplastic ECs. Restoration of PPARβ/δ expression led to maturation of microvascular structures within the tumor, suggesting a role for this receptor in maintaining the tumor vasculature (Muller-Brusselbach et al. 2007). Other studies suggest an inhibitory role for PPARβ/δ in tumor progression through inhibition of cancer cell differentiation (Yang et al. 2010), which would be expected to modify their cytokine secretory capacity. In contrast, PPARβ/δ expression has recently been implicated in the initiation of primary colorectal tumors, with subsequent loss of expression required for metastasis and progression (Yang et al. 2011). In keeping with a positive role for PPARβ/δ in tumor initiation and potentially in tumor angiogenesis, *in vitro* studies in colorectal cancer cell lines have demonstrated an ability of PPARβ/δ agonists, including PGI$_2$, to promote VEGF production (Rohrl et al. 2011).

The role of PPARβ/δ in TEC function and angiogenesis has not been investigated, although the abnormal microvasculature observed in tumors from PPARβ/δ$^{-/-}$ mice (Muller-Brusselbach et al. 2007) suggests a regulatory function in the process of tumor vascularization. There is, however, evidence from *in vitro* studies in NECs that PPARβ/δ can regulate their angiogenic potential. For example, the PPARβ/δ agonist GW501516 has been reported to increase proliferation and differentiation on Matrigel of the EAHy926 cell line, potentially through increased VEGF production, and to increase angiogenesis in *in vivo* assays (Piqueras et al. 2007). Recently, however, Meissner et al have reported that GW501516 and an additional PPARβ/δ agonist exert anti-angiogenic effects on HUVEC *in vitro*. These authors, in marked contrast to Piqueras and co-workers, showed that PPARβ/δ agonism suppressed EC migration and tubulogenesis and down-regulated VEGFR2 expression through inhibition of VEGFR2 promoter activity (Meissner et al. 2011). Thus, the importance of PPARβ/δ as a positive or negative regulator of NEC angiogenic functions remains unclear. Interestingly, in quiescent non-proliferating HUVEC PPARβ/δ agonists can promote the release of pro-inflammatory cytokines (Meissner et al. 2010) suggesting that PPARβ/δ function in ECs is likely to be influenced by cell cycle status. To date, no studies have yet defined the involvement of PPARβ/δ in TEC functions relevant for angiogenesis.

5.2.4 PPARα and PPARγ

The roles of PPARα and PPARγ in tumor growth and progression have been reviewed previously (Michalik et al. 2004; Krishnan et al. 2007; Sertznig et al. 2007). PPARα and PPARγ are negative regulators of inflammation and ligands for these receptors inhibit the growth of many cancers (Thuillier et al. 2000; Tanaka et al. 2001; Grabacka et al. 2004; Saidi et al. 2006; Panigrahy et al. 2008; Suchanek et al. 2002; Panigrahy et al. 2008). This inhibition is thought to reflect targeting of angiogenesis (Panigrahy et al. 2008) and in keeping with this studies in immortalised human dermal microvascular ECs have shown that PPARα agonists inhibit EC proliferation and migration and reduce angiogenesis both *in vivo* and *in vitro* (Varet et al. 2003). Similarly, the majority of studies have shown inhibitory properties of PPARα and γ ligands, suggesting anti-tumorigenic roles for these receptors, although their precise roles in cancer progression and tumor angiogenesis are still not entirely clear (reviewed in Gupta and Dubois 2002). Interestingly, as with

PPARβ/δ, there are conflicting reports concerning the actions of PPARγ ligands on NECs *in vitro*. For instance, PPARγ and -α agonists can promote angiogenic responses in NECs in culture and in *in vivo* models, most likely through enhanced VEGF production (Biscetti et al. 2008), whereas others have shown that VEGF-induced HUVEC migration and tubulogenesis is suppressed by PPARγ agonism and that this effect involves reduced VEGF-stimulated COX-2 expression (Scoditti et al. 2010). In support of an anti-angiogenic role, PPARγ ligands also down-regulate leukocyte VEGF production (Mattos et al. 2012) and were shown to reduce expression of mPGES-1 and PGE_2 and suppress the growth and vascularization of xenotransplanted ovarian tumors (Xin et al. 2007).

PPARα has also been proposed as a mediator of iloprost-stimulated angiogenic functions because iloprost failed to promote angiogenesis and upregulate VEGF in mice null for PPARα expression (Biscetti et al. 2009). In addition, a selective PPARα ligand (WY14643) increased inflammatory cytokine and VEGF synthesis in corneal epithelial cells which led the authors to suggest a potential role for PPARα in the pathological angiogenesis associated with ocular inflammation (Zhang and Ward 2010). Thus, individual PPARs can exert anti-or pro-angiogenic effects, depending on context. The relative levels of expression of individual PPARs and their activators within tumors, as well as variations in the cell types which express them, could account for the general lack of consensus on their roles in tumor angiogenesis. Further studies examining their significance as regulators of the angiogenic behaviors of both TECs and NECs are clearly warranted.

5.2.5   Thromboxane A_2

There is good evidence that TxA_2 and its terminal synthase (TxS) are both up-regulated in a number of cancers and that TxA_2 is able to stimulate cancer cell proliferation (Ermert et al. 2003) The importance of TxA_2 in tumor angiogenesis has been investigated through over-expression studies in mice. Animals over-expressing Tx synthase (TxS) showed increased tumor growth and died earlier than wild type mice and this effect was shown to reflect the actions of TxA_2 on tumor angiogenesis (Pradono et al. 2002). Human lung cancer cells over-expressing Tx receptor-α (TPα) induce greater tumor growth and vascularization in mice and a TPα agonist enhances cancer cell VEGF production (Wei et al. 2010), which would be expected to influence EC function through paracrine actions.

The role of thromboxane A_2 (TxA_2) in the angiogenic functions of NECs, or indeed TECs, is still not fully defined but the majority of studies indicate that this mediator is generally pro-angiogenic, in keeping with its reported ability to act as a positive regulator of tumor angiogenesis. For example, U44619, a stable TP receptor agonist, induced EC migration and tube formation *in vitro* and antagonism of the TP receptor reduced both VEGF- and bFGF-stimulated tube formation (Nie et al. 2000). Pharmacological compounds that interfere with TxA_2 synthesis and/or action also exert anti-angiogenic effects on NECs *in vitro*, principally through inhibition of

EC migration (de Leval et al. 2006). Additionally, TxA_2 has been reported to contribute to COX-2-induced EC migration and angiogenesis (Daniel et al. 1999), although this is not consistent with the observation that COX-2-inducing pro-angiogenic agonist IL-1 has no effect on TxS expression or TxA_2 generation in NECs (Caughey et al. 2001).

Several early studies demonstrated a distinct profile of prostanoid production in malignant versus normal tissue and cells (McLemore et al. 1988; Hubbard et al. 1989). Although individually most prostanoids effect ECs *in vitro* it is argued, not surprisingly, that their ratio *in vivo* is probably most important with respect to overall functional outcome (McLemore et al. 1988; Hubbard et al. 1989; Yang et al. 1998; Cathcart et al. 2010). This is particularly relevant with respect to TxA_2 and PGI_2 which generally exert opposing actions on EC function. In addition there is ample evidence that the terminal synthases responsible for catalyzing their formation from PGH_2 are differentially regulated (Helliwell et al. 2004). In this respect, it has been demonstrated that ECs of vessels located within or near lung tumors exhibit pronounced expression of both COX-2 and Tx synthase (TxS), while ECs of normal lung specimens primarily express COX-1 and PGI_2 synthase (PGIS) (Ermert et al. 2003). The ratio between TxB_2 and 6-keto-$PGF_{1}\alpha$ (stable hydrolysis products of TxA_2 and PGI_2, respectively) was also approximately two-fold higher in tumor and peri-tumor tissue, compared to control tissue (Pinto et al. 1993). Thus, there may be differential regulation of COX enzyme expression as well as terminal synthases in normal versus tumor-associated ECs which together would have consequences for the local prostanoid balance and for autocrine and paracrine control of both ECs and tumor cells. An unbalanced production of TxA_2 and PGI_2 has also been implicated in thrombosis, known to be a complication in cancer. In this regard, tissue factor is mitogenic for several cell types, including ECs, and stimulates angiogenic mediator production and tumor angiogenesis (see Rak et al. 2006).

5.2.6 PGD$_2$ and PGF$_2\alpha$

The role of PGD_2 in tumor angiogenesis is relatively unknown considering the extensive focus on PGs and cancer. However, a recent investigation reported that depletion of hematopoietic PGD synthase enhanced Lewis lung carcinoma progression, an effect which was abolished by administration of either a PGD_2 receptor agonist or the PGD_2 breakdown product 15-deoxy-$\Delta(12,14)$-PGJ_2 (Murata et al. 2012). Interestingly, PGD_2 was shown to negatively regulate $TNF\alpha$ production which could account, in part, for the anti-angiogenic and anti-tumorigenic effects of this prostanoid (Murata et al. 2011). In addition, 15-deoxy-$\Delta(12,14)$-PGJ_2 is a natural $PPAR\gamma$ ligand and has been reported to inhibit NEC proliferation (Xin et al. 1999), induce caspase-mediated HUVEC apoptosis (Bishop-Bailey and Hla 1999) and suppress endothelial-monocyte interactions (Prasad et al. 2008). Thus, overall, current evidence suggests that PGD_2 has anti-inflammatory and anti-angiogenic actions, consistent with a suppressive role in tumor angiogenesis.

Similarly, the precise role played by $PGF_2\alpha$ and its receptor (FP) in either pathological or physiological angiogenesis is not known. It has been reported that some colon cancer cell lines express FP and that exogenous $PGF_2\alpha$ stimulates cell motility (Qualtrough et al. 2007). These cells also constitutively produce $PGF_2\alpha$ (potentially through up-regulated COX-2 expression) suggesting that this mediator could also influence stromal cell function through paracrine mechanisms. The FP receptor co-localizes with CD31 (an EC marker) in endometrial adenocarcinomas and $PGF_2\alpha$ promotes cancer cell VEGF production (Sales et al. 2005). Conditioned medium from cancer cells over-expressing FP and exposed to $PGF_2\alpha$ also stimulates HUVEC tubulogenesis (Keightley et al. 2010) perhaps suggesting that $PGF_2\alpha$-FP interaction influences both tumor cells and ECs, at least in this model.

5.2.7 Resolvins, Fatty Acids and Inhibition of Angiogenesis

Factors produced by ECs and other cells that have anti-inflammatory activities are currently the subject of intense investigation. These endogenous pro-resolving eicosanoid mediators (e.g. resolvins, lipoxins and protectins) constitute several families of molecules generated from essential polyunsaturated fatty acids whose production is elevated during normal inflammatory processes to assist with resolution (see Fredman and Serhan 2011). There is currently little information regarding precisely how these various factors influence NEC or TEC functions relevant for angiogenesis, but there is clear evidence to suggest that some of them can exert direct anti-angiogenic effects on NECs. For example, AA-derived lipoxin A_4 has been reported to suppress production of pro-inflammatory cytokines and to reduce proliferation and differentiation of VEGF-stimulated HUVEC (Baker et al. 2009). Moreover, these effects were accompanied by enhanced endothelial production of IL-10 which exerts largely anti-inflammatory actions (Baker et al. 2009). IL-10 appears to have an undefined role in angiogenesis and has been reported to stimulate pathological angiogenesis (Dace et al. 2008) and to suppress VEGF production and ischemia-induced angiogenesis in mouse hindlimb (Silvestre et al. 2000). Thus, whether IL-10 directly to the anti-angiogenic actions of lipoxin A_4 (Baker et al. 2009) or other related mediators (Jin et al. 2009) remains undefined. Aspirin, which is an acetylating NSAID, switches COX-2 from generating pro-tumorigenic PGE_2 to anti-tumorigenic lipoxin A_4. The observation that the aspirin-triggered lipoxin A_4 analogue (ATLa) also directly inhibits VEGF-induced angiogenesis in a mouse model of corneal neovascularization (Jin et al. 2009) provides further support for the anti-angiogenic actions of these mediators. In addition, early work showed that lipoxins can stimulate endothelial PGI_2 synthesis (Brezinski et al. 1989). If this proves to be a significant action of lipoxins it most likely relates to the well-established cytoprotective effects of PGI_2 on NECs (see Wheeler-Jones et al. 2009) since the functional significance of this in the context of TECs and their angiogenic properties is unknown.

5.2.7.1 Omega 3 Polyunsaturated Fatty Acids

As highlighted above a family of bioactive, pro-resolving lipids (including lipoxins and resolvins) have well-documented anti-inflammatory and anti-carcinogenic actions (see Fredman and Serhan 2011). Alongside the enhanced formation of pro-angiogenic AA-derived lipid mediators (e.g. PGE_2) in tumors, insufficient production of these anti-inflammatory and potentially anti-angiogenic factors could directly contribute to the increased angiogenesis observed in the chronically inflamed tumor microenvironment.

There are now numerous publications documenting the ability of omega-3 PUFAs to suppress production of pro-angiogenic VEGF and inflammatory cytokines by tumor cells (see Wendel and Heller 2009; Fauser et al. 2011), an effect which would be expected to be accompanied by reduced angiogenesis. In this respect, a role for omega-3 PUFAs in limiting pathological angiogenesis is supported by studies in the retina where these fatty acids reduce the neo-vascularization associated with oxygen-induced retinopathy (Stahl et al. 2010). Precisely how omega-3 PUFAs protect against inflammation-associated angiogenesis in this model is not defined but retinal neovessels isolated from animals fed omega-3 PUFAs showed reduced expression of adhesion molecules (ICAM-1, VCAM-1 and E-selectin) and Ang-2 in comparison to normal vessel ECs, confirming the anti-inflammatory effects of omega-3 PUFAs in this model (Stahl et al. 2010). Collectively, the reported vasoprotective actions of omega-3 PUFAs in retinopathy are most likely mediated through PPARγ activation (Stahl et al. 2010; Sapieha et al. 2011) and involve reduced production of TNFα, but not VEGF (Stahl et al. 2010). Direct exposure of retinal microvascular ECs to docosahexaenoic acid (DHA) reduces VEGF-stimulated proliferation and sprouting and suppresses VEGF-mediated ROS production (Matesanz et al. 2010), suggesting that omega-3 PUFAs also limit pro-angiogenic signaling triggered by VEGFR activation. In addition, the beneficial actions of omega-3 PUFAs, at least in the retina, may also be mediated through 5-LOX-dependent oxidation of DHA to 4-hydroxy-DHA, a mediator which reduces EC proliferation and sprouting angiogenesis, but apparently do not depend upon COX-2 activity (Sapieha et al. 2011). This contrasts with data from *in vitro* studies where the inhibitory effects of free omega-3 PUFAs on HUVEC Ang2 and MMP-9 expression and on EC tubulogenesis were reported to depend, at least partly, on COX activities (Szymczak et al. 2008). Thus, multiple mechanisms are likely to mediate the anti-angiogenic effects of long-chain omega-3 PUFAs, and at least some of these are the result of direct actions on ECs.

5.3 Lipoxygenases and Angiogenesis

Prostaglandins and leukotrienes are generated through two distinct pathways, involving metabolism of AA via the activities of COX-2 and lipoxygenases (LOX), respectively. To date, there is comparatively less known about the functional significance of leukotrienes for angiogenesis versus that of PGs, although there is now clear evidence of a role for LOX-derived products in modifying NEC function. In inflammatory cells such as macrophages, PLA_2 and 5-LOX co-operate to generate a range of leukotrienes, including LTB_4 and LTD_4 which exert their cellular actions through a family of G protein-coupled LT receptors. Other AA-metabolizing LOX enzymes include 15-LOX and 12-LOX which, respectively, synthesize 15(S)-hydroxyeicosatetraenoic acid (15(S)HETE) and 12(S)HETE from AA. Over-expression of 15-LOX-2 in prostate cancer cells causes cell cycle arrest and apoptosis, and tumors expressing 15-LOX-2 grow more slowly than those without (Tang et al. 2009), suggesting an anti-carcinogenic role for this enzyme. In addition, intravitreal transfer of an adenoviral construct encoding 15-LOX-1 has been reported to reduce VEGF-induced neo-vascularization and associated pathology in rabbit eyes, most likely through attenuated expression of VEGFR2, supporting the ability of this enzyme to limit pathological angiogenesis (Viita et al. 2009). In contrast, enhanced expression and activities of both 5-LOX and 12-LOX are generally associated with increased angiogenesis and enhanced tumor growth (reviewed in Wang and DuBois 2010). Thus, LTD_4 directly affects NEC migration and tubulogenesis and stimulates neo-angiogenesis in the chick chorioallantoic membrane vascularization assay (Yuan et al. 2009), consistent with an ability to target EC function. Similarly, VEGF promotes 12(S)HETE production by HUVEC and exogenous 12(S)HETE stimulates EC migration and tube formation (Kim et al. 2009). Moreover, VEGF-induced angiogenic responses in HUVEC were attenuated by knockdown of 12-LOX expression and restored by addition of 12(S)HETE (Kim et al. 2009), suggesting that enzymatic products of 12-LOX activity may contribute to maintenance of the pro-angiogenic phenotype of VEGF-stimulated NECs through autocrine and paracrine actions. There is also evidence that COX-2 and LOX enzymes co-operate to generate the formation of novel eicosanoids. In this respect, Greisser and colleagues recently showed that human leukocyte preparations stimulated to enhance expression of both 5-LOX and COX-2 biosynthesize a new group of eicosanoids known as hemiketals (HK) (Griesser et al. 2011). These authors also showed that HKE_2 and HKD_2 directly increased mouse pulmonary EC migration and tubulogenesis, providing evidence for EC-directed actions (Griesser et al. 2011). The potential importance of these mediators as regulators of ECs in pathologies characterized by aberrant angiogenesis, or in normal tissue repair processes, is not yet clarified.

5.4 Phospholipase A_2 and Angiogenesis

Enzymes upstream of COX-2/LOX activities also play important roles in both EC function and cancer progression. Cytosolic phospholipase $A_2\alpha$ (cPLA$_2\alpha$) hydrolyses membrane glycerophospholipids to generate AA and lysophospholipids, initiating the cascade of enzymatic activity which is responsible for the cellular production of biologically active eicosanoids, including leukotrienes, thromboxanes and prostaglandins (see Wheeler-Jones et al. 2009). In common with COX-2 regulation, cPLA$_2\alpha$ expression and activity in ECs are increased by pro-inflammatory and angiogenic mediators. This enzyme is rapidly phosphorylated and activated, in part, through MAPK-dependent mechanisms in both cancer cells and NECs exposed to a number of pro-angiogenic cytokines and growth factors (e.g. IL-1; VEGF; thrombin) and contributes to both acute and prolonged synthesis of prostanoids with potential pro-angiogenic actions (PGI$_2$; PGE$_2$) (Wheeler-Jones et al. 1997; Houliston et al. 2001; Herbert et al. 2009). Thus, perhaps not surprisingly, there is emerging support for involvement of cPLA$_2\alpha$ in angiogenesis and tumor progression. For example, lung microvascular ECs isolated from mice deficient in cPLA$_2\alpha$ exhibit lower rates of proliferation, are less motile and show less potential for tubulogenesis than wild-type murine ECs (Linkous et al. 2010). Similarly, pharmacological blockade of cPLA$_2\alpha$ activity has been reported to suppress VEGF-mediated proliferation of retinal microvascular ECs and to reduce retinal neovascularization and levels of PGE$_2$ and VEGF in a rat model of oxygen-induced retinopathy (Barnett et al. 2010). These data are all consistent with the hypothesis that cPLA$_2\alpha$-mediated AA generation results in the downstream production of PGE$_2$ and that this exerts pro-angiogenic actions on ECs. Moreover, inhibition of cPLA$_2\alpha$ activity with CDIBA, a chemical cPLA$_2\alpha$ inhibitor, delayed tumor growth and tumors established in cPLA$_2\alpha$-null mice were less vascularised than those in wild type animals, strongly implicating cPLA$_2\alpha$ in tumor-associated angiogenesis (Linkous et al. 2010).

6   Molecular Regulation of Pro-Inflammatory Gene Expression in Endothelial Cells

Undoubtedly individual tumors will exhibit different mediator profiles depending upon the cells present and their relative cytokine generating abilities. It is clear, however, that ECs are direct targets for the majority of these mediators and that they strongly influence EC functions important for angiogenesis. Thus, the prevailing pro-inflammatory environment will ultimately support angiogenesis, as well as tumor growth and metastases. Identification of the molecular mechanisms governing cell-cell communication in tumors and the individual signaling pathways operative in the constitutive cell types remains a considerable challenge, particularly in tumor ECs.

As detailed above, angiogenesis is dependent upon cross-talk between stromal cells and ECs which strongly influences the balance of expression and release of endogenous molecules that positively or negatively regulate a spectrum of functional responses in ECs. It is clear that EC behavior during both pathological and physiological angiogenesis is directly regulated by diverse mediators generated during inflammatory processes, including growth factors, cytokines, chemokines and eicosanoids. Many of these factors can also regulate COX-2 induction and synthesis of prostanoids which, in turn, act as autocrine regulators of EC functions (Garonna et al. 2011). These intermediaries activate a vast repertoire of intracellular pathways, discussion of which is beyond the scope of this article. However, there are key signaling elements that dominate our current understanding of how these functions are regulated at the molecular level. This section will summarize current understanding of mechanisms that may couple COX-2 induction to EC angiogenic responses and how prostanoids and other products of AA metabolism influence EC functions important for angiogenesis. In considering this it must be stressed that multiple signaling pathways and transcription factors contribute to regulation of COX-2 induction downstream of growth factor receptors and GPCRs. This control occurs in an agonist- and cell type-dependent manner and is thus highly context-specific.

6.1   PI3K-Akt and MAPK Pathways

The PI3K-Akt and mitogen-activated protein kinase (MAPK; ERK, p38^{mapk}, JNK) pathways are prominent regulators of the angiogenic functions of ECs and there are now numerous reports implicating their involvement in angiogenic processes stimulated by inflammatory mediators. Akt- and MAPK-activating signals emanate from a diverse range of cell surface receptors important for stimulating inflammatory activation of ECs and several of the established players in angiogenesis (e.g. VEGF) signal prominently through PI3K-Akt and MAPK cascades to regulate EC motility, proliferation and differentiation. Similarly, a range of other angiogenic agonists generated during inflammatory processes with less well characterized mechanisms of action (e.g. leptin; thrombin), and known to be over-expressed in tumors, also control EC functions through stimulation of the PI3K-Akt axis (Garonna et al. 2011). In addition, in cancer cells these signaling cascades are generally over-active and positively regulate downstream synthesis and release of angiogenic EC-targeting factors, justifying the potential use of inhibitors of these pathways to control tumor angiogenesis and growth (Chappell et al. 2011).

   With respect to PI3K-Akt signaling, induction of constitutively active myristoylated Akt1 in ECs of normal tissues promotes the formation of structurally and functionally abnormal blood vessels, similar to those in tumors (Phung et al. 2006). Further support for overactive PI3K-Akt signaling in ECs as a determinant of tumor angiogenesis comes from chimeric tumor models where forced over-expression of p70S6 kinase-1 (a signaling intermediate downstream of PI3K-Akt) in microvascular ECs promoted tumor angiogenesis and growth whereas use of a kinase inactive mutant abrogated these effects (Liu et al. 2008). Evidence for specific upregulation of phosphorylated (and thus activated) signaling intermediates in TECs is also emerging. For example, immunohistochemical studies have shown that ECs in the vasculature of a range of mouse tumors exhibit enhanced expression of active ERK1/2 and STAT-3 compared to ECs in normal tissue vessels (Lassoued et al. 2011). In addition, upregulation of these phosphorylated signaling proteins (Lassoued et al. 2011), as well as p-Akt, was evident in tumors engineered to express VEGF (Phung et al. 2006; Lassoued et al. 2011). Whether this upregulation is accompanied by increased TEC expression of COX-2, for example, is not yet clarified.

It has been reported that VEGF-induced COX-2 induction in NECs (HUVEC) is reduced by pharmacological inhibitors of p38^{mapk} and JNK activities, but not MEK activity (Wu et al. 2006). These authors also provided evidence that separate blockade of p38^{mapk} and JNK activities suppressed EC differentiation on Matrigel. Interestingly, VEGF-stimulated COX-2 upregulation was unaffected by inhibition of MEK (Wu et al. 2006) whereas other pro-angiogenic agonists (e.g. thrombin) clearly induce EC COX-2 in a MEK-ERK-dependent manner (Syeda et al. 2006), suggesting differential use of the MEK-ERK pathway to control COX-2 expression. We have also recently shown that leptin-stimulated COX-2 induction is regulated by p38^{mapk} and Akt, and provided additional evidence that a functional endothelial p38^{mapk}/Akt/COX-2 signaling axis is required for leptin's pro-angiogenic actions on ECs, including proliferation, directional migration and capillary-like tube formation (Garonna et al. 2011). Since exposure to NS398, a COX-2-selective inhibitor, blocked VEGF- (Wu et al. 2006; Garonna et al. 2011) and leptin-mediated angiogenic responses (Garonna et al. 2011) these studies collectively suggest that COX-2-derived mediators act as autocrine modulators of ECs, and that Akt and MAPK pathways are important upstream regulators of COX-2 expression and EC functions. In further support of a role for p38^{mapk}-mediated signaling pathways in the angiogenic functions of NECs it has recently been demonstrated that ECs chronically exposed to inflammatory stimuli exhibit enhanced p38^{mapk}-dependent angiogenic behavior accompanied by increased EC expression of TNFα and that p38^{mapk} blockade in a tumor model reduced tumor growth and vessel density. These authors also suggested that an autocrine inflammatory loop maintained p38^{mapk} activity in ECs to promote angiogenesis (Rajashekhar et al. 2011).

Recent studies have also demonstrated roles for adaptor proteins recruited to RTKs upstream of PI3K-Akt activation in angiogenesis. Shc, for example, is required for VEGF-induced Akt activation and is necessary for angiogenic functions in NECs *in vitro* and for neo-angiogenesis *in vivo* (Sweet et al. 2012). Given that Shc regulates Raf-MEK-ERK cascade activation, which in turn modifies EC COX-2 expression (Syeda et al. 2006), and that Akt regulates COX-2 expression in some settings (Kurosu et al. 2011), it is tempting to speculate that this Shc-Akt pathway may also have significance for control of EC COX-2 induction and subsequent generation of pro-angiogenic prostanoids, as well as for COX-2-independent angiogenesis.

The direct effects of MEK inhibitors on agonist-stimulated EC tube formation *in vitro* (an indication of pro-angiogenic potential) have not, to our knowledge, been fully assessed but there is ample evidence to support a role for the MEK-ERK pathway in EC proliferation and sprouting angiogenesis from *in vitro* models. Studies in HUVEC, for example, showed that ERK1/2 activation enhanced EC survival and sprouting and that this resulted from down regulation of Rho kinase activity (Mavria et al. 2006). Selective over-expression of dominant negative MEK1 in the vascular compartment of mouse tumors resulting from injection of colon carcinoma cells also suppressed angiogenesis and tumor growth (Mavria et al. 2006), providing strong support for the importance of MEK-ERK signaling in TECs (Lassoued et al. 2011).

Negative regulation of pro-angiogenic kinase activities through modulation of phosphatase expression and activity offers an alternative means of regulating angiogenic processes. In this regard, the phosphatase and tensin homolog, PTEN, is a tumor suppressor mutated in many human cancers and acts as a primary physiological antagonist of pro-angiogenic signals emanating from RTKs and GPCRs (Jiang and Liu 2008). For example, PTEN phosphatase activity not only inhibits activation of the PI3K-Akt axis by preventing PIP3 generation but also promotes dephosphorylation (and thus inactivation) of several other signaling intermediates important for stimulating cell motility and tubulogenesis *in vitro* (e.g. focal adhesion kinase; Shc). PTEN has also been reported to regulate angiogenesis through phosphatase-independent mechanisms, suggesting wider significance as an anti-angiogenic factor (Tian et al. 2010). The demonstration that highly active AA metabolism (e.g. in tumors over-expressing COX-2 and/or LOXs) oxidizes PTEN and reduces its phosphatase activity (Covey et al. 2007) likely contributes to the enhanced Akt activation and elaboration of pro-angiogenic factors by cancer cells. Whether and how similar PTEN-dependent mechanisms regulate pro-angiogenic signaling and function in ECs remains to be determined.

Given that MAPKs are key regulators of EC function it is not surprising that phosphatases capable of regulating MAPK phosphorylation status also have importance as modulators of inflammatory angiogenesis. In this respect the dual-specificity phosphatases (DUSP), which target the phosphotyrosine and phosphothreonine residues on active MAPKs, are critical for determining the duration of MAPK activities. MKP-1 (DUSP-1), for example, is a nuclear expressed DUSP that has been widely studied and dephosphorylates p38^{mapk}, JNK and ERK1/2 (Haagenson and Wu 2010). Angiogenic mediators, including VEGF and thrombin, induce MKP-1 in ECs, thus prolonging MAPK activation, promoting the associated pro-angiogenic responses in ECs and contributing to neo-angiogenesis. VEGF-driven MKP-1 induction in HUVEC has been shown to selectively dephosphorylate p38^{mapk} and JNK whereas the MKP-1 induced in thrombin-stimulated ECs appears to preferentially target ERK1/2 (Kinney et al. 2008). In addition, these agonists have been reported to make differential use of non-receptor tyrosine kinases to control MKP-1 expression (Kinney et al. 2008). Thus, a complex regulation of MKP-1 expression exists which will clearly have consequences for regulation of gene expression downstream of MAPK signaling. Although forced over-expression of MKP-2 in HUVEC has been reported to inhibit TNF-induced COX-2 induction (Al-Mutairi et al. 2010), whether MKPs in general contribute significantly to the regulation of COX-2 expression or of other AA metabolizing enzymes in NECs or TECs is not yet defined.

Interestingly, it has been shown that MKP-1 over-expression reduces tube formation by human lung microvascular ECs on Matrigel and proposed that control of tubulogenesis by these cells occurs principally through the JNK pathway and not through p38^{mapk} or MEK-ERK signaling (Medhora et al. 2008). This suggests that normal microvascular ECs, in comparison to HUVEC, could differ in their utilization of MAPK pathways for controlling angiogenesis. However, it must be stressed that this study used high passage ECs and did not investigate EC differentiation by exogenous pro-angiogenic mediators, which would likely make use of several MAPK pathways to regulate motility. How these phosphatases are modulated in TECs specifically has not been investigated but it is conceivable that changes in EC expression of these enzymes contribute to upregulation of MAPK pathway activities and thus to generation of angiogenic mediators.

Overall, it is evident that angiogenic mediators regulate PI3K-Akt and MAPK pathway activation in ECs to control functional outcomes relevant for angiogenesis and that COX-2 induction, at least in some instances, couples these acute changes in signaling with altered cell behaviors. In tumor ECs these pathways are hyper-activated and COX-2 expression is up-regulated (Kurosu et al. 2011; Muraki et al. 2012). One factor that likely contributes to the enhanced COX-2 expression in TECs is the Hu antigen R (HuR) protein which stabilizes mRNAs containing AU-rich elements, and has been implicated in the regulation of inflammatory gene expression in HUVEC (Rhee et al. 2010). In the context of tumor angiogenesis it has been shown that HuR is expressed in both the cytosol and nucleus of TECs whereas it is confined to the nucleus in normal ECs (Kurosu et al. 2011). These authors demonstrated that expression of both COX-2 and VEGF was enhanced in TECs versus NECs and that knockdown of HuR inhibited COX-2 and VEGF expression in TECs as well as their motility and tube-forming capacity. Moreover, downregulation of HuR reduced the level of Akt phosphorylation in TECs which was rescued by exogenous PGE$_2$, an effect which the authors interpreted as evidence of autocrine regulation of angiogenic phenotype (Kurosu et al. 2011). Thus, the mechanisms contributing to enhanced Akt signaling and COX-2 expression in some TECs are beginning to be addressed.

It should also be noted that while COX-2 activity has an important role in mediating the functional responses of ECs to several key angiogenic factors (Wu et al. 2006; Garonna et al. 2011) there is evidence, at least *in vitro*, that a range of other pro-angiogenic molecules can stimulate angiogenic responses in ECs *in vitro* through mechanisms that do not require increased COX-2 expression or activity, suggesting that alternative and parallel

COX-2-independent pathways are also important. For example, direct exposure of NECs to a number of angiogenic metabolites, including those derived from AA, stimulates angiogenesis through mechanisms that have not, to date, been reported to involve changes in COX-2 enzyme activity (Pula et al. 2010; Fleming 2011).

6.2. NF-κB Signaling

NF-κB is a transcription factor with established importance in inflammatory signaling in ECs and other cell types and is known to regulate expression of numerous genes implicated in cell survival, proliferation and angiogenesis (Lee and Hung 2008). The most well studied signaling cascades involving NF-κB are the so-called canonical and non-canonical NF-κB pathways. The canonical NF-κB pathway is regulated by a number of pro-inflammatory, pro-angiogenic mediators including IL-1β and TNFα. In this pathway activated IKKβ directly phosphorylates IκBα (which maintains NF-κB in an inactive state), and triggers its proteasomal degradation, thus releasing the NF-κB heterodimer and facilitating its translocation to the nucleus where it binds to its target DNA sequence. Transcriptional activation of NF-κB increases expression of several genes encoding proteins with known pro-inflammatory and pro-angiogenic actions (e.g. VEGF, IL-8, IL-6, IL-1, MCP-1, COX-2) and also up-regulates genes involved in cell survival and metastases (e.g. XIAP, MMP-9) (Lee and Hung 2008; Bollrath and Greten 2009).

While the ability of pro-inflammatory cytokines to strongly activate NF-κB signaling in NECs is well established there is less known about the importance of this pathway as a regulator of the pro-angiogenic actions of growth factors (e.g. VEGF, PAR agonists) and other angiogenic mediators (e.g. prostanoids) in ECs. In this respect studies in HUVEC have shown that VEGF stimulates nuclear translocation of the p65 subunit of NF-κB, that this is mediated by degradation-independent dissociation of IκB from NF-κB, and that this pathway regulates expression of Bcl-2 and EC survival (Grosjean et al. 2006). Similarly, activators of PARs, a family of GPCRs with emerging importance in angiogenesis (Wheeler-Jones et al. 2009), promote endothelial COX-2 expression and downstream prostanoid synthesis through mechanisms mediated by IκBα-dependent NF-κB activation as well as through enhanced ERK1/2 and $p38^{mapk}$ activities (Houliston et al. 2001; Syeda et al. 2006). Activation of NF-κB transcriptional activity also seems to be essential for β_2-adrenoceptor-mediated angiogenic responses in mouse ECs (Ciccarelli et al. 2011) and for the ability of semaphorins to drive a pro-angiogenic, IL-8-secreting phenotype in HUVEC (Yang et al. 2011). Given the importance of COX-2-derived products in driving angiogenic responses in ECs (through paracrine and autocrine actions), and the ability of many growth factors to induce EC COX-2 expression it is possible that the angiogenic functions of growth factor-stimulated ECs all depend, to some extent, on NF-κB activity although this has yet to be tested directly. In this respect, whether PGE$_2$ or other mediators derived from COX-2 stimulate angiogenic responses in NECs through activation of NF-κB-dependent pathways has not, to our knowledge, been investigated.

There is also considerable interaction between NF-κB signaling and other pathways important for regulating EC function which could have relevance for angiogenesis. For example, IKKβ can directly phosphorylate FOXO3a, a forkhead transcription factor that is also phosphorylated by Akt. FOXO3a phosphorylation prevents its nuclear translocation and repression of the cell cycle, resulting in EC proliferation and suppression of apoptosis. Recent studies have also revealed that IKKβ can promote Akt phosphorylation in ECs by modifying PTEN activity (Ashida et al. 2011), thus providing additional evidence for complex signaling cross-talk mechanisms involving NF-κB with functional relevance for the angiogenic properties of NECs.

Several cancers show dysregulated NF-κB signaling and there is evidence that this pathway is constitutively active in a number of tumor types (Sakamoto et al. 2009). Conceivably, aberrant NF-κB up-regulation could reflect either over-activation of IKK activities, altered regulation of expression of genes encoding IκB proteins, or enhanced activities of components of non-canonical NF-κB signaling (e.g. IKKα; see Bollrath and Greten 2009). Because NF-κB regulates multiple pro-mitogenic and anti-apoptotic genes, persistent nuclear NF-κB would be expected to confer resistance to apoptosis and to enhance the proliferative capacity of both TECs and tumor cells themselves. Components of the NF-κB pathway could therefore offer additional targets for influencing expression of pro-angiogenic NF-κB-dependent genes (e.g. COX-2) and thus for suppressing angiogenesis. Indeed, many anti-angiogenic compounds and factors have been reported to limit the angiogenic functions of ECs by indirectly or directly modifying activation of the NF-κB pathway, which is both pro-inflammatory and anti-apoptotic (Albini et al. 2010). In addition, the recognition that IKKs and IKK-related kinases are oncogenic has fuelled strategies to target overactive NF-κB signaling through inhibition of IKKs, particularly IKKβ (Lee and Hung 2008).

While there is substantial evidence for upregulation of the NF-κB pathway within tumors there are no reports, to our knowledge, detailing expression of NF-κB or upstream elements in TECs either *in situ* or *in vitro*. However, along with the ample evidence that cancer cells from several tumor types over-express NF-κB, there is some indication from *in vitro* studies that these cells influence EC function through paracrine mechanisms. For example, colon cancer cells with constitutively active NF-κB synthesize a number of pro-angiogenic cytokines (notably IL-8 and MCP-1) and conditioned medium from these cells stimulates capillary-like tube formation by HUVEC on Matrigel (Sakamoto et al. 2009). In contrast, medium from cells in which NF-κB signaling was ablated showed reduced production of angiogenic mediators and less ability to promote EC tubulogenesis. Tumor xenografts derived from cells devoid of NF-κB also exhibited reduced vascularization in comparison to grafts from NF-κB over-expressing cells (Sakamoto et al. 2009), consistent with a model of paracrine communication.

Since upregulation of COX-2 expression has been reported in TECs (Kurosu et al. 2011; Muraki et al. 2012) it is not unreasonable to suggest that this could be accompanied by increased NF-κB pathway activity which, in turn, would enhance EC expression of other genes whose transcription is regulated through regulated through NF-κB-dependent mechanisms (e.g. IL-8, IL-6, IL-1 and TNF); the consequent production of angiogenic mediators by ECs would then facilitate bidirectional cross-talk between and tumor/stromal cells and the tumor vasculature.

Interestingly, there is evidence that MAPK pathways, which are constitutively active in some tumors, as well as NF-κB, regulate expression and activity of cPLA$_2\alpha$ and that other pro-angiogenic signaling pathways (e.g. PI-3K-Akt) are upstream regulators of cPLA$_2\alpha$ which is coupled to microvascular EC migration and tube formation (Giurdanella et al. 2011). The promoter region of the cPLA$_2\alpha$ gene, in common with that of COX-2, also contains binding sites for NF-κB, suggesting that signaling pathways involving NF-κB may also directly regulate endothelial cPLA$_2\alpha$ expression (Clark et al. 1995) as well as COX-2 induction (Syeda et al. 2006). Thus, co-ordinated induction of both AA-generating cPLA$_2\alpha$ and COX-2 in TECs would be expected to permit sustained production of pro-angiogenic prostanoids which, in turn, would influence EC functions and tumor/stromal cell behaviors through autocrine and paracrine control mechanisms.

6.3   Receptor Cross-talk Mechanisms

Inflammatory mediators are capable of influencing expression and release of a range of growth factors that operate through activation of receptor tyrosine kinases (RTKs). While most of these (e.g. VEGF, FGF) are capable of directly influencing EC signaling and function it is now becoming evident that highly complex interactions occur between families of RTKs and other receptors (e.g. GPCRs; cytokine receptors) and that some of these cross-talk mechanisms are functionally significant for the angiogenic behaviors of ECs (Garonna et al. 2011). With respect to tumor cell function it is generally accepted, for example, that cross-talk between EPs and the epidermal growth factor receptor (EGFR) contributes to the pro-tumorigenic actions of PGE$_2$ on these cells. Thus, exposure to exogenous PGE$_2$ results in rapid phosphorylation of the EGFR in both normal gastric epithelial cells and cancerous cell lines which leads to downstream activation of ERK1/2 and increased cell proliferation (Pai et al. 2002). Banu and co-workers also provided evidence that autocrine production of PGE$_2$ regulates the level of EGFR phosphorylation in colorectal carcinoma cell lines and showed that pharmacological inhibition of EGFR kinase activity inhibited cell movement (Banu et al. 2007). In contrast, the specific role of PGE$_2$-EGFR cross-talk in regulating the angiogenic functions of ECs has not been explored. However, there is evidence that PGE_2 phosphorylates fibroblast growth factor receptor (FGFR)-1 in ECs with consequent phosphorylation of FRS2α (Finetti et al. 2009), a key FGFR substrate mediating downstream activation of MEK-ERK and PI3K-Akt pathways (Sato and Gotoh 2009).

VEGFR2 is highly expressed in vascular ECs and its expression is increased in both ECs and stromal cells under conditions of active angiogenesis and, additionally, is known to be up-regulated by a range of inflammatory mediators, including IL-1 (Guo and Gonzalez-Perez 2011). Both ligand-dependent and -independent activation of VEGFR2 by pro-inflammatory and angiogenic agonists has been reported in NECs. For example, the adipokine leptin rapidly phosphorylates VEGFR2 through a mechanism that requires engagement with the leptin receptor ObRb and which triggers activation of p38[mapk] and Akt with consequent upregulation of COX-2 expression (Garonna et al. 2011). The precise molecular events involved in mediating ObRb-VEGFR2 cross-talk remain to be defined but this interaction and subsequent activation of the p38[mapk]/Akt/COX-2 axis were shown to be important for the pro-angiogenic actions of leptin on cultured ECs (proliferation, migration and differentiation) and for neovascularization in a chick chorioallantoic membrane assay, suggesting that it has functional significance. Thus, signaling through VEGFR2 contributes, at least in part, to the EC targeted angiogenic actions of a range of angiogenic and pro-inflammatory factors. Not surprisingly, in keeping with a prominent role for VEGFR2 there is substantial evidence that anti-angiogenic mediators often limit the pro-angiogenic properties of ECs by suppressing VEGFR2-dependent signaling events. Lipoxin A4, for example, reduces VEGF-induced HUVEC proliferation and capillary-like tube formation through a mechanism involving inhibition of VEGFR2 phosphorylation (Baker et al. 2009) and a range of other biologically active lipids, including omega-3 PUFAs and endocannabinoids, can attenuate VEGF-driven signaling and dependent EC functions (Pisanti et al. 2007; Matesanz et al. 2010).

Other functionally relevant relationships between VEGFR2, EGFR and FGF receptors have also recently come to light. For example, NECs completely lacking FGF signaling are unable to maintain VEGFR2 expression and lose responsiveness to VEGF, an effect which is characterized *in vivo* by impaired angiogenesis (Murakami et al. 2011). Binding of FGF7 to FGFR2b and of VEGF to VEGFR2 have also been reported to promote transactivation of the EGFR in ECs (Maretzky et al. 2011). Both these interactions are mediated by the activity of ADAM17, a cell surface metalloproteinase which releases EGFR ligands from the extracellular matrix, resulting in EGF-dependent regulation of cell motility in response to both FGF and VEGF (Maretzky et al. 2011). These findings suggest a high level of complexity and interaction between receptors important for regulating inflammation and angiogenic responses in NECs. The relevance of these interactions for EC functions in the tumor microenvironment is not yet known but a signaling complexity of this nature, likely present in TECs and cancer cells could help to explain why many tumors escape sensitivity to VEGF blockade.

6.4   Actions of PGE$_2$ on Endothelial Cells

As noted above there is substantial evidence to support the pro-angiogenic actions of PGE$_2$ in the tumor microenvironment. For example, PGE$_2$ stimulates the synthesis and release of pro-angiogenic mediators (e.g. VEGF, FGF, chemokines) from both tumor cells and stromal cells. These, in turn, directly enhance the angiogenic properties of ECs and increase their generation of angiogenic factors (e.g. IL-8). These factors are capable of acting through autocrine and paracrine mechanisms to influence EC, stromal cell and tumor cell functions.

Mechanisms that enable acquisition of invasiveness and motility will generally lead to enhanced tumor development and metastasis. In cancer cells PGE$_2$'s ability to influence cell movement is generally well described but the specific mechanisms involved in modulation of EC functions by this prostanoid are not well understood. Recent studies, however, are beginning to shed light on how PGE$_2$ influences the angiogenic properties of ECs. The molecular mechanisms underlying the direct actions of exogenous PGE$_2$ on NECs are not fully defined but several signaling pathways with known relevance for angiogenesis have been implicated. The ability of exogenous PGE$_2$ to directly affect angiogenic responses in ECs is consistent with the evidence discussed above supporting its role as a COX-2-derived autocrine mediator of EC function (Wu et al. 2006; Garonna et al. 2011).

Several recent studies have investigated signaling mechanisms responsible for mediating PGE$_2$'s actions on ECs in culture. PGE$_2$, possibly acting through EP3, was reported to synergize with FGF-2 to markedly enhance proliferation of bovine post-capillary venule ECs and murine aortic ECs (Finetti et al. 2008). This synergistic effect was not evident with combinations of PGE$_2$ and either VEGF or EGF, suggesting prominent EP3-FGF receptor (FGFR) cross-talk, at least in these EC types (Finetti et al. 2009). The effect of PGE$_2$ was shown to involve phosphorylation of FGFR1 and was associated with increased phosphorylation of phospholipase Cγ, ERK1/2 and STAT-3 (Finetti et al. 2009). The significance of this EP3-dependent pathway for PGE$_2$-mediated angiogenic responses in other EC types is not yet clear since neither mouse lung ECs nor human ECs appear to express EP3 (Dormond et al. 2002; Rao et al. 2007; Aso et al. 2012). Neonatal dermal microvascular ECs (HDMEC) have been reported to express EP3 but this receptor is not involved in mediating the pro-angiogenic effects of PGE$_2$ in this model (Zhang and Daaka 2011). Others have suggested that the pro-migratory effects of PGE$_2$ on microvascular ECs require up-regulation of CXCR4, a receptor for the chemokine stromal-cell derived factor-1 (SDF-1), and that this pro-angiogenic effect is mediated through both EP2- and EP4-signaling events (Salcedo 2003).

Most studies now point to a prominent role for EP4 as the major mediator of PGE$_2$'s angiogenic actions, at least in NECs. Thus, PGE$_2$ and EP4-selective agonists stimulate migration and differentiation on Matrigel of wild type murine lung ECs whereas these responses are absent in

EP4-null ECs (Rao et al. 2007). PGE_2-stimulated angiogenic behaviors of HDMECs seem to be mediated principally by an EP4-dependent regulation of the cyclic AMP (cAMP)-protein kinase A (PKA) pathway and consequent changes in activation of a number of downstream targets, including Rho A and GSK3β (Zhang and Daaka 2011). These findings are consistent with previous observations in HUVEC documenting pro-angiogenic affects of forskolin, a direct activator of adenylate cyclase which raises intracellular cAMP (Namkoong et al. 2009). Lung microvascular ECs exposed to PGE_2 also utilize the cAMP-PKA pathway to regulate synthesis of the pro-angiogenic and pro-inflammatory chemokine IL-8 (Aso et al. 2012). In these ECs PGE_2 also activates $p38^{mapk}$, potentially through cAMP, and blockade of $p38^{mapk}$ activity suppresses IL-8 production (Aso et al. 2011). There is also evidence for cross-talk between the PKA and PI3K-Akt pathways (Namkoong et al. 2009), suggesting further complexity in the mechanisms mediating angiogenesis in response to PGE_2 and other cAMP elevating agonists. Interestingly, PGE_2 has recently been reported to influence the angiogenic properties of NECs (HUVEC) through modification of TGFβ signaling via MMP-mediated release of active TGFβ from its latent form (Alfranca et al. 2008). In this study exogenous PGE_2 enhanced phosphorylation and nuclear translocation of Smad3 and this was shown to be dependent upon the kinase activity of the Type I TGFβ receptor, Alk5 (Alfranca et al. 2008).

PGE_2 also directly regulates the migratory capacity and differentiation of bone marrow-derived mononuclear cells in culture, increasing their ability to form capillary-like tube structures on Matrigel and up-regulating their expression of EC markers (CD31 and von Willebrand Factor) and metalloproteinases (Zhu et al. 2011). These effects appear to be mediated by EP4 and were reduced by pharmacological blockade of AMPK or by expression of a dominant-negative AMPK, but not by modifying PKA activity (Zhu et al. 2011). While AMPK is known to be involved in the proliferative and migratory responses of VEGF-A-stimulated aortic ECs (Reihill et al. 2011) and is required for angiogenesis in response to a range of other mediators, precisely what role, if any, can be attributed to AMPK in regulating prostanoid-mediated angiogenic functions of mature ECs remains to be defined. There is some evidence from studies in HUVEC using direct AMPK activators that AMPK may contribute to regulation of COX-2 induction upstream of $p38^{mapk}$ (Chang et al. 2010) but whether this pathway has any relevance for angiogenic EC functions is not defined. Recent studies suggest that AMPK activation in tumor cells themselves may be beneficial. Pharmacological activation of AMPK in mouse melanoma cells, for example, inhibits their metastatic potential through a mechanism involving reduced ERK activity and COX-2 expression (Kim et al. 2012). However, nothing is currently known about the activation state or functional significance of AMPK in TECs and whether this plays any part in positively or negatively regulating tumor angiogenesis.

Notably, in the context of tumor angiogenesis there is little understanding to date of which EC-associated EP subtypes could be responsible for mediating the angiogenic functions of TECs specifically, presumably reflecting the obvious difficulties associated with TEC isolation. However, there is growing evidence implicating several prostanoid receptor subtypes in tumor-associated angiogenesis, derived principally from studies in genetically modified mouse lines and supported by parallel investigations in NECs. EP3, for example, may be important for lung tumor angiogenesis, metastases and invasion (Amano et al. 2009). In this study the authors used EP3-deficient mice to show that EP3 up-regulated MMP9 expression in CD31-positive cells in tumors and increased endothelial VEGF-A production (Amano et al. 2009), consistent with the well documented ability of PGE_2 to enhance VEGF-A expression in a range of EC types (Clarkin et al. 2008a). Similarly, breast tumors implanted into mice devoid of EP2 expression showed reduced levels of neo-angiogenesis and lung ECs isolated from these animals had impaired migratory capacity and were more susceptible to apoptosis *in vitro* compared to wild-type ECs (Kamiyama et al. 2006).

Since it is clear that EC responses to PGE_2 are dictated to some extent by EC phenotype, which in turn will influence relative EP receptor expression, the likelihood is that ECs derived from tumors will exhibit diverse responses to PGE_2 (and to other prostanoids) and that these will be distinct, at least in part, from those evident in NECs. What is clear from the published studies in normal ECs, however, is that PGE_2 (acting mainly through EP4) is capable of triggering activation of a number of key signaling pathways and that these collectively regulate the ability of this tumorigenic prostanoid to stimulate angiogenic functions and hence contribute to processes involving neo-angiogenesis.

While the molecular actions of PGE_2 on ECs are currently being unraveled, there is comparatively little known about the molecular control of EC angiogenic responses to PGI_2. As discussed above the precise role of PGI_2 in angiogenic processes (pathological or physiological) is generally poorly defined but could be mediated through cell surface prostaglandin GPCRs and/or PPARs. In this context, cicaprost (a selective agonist at the IP receptor) has been shown to increase the rate of scratch wound closure in HUVEC monolayers and to enhance their capillary-like tube formation on Matrigel (Turner et al. 2011). These effects were shown to be mediated through the IP receptor since they were abolished by RO1138452, a specific IP antagonist, and also depended upon direct interaction of an adaptor protein PDZK1 with the IP (Turner et al. 2011). The molecular signaling events that regulate pro-angiogenic behaviors downstream of IP-PDZK1 interaction are not known and whether this mechanism has functional significance for the angiogenic properties of microvascular ECs is not yet clarified.

The vasculature of implanted tumors (lung carcinoma cells) also expresses PGD_2 receptors (DP) and exposure of bovine ECs to DP agonists improves barrier function in a cAMP-dependent manner, suggesting that, in this context, DP agonism mediates inflammation-associated vascular leakage (Murata et al. 2008).

6.5 Other Bioactive Lipid Mediators

There is now a large body of evidence to suggest that COX-2-derived mediators influence normal EC function through autocrine and paracrine signaling (e.g. Clarkin et al. 2008; Garonna et al. 2011) and some support for the hypothesis that similar regulatory mechanisms influence TEC function (Muraki et al. 2012). Alongside this, there is increasing recognition that a range of other biologically active AA metabolites derived from the activities of cytochrome P450s (reviewed in Fleming 2011) and lipoxygenases (LOX) also modify the angiogenic functions of ECs. 15(S)hydroxyeicosatetraenoic acid (15-HETE) is a product of 15-LOX activity and has been reported to stimulate microvascular EC migration and tubulogenesis and EC MMP expression through multiple signaling moieties including Rac-1 (a small Rho GTPase), Src (tyrosine kinase), PI3K-Akt and MAPK pathways (MEK-1/ERK and JNK-1) (Kundumani-Sridharan et al. 2010; Singh et al. 2010). Interestingly, 15-HETE-mediated angiogenic responses are suppressed by neutralizing FGF-2 antibodies (Kundumani-Sridharan et al. 2010) suggesting that, in common with other pro-angiogenic factors, these mediators depend to some extent upon the autocrine actions of EC-derived cytokines. More recently, Singh et al (Singh et al. 2011) have demonstrated inhibition of 15-HETE-stimulated EC motility and tube formation by the HMG-CoA reductase inhibitor simvastatin, thus providing new evidence for involvement of the cholesterol synthetic pathway in 15-LOX-dependent angiogenesis and revealing a potentially novel means of limiting pathological angiogenesis (Singh et al. 2011). The specific roles of HETEs in the functions of ECs isolated from tumors have not been studied to date but there is evidence that these factors regulate cancer cell mediator production and function (including apoptosis, motility and metastasis). The exact role of 15-LOX in controlling EC function and angiogenesis in tumors is not yet established since it has also been reported that metabolites of 15-LOX are anti-tumorigenic (Janakiram and Rao 2009).

Other lipid-derived mediators produced under inflammatory conditions also have significance for controlling the angiogenic properties of ECs. Platelet-activating factor (PAF), for example, is produced by ECs in response to several pro-angiogenic mediators (including VEGF and thrombin) and it has been reported that TECs synthesis greater quantities of PAF than normal microvascular ECs (Doublier et al. 2007). These authors showed that over-expression of PAF-acetylhydrolase, the major PAF-inactivating enzyme, in TECs suppressed their motility and tube forming capacity, strongly suggesting that autocrine effects of this mediator contribute to the pro-angiogenic phenotype of TECs (Doublier et al. 2007). Thus, multiple lipid-derived mediators have potential importance as autocrine regulators of EC angiogenic functions.

Endocannabinoids are lipid signaling mediators with emerging roles as regulators of EC function and angiogenesis and documented anti-tumor actions (Pisanti et al. 2007). Adult aortic ECs and endothelial progenitor cells are capable of synthesizing anandamide (N-arachidonoyl ethanolamide) and 2-arachidonoylglycerol and also respond to them through ligation of cannabinoid CB_1 and CB_2 receptors (Opitz et al. 2007) raising the possibility that they constitute another group of lipid-derived mediators with potential to control EC behavior through autocrine and paracrine mechanisms. 2-methyl-2'-F-anandamide, a metabolically stable anandamide analogue, has been reported to inhibit FGF-driven EC proliferation and differentiation in a CB_1 receptor-dependent manner (Pisanti et al. 2007) and knockdown of CB_1 in HUVEC suppresses FGF-stimulated migration and tubulogenesis through inhibition of MEK-ERK and Akt signaling (Pisanti et al. 2011). These studies together suggest that engagement of the CB_1 receptor facilitates EC motility and angiogenesis. However, there is also evidence that related molecules act primarily as pro-angiogenic mediators in ECs *in vitro*. N-arachidonoyl serine, for example, stimulates proliferation, migration and tube forming capacity of human microvascular ECs and enhances phosphorylation of ERK1/2 and Akt (Zhang et al. 2010). This most likely occurs in a CB_1-independent manner, suggesting that similar endogenous endocannabinoids use highly distinct mechanisms to influence the angiogenic functions of NECs. Whether the effects of these different mediators on ECs are functionally significant for angiogenesis in the tumor microenvironment and/or in physiological contexts (e.g. wound healing and repair) and the mechanisms responsible require investigation.

Since ECs are capable of producing a range of inflammatory mediators and growth factors, paracrine control of cancer cell function by EC-derived mediators is also important. It has recently been shown that prolonged incubation of cancer cells with conditioned medium from quiescent HUVEC reduced their expression of active inflammatory signaling elements (e.g. phosphorylated NF-κB and STAT-3) and inhibited their proliferation and migratory capacity (Franses et al. 2011). However, since the secretome of hyper-activated TECs is likely to be highly distinct from that of normal large vessel ECs, the significance of these findings for understanding cancer cell-EC cross-talk in the tumor environment is, as yet, unclear.

7   Anti-inflammatory Drugs as Regulators of Tumor Angiogenesis: Focus on NSAIDs

As discussed above, the tumor microenvironment is characterized by chronic inflammation and is rich in mediators that sustain inflammatory gene expression in cancer cells and stromal cells, including ECs. Thus, anti-inflammatory drugs would be expected to have benefit in suppressing tumor growth and tumor angiogenesis. This section will give a brief overview of some of the evidence supporting the inhibitory effects of NSAIDs on these processes, particularly those with selectivity for COX-2. However, it must be borne in mind that most of the existing studies focus on tumor cells themselves and that there is still very little understanding of precisely how these drugs influence the function of TECs or indeed normal ECs.

It is now well established that COX-2 promotes tumor growth and is highly up-regulated in most common cancers (Khan et al. 2011), where it is associated with neoplastic transformation, cell growth, angiogenesis, invasiveness, inhibition of apoptosis, and metastasis (Tsujii et al. 1998; Pai et al. 2003; Chang et al. 2004). The earliest direct evidence for the potential use of NSAIDs as anti-neoplastic drugs in humans was the documented regression of rectal polyps in patients treated with sulindac or indomethacin (Waddell and Loughry 1983). Since then epidemiological studies have demonstrated that the regular use of aspirin, or other NSAIDs, reduces death from colorectal cancer (CRC) (Thun et al. 2002; Huls et al. 2003) as well as limiting the recurrence of adenomas in patients with a history of CRC (Sandler et al. 2003). Not surprisingly, there is now ample evidence supporting the use of NSAIDs as anti-cancer drugs, where they have been shown to induce, or restore, apoptosis and to suppress tumor angiogenesis (Thun et al. 2002; Khan and Lee 2011).

The NSAIDs, such as aspirin and sulindac, possess chemopreventive properties through their inhibition of COX-1 and COX-2 enzyme activities. However, their regular use is associated with side effects, including gastric ulceration and coronary thrombosis, which led to the development of selective COX-2 inhibitors, collectively known as the Coxibs (e.g. celecoxib) (Khan and Lee 2011). The current approach aims to design selective COX-2 inhibitors with improved anti-cancer efficacy whilst eliminating their cardiovascular side effects.

7.1 COX-2 in Tumor Progression: Pro-apoptotic Effects of NSAIDs

The prominent role of COX-2 in tumor promotion is supported by numerous studies in colorectal tumor models. A reduction in the number and size of colonic polyps was observed in adenomatous polyposis coli $(APC)^{\Delta 716}$ mice (model of human familial adenomatous polyposis) crossed with mice harboring an inactivated *Ptgs2* (murine COX-2) gene. Furthermore, treating $APC^{\Delta 716}$ mice with a selective COX-2 inhibitor reduced the polyp number more effectively than sulindac (a dual COX-2/COX-1 inhibitor) (Oshima et al. 1996).

As discussed earlier, the COX-2-derived product, PGE_2, is a major driver of tumor progression and is now known to directly affect the behavior of tumor cells as well as ECs and other stromal cell types. The importance of PGE_2 in the tumor microenvironment was demonstrated by several key publications appearing well over a decade ago. For example, deletion of the gene encoding EP2 (but not EP1 or EP3) decreased polyps in $APC^{\Delta 716}$ mice (Sonoshita et al. 2001), providing clear evidence of a role for PGE_2 receptors in progression. Alongside these investigations *in vivo*, a series of *in vitro* studies confirmed the central role of PGE_2 in controlling cancer cell behavior. In this respect exposure of human colon cancer cells to PGE_2 was shown to increase colony forming rate (Sheng et al. 1998). In this latter study treatment with a selective COX-2 inhibitor (SC-58125) decreased colony formation and this growth inhibitory effect was reversed by exogenous PGE_2, supporting an autocrine role for COX-2-derived PGE_2 in cancer cell growth.

It is known that apoptosis maintains homeostasis in continuously replicating tissues and well established that premalignant and malignant lesions exhibit a decreased rate of apoptosis. One important mechanism that is accepted to contribute to the anti-cancer activities of NSAIDs is their ability to restore cancer cell apoptosis. Early studies in intestinal cancer cells, for example, demonstrated pro-apoptotic effects of SC-58125 (Sheng et al. 1998); PGE_2 abrogated the effect of SC-58125 and concomitantly increased expression of the anti-apoptotic protein Bcl-2 without modifying the levels of Bcl-x or Bax (Sheng et al. 1998). Thus, cytoprotection mediated by PGE_2 enhances the tumorigenic potential of intestinal epithelial cells. Subsequently, it has now become well accepted (and reviewed elsewhere) that COX-2 products influence both the extrinsic (receptor mediated) and intrinsic (mitochondrial mediated) apoptotic pathways (Li et al. 2001; Leahy et al. 2002; Iwase et al. 2006; Khan et al. 2011). In general, over-expression of COX-2 reduces Bax and Bcl-xl (pro apoptotic proteins) (Liu et al. 2001) and increases Bcl-2 and survivin expression (anti-apoptotic proteins) (Li et al. 2001; Krysan et al. 2004). Survivin is a key anti-apoptotic protein over-expressed in many human cancers (Khan 2006) and confers resistance to conventional cancer therapies, including chemotherapy and radiotherapy (Khan et al. 2010). COX-2 stabilizes survivin by maintaining ubiquitin levels, thus preventing its degradation and providing resistance to apoptosis (Barnes et al. 2006). In the absence of COX-2 activity this stabilization effect is lost, thus permitting apoptosis. In this setting, blocking COX-2 activity with NSAIDs, and therefore preventing production of PGE_2, would be predicted to allow apoptosis to take place unhindered. In this respect PGE_2 has also been reported to protect against Fas-mediated apoptosis (Casado et al. 2007) whereas COX-2 inhibition with sulindac enhanced Fas-mediated cell death (Brueggemeier and Diaz-Cruz 2006). Recent studies in colon cancer cells have also shown that celecoxib-induced apoptosis can be potentiated by ABT-737, a small molecule Bcl-2/Bcl-x(L) antagonist (Huang and Sinicrope 2010). Thus, the ability to modify the expression of proteins involved in regulating apoptosis, and therefore to reinstate the apoptotic program, is a key feature associated with the cellular actions of several NSAIDs and contributes to their anti-tumorigenic properties.

7.2 NSAIDs and Suppression of Angiogenesis

In 1998 Tsujii and co-workers used co-culture models to examine the ability of colon cancer cells to influence EC function (Tsujii et al. 1998). These authors showed that cells with constitutive over-expression of

COX-2 released a range of pro-angiogenic factors, including PGE_2, and that these cells (but not cells devoid of COX-2) promoted EC motility and tubulogenesis, raising the possibility that cancer cells control EC function through paracrine mechanisms. Both NS398 and aspirin reduced the ability of cancer cells to influence EC function, consistent with a role for COX activity. Interestingly, this early study implicated COX-1 in the paracrine effect because COX-1 antisense oligonucleotides inhibited COX-1 expression and activity and also suppressed tube formation in the co-cultures (Tsujii et al. 1998). Since expression of COX-1 does not generally change in response to pro-angiogenic factors, its potential role in angiogenesis has received little attention.

While there is emerging evidence supporting the ability of COX-2 blockade to modulate the pro-proliferative, angiogenic functions of NECs through modification of key signaling pathways (Lin et al. 2004; Wu et al. 2006; Garonna et al. 2011), information concerning the effects of NSAIDs on TEC function is sparse. However, it has recently been documented that COX-2 inhibition selectively suppresses the pro-angiogenic functions of tumor ECs (TECs) and vascular progenitor cells (Muraki et al. 2012). In this study COX-2 expression was confirmed in the vasculature of surgically resected human tumors, and studies in TECs isolated from human melanoma and oral carcinoma xenografts in mice showed that the COX-2-selective NSAID, NS398, inhibited angiogenesis by modifying EC functions. These authors also showed that COX-2 mRNA was up-regulated in TECs compared to NECs and that both migration and proliferation of TECs, but not NECs, were suppressed by exposure to NS398. COX-2 inhibition also reduced the number of circulating $CD133^+/VEGFR2^+$ cells as well as the number of these cells incorporated into tumor vessels (Muraki et al. 2012). Thus, NS398 was suggested to target both TECs and vascular progenitor cells without affecting NECs. Significantly, TECs appear to exhibit greater sensitivity to COX-2 inhibitors than NECs (in this case human dermal microvascular ECs) and tumor cells themselves (Muraki et al. 2012), suggesting that lower doses of COX-2 inhibitors could selectively target TECs and reduce tumor angiogenesis, thus potentially limiting their undesirable cardiovascular side effects. Whether other NSAIDs have similar, or distinct, effects in these models remains to be determined.

Interestingly EP4 expression in cancer cells is also reduced by exposure to some NSAIDs (e.g. sulindac sulfide), an effect which was reversed by pharmacological blockade of the MEK-ERK pathway with PD98059 and which was accompanied by increased expression of the pro-apoptotic gene NAG-1 (Kambe et al. 2009). Moreover, it has been reported that PGE_2 inhibits whereas celecoxib (COX-2-selective) up-regulates NAG-1 expression and suggested that there is, potentially, an inverse relationship between NAG-1 (pro-apoptotic) and COX-2 (pro-angiogenic) expression in tumors (Iguchi et al. 2009). These data suggest that inhibition of the expression of pro-angiogenic and pro-inflammatory receptors could represent another target for NSAIDs that would help to explain their benefits as anti-carcinogenic drugs and suppressors of tumor angiogenesis. Whether these actions extend to all NSAIDs and to TECs or NECs are not known, but warrant investigation.

A recently-described beneficial effect of celecoxib is its apparent ability to suppress epithelial to mesenchymal transition (EMT) (Bocca et al. 2011), a process important for tumor cell invasiveness and metastases and thought be involved in cancer progression. In this study the authors used human HT-29 colon cancer cells and showed that celecoxib treatment reduced hypoxia- and EGF-stimulated proliferation, invasiveness, activation of ERK1/2 and PI3K, and expression of markers of EMT. In addition, these inhibitory actions of celecoxib were not observed in SW-480 colon cancer cells (which are devoid of COX-2 expression) suggesting that the effects are likely to be 'on-target' and dependent upon reduced COX-2 activity (Bocca et al. 2011).

It should also be pointed out that there is also evidence for the development of celecoxib resistance in some settings. For instance, in a model of chronic intestinal inflammation short-term celecoxib treatment reduced COX-2 expression and PGE_2 synthesis in mouse ileum, whereas prolonged exposure was associated with much higher expression of COX-2, VEGF and IL-1β (Carothers et al. 2010). Moreover, this was accompanied by parallel changes in the microvasculature, with short term treatment reducing and longer term treatment enhancing microvessel density, consistent with the altered expression of pro-angiogenic mediators (Carothers et al. 2010).

7.3   COX-2-dependent and -independent Effects of COX-2-Selective Inhibitors

COX-2 is upregulated in a range of cancers and is thought to support tumorigenesis and tumor angiogenesis. While blockade of COX-2 activity and the consequent reduction in pro-angiogenic mediator synthesis results in reduced angiogenesis, it is now known that COX-2 inhibitors, particularly celecoxib, have additional pharmacologic activities which are independent of COX-2 inhibitory activity. In general, these actions are not mimicked by other non-selective NSAIDs at equivalent concentrations; are evident in cells devoid of COX-2 expression; and can be detected in tumor tissues from drug treated animals, suggesting that these effects are not *in vitro* artifacts and do not represent indirect consequences of COX-2 inhibition (reviewed in Ghosh et al. 2010; Schonthal 2010).

Several studies have now investigated the anti-tumorigenic effects of 2,5-dimethyl-celecoxib (DMC), a structural analogue of celecoxib that lacks ability to inhibit COX-2 enzyme activity. This drug has been reported to mimic, with greater potency, the anti-proliferative and anti-tumorigenic effects of celecoxib and therefore has promise therapeutically. A range of effects have been reported in both *in vitro* and *in vivo* models. For example, DMC inhibited cell proliferation of Burkitt's lymphoma cells *in vitro* by a mechanism involving down-regulation of cyclins A and B and subsequent loss of cyclin-dependent kinase activity. This effect was also evident *in vivo* and resulted in reduced tumor growth in mouse models (Kardosh et al. 2005). Other studies have shown that celecoxib concentration-dependently inhibited cholangiocarcinoma cell growth, causing cell cycle arrest at G_1 which was associated with increased expression of the cyclin-dependent kinase inhibitors p21(waf1/cip1) and p27(kip1). Overexpression or antisense depletion of COX-2 had no effect on expression of either p21 or p27, suggesting a COX-2-independent mechanism of action (Han et al. 2004; Mineo et al. 2010). In another investigation employing ECs (HUVECs) the anti-proliferative effects of celecoxib and DMC were attributed to inhibition of the PI3K-Akt pathway and reduced cyclin-dependent kinase activities. These authors showed that celecoxib- and DMC-driven G_1 arrest were associated with reduced retinoblastoma protein phosphorylation through inhibition of several cyclin-dependent kinases. Celecoxib and DMC have also been reported to reduce neovascularization in the chicken chorioallantoic membrane assay, suggesting the potential involvement of a COX-2-independent mechanism contributing to the *in vivo* anti-angiogenic effects of celecoxib, at least in this model (Lin et al. 2004).

Celecoxib and DMC have also both been shown to suppress survivin expression. As indicated above, survivin is an anti-apoptotic protein highly expressed in tumor cells and has been reported to confer resistance to anti-cancer therapies (Khan et al. 2010). Its expression in NECs is also regulated by a number of pro-angiogenic factors, including VEGF (Grosjean et al. 2006). COX-2 activity has been shown to prevent survivin degradation, indicating that pharmacological targeting of COX-2 may restore apoptosis (Barnes et al. 2006). However, DMC, which lacks COX-2 inhibitory function, was also able to down-regulate survivin expression (Pyrko et al. 2006) indicating that celecoxib and DMC may exert their cytotoxic anti-tumor effects through via a mechanism that does not require COX-2 inhibition. These investigators also found that suppression of survivin was specific to these two drugs, as survivin levels were not modified by equivalent concentrations of either other coxibs (valdecoxib, rofecoxib) or traditional NSAIDs (flurbiprofen, indomethacin, sulindac). When combined with irinotecan, a widely used anticancer drug, celecoxib and DMC both greatly enhanced the cytotoxic effects of this drug, in keeping with the hypothesis that reduced survivin expression may be beneficial in sensitizing cancer cells to chemotherapy. Importantly, this study also showed that these *in vitro* effects were recapitulated in tumors *in vivo* where celecoxib and DMC similarly repressed survivin, thus inducing apoptosis, and inhibited tumor growth.

There is also evidence that both celecoxib and DMC modulate the activity of the endoplasmic reticulum (ER) calcium ATPase (SERCA), a Ca^{2+}ATPase which transfers Ca^{2+} from the cytosol to the lumen of the ER and is responsible for maintaining the steep calcium gradient of calcium between these two cellular compartments (Johnson et al. 2002; Pyrko et al. 2007). Inhibition of pump activity increases cytosolic calcium and triggers the ER stress response (ESR) which in turn promotes apoptosis

(Pyrko et al. 2007; Ghosh et al. 2010). This effect is apparently not shared by other NSAIDs and is thus unique to celecoxib and its non-coxib analogue DMC. Calcium-mediated ERS is therefore a major regulator of tumor cell death induced by both celecoxib and DMC, effects which have also been confirmed in tumor tissue from animals treated *in vivo* (Tsutsumi et al. 2004; Kardosh et al. 2008).

As alluded to earlier many tumors, particularly those with PTEN deletions are characterized by chronically elevated activation of the PI3K-Akt axis that contributes to both tumor growth and chemoresistance (Kulp et al. 2004). Several studies have now shown that celecoxib directly targets and reduces the activity of 3-phosphoinositide-dependent protein kinase-1 (PDK-1), a key upstream regulator of Akt (Arico et al. 2002; Kulp et al. 2004) and PTEN, which in turn negatively regulates of PI3K-PDK1-Akt signaling. Furthermore, these studies fueled the generation of a series of COX-2 inactive celecoxib derivatives with increased potency towards PDK1, including OSU-03012 (Zhu et al. 2004). Since, as discussed earlier, PI3K-PDK1-Akt signaling plays a key role in maintaining the chronic inflammatory state of tumors and promoting tumorigenesis, these analogues could be relevant therapeutics.

Other recent studies have identified additional pathways and signaling elements targeted by celecoxib and related drugs. For instance, exposure of colon cancer cells to celecoxib also results in phosphorylation of GSK3β (a known Akt substrate) through a mechanism involving enhanced protein kinase C activity but independent of changes in Akt phosphorylation (Chen et al. 2011). This pathway was also shown to promote degradation of c-FLIP (anti-apoptotic) in celecoxib-treated cells, resulting in increased apoptosis (Chen et al. 2011). Selective induction of the expression of other genes with anti-tumorigenic activity has also been described. In this respect celecoxib induces the expression of MAGI1 (a scaffolding protein implicated in the stabilization of adherent junctions) in colon cancer line lines and this limits Wnt signaling, anchorage-independent growth, migration and invasion *in vitro* (Zaric et al. 2012). There has also been a suggestion that celecoxib can bind directly to STAT-3 (Reed et al. 2011), which is constitutively active in many cancers, and in TECs (Phung et al. 2006; Lassoued et al. 2011). Reed and co-workers have shown that celecoxib inhibits IL-6-induced STAT-3 phosphorylation, reduces expression of STAT-3-dependent genes (Bcl-2, survivin) and abrogates sarcoma cell migration (Reed et al. 2011); the authors suggested that these effects could result from celecoxib's ability to bind directly to STAT-3, although this has not yet been shown experimentally. There is also emerging evidence that NSAIDs suppress the activity of RTK-mediated signaling cascades (Schiffmann et al. 2008; Tuynman et al. 2008; Lu et al. 2009). Using peptide array analysis of drug-treated colon cancer cells Tuynman et al. (Tuynman et al. 2008) showed that celecoxib down-regulates activity of the c-Met (HGF receptor) and insulin-like growth factor receptors, resulting in decreased downstream activity of ERK1/2, the PI3K-Akt pathway and Wnt, all of which have been implicated in cancer cell growth. Thus, celecoxib is clearly capable of suppressing multiple kinase activities that normally co-operate to maintain proliferation and heighten pro-inflammatory cytokine synthesis.

As well as over-expressing COX-2, many human colorectal tumors exhibit reduced expression of 15-LOX, the product of which is 13-S-hydroxyoctadecadienoic acid (13-S-HODE), and which has growth inhibitory and pro-apoptotic actions (Shureiqi et al. 2000; Thompson et al. 2000). It has also been shown that both sulindac and NS-398 can up-regulate 15-LOX expression and activity in colon cancer cells in parallel with inhibition of proliferation and increased apoptosis, demonstrating that NSAIDs can promote apoptosis in colon cancer cells via up-regulation of 15-LOX-1 in the absence of COX-2 activity. Acetylating NSAIDs, such as aspirin, also switch COX-2 from forming PGE_2 (pro-tumorigenic and pro-angiogenic) to 15-epi-LXA_4 (anti-tumorigenic and anti-angiogenic). Thus, NSAIDs can indirectly or directly modify cellular expression of bioactive lipids to regulate tumorigenic and angiogenic potential.

DMC can also modify expression of PG synthases and hence directly influence production of pro-angiogenic PGE_2 without affecting COX activity. In this respect, Deckmann and colleagues have recently shown that exposure of HeLa cells to DMC results in formation of a signaling complex that binds to the EGR1 promoter, reduces its transcriptional activity and inhibits expression of mPGES-1 (Deckmann et al. 2012).

Beneficial effects of both celecoxib and sulindac on membrane fluidity in colon cancer cells have also been described (Vaish and Sanyal 2011; Sade et al. 2012) and these actions occur irrespective of COX-2 expression status (Sade et al. 2012), suggesting that they are independent of their COX-inhibiting activities. In addition, celecoxib activates sphingolipid biosynthesis in colon cancer cells, an effect which contributes to the established anti-proliferative, pro-apoptotic effects of this drug (Schiffmann et al. 2010). Thus, celecoxib, and potentially other NSAIDs, are likely to exert multiple effects on cell behavior through COX-dependent and -independent mechanisms.

While the majority of studies have focused on cancer cells themselves as targets for NSAIDs, it is clear that celecoxib and related drugs exert a number of off-target effects which seem to be directed towards limiting the overactive pro-inflammatory signaling that drives tumor growth and angiogenesis. Recent studies using human glioma specimens have shown that DMC triggered apoptosis of tumor-associated brain ECs and reduced their proliferation and motility, whereas normal brain ECs were unaffected (Virrey et al. 2010). These data provide evidence that DMC targets both tumor cells and tumor ECs, at least in this model.

7.4   Therapeutic Drugs and COX-2 Activity

A vast array of NSAIDs is currently undergoing use in clinical trials and in experimental research. An in-depth description of all of these drugs is beyond the scope of this article, but there are several published review articles on the topic (Taketo 1998a; Taketo 1998b; Janne and Mayer 2000; Thun et al. 2002; Huls et al. 2003). Briefly, nonselective NSAIDs such as aspirin, ibuprofen, naproxen, sulindac, and piroxicam have all been proven to inhibit carcinogenesis. However, non-selective NSAIDs suppress tumor growth to a greater extent and at lower doses when treatment is started before or coincident with exposure to carcinogen than when it is delayed until the tumor promotion/progression phase (Reddy et al. 1987; Reddy et al. 1999). Based on a need to develop safe and effective chemotherapeutic agents which are also active when treatment is delayed, selective COX-2 inhibitors have been widely investigated. For reasons discussed above, this type of drug has clear benefits and includes celecoxib (and other Coxibs, such as rofe- , valde-, etori-coxib), DMGc as well as NS398 (Jia et al. 2005; Su et al. 2006), etodolac (Chen et al. 2003), nimesulide (Su et al. 2006; Su et al. 2008) and SC-236 (Kim et al. 2005). These lists are far from exhaustive, but give a flavor of the enormous amount of ongoing research in this area.

Other novel compounds with anti-angiogenic activity have also received some recent attention because they appear to indirectly inhibit COX-2 expression and activity. For example, curcumin, a polyphenol found in dietary spice, suppresses COX-2 activity (Cheng et al. 2001; Leu and Maa 2002; Villegas et al. 2008). Organoselenium compounds, derived from the essential trace element selenium, inhibit cancer cell growth by activating AMPK and subsequently inhibiting expression of COX-2 and downstream production of PGE_2 (Hwang et al. 2006). Resveratrol, a naturally occurring polyphenolic phytoalexine found in grapes and other foods, was found to reduce tumor volume, weight and metastasis in mice with highly metastatic Lewis lung carcinoma and significantly inhibited tumor-induced neo-vascularization *in vivo* via its ability to inhibit COX-2 activity (Jang et al. 1997; Kimura and Okuda 2001).

Bereberine, an isoquinoline alkaloid, inhibits cancer cell motility through down regulation of COX-2 expression and activity, as well as EP2 and EP4 expression (Singh et al. 2011) and this inhibitory effect may be related to its ability to activate AMPK which is known to exert anti-inflammatory actions (Kim et al. 2011), in addition to contributing directly to the pro-angiogenic functions of NECs. Thus, it would appear that a number of drugs with reported anti-tumorigenic actions are capable of influencing, directly or indirectly, COX-2 activity, confirming the central importance of this enzyme in driving tumorigenesis and angiogenesis.

As discussed earlier, TECs differ from NECs in many respects (Hida and Klagsbrun 2005; Hida et al. 2008). They have an altered phenotype and cytogenetic abnormalities (St Croix et al. 2000; Akino et al. 2009) as well as expressing specific markers such as tumor endothelial markers (St Croix et al. 2000), EGF receptor (Amin et al. 2006), CD13 (Pasqualini et al. 2000) and Dkk-3 (Untergasser et al. 2008). TECs are apparently more sensitive to certain drugs including EGFR inhibitors (Amin et al. 2006) and the green tea polyphenol epigallocatechin-3 gallate (EGCG) (Ohga et al. 2009), as well as to NSAIDs (Muraki et al. 2012). In contrast, it would seem that TECs are more resistant to standard chemotherapeutic drugs, such as vincristine, than are NECs (Bussolati et al. 2003).

COX-2 inhibition is now also used as an adjunct, synergistic therapy to other chemotherapeutic agents (Dmitrovsky 2003) and cancer treatments such as radiotherapy (Soriano et al. 1999; Choy and Milas 2003). For example, EGFR inhibitor treatment often leads to acquired resistance pathways, which are now targeted with COX-2 inhibitors, in particular celecoxib (Kao et al. 2009). Notably, it has recently been demonstrated that celecoxib prevents the apoptotic effects of a range of chemotherapeutic agents in haematopoietic cell lines (Cerella et al. 2011). These authors showed that DNA-damage induced by etoposide was markedly reduced by exposure to celecoxib or its analogue DMC, implying that the anti-apoptotic effects occurred in a COX-independent manner. These data suggest that the off-target actions of NSAIDs may well be cell specific and not always beneficial. However, there are also many clinical studies in a variety of cancer types, which have examined the ability of COX-2 inhibitors to augment the effects of chemotherapy and which have shown favorable outcomes. Summaries of some of these clinical trials in different cancers and with different chemotherapeutic agents have been reviewed elsewhere (Falandry et al. 2009; Ghosh et al. 2010; Khan et al. 2011).

Interestingly, Suh et al (Suh et al. 2011) have recently noted that the combination of low-dose atorvastatin (anti-inflammatory) with either low-dose sulindac or naproxen reduced the incidence of colon adenocarcinoma in mice to a greater extent than either drug alone. In addition, tumors from these mice exhibited greatly reduced expression of a range of inflammatory genes (e.g, COX-2, IL-1β, TNF-α) and signaling elements (e.g. phosphorylated NF-κB), confirming the marked inhibition of inflammation with this drug combination. Since the pro-angiogenic effects of cytokines and COX-2-derived PGE$_2$ on ECs are relatively well established it might be predicted that these anti-inflammatory effects would be accompanied by reduced tumor angiogenesis, although this has yet to be examined directly.

The full basis for the cancer chemopreventative actions of celecoxib, its derivatives (e.g. DMC) and other NSAIDs is not yet established. Nevertheless, it is evident that these drugs are able to suppress generation of inflammatory mediators by both tumor cells and ECs and dampen activation of signaling pathways responsible for their maintained production. It is also clear that these drugs not only inhibit COX activities and reduce synthesis of pro-angiogenic PGE$_2$, but also exert a number of 'off-target' effects which have benefit in suppressing at least some of the inflammatory signaling supporting the positive feedback loops that drive tumor angiogenesis and growth and increase metastatic potential.

8 Conclusions and Perspectives

The close association between inflammation and blood vessel formation and its importance in both physiological and pathological angiogenesis is generally well accepted. Angiogenesis is clearly dependent upon the participation of multiple cells types and as a result can be influenced by numerous factors, many of which are regulated inter-dependently; this offers a wealth of potential targets for influencing the angiogenic cascade. With respect to cancer biology substantial, evidence now indicates that the tumor microenvironment is characterized by over-activity of pro-inflammatory signaling pathways, up-regulation of pro-inflammatory genes and mediators, and enhanced expression of receptors that mediate their cellular actions. Thus, chronic unresolved inflammation creates an environment in which ECs are permanently 'switched' towards a pro-angiogenic phenotype. A fuller characterization of the altered phenotypes of ECs derived from various tumor types, although challenging, is clearly warranted and could help to identify novel targets for suppressing tumor angiogenesis. It is also evident that virtually all of the key mediators and pathways involved in regulating physiological angiogenic processes are either up-regulated or over-active in the tumor microenvironment and are, therefore, drivers of pathological angiogenesis. Thus, it must be borne in mind that there could be disadvantages associated with blocking pathways that are particularly important for regulating reparative angiogenesis.

Bioactive lipid mediators have emerged as key players in the control of EC functions relevant for angiogenesis. While it is clear that COX-2-derived prostanoids play a prominent role, a fuller characterization of the repertoire of lipid-derived mediators produced by NECs and TECs and how these are influenced by products of other stromal cell populations (e.g. fibroblasts) is required. Understanding the interdependency of production of these mediators may also reveal new modes of interaction that could be therapeutically advantageous for limiting pathological angiogenesis. Given that the majority of inflammatory factors present in tumors are important regulators of EC behavior, many of them operating through altered COX-2 expression, the use of anti-inflammatory strategies to directly or indirectly modify these pathways should have benefit in managing conditions characterized by excessive angiogenesis. These strategies could include: (i) inhibiting enzyme activities responsible for producing inflammatory mediators (e.g. COX-2; cPLA$_2\alpha$; LOX), (ii) blocking receptors or co-receptors involved in generating or responding to these factors (EP4; VEGFR2), or (iii) targeting common signaling elements important for driving mediator production (MAPKs; Akt; AMPK). This latter approach may be particularly useful as some of the signaling elements involved in regulating EC function and angiogenesis in tumors may also be critical for conferring resistance to chemotherapeutics. Since multiple interacting signaling pathways control the response of ECs to paracrine and autocrine mediators it is possible that most benefit would be achieved by simultaneously targeting several of these key signaling elements alongside use of standard anti-inflammatory approaches. Indeed, COX-2-selective inhibitors show promise as anti-angiogenic drugs not only by limiting production of prostanoids (notably PGE$_2$) but also by exerting beneficial off-target effects which generally act to suppress over-active pro-inflammatory signaling pathways in both ECs and tumor cells.

Current evidence suggests the existence of highly intricate mechanisms of receptor cross-talk and interaction in ECs and bi-directional communication between ECs and tumor cells. These signaling networks are clearly important for the regulation of EC functions relevant for angiogenesis and are a central feature of tumors characterized by chronic inflammation. Given this cross-talk complexity, and the wealth of mediators and signaling pathways involved, it is not surprising that the success of anti-angiogenic therapies has been highly variable. Thus, combined blockade of key signaling 'nodes' could have benefit in suppressing inflammatory activation of ECs and would be expected to maximally disrupt the autocrine and paracrine loops responsible for maintaining excessive inflammation. Improved understanding of the complex cellular cross-talk mechanisms operative in the tumor environment should help to identify new targets for influencing tumor-associated angiogenesis.

The authors thank the British Heart Foundation and the Wellcome Trust for funding their work on endothelial cell function and angiogensis.

REFERENCES

Akino, T., K. Hida, Y. Hida, K. Tsuchiya, D. Freedman, C. Muraki, N. Ohga, K. Matsuda, K. Akiyama, T. Harabayashi, N. Shinohara, K. Nonomura, M. Klagsbrun and M. Shindoh (2009). "Cytogenetic abnormalities of tumor-associated endothelial cells in human malignant tumors." Am J Pathol 175(6): 2657-2667.

Al-Mutairi, M., S. Al-Harthi, L. Cadalbert and R. Plevin (2010). "Over-expression of mitogen-activated protein kinase phosphatase-2 enhances adhesion molecule expression and protects against apoptosis in human endothelial cells." Br J Pharmacol 161(4): 782-798.

Albini, A., S. Indraccolo, D. M. Noonan and U. Pfeffer (2010). "Functional genomics of endothelial cells treated with anti-angiogenic or angiopreventive drugs." Clin Exp Metastasis 27(6): 419-439.

Alfranca, A., J. M. Lopez-Oliva, L. Genis, D. Lopez-Maderuelo, I. Mirones, D. Salvado, A. J. Quesada, A. G. Arroyo and J. M. Redondo (2008). "PGE$_2$ induces angiogenesis via MT1-MMP-mediated activation of the TGFbeta/Alk5 signaling pathway." Blood 112(4): 1120-1128.

Amano, H., Y. Ito, T. Suzuki, S. Kato, Y. Matsui, F. Ogawa, T. Murata, Y. Sugimoto, R. Senior, H. Kitasato, I. Hayashi, Y. Satoh, S. Narumiya and M. Majima (2009). "Roles of a prostaglandin E-type receptor, EP3, in upregulation of matrix metalloproteinase-9 and vascular endothelial growth factor during enhancement of tumor metastasis." Cancer Sci 100(12): 2318-2324.

Amin, D. N., K. Hida, D. R. Bielenberg and M. Klagsbrun (2006). "Tumor endothelial cells express epidermal growth factor receptor (EGFR) but not ErbB3 and are responsive to EGF and to EGFR kinase inhibitors." Cancer Res 66(4): 2173-2180.

Aoki, N., R. Yokoyama, N. Asai, M. Ohki, Y. Ohki, K. Kusubata, B. Heissig, K. Hattori, Y. Nakagawa and T. Matsuda (2010). "Adipocyte-derived microvesicles are associated with multiple angiogenic factors and induce angiogenesis *in vivo* and *in vitro*." Endocrinology 151(6): 2567-2576.

Arico, S., S. Pattingre, C. Bauvy, P. Gane, A. Barbat, P. Codogno and E. Ogier-Denis (2002). "Celecoxib induces apoptosis by inhibiting 3-phosphoinositide-dependent protein kinase-1 activity in the human colon cancer HT-29 cell line." J Biol Chem 277(31): 27613-27621.

Arroyo, A. G. and M. L. Iruela-Arispe (2010). "Extracellular matrix, inflammation, and the angiogenic response." Cardiovasc Res 86(2): 226-235.

Ashida, N., S. Senbanerjee, S. Kodama, S. Y. Foo, M. Coggins, J. A. Spencer, P. Zamiri, D. Shen, L. Li, T. Sciuto, A. Dvorak, R. E. Gerszten, C. P. Lin, M. Karin and A. Rosenzweig (2011). "IKKbeta regulates essential functions of the vascular endothelium through kinase-dependent and -independent pathways." Nat Commun 2: 318.

Aso, H., S. Ito, A. Mori, M. Morioka, N. Suganuma, M. Kondo, K. Imaizumi and Y. Hasegawa (2012). "Prostaglandin E2 enhances interleukin-8 production via EP4 receptor in human pulmonary microvascular endothelial cells." Am J Physiol Lung Cell Mol Physiol 302(2): 266-273.

Baker, N., S. J. O'Meara, M. Scannell, P. Maderna and C. Godson (2009). "Lipoxin A4: anti-inflammatory and anti-angiogenic impact on endothelial cells." J Immunol 182(6): 3819-3826.

Banu, N., A. Buda, S. Chell, D. Elder, M. Moorghen, C. Paraskeva, D. Qualtrough and M. Pignatelli (2007). "Inhibition of COX-2 with NS-398 decreases colon cancer cell motility through blocking epidermal growth factor receptor transactivation: possibilities for combination therapy." Cell Prolif 40(5): 768-779.

Barnes, N., P. Haywood, P. Flint, W. F. Knox and N. J. Bundred (2006). "Survivin expression in in situ and invasive breast cancer relates to COX-2 expression and DCIS recurrence." Br J Cancer 94(2): 253-258.

Barnett, J. M., G. W. McCollum and J. S. Penn (2010). "Role of cytosolic phospholipase A(2) in retinal neovascularization." Invest Ophthalmol Vis Sci 51(2): 1136-1142.

Bertazza, L. and S. Mocellin (2010). "The dual role of tumor necrosis factor (TNF) in cancer biology." Curr Med Chem 17(29): 3337-3352.

Biscetti, F., E. Gaetani, A. Flex, T. Aprahamian, T. Hopkins, G. Straface, G. Pecorini, E. Stigliano, R. C. Smith, F. Angelini, J. J. Castellot, Jr. and R. Pola (2008). "Selective activation of peroxisome proliferator-activated receptor (PPAR)alpha and PPAR gamma induces neoangiogenesis through a vascular endothelial growth factor-dependent mechanism." Diabetes 57(5): 1394-1404.

Biscetti, F., E. Gaetani, A. Flex, G. Straface, G. Pecorini, F. Angelini, E. Stigliano, T. Aprahamian, R. Smith, J. Castellot and R. Pola (2009). "Peroxisome proliferator-activated receptor alpha is crucial for iloprost-induced in vivo angiogenesis and vascular endothelial growth factor upregulation." J Vasc Res 46(2): 103-108.

Bishop-Bailey, D. and T. Hla (1999). "Endothelial cell apoptosis induced by the peroxisome proliferator-activated receptor (PPAR) ligand 15-deoxy-Delta12, 14-prostaglandin J2." J Biol Chem 274(24): 17042-17048.

Bocca, C., F. Bozzo, S. Cannito, M. Parola and A. Miglietta (2011). "Celecoxib inactivates epithelial-mesenchymal transition stimulated by hypoxia and/or epidermal growth factor in colon cancer cells." Mol Carcinog.

Bollrath, J. and F. R. Greten (2009). "IKK/NF-kappaB and STAT3 pathways: central signaling hubs in inflammation-mediated tumor promotion and metastasis." EMBO Rep 10(12): 1314-1319.

Boosani, C. S., N. Nalabothula, N. Sheibani and A. Sudhakar (2010). "Inhibitory effects of arresten on bFGF-induced proliferation, migration, and matrix metalloproteinase-2 activation in mouse retinal endothelial cells." Curr Eye Res 35(1): 45-55.

Bouchentouf, M., K. A. Forner, J. Cuerquis, V. Michaud, J. Zheng, P. Paradis, E. L. Schiffrin and J. Galipeau (2010). "Induction of cardiac angiogenesis requires killer cell lectin-like receptor 1 and alpha4beta7 integrin expression by NK cells." J Immunol 185(11): 7014-7025.

Brat, D. J., A. C. Bellail and E. G. Van Meir (2005). "The role of interleukin-8 and its receptors in gliomagenesis and tumoral angiogenesis." Neuro Oncol 7(2): 122-133.

Brezinski, M. E., M. A. Gimbrone, Jr., K. C. Nicolaou and C. N. Serhan (1989). "Lipoxins stimulate prostacyclin generation by human endothelial cells." FEBS Lett 245(1-2): 167-172.

Brock, T., R. McNish and M. Peters-Golden (1999). "Arachidonic acid is preferentially metabolized by cyclooxygenase-2 to prostacyclin and prostaglandin E2." J Biol Chem 274(17): 11660-11666.

Brueggemeier, R. W. and E. S. Diaz-Cruz (2006). "Relationship between aromatase and cyclooxygenases in breast cancer: potential for new therapeutic approaches." Minerva Endocrinol 31(1): 13-26.

Bussolati, B., I. Deambrosis, S. Russo, M. C. Deregibus and G. Camussi (2003). "Altered angiogenesis and survival in human tumor-derived endothelial cells." FASEB J 17(9): 1159-1161.

Bussolati, B., M. C. Deregibus and G. Camussi (2010). "Characterization of molecular and functional alterations of tumor endothelial cells to design anti-angiogenic strategies." Curr Vasc Pharmacol 8(2): 220-232.

Carmeliet, P. (2003). "Angiogenesis in health and disease." Nat Med 9(6): 653-660.

Carmeliet, P. and R. K. Jain (2000). "Angiogenesis in cancer and other diseases." Nature 407(6801): 249-257.

Carmi, Y., E. Voronov, S. Dotan, N. Lahat, M. A. Rahat, M. Fogel, M. Huszar, M. R. White, C. A. Dinarello and R. N. Apte (2009). "The role of macrophage-derived IL-1 in induction and maintenance of angiogenesis." J Immunol 183(7): 4705-4714.

Carothers, A. M., J. S. Davids, B. C. Damas and M. M. Bertagnolli (2010). "Persistent cyclooxygenase-2 inhibition downregulates NF-{kappa}B, resulting in chronic intestinal inflammation in the min/+ mouse model of colon tumorigenesis." Cancer Res 70(11): 4433-4442.

Casado, M., B. Molla, R. Roy, A. Fernandez-Martinez, C. Cucarella, R. Mayoral, L. Bosca and P. Martin-Sanz (2007). "Protection against Fas-induced liver apoptosis in transgenic mice expressing cyclooxygenase 2 in hepatocytes." Hepatology 45(3): 631-638.

Cathcart, M.-C., J. V. Reynolds, K. J. O'Byrne and G. P. Pidgeon (2010). "The role of prostacyclin synthase and thromboxane synthase signaling in the development and progression of cancer." Biochimica et Biophysica Acta (BBA) – Reviews on Cancer 1805(2): 153-166.

Caughey, G., L. Cleland, P. Penglis, J. Gamble and M. James (2001). "Roles of cyclooxygenase (COX)-1 and COX-2 in prostanoid production by human endothelial cells: selective up-regulation of prostacyclin synthesis by COX-2." J Immunol 167(5): 2831-2838.

Cerella, C., C. Sobolewski, S. Chateauvieux, E. Henry, M. Schnekenburger, J. Ghelfi, M. Dicato and M. Diederich (2011). "COX-2 inhibitors block chemotherapeutic agent-induced apoptosis prior to commitment in hematopoietic cancer cells." Biochem Pharmacol 82(10): 1277-1290.

Chang, M. Y., F. M. Ho, J. S. Wang, H. C. Kang, Y. Chang, Z. X. Ye and W. W. Lin (2010). "AICAR induces cyclooxygenase-2 expression through AMP-activated protein kinase-transforming growth factor-beta-activated kinase 1-p38 mitogen-activated protein kinase signaling pathway." Biochem Pharmacol 80(8): 1210-1220.

Chang, S. H., C. H. Liu, R. Conway, D. K. Han, K. Nithipatikom, O. C. Trifan, T. F. Lane and T. Hla (2004). "Role of prostaglandin E2-dependent angiogenic switch in cyclooxygenase 2-induced breast cancer progression." Proc Natl Acad Sci USA 101(2): 591-596.

Chappell, W. H., L. S. Steelman, J. M. Long, R. C. Kempf, S. L. Abrams, R. A. Franklin, J. Basecke, F. Stivala, M. Donia, P. Fagone, G. Malaponte, M. C. Mazzarino, F. Nicoletti, M. Libra, D. Maksimovic-Ivanic, S. Mijatovic, G. Montalto, M. Cervello, P. Laidler, M. Milella, A. Tafuri, A. Bonati, C. Evangelisti, L. Cocco, A. M. Martelli and J. A. McCubrey (2011). "Ras/Raf/MEK/ERK and PI3K/PTEN/Akt/mTOR inhibitors: rationale and importance to inhibiting these pathways in human health." Oncotarget 2(3): 135-164.

Chen, S., W. Cao, P. Yue, C. Hao, F. R. Khuri and S. Y. Sun (2011). "Celecoxib promotes c-FLIP degradation through Akt-independent inhibition of GSK3." Cancer Res 71(19): 6270-6281.

Chen, W. S., J. H. Liu, S. J. Wei, J. M. Liu, C. Y. Hong and W. K. Yang (2003). "Colon cancer cells with high invasive potential are susceptible to induction of apoptosis by a selective COX-2 inhibitor." Cancer Sci 94(3): 253-258.

Cheng, A. L., C. H. Hsu, J. K. Lin, M. M. Hsu, Y. F. Ho, T. S. Shen, J. Y. Ko, J. T. Lin, B. R. Lin, W. Ming-Shiang, H. S. Yu, S. H. Jee, G. S. Chen, T. M. Chen, C. A. Chen, M. K. Lai, Y. S. Pu, M. H. Pan, Y. J. Wang, C. C. Tsai and C. Y. Hsieh (2001). "Phase I clinical trial of curcumin, a chemopreventive agent, in patients with high-risk or pre-malignant lesions." Anticancer Res 21(4B): 2895-2900.

Choy, H. and L. Milas (2003). "Enhancing radiotherapy with cyclooxygenase-2 enzyme inhibitors: a rational advance?" J Natl Cancer Inst 95(19): 1440-1452.

Chung, A. S., J. Lee and N. Ferrara (2010). "Targeting the tumor vasculature: insights from physiological angiogenesis." Nat Rev Cancer 10(7): 505-514.

Ciccarelli, M., D. Sorriento, E. Cipolletta, G. Santulli, A. Fusco, R. H. Zhou, A. D. Eckhart, K. Peppel, W. J. Koch, B. Trimarco and G. Iaccarino (2011). "Impaired neoangiogenesis in beta-adrenoceptor gene-deficient mice: restoration by intravascular human beta-adrenoceptor gene transfer and role of NFkappaB and CREB transcription factors." Br J Pharmacol 162(3): 712-721.

Clark, J. D., A. R. Schievella, E. A. Nalefski and L. L. Lin (1995). "Cytosolic phospholipase A2." J Lipid Mediat Cell Signal 12(2-3): 83-117.

Clarkin, C., R. Emery, A. Pitsillides and C. Wheeler-Jones (2008a). "Evaluation of VEGF-mediated signaling in primary human cells reveals a paracrine action for VEGF in osteoblast-mediated crosstalk to endothelial cells." J Cell Physiol 214(2): 537-544.

Clarkin, C., E. Garonna, A. Pitsillides and C. Wheeler-Jones (2008b). "Heterotypic contact reveals a COX-2-mediated suppression of osteoblast differentiation by endothelial cells: A negative modulatory role for prostanoids in VEGF-mediated cell: cell communication?" Exp Cell Res 314(17): 3152-3161.

Covey, T. M., K. Edes and F. A. Fitzpatrick (2007). "Akt activation by arachidonic acid metabolism occurs via oxidation and inactivation of PTEN tumor suppressor." Oncogene 26(39): 5784-5792.

Dace, D. S., A. A. Khan, J. Kelly and R. S. Apte (2008). "Interleukin-10 promotes pathological angiogenesis by regulating macrophage response to hypoxia during development." PLoS ONE 3(10): e3381.

Daniel, T. O., H. Liu, J. D. Morrow, B. C. Crews and L. J. Marnett (1999). "Thromboxane A2 Is a Mediator of Cyclooxygenase-2-dependent Endothelial Migration and Angiogenesis." Cancer Research 59(18): 4574-4577.

de Leval, X., T. Dassesse, J. M. Dogne, D. Waltregny, A. Bellahcene, V. Benoit, B. Pirotte and V. Castronovo (2006). "Evaluation of original dual thromboxane A2 modulators as antiangiogenic agents." J Pharmacol Exp Ther 318(3): 1057-1067.

Deckmann, K., F. Rorsch, G. Geisslinger and S. Grosch (2012). "Dimethylcelecoxib induces an inhibitory complex consisting of HDAC1/NF-kappaB(p65)RelA leading to transcriptional downregulation of mPGES-1 and EGR1." Cell Signal 24(2): 460-467.

Dempke, W. C. and R. Zippel (2010). "Brivanib, a novel dual VEGF-R2/bFGF-R inhibitor." Anticancer Res 30(11): 4477-4483.

Dmitrovsky, E. (2003). "Combining cytotoxic chemotherapy with cyclooxygenase-2 inhibition." J Clin Oncol 21(14): 2631-2632.

Dormond, O., M. Bezzi, A. Mariotti and C. Ruegg (2002). "Prostaglandin E2 promotes integrin alpha Vbeta 3-dependent endothelial cell adhesion, rac-activation, and spreading through cAMP/PKA-dependent signaling." J Biol Chem 277(48): 45838-45846.

Doublier, S., M. Ceretto, E. Lupia, S. Bravo, B. Bussolati and G. Camussi (2007). "The proangiogenic phenotype of tumor-derived endothelial cells is reverted by the overexpression of platelet-activating factor acetylhydrolase." Clin Cancer Res 13(19): 5710-5718.

Du, X. L., T. Jiang, W. B. Zhao, F. Wang, G. L. Wang, M. Cui and Z. Q. Wen (2008). "Gene alterations in tumor-associated endothelial cells from endometrial cancer." Int J Mol Med 22(5): 619-632.

Ermert, L., C. Dierkes and M. Ermert (2003). "Immunohistochemical Expression of Cyclooxygenase Isoenzymes and Downstream Enzymes in Human Lung Tumors." Clinical Cancer Research 9(5): 1604-1610.

Falandry, C., P. A. Canney, G. Freyer and L. Y. Dirix (2009). "Role of combination therapy with aromatase and cyclooxygenase-2 inhibitors in patients with metastatic breast cancer." Ann Oncol 20(4): 615-620.

Fauser, J. K., L. D. Prisciandaro, A. G. Cummins and G. S. Howarth (2011). "Fatty acids as potential adjunctive colorectal chemotherapeutic agents." Cancer Biol Ther 11(8): 724-731.

Ferrara, N. and R. S. Kerbel (2005). "Angiogenesis as a therapeutic target." Nature 438(7070): 967-974.

Finetti, F., S. Donnini, A. Giachetti, L. Morbidelli and M. Ziche (2009). "Prostaglandin E(2) primes the angiogenic switch via a synergic interaction with the fibroblast growth factor-2 pathway." Circ Res 105(7): 657-666.

Finetti, F., R. Solito, L. Morbidelli, A. Giachetti, M. Ziche and S. Donnini (2008). "Prostaglandin E2 regulates angiogenesis via activation of fibroblast growth factor receptor-1." J Biol Chem 283(4): 2139-2146.

Fleming, I. (2011). "The cytochrome P450 pathway in angiogenesis and endothelial cell biology." Cancer Metastasis Rev 30(3-4): 541-555.

Franses, J. W., A. B. Baker, V. C. Chitalia and E. R. Edelman (2011). "Stromal Endothelial Cells Directly Influence Cancer Progression." Science Translational Medicine 3(66): 66ra65-66ra65.

Fredman, G. and C. N. Serhan (2011). "Specialized proresolving mediator targets for RvE1 and RvD1 in peripheral blood and mechanisms of resolution." Biochem J 437(2): 185-197.

Garonna, E., K. M. Botham, G. M. Birdsey, A. M. Randi, R. R. Gonzalez-Perez and C. P. Wheeler-Jones (2011). "Vascular endothelial growth factor receptor-2 couples cyclo-oxygenase-2 with pro-angiogenic actions of leptin on human endothelial cells." PLoS ONE 6(4): e18823.

Ghosh, N., R. Chaki, V. Mandal and S. C. Mandal (2010). "COX-2 as a target for cancer chemotherapy." Pharmacol Rep 62(2): 233-244.

Giurdanella, G., C. Motta, S. Muriana, V. Arena, C. D. Anfuso, G. Lupo and M. Alberghina (2011). "Cytosolic and calcium-independent phospholipase A(2) mediate glioma-enhanced proangiogenic activity of brain endothelial cells." Microvasc Res 81(1): 1-17.

Grabacka, M., W. Placha, P. M. Plonka, S. Pajak, K. Urbanska, P. Laidler and A. Slominski (2004). "Inhibition of melanoma metastases by fenofibrate." Arch Dermatol Res 296(2): 54-58.

Grau, S., J. Thorsteinsdottir, L. von Baumgarten, F. Winkler, J. C. Tonn and C. Schichor (2011). "Bevacizumab can induce reactivity to VEGF-C and -D in human brain and tumor derived endothelial cells." J Neurooncol 104(1): 103-112.

Greenhough, A., H. J. Smartt, A. E. Moore, H. R. Roberts, A. C. Williams, C. Paraskeva and A. Kaidi (2009). "The COX-2/PGE$_2$ pathway: key roles in the hallmarks of cancer and adaptation to the tumor microenvironment." Carcinogenesis 30(3): 377-386.

Griesser, M., T. Suzuki, N. Tejera, S. Mont, W. E. Boeglin, A. Pozzi and C. Schneider (2011). "Biosynthesis of hemiketal eicosanoids by cross-over of the 5-lipoxygenase and cyclooxygenase-2 pathways." Proc Natl Acad Sci USA 108(17): 6945-6950.

Grosjean, J., S. Kiriakidis, K. Reilly, M. Feldmann and E. Paleolog (2006). "Vascular endothelial growth factor signaling in endothelial cell survival: a role for NFkappaB." Biochem Biophys Res Commun 340(3): 984-994.

Guo, P., B. Hu, W. Gu, L. Xu, D. Wang, H. J. Huang, W. K. Cavenee and S. Y. Cheng (2003). "Platelet-derived growth factor-B enhances glioma angiogenesis by stimulating vascular endothelial growth factor expression in tumor endothelia and by promoting pericyte recruitment." Am J Pathol 162(4): 1083-1093.

Guo, S. and R. R. Gonzalez-Perez (2011). "Notch, IL-1 and leptin crosstalk outcome (NILCO) is critical for leptin-induced proliferation, migration and VEGF/VEGFR-2 expression in breast cancer." PLoS ONE 6(6): e21467.

Gupta, R. A. and R. N. Dubois (2002). "Controversy: PPARgamma as a target for treatment of colorectal cancer." Am J Physiol Gastrointest Liver Physiol 283(2): G266-269.

Haagenson, K. K. and G. S. Wu (2010). "Mitogen activated protein kinase phosphatases and cancer." Cancer Biol Ther 9(5): 337-340.

Han, C., J. Leng, A. J. Demetris and T. Wu (2004). "Cyclooxygenase-2 promotes human cholangiocarcinoma growth: evidence for cyclooxygenase-2-independent mechanism in celecoxib-mediated induction of p21waf1/cip1 and p27kip1 and cell cycle arrest." Cancer Res 64(4): 1369-1376.

Hanahan, D. and R. A. Weinberg (2011). "Hallmarks of cancer: the next generation." Cell 144(5): 646-674.

Hata, A. and R. Breyer (2004). "Pharmacology and signaling of prostaglandin receptors: multiple roles in inflammation and immune modulation." Pharmacol Ther 103(2): 147-166.

He, T., T. Lu, L. V. d'Uscio, C. F. Lam, H. C. Lee and Z. S. Katusic (2008). "Angiogenic function of prostacyclin biosynthesis in human endothelial progenitor cells." Circ Res 103(1): 80-88.

Helliwell, R. J. A., L. F. Adams and M. D. Mitchell (2004). "Prostaglandin synthases: recent developments and a novel hypothesis." Prostaglandins, Leukotrienes and Essential Fatty Acids 70(2): 101-113.

Herbert, S. P., A. F. Odell, S. Ponnambalam and J. H. Walker (2009). "Activation of cytosolic phospholipase A2-{alpha} as a novel mechanism regulating endothelial cell cycle progression and angiogenesis." J Biol Chem 284(9): 5784-5796.

Hida, K., Y. Hida and M. Shindoh (2008). "Understanding tumor endothelial cell abnormalities to develop ideal anti-angiogenic therapies." Cancer Sci 99(3): 459-466.

Hida, K. and M. Klagsbrun (2005). "A new perspective on tumor endothelial cells: unexpected chromosome and centrosome abnormalities." Cancer Res 65(7): 2507-2510.

Holderfield, M. T. and C. C. Hughes (2008). "Crosstalk between vascular endothelial growth factor, notch, and transforming growth factor-beta in vascular morphogenesis." Circ Res 102(6): 637-652.

Houliston, R. A., J. D. Pearson and C. P. Wheeler-Jones (2001). "Agonist-specific cross talk between ERKs and p38(mapk) regulates PGI(2) synthesis in endothelium." Am J Physiol Cell Physiol 281(4): C1266-1276.

Huang, S. and F. A. Sinicrope (2010). "Celecoxib-induced apoptosis is enhanced by ABT-737 and by inhibition of autophagy in human colorectal cancer cells." Autophagy 6(2): 256-269.

Hubbard, W. C., M. C. Alley, G. N. Gray, K. C. Green, T. L. McLemore and M. R. Boyd (1989). "Evidence for Prostanoid Biosynthesis as a Biochemical Feature of Certain Subclasses of Non-Small Cell Carcinomas of the Lung as Determined in Established Cell Lines Derived from Human Lung Tumors." Cancer Research 49(4): 826-832.

Huls, G., J. J. Koornstra and J. H. Kleibeuker (2003). "Non-steroidal anti-inflammatory drugs and molecular carcinogenesis of colorectal carcinomas." Lancet 362(9379): 230-232.

Hwang, J. T., Y. M. Kim, Y. J. Surh, H. W. Baik, S. K. Lee, J. Ha and O. J. Park (2006). "Selenium regulates cyclooxygenase-2 and extracellular signal-regulated kinase signaling pathways by activating AMP-activated protein kinase in colon cancer cells." Cancer Res 66(20): 10057-10063.

Iguchi, G., K. Chrysovergis, S. H. Lee, S. J. Baek, R. Langenbach and T. E. Eling (2009). "A reciprocal relationship exists between non-steroidal anti-inflammatory drug-activated gene-1 (NAG-1) and cyclooxygenase-2." Cancer Lett 282(2): 152-158.

Iwase, M., S. Takaoka, M. Uchida, G. Kondo, H. Watanabe, M. Ohashi and M. Nagumo (2006). "Accelerative effect of a selective cyclooxygenase-2 inhibitor on Fas-mediated apoptosis in human neutrophils." Int Immunopharmacol 6(3): 334-341.

Jain, R. K. and M. F. Booth (2003). "What brings pericytes to tumor vessels?" J Clin Invest 112(8): 1134-1136.

Janakiram, N. B. and C. V. Rao (2009). "Role of lipoxins and resolvins as anti-inflammatory and proresolving mediators in colon cancer." Curr Mol Med 9(5): 565-579.

Jang, M., L. Cai, G. O. Udeani, K. V. Slowing, C. F. Thomas, C. W. Beecher, H. H. Fong, N. R. Farnsworth, A. D. Kinghorn, R. G. Mehta, R. C. Moon and J. M. Pezzuto (1997). "Cancer chemopreventive activity of resveratrol, a natural product derived from grapes." Science 275(5297): 218-220.

Janne, P. A. and R. J. Mayer (2000). "Chemoprevention of colorectal cancer." N Engl J Med 342(26): 1960-1968.

Jia, X. Q., N. Zhong, L. H. Han, J. H. Wang, M. Yan, F. L. Meng and S. Z. Zhang (2005). "Effect of NS-398 on colon cancer cells." World J Gastroenterol 11(3): 353-356.

Jiang, B.-H. and L.-Z. Liu (2008). "PI3K/PTEN signaling in tumorigenesis and angiogenesis." Biochimica et Biophysica Acta (BBA) – Proteins & Proteomics 1784(1): 150-158.

Jin, Y., M. Arita, Q. Zhang, D. R. Saban, S. K. Chauhan, N. Chiang, C. N. Serhan and R. Dana (2009). "Anti-angiogenesis effect of the novel anti-inflammatory and pro-resolving lipid mediators." Invest Ophthalmol Vis Sci 50(10): 4743-4752.

Johnson, A. J., A. L. Hsu, H. P. Lin, X. Song and C. S. Chen (2002). "The cyclo-oxygenase-2 inhibitor celecoxib perturbs intracellular calcium by inhibiting endoplasmic reticulum Ca^{2+}-ATPases: a plausible link with its anti-tumor effect and cardiovascular risks." Biochem J 366(Pt 3): 831-837.

Kaidi, A., D. Qualtrough, A. C. Williams and C. Paraskeva (2006). "Direct Transcriptional Up-regulation of Cyclooxygenase-2 by Hypoxia-Inducible Factor (HIF)-1 Promotes Colorectal Tumor Cell Survival and Enhances HIF-1 Transcriptional Activity during Hypoxia." Cancer Research 66(13): 6683-6691.

Kambe, A., H. Yoshioka, H. Kamitani, T. Watanabe, S. J. Baek and T. E. Eling (2009). "The cyclooxygenase inhibitor sulindac sulfide inhibits EP4 expression and suppresses the growth of glioblastoma cells." Cancer Prev Res (Phila) 2(12): 1088-1099.

Kamiyama, M., A. Pozzi, L. Yang, L. M. DeBusk, R. M. Breyer and P. C. Lin (2006). "EP2, a receptor for PGE_2, regulates tumor angiogenesis through direct effects on endothelial cell motility and survival." Oncogene 25(53): 7019-7028.

Kao, J., A. T. Sikora and S. Fu (2009). "Dual EGFR and COX-2 inhibition as a novel approach to targeting head and neck squamous cell carcinoma." Curr Cancer Drug Targets 9(8): 931-937.

Kardosh, A., E. B. Golden, P. Pyrko, J. Uddin, F. M. Hofman, T. C. Chen, S. G. Louie, N. A. Petasis and A. H. Schonthal (2008). "Aggravated endoplasmic reticulum stress as a basis for enhanced glioblastoma cell killing by bortezomib in combination with celecoxib or its non-coxib analogue, 2,5-dimethyl-celecoxib." Cancer Res 68(3): 843-851.

Kardosh, A., W. Wang, J. Uddin, N. A. Petasis, F. M. Hofman, T. C. Chen and A. H. Schonthal (2005). "Dimethyl-celecoxib (DMC), a derivative of celecoxib that lacks cyclooxygenase-2-inhibitory function, potently mimics the anti-tumor effects of celecoxib on Burkitt's lymphoma *in vitro* and *in vivo*." Cancer Biol Ther 4(5): 571-582.

Keightley, M. C., P. Brown, H. N. Jabbour and K. J. Sales (2010). "F-Prostaglandin receptor regulates endothelial cell function via fibroblast growth factor-2." BMC Cell Biol 11: 8.

Keith, R. L., Y. E. Miller, T. M. Hudish, C. E. Girod, S. Sotto-Santiago, W. A. Franklin, R. A. Nemenoff, T. H. March, S. P. Nana-Sinkam and M. W. Geraci (2004). "Pulmonary prostacyclin synthase overexpression chemoprevents tobacco smoke lung carcinogenesis in mice." Cancer Res 64(16): 5897-5904.

Khan, M. N. and Y. S. Lee (2011). "Cyclooxygenase inhibitors: scope of their use and development in cancer chemotherapy." Med Res Rev 31(2): 161-201.

Khan, Z., Bhadouria, P., Radha Gupta, Bisen, PS (2006). "Tumor control by manipulation of the human anti-apoptotic survivin gene." Curr Cancer Ther Rev 2: 73-79.

Khan, Z., N. Khan, R. P. Tiwari, I. K. Patro, G. B. K. S. Prasad and P. S. Bisen (2010). "Down-regulation of survivin by oxaliplatin diminishes radioresistance of head and neck squamous carcinoma cells." Radiother Oncol 96(2): 267-273.

Khan, Z., N. Khan, R. P. Tiwari, N. K. Sah, G. B. Prasad and P. S. Bisen (2011). "Biology of Cox-2: An Application in Cancer Therapeutics." Curr Drug Targets 12(7): 1082-1093.

Kim, G. Y., J. W. Lee, S. H. Cho, J. M. Seo and J. H. Kim (2009). "Role of the low-affinity leukotriene B4 receptor BLT2 in VEGF-induced angiogenesis." Arterioscler Thromb Vasc Biol 29(6): 915-920.

Kim, H. S., M. J. Kim, E. J. Kim, Y. Yang, M. S. Lee and J. S. Lim (2012). "Berberine-induced AMPK activation inhibits the metastatic potential of melanoma cells via reduction of ERK activity and COX-2 protein expression." Biochem Pharmacol 83(3): 385-394.

Kim, S. J., H. J. Jeong, I. Y. Choi, K. M. Lee, R. K. Park, S. H. Hong and H. M. Kim (2005). "Cyclooxygenase-2 inhibitor SC-236 [4-[5-(4-chlorophenyl)-3-(trifluoro-methyl)-1-pyrazol-1-l] benzenesulfonamide] suppresses nuclear factor-kappaB activation and phosphorylation of p38 mitogen-activated protein kinase, extracellular signal-regulated kinase, and c-Jun N-terminal kinase in human mast cell line cells." J Pharmacol Exp Ther 314(1): 27-34.

Kimura, Y. and H. Okuda (2001). "Resveratrol isolated from Polygonum cuspidatum root prevents tumor growth and metastasis to lung and tumor-induced neovascularization in Lewis lung carcinoma-bearing mice." J Nutr 131(6): 1844-1849.

Kinney, C. M., U. M. Chandrasekharan, L. Mavrakis and P. E. DiCorleto (2008). "VEGF and thrombin induce MKP-1 through distinct signaling pathways: role for MKP-1 in endothelial cell migration." Am J Physiol Cell Physiol 294(1): C241-250.

Krishnan, A., S. A. Nair and M. R. Pillai (2007). "Biology of PPAR gamma in cancer: a critical review on existing lacunae." Curr Mol Med 7(6): 532-540.

Krysan, K., F. H. Merchant, L. Zhu, M. Dohadwala, J. Luo, Y. Lin, N. Heuze-Vourc'h, M. Pold, D. Seligson, D. Chia, L. Goodglick, H. Wang, R. Strieter, S. Sharma and S. Dubinett (2004). "COX-2-dependent stabilization of survivin in non-small cell lung cancer." FASEB J 18(1): 206-208.

Kulp, S. K., Y. T. Yang, C. C. Hung, K. F. Chen, J. P. Lai, P. H. Tseng, J. W. Fowble, P. J. Ward and C. S. Chen (2004). "3-phosphoinositide-dependent protein kinase-1/Akt signaling represents a major cyclooxygenase-2-independent target for celecoxib in prostate cancer cells." Cancer Res 64(4): 1444-1451.

Kundumani-Sridharan, V., J. Niu, D. Wang, D. Van Quyen, Q. Zhang, N. K. Singh, J. Subramani, S. Karri and G. N. Rao (2010). "15(S)-hydroxyeicosatetraenoic acid-induced angiogenesis requires Src-mediated Egr-1-dependent rapid induction of FGF-2 expression." Blood 115(10): 2105-2116.

Kurosu, T., N. Ohga, Y. Hida, N. Maishi, K. Akiyama, W. Kakuguchi, T. Kuroshima, M. Kondo, T. Akino, Y. Totsuka, M. Shindoh, F. Higashino and K. Hida (2011). "HuR keeps an angiogenic switch on by stabilising mRNA of VEGF and COX-2 in tumor endothelium." Br J Cancer 104(5): 819-829.

Lassoued, W., D. Murphy, J. Tsai, R. Oueslati, G. Thurston and W. M. Lee (2011). "Effect of VEGF and VEGF Trap on vascular endothelial cell signaling in tumors." Cancer Biol Ther 10(12): 1326-1333.

Leahy, K. M., R. L. Ornberg, Y. Wang, B. S. Zweifel, A. T. Koki and J. L. Masferrer (2002). "Cyclooxygenase-2 inhibition by celecoxib reduces proliferation and induces apoptosis in angiogenic endothelial cells *in vivo*." Cancer Res 62(3): 625-631.

Lee, D. F. and M. C. Hung (2008). "Advances in targeting IKK and IKK-related kinases for cancer therapy." Clin Cancer Res 14(18): 5656-5662.

Leu, T. H. and M. C. Maa (2002). "The molecular mechanisms for the antitumorigenic effect of curcumin." Curr Med Chem Anticancer Agents 2(3): 357-370.

Lewis, C. E. and J. W. Pollard (2006). "Distinct Role of Macrophages in Different Tumor Microenvironments." Cancer Research 66(2): 605-612.

Li, A., S. Dubey, M. L. Varney, B. J. Dave and R. K. Singh (2003). "IL-8 directly enhanced endothelial cell survival, proliferation, and matrix metalloproteinases production and regulated angiogenesis." J Immunol 170(6): 3369-3376.

Li, J. L. and A. L. Harris (2007). "The potential of new tumor endothelium-specific markers for the development of antivascular therapy." Cancer Cell 11(6): 478-481.

Li, M., X. Wu and X. C. Xu (2001). "Induction of apoptosis in colon cancer cells by cyclooxygenase-2 inhibitor NS398 through a cytochrome c-dependent pathway." Clin Cancer Res 7(4): 1010-1016.

Ligresti, G., A. C. Aplin, P. Zorzi, A. Morishita and R. F. Nicosia (2011). "Macrophage-Derived Tumor Necrosis Factor-{alpha} Is an Early Component of the Molecular Cascade Leading to Angiogenesis in Response to Aortic Injury." Arterioscler Thromb Vasc Biol 31(5): 1151-1159.

Lin, H. P., S. K. Kulp, P. H. Tseng, Y. T. Yang, C. C. Yang and C. S. Chen (2004). "Growth inhibitory effects of celecoxib in human umbilical vein endothelial cells are mediated through G_1 arrest via multiple signaling mechanisms." Mol Cancer Ther 3(12): 1671-1680.

Linkous, A. G., E. M. Yazlovitskaya and D. E. Hallahan (2010). "Cytosolic phospholipase A2 and lysophospholipids in tumor angiogenesis." J Natl Cancer Inst 102(18): 1398-1412.

Liu, C. H., S.-H. Chang, K. Narko, O. C. Trifan, M.-T. Wu, E. Smith, C. Haudenschild, T. F. Lane and T. Hla (2001). "Overexpression of Cyclooxygenase-2 Is Sufficient to Induce Tumorigenesis in Transgenic Mice." Journal of Biological Chemistry 276(21): 18563-18569.

Liu, C. H., S. H. Chang, K. Narko, O. C. Trifan, M. T. Wu, E. Smith, C. Haudenschild, T. F. Lane and T. Hla (2001). "Overexpression of cyclooxygenase-2 is sufficient to induce tumorigenesis in transgenic mice." J Biol Chem 276(21): 18563-18569.

Liu, L. Z., J. Z. Zheng, X. R. Wang and B. H. Jiang (2008). "Endothelial p70 S6 kinase 1 in regulating tumor angiogenesis." Cancer Res 68(19): 8183-8188.

Liu, Q., G. Li, R. Li, J. Shen, Q. He, L. Deng, C. Zhang and J. Zhang (2010). "IL-6 promotion of glioblastoma cell invasion and angiogenesis in U251 and T98G cell lines." J Neurooncol 100(2): 165-176.

Lu, W., H. N. Tinsley, A. Keeton, Z. Qu, G. A. Piazza and Y. Li (2009). "Suppression of Wnt/beta-catenin signaling inhibits prostate cancer cell proliferation." Eur J Pharmacol 602(1): 8-14.

Maretzky, T., A. Evers, W. Zhou, S. L. Swendeman, P. M. Wong, S. Rafii, K. Reiss and C. P. Blobel (2011). "Migration of growth factor-stimulated epithelial and endothelial cells depends on EGFR transactivation by ADAM17." Nat Commun 2: 229.

Martinez, F. O., A. Sica, A. Mantovani and M. Locati (2008). "Macrophage activation and polarization." Front Biosci 13: 453-461.

Matesanz, N., G. Park, H. McAllister, W. Leahey, A. Devine, G. E. McVeigh, T. A. Gardiner and D. M. McDonald (2010). "Docosahexaenoic acid improves the nitroso-redox balance and reduces VEGF-mediated angiogenic signaling in microvascular endothelial cells." Invest Ophthalmol Vis Sci 51(12): 6815-6825.

Matsuo, Y., H. Sawai, J. Ma, D. Xu, N. Ochi, A. Yasuda, H. Takahashi, H. Funahashi and H. Takeyama (2009). "IL-1alpha secreted by colon cancer cells enhances angiogenesis: the relationship between IL-1alpha release and tumor cells' potential for liver metastasis." J Surg Oncol 99(6): 361-367.

Mattos, R. T., A. A. Bosco and J. A. Nogueira-Machado (2012). "Rosiglitazone, a PPAR-gamma agonist, inhibits VEGF secretion by peripheral blood mononuclear cells and ROS production by human leukocytes." Inflamm Res 61(1): 37-41.

Mavria, G., Y. Vercoulen, M. Yeo, H. Paterson, M. Karasarides, R. Marais, D. Bird and C. J. Marshall (2006). "ERK-MAPK signaling opposes Rho-kinase to promote endothelial cell survival and sprouting during angiogenesis." Cancer Cell 9(1): 33-44.

Mawatari, M., K. Kohno, H. Mizoguchi, T. Matsuda, K. Asoh, J. Van Damme, H. G. Welgus and M. Kuwano (1989). "Effects of tumor necrosis factor and epidermal growth factor on cell morphology, cell surface receptors, and the production of tissue inhibitor of metalloproteinases and IL-6 in human microvascular endothelial cells." J Immunol 143(5): 1619-1627.

McLemore, T. L., W. C. Hubbard, C. L. Litterst, M. C. Liu, S. Miller, N. A. McMahon, J. C. Eggleston and M. R. Boyd (1988). "Profiles of Prostaglandin Biosynthesis in Normal Lung and Tumor Tissue from Lung Cancer Patients." Cancer Research 48(11): 3140-3147.

Medhora, M., A. Dhanasekaran, P. F. Pratt, Jr., C. R. Cook, L. K. Dunn, S. K. Gruenloh and E. R. Jacobs (2008). "Role of JNK in network formation of human lung microvascular endothelial cells." Am J Physiol Lung Cell Mol Physiol 294(4): L676-685.

Meissner, M., I. Hrgovic, M. Doll and R. Kaufmann (2011). "PPARdelta agonists suppress angiogenesis in a VEGFR2-dependent manner." Arch Dermatol Res 303(1): 41-47.

Meissner, M., I. Hrgovic, M. Doll, J. Naidenow, G. Reichenbach, T. Hailemariam-Jahn, D. Michailidou, J. Gille and R. Kaufmann (2010). "Peroxisome Proliferator-activated Receptor Activators Induce IL-8 Expression in Nonstimulated Endothelial Cells in a Transcriptional and Posttranscriptional Manner." Journal of Biological Chemistry 285(44): 33797-33804.

Michalik, L., B. Desvergne and W. Wahli (2004). "Peroxisome-proliferator-activated receptors and cancers: complex stories." Nat Rev Cancer 4(1): 61-70.

Mineo, T. C., V. Ambrogi, M. E. Cufari and E. Pompeo (2010). "May cyclooxygenase-2 (COX-2), p21 and p27 expression affect prognosis and therapeutic strategy of patients with malignant pleural mesothelioma?" Eur J Cardiothorac Surg 38(3): 245-252.

Muller-Brusselbach, S., M. Komhoff, M. Rieck, W. Meissner, K. Kaddatz, J. Adamkiewicz, B. Keil, K. J. Klose, R. Moll, A. D. Burdick, J. M. Peters and R. Muller (2007). "Deregulation of tumor angiogenesis and blockade of tumor growth in PPARbeta-deficient mice." EMBO J 26(15): 3686-3698.

Müller-Decker, K., G. Neufang, I. Berger, M. Neumann, F. Marks and G. Fürstenberger (2002). "Transgenic cyclooxygenase-2 overexpression sensitizes mouse skin for carcinogenesis." Proceedings of the National Academy of Sciences 99(19): 12483-12488.

Murakami, M. and I. Kudo (2006). "Prostaglandin E synthase: a novel drug target for inflammation and cancer." Curr Pharm Des 12(8): 943-954.

Murakami, M., L. T. Nguyen, K. Hatanaka, W. Schachterle, P. Y. Chen, Z. W. Zhuang, B. L. Black and M. Simons (2011). "FGF-dependent regulation of VEGF receptor 2 expression in mice." J Clin Invest 121(7): 2668-2678.

Muraki, C., N. Ohga, Y. Hida, H. Nishihara, Y. Kato, K. Tsuchiya, K. Matsuda, Y. Totsuka, M. Shindoh and K. Hida (2012). "Cyclooxygenase-2 inhibition causes antiangiogenic effects on tumor endothelial and vascular progenitor cells." Int J Cancer 130(1): 59-70.

Murata, T., K. Aritake, S. Matsumoto, S. Kamauchi, T. Nakagawa, M. Hori, E. Momotani, Y. Urade and H. Ozaki (2011). "Prostagladin D2 is a mast cell-derived antiangiogenic factor in lung carcinoma." Proc Natl Acad Sci USA 108(49): 19802-19807.

Murata, T., M. I. Lin, K. Aritake, S. Matsumoto, S. Narumiya, H. Ozaki, Y. Urade, M. Hori and W. C. Sessa (2008). "Role of prostaglandin D2 receptor DP as a suppressor of tumor hyperpermeability and angiogenesis in vivo." Proc Natl Acad Sci USA 105(50): 20009-20014.

Nagy, J. A., S. H. Chang, A. M. Dvorak and H. F. Dvorak (2009). "Why are tumor blood vessels abnormal and why is it important to know?" Br J Cancer 100(6): 865-869.

Nakanishi, M., D. C. Montrose, P. Clark, P. R. Nambiar, G. S. Belinsky, K. P. Claffey, D. Xu and D. W. Rosenberg (2008). "Genetic deletion of mPGES-1 suppresses intestinal tumorigenesis." Cancer Res 68(9): 3251-3259.

Nakao, S., T. Kuwano, C. Tsutsumi-Miyahara, S. Ueda, Y. N. Kimura, S. Hamano, K. H. Sonoda, Y. Saijo, T. Nukiwa, R. M. Strieter, T. Ishibashi, M. Kuwano and M. Ono (2005). "Infiltration of COX-2-expressing macrophages is a prerequisite for IL-1 beta-induced neovascularization and tumor growth." J Clin Invest 115(11): 2979-2991.

Namkoong, S., C. K. Kim, Y. L. Cho, J. H. Kim, H. Lee, K. S. Ha, J. Choe, P. H. Kim, M. H. Won, Y. G. Kwon, E. B. Shim and Y. M. Kim (2009). "Forskolin increases angiogenesis through the coordinated cross-talk of PKA-dependent VEGF expression and Epac-mediated PI3K/Akt/eNOS signaling." Cell Signal 21(6): 906-915.

Nemenoff, R., A. M. Meyer, T. M. Hudish, A. B. Mozer, A. Snee, S. Narumiya, R. S. Stearman, R. A. Winn, M. Weiser-Evans, M. W. Geraci and R. L. Keith (2008). "Prostacyclin prevents murine lung cancer independent of the membrane receptor by activation of peroxisomal proliferator-activated receptor gamma." Cancer Prev Res (Phila) 1(5): 349-356.

Nie, D., M. Lamberti, A. Zacharek, L. Li, K. Szekeres, K. Tang, Y. Chen and K. V. Honn (2000). "Thromboxane A2 Regulation of Endothelial Cell Migration, Angiogenesis, and Tumor Metastasis." Biochemical and Biophysical Research Communications 267(1): 245-251.

Niu, Y. N. and S. J. Xia (2009). "Stroma-epithelium crosstalk in prostate cancer." Asian J Androl 11(1): 28-35.

Ohga, N., K. Hida, Y. Hida, C. Muraki, K. Tsuchiya, K. Matsuda, Y. Ohiro, Y. Totsuka and M. Shindoh (2009). "Inhibitory effects of epigallocatechin-3 gallate, a polyphenol in green tea, on tumor-associated endothelial cells and endothelial progenitor cells." Cancer Sci 100(10): 1963-1970.

Okamura, K., Y. Sato, T. Matsuda, R. Hamanaka, M. Ono, K. Kohno and M. Kuwano (1991). "Endogenous basic fibroblast growth factor-dependent induction of collagenase and interleukin-6 in tumor necrosis factor-treated human microvascular endothelial cells." J Biol Chem 266(29): 19162-19165.

Olsen, C. L., P. P. Hsu, J. Glienke, G. M. Rubanyi and A. R. Brooks (2004). "Hedgehog-interacting protein is highly expressed in endothelial cells but down-regulated during angiogenesis and in several human tumors." BMC Cancer 4: 43.

Ono, M. (2008). "Molecular links between tumor angiogenesis and inflammation: inflammatory stimuli of macrophages and cancer cells as targets for therapeutic strategy." Cancer Sci 99(8): 1501-1506.

Opitz, C. A., N. Rimmerman, Y. Zhang, L. E. Mead, M. C. Yoder, D. A. Ingram, J. M. Walker and J. Rehman (2007). "Production of the endocannabinoids anandamide and 2-arachidonoylglycerol by endothelial progenitor cells." FEBS Lett 581(25): 4927-4931.

Oshima, M., J. E. Dinchuk, S. L. Kargman, H. Oshima, B. Hancock, E. Kwong, J. M. Trzaskos, J. F. Evans and M. M. Taketo (1996). "Suppression of intestinal polyposis in APC$^{\Delta716}$ knockout mice by inhibition of cyclooxygenase 2 (COX-2)." Cell 87(5): 803-809.

Pai, R., T. Nakamura, W. S. Moon and A. S. Tarnawski (2003). "Prostaglandins promote colon cancer cell invasion; signaling by cross-talk between two distinct growth factor receptors." FASEB J 17(12): 1640-1647.

Pai, R., B. Soreghan, I. L. Szabo, M. Pavelka, D. Baatar and A. S. Tarnawski (2002). "Prostaglandin E2 transactivates EGF receptor: a novel mechanism for promoting colon cancer growth and gastrointestinal hypertrophy." Nat Med 8(3): 289-293.

Panigrahy, D., A. Kaipainen, S. Huang, C. E. Butterfield, C. M. Barnes, M. Fannon, A. M. Laforme, D. M. Chaponis, J. Folkman and M. W. Kieran (2008). "PPARalpha agonist fenofibrate suppresses tumor growth through direct and indirect angiogenesis inhibition." Proc Natl Acad Sci USA 105(3): 985-990.

Pasqualini, R., E. Koivunen, R. Kain, J. Lahdenranta, M. Sakamoto, A. Stryhn, R. A. Ashmun, L. H. Shapiro, W. Arap and E. Ruoslahti (2000). "Aminopeptidase N is a receptor for tumor-homing peptides and a target for inhibiting angiogenesis." Cancer Res 60(3): 722-727.

Phung, T. L., K. Ziv, D. Dabydeen, G. Eyiah-Mensah, M. Riveros, C. Perruzzi, J. Sun, R. A. Monahan-Earley, I. Shiojima, J. A. Nagy, M. I. Lin, K. Walsh, A. M. Dvorak, D. M. Briscoe, M. Neeman, W. C. Sessa, H. F. Dvorak and L. E. Benjamin (2006). "Pathological angiogenesis is induced by sustained Akt signaling and inhibited by rapamycin." Cancer Cell 10(2): 159-170.

Pinto, S., L. Gori, O. Gallo, S. Boccuzzi, R. Paniccia and R. Abbate (1993). "Increased thromboxane A2 production at primary tumor site in metastasizing squamous cell carcinoma of the larynx." Prostaglandins Leukot Essent Fatty Acids 49(1): 527-530.

Piqueras, L., A. R. Reynolds, K. M. Hodivala-Dilke, A. Alfranca, J. M. Redondo, T. Hatae, T. Tanabe, T. D. Warner and D. Bishop-Bailey (2007). "Activation of PPARbeta/delta induces endothelial cell proliferation and angiogenesis." Arterioscler Thromb Vasc Biol 27(1): 63-69.

Pisanti, S., C. Borselli, O. Oliviero, C. Laezza, P. Gazzerro and M. Bifulco (2007). "Antiangiogenic activity of the endocannabinoid anandamide: correlation to its tumor-suppressor efficacy." J Cell Physiol 211(2): 495-503.

Pisanti, S., P. Picardi, L. Prota, M. C. Proto, C. Laezza, P. G. McGuire, L. Morbidelli, P. Gazzerro, M. Ziche, A. Das and M. Bifulco (2011). "Genetic and pharmacologic inactivation of cannabinoid CB_1 receptor inhibits angiogenesis." Blood 117(20): 5541-5550.

Pradono, P., R. Tazawa, M. Maemondo, M. Tanaka, K. Usui, Y. Saijo, K. Hagiwara and T. Nukiwa (2002). "Gene Transfer of Thromboxane A2 Synthase and Prostaglandin I2 Synthase Antithetically Altered Tumor Angiogenesis and Tumor Growth." Cancer Research 62(1): 63-66.

Prasad, R., S. Giri, A. K. Singh and I. Singh (2008). "15-deoxy-delta12,14-prostaglandin J2 attenuates endothelial-monocyte interaction: implication for inflammatory diseases." J Inflamm (Lond) 5: 14.

Pula, G., E. Garonna, W. B. Dunn, M. Hirano, G. Pizzorno, M. Campanella, E. L. Schwartz, M. H. El Kouni and C. P. Wheeler-Jones (2010). "Paracrine stimulation of endothelial cell motility and angiogenesis by platelet-derived deoxyribose-1-phosphate." Arterioscler Thromb Vasc Biol 30(12): 2631-2638.

Pupo, E., A. F. Pla, D. Avanzato, F. Moccia, J. E. Cruz, F. Tanzi, A. Merlino, D. Mancardi and L. Munaron (2011). "Hydrogen sulfide promotes calcium signals and migration in tumor-derived endothelial cells." Free Radic Biol Med 51(9): 1765-1773.

Pyrko, P., A. Kardosh, Y. T. Liu, N. Soriano, W. Xiong, R. H. Chow, J. Uddin, N. A. Petasis, A. K. Mircheff, R. A. Farley, S. G. Louie, T. C. Chen and A. H. Schonthal (2007). "Calcium-activated endoplasmic reticulum stress as a major component of tumor cell death induced by 2,5-dimethyl-celecoxib, a non-coxib analogue of celecoxib." Mol Cancer Ther 6(4): 1262-1275.

Pyrko, P., N. Soriano, A. Kardosh, Y. T. Liu, J. Uddin, N. A. Petasis, F. M. Hofman, C. S. Chen, T. C. Chen and A. H. Schonthal (2006). "Downregulation of survivin expression and concomitant induction of apoptosis by celecoxib and its non-cyclooxygenase-2-inhibitory analog, dimethyl-celecoxib (DMC), in tumor cells *in vitro* and *in vivo*." Mol Cancer 5: 19.

Qualtrough, D., A. Kaidi, S. Chell, H. N. Jabbour, A. C. Williams and C. Paraskeva (2007). "Prostaglandin F(2alpha) stimulates motility and invasion in colorectal tumor cells." Int J Cancer 121(4): 734-740.

Raica, M., A. M. Cimpean and D. Ribatti (2009). "Angiogenesis in pre-malignant conditions." Eur J Cancer 45(11): 1924-1934.

Rajashekhar, G., M. Kamocka, A. Marin, M. A. Suckow, W. R. Wolter, S. Badve, A. R. Sanjeevaiah, K. Pumiglia, E. Rosen and M. Clauss (2011). "Pro-inflammatory angiogenesis is mediated by p38 MAP kinase." J Cell Physiol 226(3): 800-808.

Rak, J., C. Milsom, L. May, P. Klement and J. Yu (2006). "Tissue factor in cancer and angiogenesis: the molecular link between genetic tumor progression, tumor neovascularization, and cancer coagulopathy." Semin Thromb Hemost 32(1): 54-70.

Rao, R., R. Redha, I. Macias-Perez, Y. Su, C. Hao, R. Zent, M. D. Breyer and A. Pozzi (2007). "Prostaglandin E2-EP4 receptor promotes endothelial cell migration via ERK activation and angiogenesis *in vivo*." J Biol Chem 282(23): 16959-16968.

Reddy, B. S., T. Kawamori, R. A. Lubet, V. E. Steele, G. J. Kelloff and C. V. Rao (1999). "Chemopreventive efficacy of sulindac sulfone against colon cancer depends on time of administration during carcinogenic process." Cancer Res 59(14): 3387-3391.

Reddy, B. S., H. Maruyama and G. Kelloff (1987). "Dose-related inhibition of colon carcinogenesis by dietary piroxicam, a nonsteroidal antiinflammatory drug, during different stages of rat colon tumor development." Cancer Res 47(20): 5340-5346.

Reed, S., H. Li, C. Li and J. Lin (2011). "Celecoxib inhibits STAT3 phosphorylation and suppresses cell migration and colony forming ability in rhabdomyosarcoma cells." Biochem Biophys Res Commun 407(3): 450-455.

Reihill, J. A., M. A. Ewart and I. P. Salt (2011). "The role of AMP-activated protein kinase in the functional effects of vascular endothelial growth factor-A and -B in human aortic endothelial cells." Vasc Cell 3: 9.

Rhee, W. J., C. W. Ni, Z. Zheng, K. Chang, H. Jo and G. Bao (2010). "HuR regulates the expression of stress-sensitive genes and mediates inflammatory response in human umbilical vein endothelial cells." Proc Natl Acad Sci USA 107(15): 6858-6863.

Rohrl, C., U. Kaindl, I. Koneczny, X. Hudec, D. M. Baron, J. S. Konig and B. Marian (2011). "Peroxisome-proliferator-activated receptors gamma and beta/delta mediate vascular endothelial growth factor production in colorectal tumor cells." J Cancer Res Clin Oncol 137(1): 29-39.

Ruoslahti, E., S. N. Bhatia and M. J. Sailor (2010). "Targeting of drugs and nanoparticles to tumors." J Cell Biol 188(6): 759-768.

Sade, A., S. Tuncay, I. Cimen, F. Severcan and S. Banerjee (2012). "Celecoxib reduces fluidity and decreases metastatic potential of colon cancer cell lines irrespective of COX-2 expression." Biosci Rep 32(1): 35-44.

Saidi, A., M. Hagedorn, N. Allain, C. Verpelli, C. Sala, L. Bello, A. Bikfalvi and S. Javerzat (2009). "Combined targeting of interleukin-6 and vascular endothelial growth factor potently inhibits glioma growth and invasiveness." Int J Cancer 125(5): 1054-1064.

Saidi, S. A., C. M. Holland, D. S. Charnock-Jones and S. K. Smith (2006). "*In vitro* and *in vivo* effects of the PPAR-alpha agonists fenofibrate and retinoic acid in endometrial cancer." Mol Cancer 5: 13.

Sakamoto, K., S. Maeda, Y. Hikiba, H. Nakagawa, Y. Hayakawa, W. Shibata, A. Yanai, K. Ogura and M. Omata (2009). "Constitutive NF-kappaB activation in colorectal carcinoma plays a key role in angiogenesis, promoting tumor growth." Clin Cancer Res 15(7): 2248-2258.

Salcedo, R. (2003). "Angiogenic effects of prostaglandin E2 are mediated by up-regulation of CXCR4 on human microvascular endothelial cells." Blood 102(6): 1966-1977.

Sales, K. J., T. List, S. C. Boddy, A. R. Williams, R. A. Anderson, Z. Naor and H. N. Jabbour (2005). "A novel angiogenic role for prostaglandin F2alpha-FP receptor interaction in human endometrial adenocarcinomas." Cancer Res 65(17): 7707-7716.

Salvado, M. D., A. Alfranca, A. Escolano, J. Z. Haeggstrom and J. M. Redondo (2009). "COX-2 Limits Prostanoid Production in Activated HUVECs and Is a Source of PGH2 for Transcellular Metabolism to PGE$_2$ by Tumor Cells." Arterioscler Thromb Vasc Biol 29(7): 1131-1137.

Sandler, R. S., S. Halabi, J. A. Baron, S. Budinger, E. Paskett, R. Keresztes, N. Petrelli, J. M. Pipas, D. D. Karp, C. L. Loprinzi, G. Steinbach and R. Schilsky (2003).

"A randomized trial of aspirin to prevent colorectal adenomas in patients with previous colorectal cancer." N Engl J Med 348(10): 883-890.

Sapieha, P., A. Stahl, J. Chen, M. R. Seaward, K. L. Willett, N. M. Krah, R. J. Dennison, K. M. Connor, C. M. Aderman, E. Liclican, A. Carughi, D. Perelman, Y. Kanaoka, J. P. Sangiovanni, K. Gronert and L. E. Smith (2011). "5-Lipoxygenase metabolite 4-HDHA is a mediator of the antiangiogenic effect of omega-3 polyunsaturated fatty acids." Sci Transl Med 3(69): 69ra12.

Sasi, S. P., X. Yan, H. Enderling, D. Park, H. Y. Gilbert, C. Curry, C. Coleman, L. Hlatky, G. Qin, R. Kishore and D. A. Goukassian (2012). "Breaking the 'harmony' of TNF-alpha signaling for cancer treatment." Oncogene 31(37): 4117-4127.

Sato, T. and N. Gotoh (2009). "The FRS2 family of docking/scaffolding adaptor proteins as therapeutic targets of cancer treatment." Expert Opin Ther Targets 13(6): 689-700.

Schiffmann, S., T. J. Maier, I. Wobst, A. Janssen, H. Corban-Wilhelm, C. Angioni, G. Geisslinger and S. Grosch (2008). "The anti-proliferative potency of celecoxib is not a class effect of coxibs." Biochem Pharmacol 76(2): 179-187.

Schiffmann, S., S. Ziebell, J. Sandner, K. Birod, K. Deckmann, D. Hartmann, S. Rode, H. Schmidt, C. Angioni, G. Geisslinger and S. Grosch (2010). "Activation of ceramide synthase 6 by celecoxib leads to a selective induction of C16:0-ceramide." Biochem Pharmacol 80(11): 1632-1640.

Schonthal, A. H. (2010). "Exploiting cyclooxygenase-(in)dependent properties of COX-2 inhibitors for malignant glioma therapy." Anticancer Agents Med Chem 10(6): 450-461.

Scoditti, E., M. Massaro, M. A. Carluccio, A. Distante, C. Storelli and R. De Caterina (2010). "PPARgamma agonists inhibit angiogenesis by suppressing PKCalpha- and CREB-mediated COX-2 expression in the human endothelium." Cardiovasc Res 86(2): 302-310.

Sertznig, P., M. Seifert, W. Tilgen and J. Reichrath (2007). "Present concepts and future outlook: function of peroxisome proliferator-activated receptors (PPARs) for pathogenesis, progression, and therapy of cancer." J Cell Physiol 212(1): 1-12.

Sheng, H., J. Shao, J. D. Morrow, R. D. Beauchamp and R. N. DuBois (1998). "Modulation of apoptosis and Bcl-2 expression by prostaglandin E2 in human colon cancer cells." Cancer Res 58(2): 362-366.

Shureiqi, I., D. Chen, R. Lotan, P. Yang, R. A. Newman, S. M. Fischer and S. M. Lippman (2000). "15-Lipoxygenase-1 mediates nonsteroidal anti-inflammatory drug-induced apoptosis independently of cyclooxygenase-2 in colon cancer cells." Cancer Res 60(24): 6846-6850.

Silvestre, J. S., Z. Mallat, M. Duriez, R. Tamarat, M. F. Bureau, D. Scherman, N. Duverger, D. Branellec, A. Tedgui and B. I. Levy (2000). "Antiangiogenic effect of interleukin-10 in ischemia-induced angiogenesis in mice hindlimb." Circ Res 87(6): 448-452.

Singh, N. K., V. Kundumani-Sridharan and G. N. Rao (2011). "12/15-Lipoxygenase gene knockout severely impairs ischemia-induced angiogenesis due to lack of Rac1 farnesylation." Blood 118(20): 5701-5712.

Singh, N. K., D. V. Quyen, V. Kundumani-Sridharan, P. C. Brooks and G. N. Rao (2010). "AP-1 (Fra-1/c-Jun)-mediated induction of expression of matrix metallo-proteinase-2 is required for 15S-hydroxyeicosatetraenoic acid-induced angiogenesis." J Biol Chem 285(22): 16830-16843.

Singh, T., M. Vaid, N. Katiyar, S. Sharma and S. K. Katiyar (2011). "Berberine, an isoquinoline alkaloid, inhibits melanoma cancer cell migration by reducing the expressions of cyclooxygenase-2, prostaglandin E and prostaglandin E receptors." Carcinogenesis 32(1): 86-92.

Smartt, H. J., A. Greenhough, P. Ordonez-Moran, E. Talero, C. A. Cherry, C. A. Wallam, L. Parry, M. Al Kharusi, H. R. Roberts, J. M. Mariadason, A. R. Clarke, J. Huelsken, A. C. Williams and C. Paraskeva (2012). "beta-catenin represses expression of the tumor suppressor 15-prostaglandin dehydrogenase in the normal intestinal epithelium and colorectal tumor cells." Gut 61(9): 1306-1314.

Smith, W., L. Marnett and D. DeWitt (1991). "Prostaglandin and thromboxane biosynthesis." Pharmacol Ther 49(3): 153-179.

Sonoshita, M., K. Takaku, N. Sasaki, Y. Sugimoto, F. Ushikubi, S. Narumiya, M. Oshima and M. M. Taketo (2001). "Acceleration of intestinal polyposis through prostaglandin receptor EP2 in (APC)$^{\Delta716}$ knockout mice." Nat Med 7(9): 1048-1051.

Soriano, A. F., B. Helfrich, D. C. Chan, L. E. Heasley, P. A. Bunn, Jr. and T. C. Chou (1999). "Synergistic effects of new chemopreventive agents and conventional cytotoxic agents against human lung cancer cell lines." Cancer Res 59(24): 6178-6184.

St Croix, B., C. Rago, V. Velculescu, G. Traverso, K. E. Romans, E. Montgomery, A. Lal, G. J. Riggins, C. Lengauer, B. Vogelstein and K. W. Kinzler (2000). "Genes expressed in human tumor endothelium." Science 289(5482): 1197-1202.

Stahl, A., P. Sapieha, K. M. Connor, J. P. Sangiovanni, J. Chen, C. M. Aderman, K. L. Willett, N. M. Krah, R. J. Dennison, M. R. Seaward, K. I. Guerin, J. Hua and L. E. Smith (2010). "Short communication: PPAR gamma mediates a direct antiangiogenic effect of omega 3-PUFAs in proliferative retinopathy." Circ Res 107(4): 495-500.

Su, B., M. V. Darby and R. W. Brueggemeier (2008). "Synthesis and biological evaluation of novel sulfonanilide compounds as antiproliferative agents for breast cancer." J Comb Chem 10(3): 475-483.

Su, B., E. S. Diaz-Cruz, S. Landini and R. W. Brueggemeier (2006). "Novel sulfonanilide analogues suppress aromatase expression and activity in breast cancer cells independent of COX-2 inhibition." J Med Chem 49(4): 1413-1419.

Suchanek, K. M., F. J. May, J. A. Robinson, W. J. Lee, N. A. Holman, G. R. Monteith and S. J. Roberts-Thomson (2002). "Peroxisome proliferator-activated receptor alpha in the human breast cancer cell lines MCF-7 and MDA-MB-231." Mol Carcinog 34(4): 165-171.

Suh, N., B. S. Reddy, A. DeCastro, S. Paul, H. J. Lee, A. K. Smolarek, J. Y. So, B. Simi, C. X. Wang, N. B. Janakiram, V. Steele and C. V. Rao (2011). "Combination of atorvastatin with sulindac or naproxen profoundly inhibits colonic adeno-carcinomas by suppressing the p65/beta-catenin/cyclin D1 signaling pathway in rats." Cancer Prev Res (Phila) 4(11): 1895-1902.

Sweet, D. T., Z. Chen, D. M. Wiley, V. L. Bautch and E. Tzima (2012). "The adaptor protein Shc integrates growth factor and ECM signaling during postnatal angiogenesis." Blood 119(8): 1946-1955.

Syeda, F., J. Grosjean, R. A. Houliston, R. J. Keogh, T. D. Carter, E. Paleolog and C. P. Wheeler-Jones (2006). "Cyclooxygenase-2 induction and prostacyclin release by protease-activated receptors in endothelial cells require cooperation between mitogen-activated protein kinase and NF-kappaB pathways." J Biol Chem 281(17): 11792-11804.

Szymczak, M., M. Murray and N. Petrovic (2008). "Modulation of angiogenesis by omega-3 polyunsaturated fatty acids is mediated by cyclooxygenases." Blood 111(7): 3514-3521.

Taeger, J., C. Moser, C. Hellerbrand, M. E. Mycielska, G. Glockzin, H. J. Schlitt, E. K. Geissler, O. Stoeltzing and S. A. Lang (2011). "Targeting FGFR/PDGFR/VEGFR impairs tumor growth, angiogenesis, and metastasis by effects on tumor cells, endothelial cells, and pericytes in pancreatic cancer." Mol Cancer Ther 10(11): 2157-2167.

Taketo, M. M. (1998a). "Cyclooxygenase-2 inhibitors in tumorigenesis (part I)." J Natl Cancer Inst 90(20): 1529-1536.

Taketo, M. M. (1998b). "Cyclooxygenase-2 inhibitors in tumorigenesis (Part II)." J Natl Cancer Inst 90(21): 1609-1620.

Tanaka, T., H. Kohno, S. Yoshitani, S. Takashima, A. Okumura, A. Murakami and M. Hosokawa (2001). "Ligands for peroxisome proliferator-activated receptors alpha and gamma inhibit chemically induced colitis and formation of aberrant crypt foci in rats." Cancer Res 61(6): 2424-2428.

Tang, Y., M. T. Wang, Y. Chen, D. Yang, M. Che, K. V. Honn, G. D. Akers, S. R. Johnson and D. Nie (2009). "Downregulation of vascular endothelial growth factor and induction of tumor dormancy by 15-lipoxygenase-2 in prostate cancer." Int J Cancer 124(7): 1545-1551.

Tennis, M. A., M. Van Scoyk, L. E. Heasley, K. Vandervest, M. Weiser-Evans, S. Freeman, R. L. Keith, P. Simpson, R. A. Nemenoff and R. A. Winn (2010). "Prostacyclin inhibits non-small cell lung cancer growth by a frizzled 9-dependent pathway that is blocked by secreted frizzled-related protein 1." Neoplasia 12(3): 244-253.

Thompson, W. J., G. A. Piazza, H. Li, L. Liu, J. Fetter, B. Zhu, G. Sperl, D. Ahnen and R. Pamukcu (2000). "Exisulind induction of apoptosis involves guanosine 3′,5′-cyclic monophosphate phosphodiesterase inhibition, protein kinase G activation, and attenuated beta-catenin." Cancer Res 60(13): 3338-3342.

Thuillier, P., G. J. Anchiraico, K. P. Nickel, R. E. Maldve, I. Gimenez-Conti, S. J. Muga, K. L. Liu, S. M. Fischer and M. A. Belury (2000). "Activators of peroxisome proliferator-activated receptor-alpha partially inhibit mouse skin tumor promotion." Mol Carcinog 29(3): 134-142.

Thun, M. J., S. J. Henley and C. Patrono (2002). "Nonsteroidal anti-inflammatory drugs as anticancer agents: mechanistic, pharmacologic, and clinical issues." J Natl Cancer Inst 94(4): 252-266.

Tian, T., K. J. Nan, S. H. Wang, X. Liang, C. X. Lu, H. Guo, W. J. Wang and Z. P. Ruan (2010). "PTEN regulates angiogenesis and VEGF expression through phosphatase-dependent and -independent mechanisms in HepG2 cells." Carcinogenesis 31(7): 1211-1219.

Torisu, H., M. Ono, H. Kiryu, M. Furue, Y. Ohmoto, J. Nakayama, Y. Nishioka, S. Sone and M. Kuwano (2000). "Macrophage infiltration correlates with tumor stage and angiogenesis in human malignant melanoma: Possible involvement of TNFα and IL-1α." Int J Cancer 85(2): 182-188.

Tsujii, M., S. Kawano, S. Tsuji, H. Sawaoka, M. Hori and R. N. DuBois (1998). "Cyclooxygenase regulates angiogenesis induced by colon cancer cells." Cell 93(5): 705-716.

Tsutsumi, S., T. Gotoh, W. Tomisato, S. Mima, T. Hoshino, H. J. Hwang, H. Takenaka, T. Tsuchiya, M. Mori and T. Mizushima (2004). "Endoplasmic reticulum stress response is involved in nonsteroidal anti-inflammatory drug-induced apoptosis." Cell Death Differ 11(9): 1009-1016.

Turner, E. C., E. P. Mulvaney, H. M. Reid and B. T. Kinsella (2011). "Interaction of the human prostacyclin receptor with the PDZ adapter protein PDZK1: role in endothelial cell migration and angiogenesis." Mol Biol Cell 22(15): 2664-2679.

Tuynman, J. B., L. Vermeulen, E. M. Boon, K. Kemper, A. H. Zwinderman, M. P. Peppelenbosch and D. J. Richel (2008). "Cyclooxygenase-2 inhibition inhibits c-Met kinase activity and Wnt activity in colon cancer." Cancer Res 68(4): 1213-1220.

Untergasser, G., M. Steurer, M. Zimmermann, M. Hermann, J. Kern, A. Amberger, G. Gastl and E. Gunsilius (2008). "The Dickkopf-homolog 3 is expressed in tumor endothelial cells and supports capillary formation." Int J Cancer 122(7): 1539-1547.

Vaish, V. and S. N. Sanyal (2011). "Non steroidal anti-inflammatory drugs modulate the physicochemical properties of plasma membrane in experimental colorectal cancer: a fluorescence spectroscopic study." Mol Cell Biochem 358(1-2): 161-171.

Valacchi, G., A. Pecorelli, C. Sticozzi, C. Torricelli, M. Muscettola, C. Aldinucci and E. Maioli (2011). "Rottlerin exhibits antiangiogenic effects *in vitro*." Chem Biol Drug Des 77(6): 460-470.

van Beijnum, J. R., R. P. Dings, E. van der Linden, B. M. Zwaans, F. C. Ramaekers, K. H. Mayo and A. W. Griffioen (2006). "Gene expression of tumor angiogenesis dissected: specific targeting of colon cancer angiogenic vasculature." Blood 108(7): 2339-2348.

van Moorselaar, R. J. and E. E. Voest (2002). "Angiogenesis in prostate cancer: its role in disease progression and possible therapeutic approaches." Mol Cell Endocrinol 197(1-2): 239-250.

Varet, J., L. Vincent, P. Mirshahi, J. V. Pille, E. Legrand, P. Opolon, Z. Mishal, J. Soria, H. Li and C. Soria (2003). "Fenofibrate inhibits angiogenesis *in vitro* and *in vivo*." Cell Mol Life Sci 60(4): 810-819.

Viita, H., K. Kinnunen, E. Eriksson, J. Lahteenvuo, M. Babu, G. Kalesnykas, T. Heikura, S. Laidinen, T. Takalo and S. Yla-Herttuala (2009). "Intravitreal adenoviral 15-lipoxygenase-1 gene transfer prevents vascular endothelial growth factor A-induced neovascularization in rabbit eyes." Hum Gene Ther 20(12): 1679-1686.

Villegas, I., S. Sanchez-Fidalgo and C. Alarcon de la Lastra (2008). "New mechanisms and therapeutic potential of curcumin for colorectal cancer." Mol Nutr Food Res 52(9): 1040-1061.

Virrey, J. J., Z. Liu, H. Y. Cho, A. Kardosh, E. B. Golden, S. G. Louie, K. J. Gaffney, N. A. Petasis, A. H. Schonthal, T. C. Chen and F. M. Hofman (2010). "Antiangiogenic activities of 2,5-dimethyl-celecoxib on the tumor vasculature." Mol Cancer Ther 9(3): 631-641.

Voronov, E., D. S. Shouval, Y. Krelin, E. Cagnano, D. Benharroch, Y. Iwakura, C. A. Dinarello and R. N. Apte (2003). "IL-1 is required for tumor invasiveness and angiogenesis." Proc Natl Acad Sci USA 100(5): 2645-2650.

Waddell, W. R. and R. W. Loughry (1983). "Sulindac for polyposis of the colon." J Surg Oncol 24(1): 83-87.

Wang, D. and R. N. DuBois (2010). "Eicosanoids and cancer." Nat Rev Cancer 10(3): 181-193.

Wei, J., W. Yan, X. Li, Y. Ding and H. H. Tai (2010). "Thromboxane receptor alpha mediates tumor growth and angiogenesis via induction of vascular endothelial growth factor expression in human lung cancer cells." Lung Cancer 69(1): 26-32.

Weis, S. M. (2008). "Vascular permeability in cardiovascular disease and cancer." Curr Opin Hematol 15(3): 243-249.

Wendel, M. and A. R. Heller (2009). "Anticancer actions of omega-3 fatty acids-current state and future perspectives." Anticancer Agents Med Chem 9(4): 457-470.

Wheeler-Jones, C., R. Abu-Ghazaleh, R. Cospedal, R. A. Houliston, J. Martin and I. Zachary (1997). "Vascular endothelial growth factor stimulates prostacyclin production and activation of cytosolic phospholipase A2 in endothelial cells via p42/p44 mitogen-activated protein kinase." FEBS Lett 420(1): 28-32.

Wheeler-Jones, C., C. Farrar and E. Garonna (2009). "Protcase-activated receptors, cyclo-oxygenases and pro-angiogenic signaling in endothelial cells." Biochem Soc Trans 37(Pt 6): 1179-1183.

Wu, C. C., R. Y. Shyu, J. M. Chou, S. W. Jao, P. C. Chao, J. C. Kang, S. T. Wu, S. L. Huang and S. Y. Jiang (2006). "RARRES1 expression is significantly related to tumor differentiation and staging in colorectal adenocarcinoma." Eur J Cancer 42(4): 557-565.

Wu, G., A. Mannam, J. Wu, S. Kirbis, J. Shie, C. Chen, R. Laham, F. Sellke and J. Li (2003). "Hypoxia induces myocyte-dependent COX-2 regulation in endothelial cells: role of VEGF." Am J Physiol Heart Circ Physiol 285(6): H2420-2429.

Xin, B., Y. Yokoyama, T. Shigeto, M. Futagami and H. Mizunuma (2007). "Inhibitory effect of meloxicam, a selective cyclooxygenase-2 inhibitor, and ciglitazone, a peroxisome proliferator-activated receptor gamma ligand, on the growth of human ovarian cancers." Cancer 110(4): 791-800.

Xin, X., S. Yang, J. Kowalski and M. E. Gerritsen (1999). "Peroxisome proliferator-activated receptor gamma ligands are potent inhibitors of angiogenesis *in vitro* and *in vivo*." J Biol Chem 274(13): 9116-9121.

Yadav, A., B. Kumar, T. N. Teknos and P. Kumar (2011). "Sorafenib enhances the antitumor effects of chemoradiation treatment by downregulating ERCC-1 and XRCC-1 DNA repair proteins." Mol Cancer Ther 10(7): 1241-1251.

Yakes, F. M., J. Chen, J. Tan, K. Yamaguchi, Y. Shi, P. Yu, F. Qian, F. Chu, F. Bentzien, B. Cancilla, J. Orf, A. You, A. D. Laird, S. Engst, L. Lee, J. Lesch, Y. C. Chou and A. H. Joly (2011). "Cabozantinib (XL184), a Novel MET and VEGFR2 Inhibitor, Simultaneously Suppresses Metastasis, Angiogenesis, and Tumor Growth." Mol Cancer Ther 10(12): 2298-2308.

Yang, L., B. Olsson, D. Pfeifer, J. I. Jonsson, Z. G. Zhou, X. Jiang, B. A. Fredriksson, H. Zhang and X. F. Sun (2010). "Knockdown of peroxisome proliferator-activated receptor-beta induces less differentiation and enhances cell-fibronectin adhesion of colon cancer cells." Oncogene 29(4): 516-526.

Yang, L., H. Zhang, Z. G. Zhou, H. Yan, G. Adell and X. F. Sun (2011). "Biological function and prognostic significance of peroxisome proliferator-activated receptor delta in rectal cancer." Clin Cancer Res 17(11): 3760-3770.

Yang, V. W., J. M. Shields, S. R. Hamilton, E. W. Spannhake, W. C. Hubbard, L. M. Hylind, C. R. Robinson and F. M. Giardiello (1998). "Size-dependent Increase in Prostanoid Levels in Adenomas of Patients with Familial Adenomatous Polyposis." Cancer Research 58(8): 1750-1753.

Yang, Y. H., H. Zhou, N. O. Binmadi, P. Proia and J. R. Basile (2011). "Plexin-B1 activates NF-kappaB and IL-8 to promote a pro-angiogenic response in endothelial cells." PLoS ONE 6(10): e25826.

Youssef, J. and M. Badr (2011). "Peroxisome proliferator-activated receptors and cancer: challenges and opportunities." Br J Pharmacol 164(1): 68-82.

Yuan, Y. M., S. H. Fang, X. D. Qian, L. Y. Liu, L. H. Xu, W. Z. Shi, L. H. Zhang, Y. B. Lu, W. P. Zhang and E. Q. Wei (2009). "Leukotriene D4 stimulates the migration but not proliferation of endothelial cells mediated by the cysteinyl leukotriene cyslt(1) receptor via the extracellular signal-regulated kinase pathway." J Pharmacol Sci 109(2): 285-292.

Zaric, J., J. M. Joseph, S. Tercier, T. Sengstag, L. Ponsonnet, M. Delorenzi and C. Ruegg (2012). "Identification of MAGI1 as a tumor-suppressor protein induced by cyclooxygenase-2 inhibitors in colorectal cancer cells." Oncogene 31(1): 48-59.

Zhang, J. Z. and K. W. Ward (2010). "WY-14 643, a selective PPAR{alpha} agonist, induces proinflammatory and proangiogenic responses in human ocular cells." Int J Toxicol 29(5): 496-504.

Zhang, X., Y. Maor, J. F. Wang, G. Kunos and J. E. Groopman (2010). "Endocannabinoid-like N-arachidonoyl serine is a novel pro-angiogenic mediator." Br J Pharmacol 160(7): 1583-1594.

Zhang, Y. and Y. Daaka (2011). "PGE_2 promotes angiogenesis through EP4 and PKA Cgamma pathway." Blood 118(19): 5355-5364.

Zhu, B. H., H. Y. Chen, W. H. Zhan, C. Y. Wang, S. R. Cai, Z. Wang, C. H. Zhang and Y. L. He (2011). "(-)-Epigallocatechin-3-gallate inhibits VEGF expression induced by IL-6 via Stat3 in gastric cancer." World J Gastroenterol 17(18): 2315-2325.

Zhu, J., J. W. Huang, P. H. Tseng, Y. T. Yang, J. Fowble, C. W. Shiau, Y. J. Shaw, S. K. Kulp and C. S. Chen (2004). "From the cyclooxygenase-2 inhibitor celecoxib to a novel class of 3-phosphoinositide-dependent protein kinase-1 inhibitors." Cancer Res 64(12): 4309-4318.

Zhu, Z., C. Fu, X. Li, Y. Song, C. Li, M. Zou, Y. Guan and Y. Zhu (2011). "Prostaglandin E2 promotes endothelial differentiation from bone marrow-derived cells through AMPK activation." PLoS ONE 6(8): e23554.

Chapter 6

Tumor Angiogenesis
Regulation by Non-classical Factors
Regulation by Non-classical Factors:
Peptide, Proteins, Lipids, Steroids and other small molecules

Gary L. Sanford[1*], Ravindra Kumar[2] and Tanisha McGlothen[3]

[1]*Professor, Department of Microbiology, Biochemistry & Immunology, Morehouse School of Medicine 720 Westview Drive, S.W., Atlanta, Georgia 30349. Email: gsanford@msm.edu*

[2]*Instructor, Department of Microbiology, Biochemistry & Immunology, Morehouse School of Medicine 720 Westview Drive, S.W., Atlanta, Georgia 30349. Email: ravi_kumar4@hotmail.com*

[3]*Ph.D. in Biomedical Sciences Program, Morehouse School of Medicine, 720 Westview Drive, S.W. Atlanta, Georgia 30349. Email: tmcglothen@msm.edu*

CHAPTER OUTLINE

▶ Tumor angiogenesis – general introduction & focus of this chapter
▶ Factors regulating tumor angiogenesis
▶ Regulation by cyclooxygenase and prostaglandins
▶ Regulation by nitric oxide and inducible nitric oxide synthase (iNOS)
▶ Regulation by galectins
▶ Regulation by steroid hormones, calcitriol and retinoic acid
▶ Crosstalk between non-classical pro-angiogenic factors
▶ Conclusions and perspectives
▶ References

ABSTRACT

Angiogenesis is controlled by a number of pro-angiogenic and anti-angiogenic factors, which can also be regulated by the oxygen-tension of the microenvironment. This phenomenon has been termed the "angiogenic switch" (Hanahan and Folkman 1996). These classical angiogenesis regulation pathways have been well-described

Correspondence Author Department of Microbiology, Biochemistry & Immunology, Morehouse School of Medicine, 720 Westview Drive, S.W., Atlanta, Georgia 30349. Email: gsanford@msm.edu

in numerous studies. Equally, it is well-established that tumors express and secrete several factors that stimulate angiogenesis, including the classical growth factors (VEGF and bFGF), thrombospondin and angiopoetin. However, tumors produce a number of non-classical factors capable of modulating the angiogenesis process. These factors can be classified as 1) peptides or proteins (e.g., Cox-2, galectins, iNOS), 2) lipids and steroids (e.g., steroid hormones and retinoic acid), and 3) other small molecules (e.g., nitric oxide). This chapter will describe these factors, provide the latest research findings concerning the roles they play in angiogenesis, and explore possible mechanisms for the effects of these factors.

Key words: tumor angiogenesis, COX-2, galectins, nitric oxide, nitric oxide synthase, prostaglandins and receptors, steroids

1   Tumor Angiogenesis—General Introduction & Focus of this Chapter

Angiogenesis is a multistep process leading to the formation of new blood vessels through endothelial cell sprouting from pre-existing blood vessels. This process is controlled by the expression or activation of opposing regulatory factors that modulate endothelial cell behavior necessary for the assembly of endothelial cells into blood vessel-like structures (Detmar 2000; McColl et al. 2004; Nagy et al. 2007; Ferrara 2010); this process has been termed the "angiogenic switch"(Hanahan and Folkman 1996) which is shown in Fig. 1. There is considerable data to suggest that solid tumors need to induce an infusion of new blood vessel growth in order to receive the nutrients needed for sustained tumor growth, survival and metastasis. Nearly all solid tumors express a number of pro-angiogenic factors, some (like VEGF) at elevated levels that are used to promote and enhance an angiogenic response resulting in a network of new blood vessels supplying the tumor mass with oxygen and nutrients. It is well-established that classical regulation of the angiogenic process is controlled by a balance between the expression of pro- (these growth factors and other factors) and anti-angiogenic factors, which in turn appear to be regulated by a number of molecules, as well as, the oxygen tension of the microenvironment. This has been the subject of several reviews and only briefly described in this chapter. However, tumors produce a number of non-classical factors that will modulate the angiogenesis process. These can be classified as 1) peptides or proteins (e.g., COX-2, galectins, iNOS), 2) lipids and steroids (e.g., steroid hormones and retinoic acid), or 3) other small molecules (e.g., nitric oxide). This chapter will describe these factors, provide the latest research findings concerning the roles they play in angiogenesis, and explore possible mechanisms for the effects of these factors.

2   Factors Regulating Tumor Angiogenesis

Numerous studies have provided evidence that most human tumors express all or some of the classical pro-angiogenic growth factors

THE BALANCE HYPOTHESIS FOR THE ANGIOGENIC SWITCH

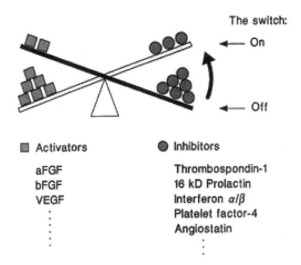

Figure 1   The balance hypothesis for the angiogenic switch. This hypothesis suggests that changes in the relative balance of pro-angiogenic factors (activators or inducers) and anti-angiogenic factors (inhibitors) control the activation of angiogenesis. In many normal tissues, the lack of pro-angiogenic factors may keep the switch off, while in others (such as cancer) pro-angiogenic factors are overexpressed which cannot be held in check by the levels of anti-angiogenic factors. The latter will result in the activation of the switch, initiating an angiogenic response (the growth of new blood vessels).

Reprinted from Cell 86(3), Hanahan D and Folkman J, Patterns and emerging mechanisms of the angiogenic switch during tumorigenesis, Page 353-364, 1996 with permission from Elsevier Ltd.

(VEGF, bFGF, PDGF and TGF-β). Many are further regulated by the hypoxic environment that exist in the microenvironment of solid tumors, through the up-regulation of hypoxia-inducible factor-1-alpha (HIF-1-α), a transcription factor that modulated the expression of several genes required for an angiogenic response. The VEGF gene family is the most studied classical pro-angiogenic growth factors; this family includes VEGF-A (often referred to as just VEGF), VEGF-B, VEGF-C, VEGF-D, and PlGF. VEGF-A, the major modulator of angiogenesis in this family, binds to two trans-membrane tyrosine kinase receptors, VEGFR1 (Flt-1) and VEGFR2 (Flk-1/KDR). Both receptors are expressed in endothelial cells, and to a lesser degree in tumor cells. Recent studies have shown that blocking VEGF or VEGFR1/2 with specific antibodies severely inhibited tumor growth and induction of angiogenesis, and suggested that this approach as a possible therapeutic treatment for cancer. The FDA has approved bevacizumab, a humanized variant of an anti-VEGF-A neutralizing antibody for combination therapy for metastatic colon, breast non-small-cell lung cancer. Similarly, receptor tyrosine kinase inhibitors that target VEGF signaling pathways in anti-cancer therapy are undergoing clinical trials (Ferrara 2010)

The fibroblast growth factor (FGF) family also seems to play a role in angiogenesis that occur in various pathologic processes, including cancer (Kos and Dabrowski 2002). This family includes 18 members that bind to four FGF receptors. FGFs have been reported to promote angiogenesis independent of VEGF (Bogin and Degani 2002). Treatment with FGF-trap (FGFR-Fc fusion peptide) in combination with VEGF inhibitors decreased tumor growth and reduced tumor angiogenesis. The delta-like ligand 4 (DLL4) is a member of the Delta/Jagged family of transmembrane ligands that binds to Notch receptors, and mediates cell–cell communication, and vascular development (Cao et al. 2008; Thurston and Kitajewski 2008). Two groups have demonstrated independently that inhibiting Dll4 leads to tumor growth suppression by deregulating angiogenesis, resulting in increased, but non-functional vessels (Noguera-Troise et al. 2006; Ridgway et al. 2006). Importantly, this strategy is also effective in slowing the growth of tumors that are relative resistant to anti-VEGF therapy and exhibit additive effects with anti-VEGF in slow resistant tumor growth (Noguera-Troise et al. 2006; Ridgway et al. 2006).

3   Regulation by Cyclooxygenase and Prostaglandins

3.1   COX-2

One factor correlated with aggressive and invasive potential of many different tumor cells is the overexpression of cyclo-oxygenase-2 (COX-2), which may play an essential role in tumor-induced angiogenesis. Presumably this role occurs through increased production of pro-angiogenic factors, such as VEGF. There are three COX isoenzymes – COX-1, COX-2 and COX-3 (a splice variant of COX-1). COX-1 and COX-2 differ mainly in their patterns of expressions; COX-1 is produced constitutively, whereas COX-2 is inducible. COX-1 is widely found in most mammalian cells and tissues, such as gastric mucosa (Chu et al. 2003). The earliest expression of COX-2 is detected in stromal cells, but it also can be found in multiple cells, like epithelial, endothelial, and stromal cells in several types of tumors. Abnormal or increased expression of COX-2 has been reported for a number of different cancers, including colon, pancreatic (Chu et al. 2003; Kasper et al. 2010), lung (Koch et al. 2011), gastric (Chen et al. 2006), breast (Singh and Lucci 2002), head and neck (Cao et al. 2008) and prostate cancers (Richardsen et al. 2010). Tumor expression of COX-2 may be stimulated by oncogenes, such as, ras (Araki et al. 2003; Cho et al. 2010), but down-regulated by tumor suppressors, such as, p53 (Chiarugi et al. 1998).

High COX-2 expression seems to enhance the metastatic potential of cancer cells, particularly colon cancer. In this case, the over-expression of COX-2 increased matrix metalloproteinase expression allowing for enhanced invasion of cancer cells (Larkins et al. 2006). Two research studies provide strong evidence that link COX-2 expression with an angiogenic response. In a study by Chu and coworkers (Chu et al. 2003) the angiogenic characteristics of a COX-2-positive and COX-2-negative pancreatic tumor cell lines were compared. Culture medium from the COX-2 positive cells was able to increase endothelial cell migration and induced an angiogenic response both *in vitro* and *in vivo*. In contrast, culture medium from the non-COX-2 cells produced only a modest effect on endothelial cell migration and angiogenesis *in vivo*. Pretreatment of cells with a selective COX-2 inhibitor (NS-398) blunted the angiogenic response of endothelial cells; SC-560 (a selective COX-1 inhibitor) had no effect. These results suggest that COX-2 may be involved in the control of tumor-dependent angiogenesis for certain pancreatic cancers. The second study, by Yao and coworkers (Yao et al. 2011), found that showed that down-regulating COX-2 expression using COX-2 siRNA resulted in significant suppression of endothelial cell migration and tube formation of human umbilical vein endothelial cells (angiogenesis), consistent with previous reports. This study also found that the down-regulation of COX-2 resulted in the differential expression of over 20 genes, many of which are associated with the angiogenesis response, including FGF4, TGFB2, TGFBR1, VEGF, Flt1, Flk-1, angiopoietin, Tie2, MMP2, and THBS2. These researchers confirmed the effect of COX-2 down-regulation on VEGF, Flt-1, Flk-1, angiopoietin, tie-2, and MMP2 expression using RT-PCR and western blotting. This study indicates that COX-2 is necessary for tumor expression of a number of pro-angiogenic factors. Hence, both studies provide additional evidence that COX-2 may mediate tumor angiogenesis and growth of different cancer cells.

Additionally, Lavalle et al (Lavalle et al. 2009) evaluated Cox-2 expression and microvessel density in canine mammary carcinomas. They demonstrated that increased microvessel density and increased Cox-2 expression were linearly related in the canine mammary tumors, suggesting that Cox-2 inhibitors could be an alternative for the treatment and control of advanced neoplasms. In a study by Basu and coworkers (Basu et al. 2006), invasive human breast cancer cells that overexpress COX-2 was reported to have the unique ability to differentiate into extracellular-matrix-rich vascular channels, also known as vascular mimicry. Vascular channels have been associated with angiogenesis without involvement of endothelial cells. They found that invasive human breast cancer cells that over-express COX-2 develop vascular channels when plated on three-dimensional matrigel cultures, whereas non-invasive cell lines that express low levels of COX-2 did not develop such channels. Similarly, vascular channels were identified in high grade invasive ductal carcinoma of the breast over-expressing COX-2, but not in low-grade breast tumors. Vascular channel formation was significantly suppressed when cells were treated with celecoxib or COX-2 siRNA. Inhibition of channel formation was abrogated by addition of exogenous prostaglandin E2. *In vitro* results were corroborated *in vivo* in tumor-bearing mice treated with celecoxib. These studies indicate that COX-2 not only mediate tumor angiogenesis but the vascular mimicry process as well.

Lastly, what the exact mechanism, including which are primary or which are secondary, for the effects of COX-2 on tumor angiogenesis has not been determined. However, it is also becoming clear that a complex regulatory network, rather than simple signaling pathways modulates tumor angiogenesis; these networks may be connected in different ways in different tumor cells. However, it is also becoming clear that a complex regulatory network, rather than simple signaling pathways modulates tumor angiogenesis. These networks may be connected in different ways in different tumor cells (Yao et al. 2011).

3.2 Prostaglandins

The COX pathway generates five primary prostanoids from arachidonic acid (Fig. 2): prostaglandin D2 (PGD2), prostaglandin E2 (PGE2), prostaglandin F2α (PGF2α), prostaglandin I2 (PGI2) and thromboxane A2 (TXA2). Several of these COX metabolites (PGE2, PGI2, and TXA2) have been shown to increase endothelial cell migration and experimental angiogenesis. In

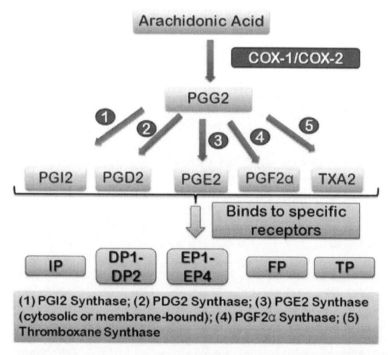

Figure 2 The downstream products (prostaglandins) formed by cyclo-oxygenase (COX1/2) activity, and their specific receptors.

fact, PGE2 was reported to support tumor growth mainly by inducing angiogenesis. In addition, PGE2 can induce the expression of matrix metalloproteinases (MMPs), VEGF, and bFGF, all of which are important mediators of angiogenesis. Using the VEGF-induced mouse corneal model of angiogenesis, it was clearly demonstrated that the selective COX-2 inhibitor NS-398 inhibited VEGF-induced angiogenesis (Hernandez et al. 2001). This inhibitory effect was reversed by PGE2 treatment, indicating that the VEGF-induced angiogenesis in this model is mediated by COX-2 through PGE2.

Each of the prostanoid metabolites are synthesized downstream of COX-1/COX-2 by specific prostaglandin synthases (Fig. 2). PGE2 synthase (PGES) is found in two forms: cytosolic PGES (cPGES) and microsomal or membrane associated PGES (mPGES), which seems to play different roles. The cPGES is constitutively expressed in most cells and tissues but is coupled to COX-1 rather than COX-2. On the other hand, mPGES is more likely to be linked to inducible COX-2 generating PGE2 downstream of COX-2. COX-2 and mPGES co-localizes in the perinuclear membrane, which allows for efficient transfer of COX-2-generated unstable PGH2 to mPGES where it is converted to PGE2. Several recent studies provide evidence that mPGES mediates COX-2 in different diseases, including cancer. Co-transfection of COX-2 and mPGES in HEK293 cells resulted in the production of high levels of PGE2, compared to cells transfected with either COX-2 or mPGES alone. In addition to PGE2, TXA2 seems to play an important role in angiogenesis. TXA2 receptor antagonist (SQ29548) inhibited both endothelial cell migration and corneal angiogenesis. When COX-2 was inhibited, the TXA2 agonist U46619 restored endothelial cell migration and corneal angiogenesis.

The mechanism by which PGE2 regulates prostate cancer, and other cancers, progression and angiogenesis is not well defined. Researchers have demonstrated that PGE2 may induce the phosphorylation of Akt, ERK1/2, and eNOS in endothelial cells; the effects of PGE2 seems to require the expression of mTORC2 (Namkoong et al. 2005; Dada et al. 2008; Sakurai et al. 2011). It is established that PGE2 induces cellular responses through the EP family of receptors, consisting of four different subtypes (EP1–EP4). The expression of two of these receptors (EP2 and EP4) was enhanced in prostate cancer, as well as, endothelial cells (Kamiyama et al. 2006; Donnini et al. 2007). A recent study showed that both exogenous and tumor-derived PGE2 increased endothelial cell motility, but blocking endothelial EP2 and EP4 receptors using specific receptor antagonists reduced endothelial cell motility toward tumor cells. Additionally, PGE2 was shown to induce angiogenesis in the *in vivo* Matrigel plug angiogenesis assay, which was attenuated by blocking the EP2 and EP4 receptors; similar findings have been reported for tumor-derived PGE2. This combination of findings shows that PGE2 via EP2/EP4 receptor promotes *in vivo* angiogenesis. It has also been reported that PGE2 induces cyclic-AMP (cAMP) production, increases cellular growth, and regulates differentiated cell functions by promoting the activation of cAMP-dependent protein kinase A (PKA) (Sheng et al. 2001; Diaz-Munoz et al. 2012). The PKA-mediated phosphorylation of cAMP-responsive element binding protein (CREB) and regulation of transcription via interaction between cAMP-response elements with CREB are considered as the major pathways that alter gene expression in cancer cells (Fujino et al. 2005; Bidwell et al. 2010). Two studies found that PGE2 also activates transcription factor 4 (ATF-4; also called CREB-2), which regulates the expression of several genes involved in angiogenesis (Shao et al. 2005; Jain et al. 2008), suggesting that PGE2 plays an important role in the regulation of tumor angiogenesis.

4 Regulation by Nitric Oxide and Inducible Nitric Oxide Synthase (iNOS)

The role of NO in tumor progression and angiogenesis has been controversial at best, having both pro- and anti-tumor effects (Morbidelli et al. 2004). There have been a number of reports that NO increases VEGF expression, which would enhance the angiogenic response. There have also been an almost equal number of contradictory reports, some of which found that NO donors (e.g., sodium nitroprusside, SNP) decreases VEGF expression or inhibited hypoxia-induced VEGF expression. Tsurumi, et al., demonstrated that SNP, a NO donor, down-regulated the VEGF promoter activity and its synthesis in vascular smooth muscle cells (VSMC) by interfering binding of the AP-1 transcription factor (Tsurumi et al. 1997). Other reports showed that NO inhibit hypoxic induction of the VEGF gene through attenuation of HIF-1 binding activity by abrogating accumulation of HIF-1α protein in VSMC and tumor cell lines (Liu et al. 1998; Sogawa et al. 1998; Huang et al. 1999). Recent studies that used SNP, found that this NO donor has distinctly different effect compared to other non-SNP NO donors, such as SNAP. Even a small amount of SNP inhibited VEGF expression, and its inhibitory effect is not ascribed to NO production (Kimura et al. 2002; Dulak and Jozkowicz 2003). NO induced VEGF expression with the non-SNP NO donors by modulating hypoxia inducible factor-1α (HIF-1α) activity in tumor cells, independent of other pathways.

Endogenous NO is produced by the action of nitric oxide synthase (NOS) on L-arginine, as shown in Fig 3. There are three NOS isotypes: nNOS (NOS1), iNOS (NOS2) and eNOS (NOS3). The nNOS (found in neural and smooth muscle cells) and eNOS (found in endothelial cells) are constitutively expressed and calcium-dependent, whereas, iNOS (inducible isoform found in a number of tissues and pathologies) is calcium-independent. Both eNOS and nNOS produce low level bursts (10-30 nmole/L) of NO, but phosphorylation of eNOS results in sustained low levels (1-10 nmole/L) of NO. These low NO levels activate cGMP and ERK signaling associated with endothelial cell migration and proliferation.

In contrast, the iNOS isotype is inducible under certain conditions, and may be over-expressed in a number of different tumors, can produce a high and sustained level of NO (>300 nmole/L). NO levels this high may promote the phosphorylation of Ser[15] on p53, as well as, inactivation of mitogen kinase phosphatase-1, leading to endothelial cell growth arrest and possibly cell death. Additionally, high levels of NO may have anti-angiogenic responses (Kimura and Esumi 2003). However, high levels of NO accompanying the overexpression of iNOS in cancer correlates with tumor growth and metastasis using *in vitro* models. In addition, a number of studies found increased expression of iNOS in different human tumors, including laryngeal neoplasia, gastric, hepatocellular, skin, and colorectal cancers (Franchi et al. 2002; Kimura and Esumi 2003; Peng et al. 2003; Massi et al. 2009; Zhang et al. 2011) and correlates with aggressive tumor behavior and poor prognosis (Zhang et al. 2011).

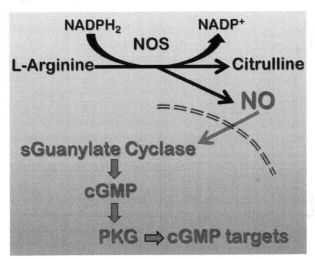

Figure 3   Nitric oxide (NO) signaling. The mechanism for generating nitric oxide (NO) is shown, along with the subsequence cellular signaling.

NOS may contribute to tumor progression by promoting angiogenesis, which has been suggested by a number of studies. Jenkins et al (Jenkins et al. 1995) demonstrated that human carcinoma cell lines that were transfected with a murine iNOS cDNA exhibited accelerated growth and increased neovascularization in a nude mouse/xenograft model. Cianchi et al (Cianchi et al. 2010) reported an up-regulation of the iNOS gene and protein in colon cancer tissues, compared with adjacent normal mucosa. They also found a positive correlation between iNOS expression and tumor angiogenesis, as reflected by tumor microvessel density. Shang and Li (Shang and Li 2005) found a correlation between microvessel density and the expression of iNOS and VEGF in oral squamous cell carcinoma.

Similarly, Li et al (Li et al. 2005) found that iNOS expression in endometrial cancer was associated with increased tumor invasion and vascularization. Hellmuth et al (Hellmuth et al. 2004) studied the effect of nitric oxide on angiogenesis in a colon cancer cell line. They found that nitric oxide up-regulated the expression of the pro-angiogenic factors interleukin-8 and VEGF and down-regulated the expression of two antiangiogenic chemokines, interferon-inducible protein-10 and monokine-induced by interferon-gamma. It has also been demonstrated that VEGF-mediated angiogenesis is dependent on nitric oxide production and requires the activation of the nitric oxide/cyclic guanosine monophosphate pathway within the endothelial compartment. Malone et al (2006) reported that with hypoxia exposure, a known angiogenesis stimulus, ovarian cancer cells showed increased expression of iNOS and VEGF and increased secretion of VEGF in the cell culture supernatants, which were also associated with a rich and thick tubular network in an *in vitro* angiogenesis model (Malone et al. 2006). Treatment with an iNOS inhibitor resulted in a significant reduction in VEGF production and inhibition of tube formation, but treatment with SNAP (an NO donor) stimulated VEGF production and tube formation. These *in vitro* data support an active role for iNOS in modulating VEGF production and angiogenesis in ovarian cancer.

5 Regulation by Galectins

5.1 Overview

Evidence is increasing that galectins may play critical roles in how endothelial cell behavior is altered in associated with tumor angiogenesis (Thijssen et al. 2008), especially, galectin-1, -3 and -8. Galectins are a family of soluble carbohydrate binding proteins (S-type or S-lac) that have carbohydrate recognition domains (CRD) that recognize and bind β-galactoside moieties. The galectin family is commonly defined by a shared consensus amino acid sequences and affinity for β-galactoside-containing oligosaccharides (Liu and Rabinovich 2005). This is an evolutionarily highly conserved family and homologues exist in lower organisms such as nematodes and sponges (Houzelstein et al. 2004). Fifteen galectins have been described and subdivided into: a) prototype (galectin-1, -2, -5, -7, -10, -11, -13, -14, and -15) that exist as monomers or homodimers consisting of one carbohydrate-recognition domain (CRD); b) chimera type (galectin-3) that have a non-lectin part connected to a CRD; and c) tandem repeat type (galectin-4, -6, -8, -9 and -12) that have two distinct but homologous CRDs in a single polypeptide chain. Additionally, there are proteins that have sequence similarities with galectins but do have the carbohydrate recognition domain; e.g., griffin (galectin related inter-fiber protein), which has been identified in the lens of the eye. Hence the galectin super family has members that may have or not have carbohydrate binding affinity. Three galectins are expressed in different types of cancers and have been associated with some aspect of angiogenesis (Table 1).

Table 1 Distribution and biological function of galectins in different cancer types

Galectins	Cancer types	Biological functions	Mechanisms
Galectin-1	colon, oral, head & neck, pancreas, prostate	Cell migration Angiogenesis Chemo-/Radio-resistance	• Modulation of cytoskeleton organization • Modulation of RhoA expression • Modulation of integrin β1 recycling • Regulation of VEGF secretion via the regulation of ORP150 expression
Galectin-3	breast, lung, prostate, colon and others	Cell migration Angiogenesis	• Modulation of cytoskeleton organization • Formation of a NG2–galectin-3–α3β1 integrin complex • Activation of Fak • ~22 kDa fragment of MMP-cleaved galectin-3
Galectin-8	brain, breast, colon, prostate, uterus, lung, and others	Cell migration Angiogenesis	• Integrin (α3β1, α6β1)-mediated signaling • Activation of Fak & Paxillin

Modified from Le Mercier M et al Galectins and Gliomas, Brain Pathology 20 (2010) 17–27.

Galectins function both extracellularly, interacting with both cell surface proteins and components of the extracellular matrix, and intracellularly, modulating soluble and nuclear proteins signaling pathways. Research to date indicates that galectins play important and possibly crucial roles in different cancer types ranging from tumor transformation, tumor cell survival, metastasis and angiogenesis. Galectins can also alter immune and inflammatory responses, and may be essential to how tumor cells avoid immune detection.

5.2   Galectin-1

The expression of galectin-1 by cancer cells is thought to promote tumor progression. However, the data for Galectin-1 playing a role as a pro-angiogenic factor is in debate. Currently, there is considerable evidence that link galectin-1 expression to endothelial cell adaptations to tumor derived factors and microenvironment that ultimately results in an angiogenic response. When galectin-1 was knocked-down, endothelial cell showed reduced migration and proliferation. Additionally, both tumor growth and angiogenesis was reduced in galectin-1 knockout mice (gal-1$^{-/-}$), but was less effective when the tumor expressed high levels of galectin-1 (Thurston and Kitajewski 2008). A similar finding has been reported galectin-1 knockdown in glioblastoma tumor cells resulted in reduced tumor vessels following the injection of these cells into mice (Le Mercier et al. 2008).

Thiodigalactoside (TDG) as a non-metabolized disaccharide that have been shown to suppress tumor growth. Ito et al (Ito et al. 2011) showed that TDG treatment of tumors in Balb/c nude mice reduced angiogenesis and slowed tumor growth. Further, knocking down galectin-1 expression significantly impeded tumor growth and the sensitivity of the resulting tumors to TDG was severely reduced, highlighting a specific role for galectin-1. Lastly, these researchers found that endothelial cells were protected by galectin-1 from oxidative stress-induced apoptosis induced by H_2O_2, but TDG inhibited this antioxidant protective effect of galectin-1 and reduced tube forming activity in angiogenic assays.

Galectin-1 may indeed function as a chemoattractant to endothelial cells, as well as, promote endothelial viability and adhesion to matrix components; galectin-1 inhibitors (lactulose, thiodigalactoside) reduced endothelial tube formation (Rabinovich et al. 2006). In fact, tumor endothelial cells will take up galectin-1 from galectin-1 secreting tumors. *In vitro* studies with endothelial cells found that exogenous galectin-1 stimulated endothelial cell migration and proliferation, which is attenuated by galectin-1 inhibitors. This study also showed that galectin-1 activated H-ras signaling leading to the activation of the MEK/ERK pathway.

5.3   Galectin-3

A large body of data shows that galectin-3 is expressed in a wide range of tumor cells and is strongly associated with tumor invasion and metastasis. Galectin-3 seems to be a multifunctional oncogenic protein that been reported to modulate cell growth, adhesion, apoptosis and angiogenesis. Figure 4 shows the extracellular roles played by galectin-3 (Nangia-Makker et al. 2008), including angiogenesis by promoting endothelial cell migration and formation of vessel like structures. This figure also shows that galectin-3 binds to cell surface proteins through glycoconjugates and inducing signaling pathways. For example, secreted galectin-3 binds to α5β3 and α3β1 integrins resulting in the clustering of these integrins and activating focal adhesion kinase. Galectin-3 can also bind to EFG, TGFβ, and VEGF receptors, retaining them on the cell surface and modulate their signal transduction. Markoska and coworkers (Markowska et al. 2010) found that galctin-3 can induce the phosphorylation of the VEGF receptor (VEGFR2) in endothelial cells; knockdown of galectin-3 attenuated VEGF-A mediated angiogenesis. There was increased uptake of VEGFR2 in the galectin-3 knockdown cells, supporting the idea that galectin-3 retain growth factor receptors on the cell surface. In a separate study by these researchers, galectin-3 knockdown, as well as, galectin-3 inhibitors reduced both VEGF and FGF induced angiogenesis. Secreted extracellular galectin-3 has been reported to induce endothelial cell morphogenesis *in vitro* and angiogenesis *in vivo*.

Intracellular galectin-3 has also been associated with promoting cell behavior necessary for angiogenesis. Califice et al (Califice et al. 2004) reported that galectin-3 exerts opposing effects depending on whether it is found in the cytoplasm or nucleus of prostate cancer cells. Cytoplasmic galectin-3 promoted aggressive cell behavior including increased cell invasion through Matrigel, anchorage independent growth, *in vivo* tumor

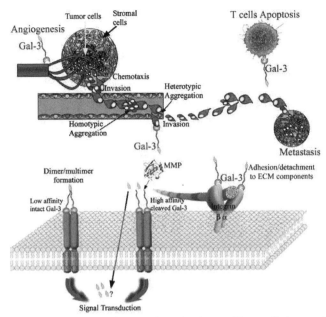

Figure 4 Angiogenesis actions of galectin-3. Extracellular galectin-3 can promote angiogenesis by inducing the migration, chemotaxis and morphogenesis of endothelial cells. Extracellular galectin-3 also crosslinks cell surface glyco-conjugates, form dimers and multimers, that delivers signals inside the cell. The ~22 kDa fragment of MMP cleaved galectin-3 binds to the glycan receptors more efficiently than the intact protein, however the significance of the smaller fragments are not yet known.

Color image of this figure appears in the color plate section at the end of the book.

Reprinted with kind permission from Springer Science+Business Media: Cancer Microenvironment, Regulation of Tumor Progression by Extracellular Galectin-3, volume 1(2008): 43–51, Nangia-Makker P, Balan V and Raz A, figure 1.

growth and angiogenesis. On the other hand nuclear galectin-3 significantly reduced malignant behavior of cancer cells and was generally associated with reduced angiogenesis and tumor growth.

5.4   Galectin-8

One galectin believed to be involved in cell-cell and cell-matrix interactions is galectin-8, which directly associates with fibronectin and other matrix glycoproteins. Galectin-8 has two CRD separated by a variable length linker peptide, making it a tandem repeat type galectin (Bidon-Wagner and Le Pennec 2004; Thijssen et al. 2008). The variable length linker peptide is formed by alternative splicing mechanisms and gives rise to several isoforms of galectin-8 in normal and tumor cells. Galectin-8 may be a potent pro-angiogenic factor that function at concentrations 10 times lower than the levels necessary for galectin-1 or -3 to modulate cell behavior. These low levels of galectin-8 may be similar to matrix levels, but may be able to induce endothelial morphogenesis *in vitro*. Galectin-8 is secreted by non-small cell lung carcinoma cells; where it can bind to and become immobilized by matrix glycoconjugates. Delgado et al (Delgado et al. 2011) found that galectin-8 immobilized in Matrigel was able to induce an angiogenic response by endothelial cells. Additionally, galectin-8-supplemented Matrigel was shown to promote endothelial cell migration and *in vivo* angiogenesis in mice. The mechanism underlying these effects of galectin-8 is not clear. However, galectin-8 can bind a number of integrins such as α3β1 found on endothelial cells, CD166 (implicated in modulating vascular tube formation), and pro-metalloproteinase-9, which may activate several cellular signaling pathways.

6   Regulation by Steroid Hormones, Calcitriol and Retinoic Acid

6.1   Estrogens

This class of sex steroids is a major risk factor in breast cancer progression. In fact breast cancer maintains higher levels of estradiol than normal breast tissue. The major enzymes involved in estrogen synthesis (aromatase, sulphatase, and 17β-hydroxysteroid dehydrogenase) are also found in breast cancer. Additionally, data has amassed suggesting that estrogen may directly regulate angiogenesis, primarily through modulation of VEGF in tumor cells and endothelial cells. The research data in this area is controversial and the underlying mechanisms for the reported finding are not entirely clear. A number of studies have reported an association between estrogen, endothelial estrogen receptor expression breast cancer, angiogenesis, and/or tumor invasiveness (Florian et al. 2008). A recent study found that extracellular levels of VEGF were significantly higher in tumors than in normal breast tissue *in vivo*; it was positively correlated with plasma levels of estradiol (Saarinen et al. 2010). 17β-Estradiol was reported to induce VEGF expression in several target cells, including epithelial cells, fibroblasts and smooth muscle cells, which again suggest a role for estrogens in modulating angiogenesis.

One mechanism that may be involved in mediating the effects of estradiol on VEGF expression is c-Myc. Dadiani et al (Dadiani et al. 2009) reported that c-Myc up-regulation by estrogen is necessary for the transient induction of VEGF transcription, however, over expression of c-Myc alone is not sufficient for this induction. These researchers also showed that both c-Myc and activated estrogen receptor-α co-bind to the VEGF promoter in close proximity, which may be a novel mechanism for estrogen regulation of VEGF. Long-term studies of estrogen treatment of tumors overexpressing c-myc resulted in stable levels of VEGF expression *in vitro* and *in vivo*, maintaining steady vascular permeability in tumors. On the other hand, when estrogen treatment stopped, there was increased VEGF and elevated vascular permeability. Hence, these studies show an apparent cooperative role for the estrogen receptor-α and c-Myc mediating estrogen regulation of VEGF and the ability of c-Myc to partially mimic estrogen regulation of angiogenesis.

Other researchers have shown that estradiol regulates the expression of VEGF in murine mammary cancer cells (Dabrosin et al. 2003) and human breast cancer cells (Garvin et al. 2005). An estrogen-responsive element was described for both the rat and human VEGF gene (Buteau-Lozano et al. 2002; Stoner et al. 2004) it interacts with both the estrogen receptor-α and -β. However, there have been reports that describe estradiol inhibition of VEGF mRNA expression in breast cancer cells (Bogin and Degani 2002); these results may be a consequence of using very high levels of estradiol. Buteau-Lozano et al. (Buteau-Lozano et al. 2002) showed that transient transfection of estrogen receptor-β repressed the VEGF promoter in breast cancer cells, although it is not clear that such an effect would occur in cell lines that expresses high levels of estrogen receptor-β. Additional researchers have found evidence that estrogen will induce VEGF in breast cancer cells (Sengupta et al. 2003) and in an *in vivo* mammary tumor model (Nakamura et al. 1996; Kazi et al. 2005) proposed that HIF-1α may influence estrogen-dependent increases in VEGF message in the rat uterus; a similar mechanism may exist in breast cancer cells. Instability of this factor (or the estrogen receptor itself) may dictate estrogen-dependent induction VEGF (Coradini et al. 2004). Interestingly, there have been suggestions that post-transcriptional events might also influence estrogen-dependent release of VEGF from human breast cancer cells; this effect is attenuated by tamoxifen (Garvin and Dabrosin 2003). Multiple researchers have showed that estrogen can act as an "angiogenic switch" in breast cancer cells by down-regulating soluble VEGFR-1, which may sequester VEGF (Elkin et al. 2004; Garvin et al. 2005). The loss of soluble VEGFR-1 may provide an increase in the net balance towards estrogen-dependent angiogenesis in breast cancer cells.

Hyder et al (Hyder et al. 2009) provided evidence that estradiol may induce the expression of thrombospondin-1 receptor in breast cancer cells *in vitro*; this induction could be blocked by ICI 182,780 (estrogen antagonist), suggesting a requirement for the estrogen receptor. Thrombospondin-1 is an extracellular matrix protein that inhibits angiogenesis. More recently, the role of thrombospondin-1 as an antiangiogenic factor has come into question and data now suggest that it plays a proliferative role in breast cancer (Tischer et al. 1991; Tsurumi et al. 1997; Sogawa et al. 1998; Stoner et al. 2004; Thijssen et al. 2008; Thurston and Kitajewski 2008); this role seems to vary with cell type and level. Transfection studies with deletion constructs of the thrombospondin-1 promoter found an estrogen-responsive region upstream of the transcription start site. Hence, thrombospondin-1 production seems to be directly controlled by estrogens in estrogen receptor-positive breast cancer cells.

6.2   Progesterone

The research to date suggests that sex steroids may modulate VEGF production in normal and tumor breast tissue, although it is not yet clear whether this is an effect of estradiol alone or whether progesterone is also involved (Dabrosin et al. 2003; Hyder 2006). Addition of progesterone to estradiol in tissue culture experiments with normal breast biopsies did not alter the estrogen-induced VEGF levels (Dabrosin et al. 2003). Importantly, it has not been proven whether VEGF in the normal breast can be associated with the initiation or progression of tumors. The effects of progesterone on angiogenesis and VEGF expression in breast cancer are poorly understood. Consequently, there is little information relating to the possible role of progesterone in angiogenesis and/or the regulation of VEGF expression.

Several reports demonstrated that both natural and synthetic progestins induced VEGF in breast cancer cells (Hyder and Stancel 2002; Liang and Hyder 2005; Mirkin et al. 2005). The human VEGF promoter was found to have progesterone response elements, which seems to be cell specific for modulating VEGF synthesis (Mueller et al. 2003; Wu et al. 2005). In breast cancer cells, the region that seems to induce VEGF expression requires a Sp-1 region since a mutation in this region eliminates the ability of progestin to induce the VEGF promoter (Wu et al. 2005). It is possible that Sp-1 and the progesterone receptor might work together to produce a response, which was reported for estrogen-dependent VEGF expression in breast cancer cells (Stoner et al. 2004). There is evidence that progesterone receptor isoforms may be important factors for the induction of VEGF in breast cancer cells. Progesterone receptor-B is a more potent inducer of VEGF than progesterone receptor-A (Wu et al. 2004), which suggests that progesterone receptor-B-rich tumors will be faster growing and more vascularized than tumors containing predominantly progesterone receptor-A. The study by Sortorius et al (Sartorius et al. 2003) provides additional support for this idea; this study reported that larger tumors could be produced when progesterone receptor-B-containing cells were grown in. Furthermore, both anti-estrogens and anti-progestins appear to induce VEGFs by breast cancer cells are rich in progesterone receptor-B, implying that anti-hormones are seen differently depending on which progesterone receptor isoform is present (Wu et al. 2004; Wu et al. 2005). All synthetic progesterones investigated were able to induce VEGF in breast cancer cells in a mechanism that involves both PI3K and MAPK pathways (Hyder and Stancel 2002; Liang and Hyder 2005). Interestingly, Wu et al reported that the regulation of the VEGF gene occurred in a either a cell or ligand-dependent manner (Wu et al. 2005). Treatment with PI3K inhibitors attenuated both the cellular expression of VEGF and secretion of VEGF protein. Treatment with MAP kinase inhibitors, on the other hand, did not inhibit gene transcription in certain cells, but did attenuate protein secretion; in this case suggesting that progesterone can regulate VEGF expression at both at the levels of transcription and post-transcription. Interestingly, progesterone-dependent VEGF expression could be blocked at both the protein and transcriptional level, by anti-estrogen therapy using ICI 182,780, supporting the idea that there may be crosstalk mechanisms involved in regulating these two receptors (Hyder and Stancel 2002). The picture emerging is that with breast cancer cells, the regulation of angiogenic pathway involving VEGF and VEGF receptors is under the control of both estrogens and progestins. Still, very little has been established detailing the molecular mechanism(s) involved in regulating this expression, or the cellular consequences resulting from inducting or suppressing VEGF and VEGF receptors in breast tissues.

6.3 Glucocorticoids

Steroid hormones such as prednisone, hydrocortisone, and dexamethasone, are known to produce some clinical benefit for patients with hormone-refractory prostate cancer (HRPC).

How glucocorticoid exert these effects are not clearly established, but may involve direct inhibition of angiogenesis. When glucocorticoid, such as dexamethasone bind to the glucocorticoid receptor two major angiogenic factors are down-regulation, VEGF and interleukin-8. Researchers have shown that dexamethasone and hydrocortisone down-regulated VEGF and IL-8 expression, which were completely reversed by the glucocorticoid receptor antagonist RU486 (Yano et al. 2006). In DU145 prostate cancer cell xenografts, treatment with dexamethasone resulted in decreased tumor volume and microvessel density, as well as, the down-regulation of VEGF and IL-8 expression. However, dexamethasone treatment did not affect the proliferation of the prostate cancer cells *in vitro*. Hence, glucocorticoids seem to suppress androgen-independent prostate cancer growth by inhibiting VEGF and IL-8 production resulting in decreased tumor-associated angiogenesis.

Iwai et al (Iwai et al. 2004) also found that that low doses of glucocorticoids markedly down-regulated VEGF expression and secretion in renal carcinoma cells possibly, again through the glucocorticoid receptor pathway. These researchers found that among the steroid hormones only glucocorticoids had an effect on VEGF expression; these effects were completely blocked by the specific glucocorticoid receptor antagonist RU486. Since the VEGF promoter does not contain a glucocorticoid-responsive element, possible mechanisms by which glucocorticoids down-regulate VEGF may include modulating the HIF-1α, AP-1, or NF-κB pathways. One candidate, AP-1 has four binding sites within the VEGF promoter (Tischer et al. 1991), and has been suggested for a number of genes repressed by glucocorticoids, but did not contain a glucocorticoid-responsive element (Webster and Cidlowski 1999). Another possibility by which glucocorticoids down-regulate VEGF is through the NF-κB pathway, which has previously been implicated as a regulator of cytokine genes, which is also inhibited by glucocorticoids. More research is required to determine whether or not the glucocorticoids down-regulate VEGF expression by destabilizing the HIF-1α pathway.

6.4 Calcitriol (Vitamin D3)

Calcitriol (1,25-dihydroxycholecalciferol, vitamin D3), the major active form of vitamin D, is anti-proliferative in tumor cells and tumor-derived endothelial cells The actions of calcitriol are mediated, in part by vitamin D receptor (VDR), which is expressed in many tissues including endothelial cells. Chung et al (Chung et al. 2009) examined the role of VDR in mediating the effects of calcitriol on the tumor vasculature using *in vivo* tumor model established in either VDR wild type (WT) or VDR-knockout (KO) mice. These researchers found that calcitriol treatment produced growth inhibition of tumor-derived endothelial cells expressing VDR, but those from VDR-KO mice were relatively resistant. These researchers also demonstrated that blood vessels in VDR-KO mice were enlarged and had less pericyte coverage compared to WT. There was increased expression of HIF-1α, VEGF, Ang-1 and PDGF-BB levels by tumors from VDR-KO mice. Collectively, these results suggest that calcitriol attenuation of tumor angiogenesis is VDR dependent; loss of VDR may lead to abnormal tumor angiogenesis.

Levine and Teegarden (Levine and Teegarden 2004) showed that 1,25-dihydroxycholecalciferol altered the angiogenic phenotype of the cells. The VEGF promoter was activated by 1,25-dihydroxycholecalciferol in a dose-dependent leading to the induction of VEGF mRNA expression, and secretion of VEGF protein. Thus, these data provide evidence that 1,25-dihydroxycholecalciferol modulates angiogenesis in normal and disease states. Since 1,25-dihydroxycholecalciferol affects multiple cellular signaling pathways and modulate a variety of cellular processes including proliferation, differentiation, and apoptosis in multiple cell types, it is possible that complex signaling could also affect angiogenesis (Krishnan et al. 2003). Research in chondrocytes and osteoblasts demonstrated that in addition to inducing proliferation and growth, 1,25-dihydroxycholecalciferol increases VEGF mRNA and secretion, subsequently increasing angiogenesis in bone, which is critical to normal bone development, maintenance, and fracture healing. VEGF secretion was also enhanced by 1,25-dihydroxycholecalciferol treatment of vascular smooth muscle cells, the primary source of circulating VEGF (Levine and Teegarden 2004). Conversely, 1,25-dihydroxycholecalciferol suppresses angiogenesis, tumor size and number, and VEGF staining in several animal models of cancer (Ben-Shoshan et al. 2007; Chung et al. 2009; Gonzalez-Pardo et al. 2010). In this case, 1,25-dihydroxycholecalciferol suppresses endothelial cell migration. Therefore, the suppression of angiogenesis in some animal models of cancer may be partially explained by reduction of the proliferating tumor cell population, thus decreasing the source of VEGF. There is evidence that 1,25-dihydroxycholecalciferol regulates angiogenesis and VEGF production, although the exact targets of 1,25-dihydroxycholecalciferol remains unclear. The effects of 1,25-dihydroxycholecalciferol on VEGF, and therefore angiogenesis, may depend on the cell type and model system studied.

6.5   Retinoic Acid

All-trans retinoic acid (ATRA) regulates a variety of important cellular functions via retinoic acid receptor (RAR). However, its effects on angiogenesis remain controversial. ATRA, as well as, the RAR agonist Am80 significantly induced capillary-like tube formation. The ATRA-induced tube formation was inhibited by co-incubation with RAR antagonist LE540/LE135. HUVEC proliferation, but not its migration, was also induced by ATRA. The ATRA-induced tube formation was completely abolished by co-incubation with VEGF antibody or with VEGF receptor (VEGFR)-2 neutralizing antibody, but not VEGFR-1 antibody. VEGF secretion was stimulated by both ATRA and Am80 by endothelial cell/dermal fibroblast cells coculture, and VEGF expression and secretion by dermal fibroblast cells. Additionally, dermal fibroblast's VEGF gene promoter transcription activity of human in was enhanced by ATRA when RAR was overexpressed. ATRA also induced VEGFR-2 (KDR) transcriptional expression in human umbilical vein endothelial cells. Moreover, ATRA induced the secretion of hepatocyte growth factor and angiopoietin-2 in co-cultures. In summary, ATRA treatment resulted in an angiogenic response mediated by the RAR through the stimulation of human umbilical vein endothelial cell proliferation. This mechanism may include enhanced endogenous VEGF signaling and, in part, increased hepatocyte growth factor and angiopoietin-2 production (Saito et al. 2007) .

In contrast, Hoffmann et al (Hoffmann et al. 2007) described anti-proliferative effects of retinoic acid (RA) in patients with advanced and poorly differentiated thyroid cancer. These patients responded to RA with increased uptake of radioiodine. It has been suggested that these effects may be caused by re-differentiation. Presently, little is known about the effects of RA on tumor angiogenesis, a prerequisite for growth and metastatic spread. *In vitro*, thyroid cancer cell lines responded to RA with a 34% reduced proliferation and with 80% reduced secretion of VEGF. *In vivo*, tumor volumes of animals receiving RA were reduced by 33% *in vivo*, VEGF-expression and apoptosis were not significantly affected by RA. *In vitro*, proliferation of HUVEC was inhibited by conditioned medium of pretreated with RA and by administration of RA. This report suggest that reduced angiogenesis may be an important mechanism responsible

7   Crosstalk between Non-classical Pro-angiogenic Factors

Nearly all of the non-classical pro-angiogenic factors modulate the expression of VEGF, VEGF receptors, or both in promoting angiogenesis. The question that has not been adequately addressed is whether there is crosstalk between pathways induced by different non-classical pro-angiogenic factors that may lead to enhanced or attenuated angiogenesis. For example, current data suggests that there is a reciprocal relationship between NOS activity and VEGF expression. There is also limited studies that examined possible crosstalk between estrogen (and other steroid hormone), COX2 (and prostaglandins), NOS (NO) and VEGF pathways in mediating tumor angiogenesis. Kim and coworkers (Kim et al. 2008) have reported steroid receptor signaling through the activation of eNOS in endothelial cells. These researchers found estrogen receptor-46, a shortened isoform of estrogen receptor-α, which mediates rapid responses to estradiol by forming a complex with c-Src, PI3K, Akt and eNOS within caveolar membranes of endothelial cells leading to the activation of eNOS. This results in increase production of NO. Kimura and Esumi (Kimura and Esumi 2003) reported that some NO donors, other than SNP, was able to induce VEGF expression by stabilizing HIF-1 levels in tumor cell lines, independent from the cGMP pathway.

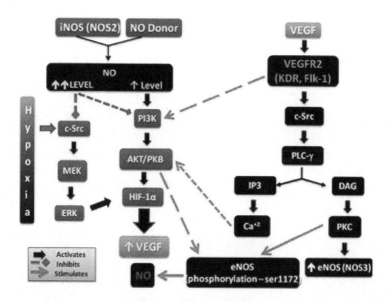

Figure 5 Crosstalk between NO, VEGF and hypoxia. This crosstalk triggers a complex signaling pathway showing that each pathway can regulate the others.

On the other hand, the ability of VEGF to stimulate angiogenesis can be mediated by NO formed by VEGF-activated eNOS in endothelial cells (Kimura and Esumi 2003). The mechanism for this effect is that VEGF bind to VEGFR-2, activating autophosphorylation, triggering an increased in Ca+2 levels, which bind to calmodulin and this complex associates with eNOS, activating this enzyme (as shown in Fig. 5).

In general, angiogenic growth factors bind to membrane receptors that have tyrosine kinase activity, which give rise to complex signal transduction pathways (Fig. 5). Besides receptor signaling through tyrosine kinase activation, there may be sideways cross talk involving other membrane components play important roles. Crosstalk between growth factor receptors and neuropilins, integrins, and VE-cadherin seem to be vital for VEGF receptor signaling. In addition, heparan sulfate proteoglycans and syndecans also play important roles in

A complex mechanism may mediate VEGF-activation of eNOS, including three autophosphorylation of tyrosine (Tyr) residues of VEGFR2. Phosphorylation of Tyr951 recruits SH2 domain-containing protein 2A, which then recruits of Src kinase26. Phosphorylation of Tyr1175 recruits phospholipase Cγ and Shb; phosphorylation of Tyr801 is required to recruit PI 3-kinase and consequent activation of Akt86. These targets in turn activate several downstream targets, altering endothelial proliferation, motility, and permeability. All three of these pathways are implicated in VEGF-mediated activation of eNOS (Fig. 4). In mice lacking eNOS, VEGF produced decreased angiogenesis compared to wild type mice. In contrast, the iNOS null mice had less of impairment in VEGF-induced angiogenesis, suggesting that eNOS may be a major mediator of VEGF-induced angiogenesis *in vivo*.

However, some discrepancies still exist in the literature reporting studies of NO mediating VEGF driven angiogenesis. These studies employ NO donating molecules at high non-physiological concentrations, as well as, NO donors that have reactive properties on top of releasing NO. Other studies indicate that NO has a biphasic reliance on its level in order to produce a number of different aspects of cell proliferation and migration. However, the responses triggered by NO donors, present at appropriate levels to activate soluble guanylate cycles, are consistently pro-angiogenic. In addition to VEGF, other angiogenic growth factor, such as, angiopoietin-1 may be mediated by the activation of eNOS. Angiopoietin-1 induced a decreased angiogenic response in eNOS-null mice.

Lastly, the anti-angiogenic activity of thrombospondin-1 may also involve the NO/cGMP pathway in endothelial cells. Miller et al (Miller et al. 2009) found that murine, porcine and human endothelial cells accumulation of NO-stimulated cGMP is attenuated by thrombospondin-1 signaling. These researchers showed that thrombospondin-1 modulates NO/cGMP signaling in vascular cells at three distinct levels: eNOS activation, soluble guanylate cyclase activation, and downstream at the level of the cGMP-dependent protein kinase (Fig. 3). This should enable thrombospondin-1 to be a highly effective physiological antagonist of NO signaling and decreasing VEGF induced angiogenesis.

8   Conclusions and Perspectives

Angiogenesis is a complex process controlled by the expression or activation of opposing regulatory factors that modulate endothelial cell behavior necessary for the assembly of endothelial cells into blood vessel-like structures. Tumor angiogenesis results from the induction and secretion of pro-angiogenic factors such as VEGF, which induces changes in vascular endothelial cell signaling and results in enhanced endothelial cell proliferation, migration and vascular tube formation. This chapter describes several non-classical factors that modulate the expression and activity pro- and anti-angiogenic factors and contribute significantly to enhanced angiogenesis required for continue growth of the tumor. One factor that is overexpressed by tumor cells and seems to play a critical role in angiogenesis is COX-2. Prostaglandins, the downstream metabolites of COX-2, are form by prostaglandin synthases, released from the cells, and bind to prostaglandin receptors to signal endothelial cells to increase proliferation and migration, as well as, to induce other cellular changes necessary for angiogenesis. Prostaglandins also induce the expression of VEGF and other pro-angiogenic factors. Tumor cells also overexpress iNOS and have high production of NO, which can induce the expression of pro-angiogenic factors, as well as, promote endothelial behavior necessary for angiogenesis.

Galectins are a family of multifunctional carbohydrate-binding proteins that play numerous intracellular and extracellular roles. Galectin-1, -3 and -8 have been shown to be expressed by tumor cells and play roles in the angiogenic process. Galectin-3 interacts with several growth factor receptors, including VEGFR-2, stabilizing receptor complexes in plasma membranes, which activate several signal transduction mechanisms. Galectin-1 expression has been linked to the endothelial cell adaptations to tumor derived factors and microenvironment that ultimately results in an angiogenic response. Galectin-8 may be a potent pro-angiogenic factor that function at concentrations 10 times lower than the levels necessary for galectin-1 or -3 to modulate endothelial cell behavior and morphogenesis.

Steroids such as estrogens, progestins, glucocorticoids, and calcitriols have all been linked to modulated tumor cell growth and gene expression. Steroids can also regulate the expression of pro-angiogenic factors including VEGF and VEGFR-2, and promote or inhibit tumor angiogenesis. There may be considerable crosstalk between COX-2, NOS (NO), steroids and VEGF pathways giving rise to multiple regulatory mechanisms that function to either activate or inactivate the "angiogenic switch" in tumor cells and tumor-derived endothelial cells.

REFERENCES

Araki, Y., S. Okamura, S. P. Hussain, M. Nagashima, P. He, M. Shiseki, K. Miura and C. C. Harris (2003). "Regulation of cyclooxygenase-2 expression by the Wnt and ras pathways." Cancer Res 63(3): 728-734.

Basu, G. D., W. S. Liang, D. A. Stephan, L. T. Wegener, C. R. Conley, B. A. Pockaj and P. Mukherjee (2006). "A novel role for cyclooxygenase-2 in regulating vascular channel formation by human breast cancer cells." Breast Cancer Res 8(6): R69.

Ben-Shoshan, M., S. Amir, D. T. Dang, L. H. Dang, Y. Weisman and N. J. Mabjeesh (2007). "1alpha,25-dihydroxyvitamin D3 (Calcitriol) inhibits hypoxia-inducible factor-1/vascular endothelial growth factor pathway in human cancer cells." Mol Cancer Ther 6(4): 1433-1439.

Bidon-Wagner, N. and J. P. Le Pennec (2004). "Human galectin-8 isoforms and cancer." Glycoconj J 19(7-9): 557-563.

Bidwell, P., K. Joh, H. A. Leaver and M. T. Rizzo (2010). "Prostaglandin E2 activates cAMP response element-binding protein in glioma cells via a signaling pathway involving PKA-dependent inhibition of ERK." Prostaglandins Other Lipid Mediat 91(1-2): 18-29.

Bogin, L. and H. Degani (2002). "Hormonal regulation of VEGF in orthotopic MCF7 human breast cancer." Cancer Res 62(7): 1948-1951.

Buteau-Lozano, H., M. Ancelin, B. Lardeux, J. Milanini and M. Perrot-Applanat (2002). "Transcriptional regulation of vascular endothelial growth factor by estradiol and tamoxifen in breast cancer cells: a complex interplay between estrogen receptors alpha and beta." Cancer Res 62(17): 4977-4984.

Califice, S., V. Castronovo, M. Bracke and F. van den Brule (2004). "Dual activities of galectin-3 in human prostate cancer: tumor suppression of nuclear galectin-3 vs tumor promotion of cytoplasmic galectin-3." Oncogene 23(45): 7527-7536.

Cao, Y., R. Cao and E. M. Hedlund (2008). "R Regulation of tumor angiogenesis and metastasis by FGF and PDGF signaling pathways." J Mol Med (Berl) 86(7): 785-789.

Chen, C. N., F. J. Hsieh, Y. M. Cheng, K. J. Chang and P. H. Lee (2006). "Expression of inducible nitric oxide synthase and cyclooxygenase-2 in angiogenesis and clinical outcome of human gastric cancer." J Surg Oncol 94(3): 226-233.

Chiarugi, V., L. Magnelli and O. Gallo (1998). "Cox-2, iNOS and p53 as play-makers of tumor angiogenesis (review)." Int J Mol Med 2(6): 715-719.

Cho, S. O., J. W. Lim, K. H. Kim and H. Kim (2010). "Involvement of Ras and AP-1 in *Helicobacter pylori*-induced expression of COX-2 and iNOS in gastric epithelial AGS cells." Dig Dis Sci 55(4): 988-996.

Chu, J., F. L. Lloyd, O. C. Trifan, B. Knapp and M. T. Rizzo (2003). "Potential involvement of the cyclooxygenase-2 pathway in the regulation of tumor-associated angiogenesis and growth in pancreatic cancer." Mol Cancer Ther 2(1): 1-7.

Chung, I., G. Han, M. Seshadri, B. M. Gillard, W. D. Yu, B. A. Foster, D. L. Trump and C. S. Johnson (2009). "Role of vitamin D receptor in the antiproliferative effects of calcitriol in tumor-derived endothelial cells and tumor angiogenesis *in vivo*." Cancer Res 69(3): 967-975.

Cianchi, F., S. Cuzzocrea, M. C. Vinci, L. Messerini, C. E. Comin, G. Navarra, G. Perigli, T. Centorrino, S. Marzocco, E. Lenzi, N. Battisti, G. Trallori and E. Masini (2010). "Heterogeneous expression of cyclooxygenase-2 and inducible nitric oxide synthase within colorectal tumors: correlation with tumor angiogenesis." Dig Liver Dis 42(1): 20-27.

Coradini, D., C. Pellizzaro, A. Speranza and M. G. Daidone (2004). "Hypoxia and estrogen receptor profile influence the responsiveness of human breast cancer cells to estradiol and antiestrogens." Cell Mol Life Sci 61(1): 76-82.

Dabrosin, C., P. J. Margetts and J. Gauldie (2003). "Estradiol increases extracellular levels of vascular endothelial growth factor *in vivo* in murine mammary cancer." Int J Cancer 107(4): 535-540.

Dada, S., N. Demartines and O. Dormond (2008). "mTORC2 regulates PGE2-mediated endothelial cell survival and migration." Biochem Biophys Res Commun 372(4): 875-879.

Dadiani, M., D. Seger, T. Kreizman, D. Badikhi, R. Margalit, R. Eilam and H. Degani (2009). "Estrogen regulation of vascular endothelial growth factor in breast cancer *in vitro* and *in vivo*: the role of estrogen receptor alpha and c-Myc." Endocr Relat Cancer 16(3): 819-834.

Delgado, V. M., L. G. Nugnes, L. L. Colombo, M. F. Troncoso, M. M. Fernandez, E. L. Malchiodi, I. Frahm, D. O. Croci, D. Compagno, G. A. Rabinovich, C. Wolfenstein-Todel and M. T. Elola (2011). "Modulation of endothelial cell migration and angiogenesis: a novel function for the "tandem-repeat" lectin galectin-8." FASEB J 25(1): 242-254.

Detmar, M. (2000). "The role of VEGF and thrombospondins in skin angiogenesis." J Dermatol Sci 24 Suppl 1: S78-84.

Diaz-Munoz, M. D., I. C. Osma-Garcia, M. Fresno and M. A. Iniguez (2012). "Involvement of PGE2 and the cAMP signalling pathway in the up-regulation of COX-2 and mPGES-1 expression in LPS-activated macrophages." Biochem J 443(2): 451-461.

Donnini, S., F. Finetti, R. Solito, E. Terzuoli, A. Sacchetti, L. Morbidelli, P. Patrignani and M. Ziche (2007). "EP2 prostanoid receptor promotes squamous cell carcinoma growth through epidermal growth factor receptor transactivation and iNOS and ERK1/2 pathways." FASEB J 21(10): 2418-2430.

Dulak, J. and A. Jozkowicz (2003). "Regulation of vascular endothelial growth factor synthesis by nitric oxide: facts and controversies." Antioxid Redox Signal 5(1): 123-132.

Elkin, M., A. Orgel and H. K. Kleinman (2004). "An angiogenic switch in breast cancer involves estrogen and soluble vascular endothelial growth factor receptor 1." J Natl Cancer Inst 96(11): 875-878.

Ferrara, N. (2010). "Pathways mediating VEGF-independent tumor angiogenesis." Cytokine Growth Factor Rev 21(1): 21-26.

Florian, M., L. Florianova, S. Hussain and S. Magder (2008). "Interaction of estrogen and tumor necrosis factor alpha in endothelial cell migration and early stage of angiogenesis." Endothelium 15(5-6): 265-275.

Franchi, A., O. Gallo, M. Paglierani, I. Sardi, L. Magnelli, E. Masini and M. Santucci (2002). "Inducible nitric oxide synthase expression in laryngeal neoplasia: correlation with angiogenesis." Head Neck 24(1): 16-23.

Fujino, H., S. Salvi and J. W. Regan (2005). "Differential regulation of phosphorylation of the cAMP response element-binding protein after activation of EP2 and EP4 prostanoid receptors by prostaglandin E2." Mol Pharmacol 68(1): 251-259.

Garvin, S. and C. Dabrosin (2003). "Tamoxifen inhibits secretion of vascular endothelial growth factor in breast cancer *in vivo*." Cancer Res 63(24): 8742-8748.

Garvin, S., U. W. Nilsson and C. Dabrosin (2005). "Effects of oestradiol and tamoxifen on VEGF, soluble VEGFR-1, and VEGFR-2 in breast cancer and endothelial cells." Br J Cancer 93(9): 1005-1010.

Gonzalez-Pardo, V., D. Martin, J. S. Gutkind, A. Verstuyf, R. Bouillon, A. R. de Boland and R. L. Boland (2010). "1 Alpha,25-dihydroxyvitamin D3 and its TX527 analog inhibit the growth of endothelial cells transformed by Kaposi sarcoma-associated herpes virus G protein-coupled receptor *in vitro* and *in vivo*." Endocrinology 151(1): 23-31.

Hanahan, D. and J. Folkman (1996). "Patterns and emerging mechanisms of the angiogenic switch during tumorigenesis." Cell 86(3): 353-364.

Hellmuth, M., J. Paulukat, J. Ninic, J. Pfeilschifter and H. Muhl (2004). "Nitric oxide differentially regulates pro- and anti-angiogenic markers in DLD-1 colon carcinoma cells." FEBS Lett 563(1-3): 98-102.

Hernandez, G. L., O. V. Volpert, M. A. Iniguez, E. Lorenzo, S. Martinez-Martinez, R. Grau, M. Fresno and J. M. Redondo (2001). "Selective inhibition of vascular endothelial growth factor-mediated angiogenesis by cyclosporin A: roles of the nuclear factor of activated T cells and cyclooxygenase 2." J Exp Med 193(5): 607-620.

Hoffmann, S., A. Rockenstein, A. Ramaswamy, I. Celik, A. Wunderlich, S. Lingelbach, L. C. Hofbauer and A. Zielke (2007). "Retinoic acid inhibits angiogenesis and tumor growth of thyroid cancer cells." Mol Cell Endocrinol 264(1-2): 74-81.

Houzelstein, D., I. R. Goncalves, A. J. Fadden, S. S. Sidhu, D. N. Cooper, K. Drickamer, H. Leffler and F. Poirier (2004). "Phylogenetic analysis of the vertebrate galectin family." Mol Biol Evol 21(7): 1177-1187.

Huang, L. E., W. G. Willmore, J. Gu, M. A. Goldberg and H. F. Bunn (1999). "Inhibition of hypoxia-inducible factor 1 activation by carbon monoxide and nitric oxide. Implications for oxygen sensing and signaling." J Biol Chem 274(13): 9038-9044.

Hyder, S. M. (2006). "Sex-steroid regulation of vascular endothelial growth factor in breast cancer." Endocr Relat Cancer 13(3): 667-687.

Hyder, S. M., Y. Liang and J. Wu (2009). "Estrogen regulation of thrombospondin-1 in human breast cancer cells." Int J Cancer 125(5): 1045-1053.

Hyder, S. M. and G. M. Stancel (2002). "Inhibition of progesterone-induced VEGF production in human breast cancer cells by the pure antiestrogen ICI 182,780." Cancer Lett 181(1): 47-53.

Ito, K., S. A. Scott, S. Cutler, L. F. Dong, J. Neuzil, H. Blanchard and S. J. Ralph (2011). "Thiodigalactoside inhibits murine cancers by concurrently blocking effects of galectin-1 on immune dysregulation, angiogenesis and protection against oxidative stress." Angiogenesis 14(3): 293-307.

Iwai, A., Y. Fujii, S. Kawakami, R. Takazawa, Y. Kageyama, M. A. Yoshida and K. Kihara (2004). "Down-regulation of vascular endothelial growth factor in renal cell carcinoma cells by glucocorticoids." Mol Cell Endocrinol 226(1-2): 11-17.

Jain, S., G. Chakraborty, R. Raja, S. Kale and G. C. Kundu (2008). "Prostaglandin E2 regulates tumor angiogenesis in prostate cancer." Cancer Res 68(19): 7750-7759.

Jenkins, D. C., I. G. Charles, L. L. Thomsen, D. W. Moss, L. S. Holmes, S. A. Baylis, P. Rhodes, K. Westmore, P. C. Emson and S. Moncada (1995). "Roles of nitric oxide in tumor growth." Proc Natl Acad Sci USA 92(10): 4392-4396.

Kamiyama, M., A. Pozzi, L. Yang, L. M. DeBusk, R. M. Breyer and P. C. Lin (2006). "EP2, a receptor for PGE2, regulates tumor angiogenesis through direct effects on endothelial cell motility and survival." Oncogene 25(53): 7019-7028.

Kasper, H. U., E. Konze, H. P. Dienes, D. L. Stippel, P. Schirmacher and M. Kern (2010). "COX-2 expression and effects of COX-2 inhibition in colorectal carcinomas and their liver metastases." Anticancer Res 30(6): 2017-2023.

Kazi, A. A., J. M. Jones and R. D. Koos (2005). "Chromatin immunoprecipitation analysis of gene expression in the rat uterus *in vivo*: estrogen-induced recruitment of both estrogen receptor alpha and hypoxia-inducible factor 1 to the vascular endothelial growth factor promoter." Mol Endocrinol 19(8): 2006-2019.

Kim, K. H., K. Moriarty and J. R. Bender (2008). "Vascular cell signaling by membrane estrogen receptors." Steroids 73(9-10): 864-869.

Kimura, H. and H. Esumi (2003). "Reciprocal regulation between nitric oxide and vascular endothelial growth factor in angiogenesis." Acta Biochim Pol 50(1): 49-59.

Kimura, H., T. Ogura, Y. Kurashima, A. Weisz and H. Esumi (2002). "Effects of nitric oxide donors on vascular endothelial growth factor gene induction." Biochem Biophys Res Commun 296(4): 976-982.

Koch, A., B. Gustafsson, H. Fohlin and S. Sorenson (2011). "Cyclooxygenase-2 expression in lung cancer cells evaluated by immunocytochemistry." Diagn Cytopathol 39(3): 188-193.

Kos, M. and A. Dabrowski (2002). "Tumour's angiogenesis – the function of VEGF and bFGF in colorectal cancer." Ann Univ Mariae Curie Sklodowska Med 57(2): 556-561.

Krishnan, A. V., D. M. Peehl and D. Feldman (2003). "The role of vitamin D in prostate cancer." Recent Results Cancer Res 164: 205-221.

Larkins, T. L., M. Nowell, S. Singh and G. L. Sanford (2006). "Inhibition of cyclooxygenase-2 decreases breast cancer cell motility, invasion and matrix metalloproteinase expression." BMC Cancer 6: 181.

Lavalle, G. E., A. C. Bertagnolli, W. L. Tavares and G. D. Cassali (2009). "Cox-2 expression in canine mammary carcinomas: correlation with angiogenesis and overall survival." Vet Pathol 46(6): 1275-1280.

Le Mercier, M., F. Lefranc, T. Mijatovic, O. Debeir, B. Haibe-Kains, G. Bontempi, C. Decaestecker, R. Kiss and V. Mathieu (2008). "Evidence of galectin-1 involvement in glioma chemoresistance." Toxicol Appl Pharmacol 229(2): 172-183.

Levine, M. J. and D. Teegarden (2004). "1alpha,25-dihydroxycholecalciferol increases the expression of vascular endothelial growth factor in C3H10T1/2 mouse embryo fibroblasts." J Nutr 134(9): 2244-2250.

Li, W., R. J. Xu, L. H. Jiang, J. Shi, X. Long and B. Fan (2005). "Expression of cyclooxygenase-2 and inducible nitric oxide synthase correlates with tumor angiogenesis in endometrial carcinoma." Med Oncol 22(1): 63-70.

Liang, Y. and S. M. Hyder (2005). "Proliferation of endothelial and tumor epithelial cells by progestin-induced vascular endothelial growth factor from human breast cancer cells: paracrine and autocrine effects." Endocrinology 146(8): 3632-3641.

Liu, F. T. and G. A. Rabinovich (2005). "Galectins as modulators of tumour progression." Nat Rev Cancer 5(1): 29-41.

Liu, Y., H. Christou, T. Morita, E. Laughner, G. L. Semenza and S. Kourembanas (1998). "Carbon monoxide and nitric oxide suppress the hypoxic induction of vascular endothelial growth factor gene via the 5' enhancer." J Biol Chem 273(24): 15257-15262.

Markowska, A. I., F. T. Liu and N. Panjwani (2010). "Galectin-3 is an important mediator of VEGF- and bFGF-mediated angiogenic response." J Exp Med 207(9): 1981-1993.

Massi, D., M. C. De Nisi, A. Franchi, V. Mourmouras, G. Baroni, J. Panelos, M. Santucci and C. Miracco (2009). "Inducible nitric oxide synthase expression in melanoma: implications in lymphangiogenesis." Mod Pathol 22(1): 21-30.

McColl, B. K., S. A. Stacker and M. G. Achen (2004). "Molecular regulation of the VEGF family – inducers of angiogenesis and lymphangiogenesis." APMIS 112 (7-8): 463-480.

Miller, T. W., J. S. Isenberg and D. D. Roberts (2009). "Molecular regulation of tumor angiogenesis and perfusion via redox signaling." Chem Rev 109(7): 3099-3124.

Mirkin, S., B. C. Wong and D. F. Archer (2005). "Effect of 17 beta-estradiol, progesterone, synthetic progestins, tibolone, and tibolone metabolites on vascular endothelial growth factor mRNA in breast cancer cells." Fertil Steril 84(2): 485-491.

Morbidelli, L., S. Donnini and M. Ziche (2004). "Role of nitric oxide in tumor angiogenesis." Cancer Treat Res 117: 155-167.

Mueller, M. D., J. L. Vigne, E. A. Pritts, V. Chao, E. Dreher and R. N. Taylor (2003). "Progestins activate vascular endothelial growth factor gene transcription in endometrial adenocarcinoma cells." Fertil Steril 79(2): 386-392.

Nagy, J. A., A. M. Dvorak and H. F. Dvorak (2007). "VEGF-A and the induction of pathological angiogenesis." Annu Rev Pathol 2: 251-275.

Nakamura, J., A. Savinov, Q. Lu and A. Brodie (1996). "Estrogen regulates vascular endothelial growth/permeability factor expression in 7,12-dimethylbenz(a) anthracene-induced rat mammary tumors." Endocrinology 137(12): 5589-5596.

Namkoong, S., S. J. Lee, C. K. Kim, Y. M. Kim, H. T. Chung, H. Lee, J. A. Han, K. S. Ha and Y. G. Kwon (2005). "Prostaglandin E2 stimulates angiogenesis by activating the nitric oxide/cGMP pathway in human umbilical vein endothelial cells." Exp Mol Med 37(6): 588-600.

Nangia-Makker, P., V. Balan and A. Raz (2008). "Regulation of tumor progression by extracellular galectin-3." Cancer Microenviron 1(1): 43-51.

Noguera-Troise, I., C. Daly, N. J. Papadopoulos, S. Coetzee, P. Boland, N. W. Gale, H. C. Lin, G. D. Yancopoulos and G. Thurston (2006). "Blockade of Dll4 inhibits tumour growth by promoting non-productive angiogenesis." Nature 444(7122): 1032-1037.

Peng, J. P., S. Zheng, Z. X. Xiao and S. Z. Zhang (2003). "Inducible nitric oxide synthase expression is related to angiogenesis, bcl-2 and cell proliferation in hepatocellular carcinoma." J Zhejiang Univ Sci 4(2): 221-227.

Rabinovich, G. A., A. Cumashi, G. A. Bianco, D. Ciavardelli, I. Iurisci, M. D'Egidio, E. Piccolo, N. Tinari, N. Nifantiev and S. Iacobelli (2006). "Synthetic lactulose amines: novel class of anticancer agents that induce tumor-cell apoptosis and inhibit galectin-mediated homotypic cell aggregation and endothelial cell morphogenesis." Glycobiology 16(3): 210-220.

Richardsen, E., R. D. Uglehus, J. Due, C. Busch and L. T. Busund (2010). "COX-2 is overexpressed in primary prostate cancer with metastatic potential and may predict survival. A comparison study between COX-2, TGF-beta, IL-10 and Ki67." Cancer Epidemiol 34(3): 316-322.

Ridgway, J., G. Zhang, Y. Wu, S. Stawicki, W. C. Liang, Y. Chanthery, J. Kowalski, R. J. Watts, C. Callahan, I. Kasman, M. Singh, M. Chien, C. Tan, J. A. Hongo, F. de Sauvage, G. Plowman and M. Yan (2006). "Inhibition of Dll4 signalling inhibits tumour growth by deregulating angiogenesis." Nature 444(7122): 1083-1087.

Saarinen, N. M., A. Abrahamsson and C. Dabrosin (2010). "Estrogen-induced angiogenic factors derived from stromal and cancer cells are differently regulated by enterolactone and genistein in human breast cancer *in vivo*." Int J Cancer 127(3): 737-745.

Saito, A., A. Sugawara, A. Uruno, M. Kudo, H. Kagechika, Y. Sato, Y. Owada, H. Kondo, M. Sato, M. Kurabayashi, M. Imaizumi, S. Tsuchiya and S. Ito (2007). "All-trans retinoic acid induces *in vitro* angiogenesis via retinoic acid receptor: possible involvement of paracrine effects of endogenous vascular endothelial growth factor signaling." Endocrinology 148(3): 1412-1423.

Sakurai, T., K. Suzuki, M. Yoshie, K. Hashimoto, E. Tachikawa and K. Tamura (2011). "Stimulation of tube formation mediated through the prostaglandin EP2 receptor in rat luteal endothelial cells." J Endocrinol 209(1): 33-43.

Sartorius, C. A., T. Shen and K. B. Horwitz (2003). "Progesterone receptors A and B differentially affect the growth of estrogen-dependent human breast tumor xenografts." Breast Cancer Res Treat 79(3): 287-299.

Sengupta, K., S. Banerjee, N. Saxena and S. K. Banerjee (2003). "Estradiol-induced vascular endothelial growth factor-A expression in breast tumor cells is biphasic and regulated by estrogen receptor-alpha dependent pathway." Int J Oncol 22(3): 609-614.

Shang, Z. J. and J. R. Li (2005). "Expression of endothelial nitric oxide synthase and vascular endothelial growth factor in oral squamous cell carcinoma: its correlation with angiogenesis and disease progression." J Oral Pathol Med 34(3): 134-139.

Shao, J., C. Jung, C. Liu and H. Sheng (2005). "Prostaglandin E2 Stimulates the beta-catenin/T cell factor-dependent transcription in colon cancer." J Biol Chem 280(28): 26565-26572.

Sheng, H., J. Shao, M. K. Washington and R. N. DuBois (2001). "Prostaglandin E2 increases growth and motility of colorectal carcinoma cells." J Biol Chem 276(21): 18075-18081.

Singh, B. and A. Lucci (2002). "Role of cyclooxygenase-2 in breast cancer." J Surg Res 108(1): 173-179.

Sogawa, K., K. Numayama-Tsuruta, M. Ema, M. Abe, H. Abe and Y. Fujii-Kuriyama (1998). "Inhibition of hypoxia-inducible factor 1 activity by nitric oxide donors in hypoxia." Proc Natl Acad Sci USA 95(13): 7368-7373.

Stoner, M., M. Wormke, B. Saville, I. Samudio, C. Qin, M. Abdelrahim and S. Safe (2004). "Estrogen regulation of vascular endothelial growth factor gene expression in ZR-75 breast cancer cells through interaction of estrogen receptor alpha and SP proteins." Oncogene 23(5): 1052-1063.

Thijssen, V. L., S. Hulsmans and A. W. Griffioen (2008). "The galectin profile of the endothelium: altered expression and localization in activated and tumor endothelial cells." Am J Pathol 172(2): 545-553.

Thurston, G. and J. Kitajewski (2008). "VEGF and Delta-Notch: interacting signalling pathways in tumour angiogenesis." Br J Cancer 99(8): 1204-1209.

Tischer, E., R. Mitchell, T. Hartman, M. Silva, D. Gospodarowicz, J. C. Fiddes and J. A. Abraham (1991). "The human gene for vascular endothelial growth factor. Multiple protein forms are encoded through alternative exon splicing." J Biol Chem 266(18): 11947-11954.

Tsurumi, Y., T. Murohara, K. Krasinski, D. Chen, B. Witzenbichler, M. Kearney, T. Couffinhal and J. M. Isner (1997). "Reciprocal relation between VEGF and NO in the regulation of endothelial integrity." Nat Med 3(8): 879-886.

Webster, J. C. and J. A. Cidlowski (1999). "Mechanisms of Glucocorticoid-receptor-mediated Repression of Gene Expression." Trends Endocrinol Metab 10(10): 396-402.

Wu, J., S. Brandt and S. M. Hyder (2005). "Ligand- and cell-specific effects of signal transduction pathway inhibitors on progestin-induced vascular endothelial growth factor levels in human breast cancer cells." Mol Endocrinol 19(2): 312-326.

Wu, J., Y. Liang, Z. Nawaz and S. M. Hyder (2005). "Complex agonist-like properties of ICI 182,780 (Faslodex) in human breast cancer cells that predominantly express progesterone receptor-B: implications for treatment resistance." Int J Oncol 27(6): 1647-1659.

Wu, J., J. Richer, K. B. Horwitz and S. M. Hyder (2004). "Progestin-dependent induction of vascular endothelial growth factor in human breast cancer cells: preferential regulation by progesterone receptor B." Cancer Res 64(6): 2238-2244.

Yano, A., Y. Fujii, A. Iwai, Y. Kageyama and K. Kihara (2006). "Glucocorticoids suppress tumor angiogenesis and *in vivo* growth of prostate cancer cells." Clin Cancer Res 12(10): 3003-3009.

Yao, L., F. Liu, L. Hong, L. Sun, S. Liang, K. Wu and D. Fan (2011). "The function and mechanism of COX-2 in angiogenesis of gastric cancer cells." J Exp Clin Cancer Res 30: 13.

Zhang, W., X. J. He, Y. Y. Ma, H. J. Wang, Y. J. Xia, Z. S. Zhao, Z. Y. Ye and H. Q. Tao (2011). "Inducible nitric oxide synthase expression correlates with angiogenesis, lymphangiogenesis, and poor prognosis in gastric cancer patients." Hum Pathol 42(9): 1275-1282.

The Role of Cyclooxygenases and Lipoxygenases in the Regulation of Tumor Angiogenesis

Graham Pidgeon

Dept. Surgery, Institute of Molecular Medicine, Trinity Centre for Health Sciences, St. James's Hospital, Dublin 8/TCD. Email: pidgeong@tcd.ie

CHAPTER OUTLINE

▶ Introduction
▶ Cyclooxygenase and lipoxygenase signaling
▶ Cyclooxygenase and lipoxygenase pathways in cancer
▶ Cylooxygenases, lipoxygenases and tumor angiogenesis
▶ Controversial data/concepts
▶ Anti-angiogenic targeting of COX/LOX pathways: Pharmacological inhibitors, natural agents and dual targeting approaches
▶ Novel methodological approaches for anti-angiogenic drug discovery/development
▶ Conclusions and perspectives
▶ References

ABSTRACT

The metabolism of arachidonic acid through both cyclooxygenase (COX) and lipoxygenase (LOX) pathways leads to the generation of various biologically active eicosanoids. The expression of these enzymes varies across the progressive sequence from metaplasia-dysplasia to cancer, and thereby has been shown to regulate various aspects of tumor development. Pharmacologic and natural inhibitors of COX and LOXs have been shown to suppress carcinogenesis and tumor growth in a number of experimental models. Arachidonic acid metabolism potentially effect diverse biological phenomenon regulating processes such as cell growth, cell survival, angiogenesis, cell invasion and metastatic potential. The influence of distinct COX and LOX isoforms, and their downstream metabolites, differ considerably with respect to their effect on both the individual mechanisms described and the tumor being examined. COX-2, 5-LOX and platelet type 12-LOX are generally considered pro-angiogenic, while 15-LOX-2 suppresses angiogenesis. In this chapter, we focus on the molecular mechanisms of angiogenesis and metastasis regulated by COX and LOX metabolism in some of the major cancers. We discuss the effects of downstream mediators of COX and LOX pathways on angiogenesis, focusing on the various steps regulating this process. Understanding the molecular mechanisms underlying the anti-angiogenic effect of specific or general COX and LOX inhibitors may lead to the design of biologically and pharmacologically targeted therapeutic strategies inhibiting these biologically active metabolites, which may ultimately prove useful in the treatment of cancer, either alone or in combination with conventional therapies.

Key words: Angiogenesis, Cancer, Metastasis, Cyclooxygenase, Lipoxygenase, Thromboxane Synthase, Prostaglandins, Pharmacological inhibitors, Natural agents

1   Introduction

The functional relationship between polyunsaturated fatty acid metabolism, inflammation and carcinogenesis has been extensively examined in numerous molecular studies, revealing potential novel targets for the treatment and chemoprevention of a number of different cancers (Klurfeld and Bull 1997; Cuendet and Pezzuto 2000; Greene et al. 2011). Three types of enzymes; cyclooxygenases, epoxygenases (cytochrome P-450), and lipoxygenases can metabolize arachidonic acid to biologically active eicosanoids (Fig. 1). The various lipid peroxides and bioactive lipids generated by this metabolism can regulate cellular proliferation, apoptosis, differentiation and senescence. The role of cyclooxygenase isoforms have been widely demonstrated in a number of studies and COX-2 in particular is regarded as a promising target for chemoprevention and treatment of several cancers (Krysan et al. 2004; Furstenberger et al. 2006). Accumulating evidence links COX-2 with angiogenesis suggesting that drugs which target the COX-2 signaling pathway could be used as anti-angiogenic agents. While selective COX-2 inhibition has demonstrated some promise for the treatment of cancer, enthusiasm for this class of drugs was dampened by their association with significantly increased cardiovascular risk. The effect of COX-2 on tumor angiogenesis has been attributed to the action of its downstream prostanoids, with significant interest in the role of these agents in tumor development and progression in the last number of years. Additionally, in the last decade there has been a significant rise in the number of reports examining the functional links between various LOX isoforms and mechanisms controlling cancer development and progression (Nie 2007). In this chapter, we review the role of cyclooxygenases and lipoxygenases in the most commonly diagnosed cancers with a focus on their effects of angiogenesis. The regulation of processes from angiogenic factor secretion to endothelial cell proliferation and migration, to vessel stability and tumor invasion is discussed. The anti-angiogenic mechanisms

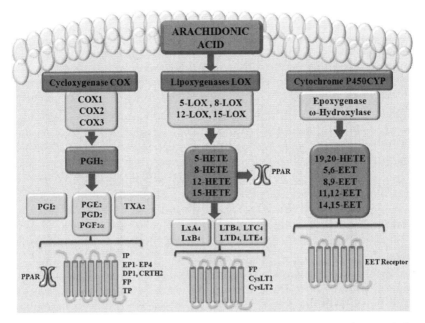

Figure 1   The arachidonic acid signaling pathway.   Arachidonic acid is metabolized to eicosanoids by one of three major pathways: the cyclooxygenase (COX), the lipoxygenase (LOX) and the cytochrome P-450 epoxygenase pathway (EPOX). In the COX pathway, COX enzymes convert arachidonic acid to prostaglandin G_2 (PGG$_2$), which is then reduced to an intermediate PGH$_2$ by the peroxidase activity of COX. PGH$_2$ is then further converted to prostanoids, including prostaglandins (PGs) and thromboxanes (TXs) by the action of cell specific synthases and isomerases. The LOX isoforms convert arachidonic acid into biologically active metabolites such as leukotrienes and hydroxyeicosatetraenoic acids (HETEs), while cytochrome P-450 metabolizes arachidonic acid into epoxyeicosatrienoic acids (EETs), HETEs and hydroperoxyeicosatetraenoic acids (HPETEs).

of action of particular agents, both naturally occurring and pharmacological inhibitors, targeting the COX and LOX pathways of arachidonic acid metabolism are discussed. The use of novel screening strategies applicable to anti-angiogenic drug discovery for COX/LOX pathway inhibitors is also described. It is likely that in the future, treatments aimed at inhibiting these pathways, either alone or in combination with other molecular targeted therapies, should improve the anti-tumor and anti-angiogenic efficacy of conventional treatments such as chemotherapy and/or radiation therapy.

2   Cyclooxygenase and Lipoxygenase Signaling

2.1   Cyclooxygenase Signaling

Cyclooxygenase (COX) is a rate limiting enzyme in the synthesis of the prostanoids from arachidonic acid (AA). It catalyses the conversion of AA to prostaglandin G_2 (PGG$_2$), and subsequently to prostaglandin H_2 (PGH$_2$), which is then further converted to the prostanoids by a range of cell specific synthases (prostaglandin-D-synthase, prostaglandin-E-synthase, prostaglandin-F-synthase, prostacyclin synthase and thromboxane synthase). The resulting prostanoids include PGD$_2$, PGE$_2$, PGF$_{2\alpha}$, PGI$_2$ (prostacyclin) and TXA$_2$ (thromboxane A$_2$). At least two isoforms of COX have been identified to date; COX-1 and COX-2. COX-1 is constitutively expressed in most tissue types. Prostanoids produced by this isoform generally mediate 'housekeeping' functions such as cytoprotections of the gastric mucosa, regulation of renal blood flow, and platelet aggregation. However, COX-1 has more recently also been shown to be inducible, particularly at sites of inflammation (Dirig et al. 1998; Gilroy et al. 1998; Morteau et al. 2000). Many studies have demonstrated that expression of COX-1 may be induced in endothelial cells in response to a range of stimuli (Okahara et al. 1998; Murphy and Fitzgerald 2001; Schwab et al. 2002). COX-1 is the only isoform expressed within platelets (Patrignani et al. 1994; Funk 2001) and is an important mediator of platelet aggregation. COX-2 on the other hand is a highly inducible enzyme under physiological conditions, although it is constitutively expressed in some tissues such as the brain, spinal cord, (Hoffmann 2000), and kidneys (Harris et al. 1994). While COX-2 expression is highly restricted under basal conditions, its expression is dramatically upregulated at sites of inflammation in response to cytokines such as interferon γ, TNFα and IL-1β (Jones et al. 1993; Asano et al. 1997), hormones (Shemesh et al. 1997), growth factors (Jones et al. 1999; Di Popolo et al. 2000; Tamura et al. 2002) and hypoxia (Lukiw et al. 2003; Pichiule et al. 2004). COX-2 expression has also been observed in neoplastic and endothelial cells within many different tumors (Masferrer et al. 2000). Recently, a third isoform of COX was discovered by Chandrasekharan et al, and termed COX-3. However, this isoform is unlikely to have prostaglandin-producing activity in human tissues, however (Stamford et al. 1986).

The COX-derived prostanoid (prostaglandins and thromboxanes) products are unstable compounds and are therefore rapidly metabolised *in vivo* (Wang and DuBois 2007). These prostanoids are readily generated by a number of cell types. Platelets, mast cells and monocytes/macrophages synthesize TXA$_2$, PGD$_2$, PGE$_2$ and PGF$_{2\delta}$, while endothelium is the major source of PGI$_2$ (Bogatcheva et al. 2005) (Fig. 1). Arachidonic acid-derived prostanoids, are biologically active lipid mediators involved in a wide range of physiological processes such as modulation of vascular tone, the inflammatory response and gastric cytoprotection. However, they have also been implicated in various disease states such as arthritis, heart disease, pulmonary hypertension and cancer (Smyth et al. 2009). The prostanoids exert their cellular functions by binding to cell surface receptors belonging to a family of seven transmembrane domain G-protein-coupled receptors (Hata and Breyer 2004). The prostanoid receptor nomenclature is assigned based on the ligand bound by as opposed to genetic or functional relationships. TP$_\alpha$/TP$_\beta$ binds TXA$_2$, DP binds PGD$_2$, EP1-4 binds PGE$_2$, FP binds PGF$_{2\delta}$, while IP binds PGI_2. In some cases, prostanoids and their metabolites bind their nuclear receptors such as peroxisome proliferator-activated receptors (PPARs) (Alfranca et al. 2006). Three distinct isotypes of PPARs have been identified to date and may be generally designated as PPARα, PPARβ/δ and PPARγ (Nolte et al. 1998; Willson et al. 2000). The prostanoid receptors subsequently activate a number of intracellular signaling pathways to mediate their cellular effects (Ricciotti and FitzGerald 2011).

2.2 Lipoxygenase Signaling

Lipoxygenases constitute a family of non-heme iron dioxygenases that insert molecular oxygen into free and/or esterified polyunsaturated fatty acids with regional specificity, and are designated 5-, 8-, 12- and 15-lipoxygenase (5-LOX, 8-LOX, 12-LOX, 15-LOX) accordingly (Funk 2001). The most studied LOX enzymes are 5-LOX, leukocyte and platelet type 12-LOX and reticulocyte-type 15-LOX-1; however the mammalian family of LOX isozymes has grown in the last number of years to include isozymes preferentially expressed in both human and murine epidermis. These isozymes include epidermal 12-LOX and 12R-LOX (producing products with R chirality), mouse 8-LOX and its human ortholog 15-LOX-2, and epidermal LOX-3 exhibiting hydroperoxide isomerase activity (Krieg et al. 2002; Yu et al. 2003). The LOX enzymes metabolize arachidonic acid (AA) to the biologically active metabolites hydroperoxy–eicosatetraenoic acids (HPETEs), which upon reduction result in the formation of corresponding HETEs (Fig. 1), while the metabolism of linoleic acid preferentially results in the formation of hydroxy–octadecadienoic acids (HODEs) (Funk 1993). Therefore, 5-, 8-, 12- and 15-HETE are the major AA metabolites formed by mammalian LOXs, and 9- and 13-HODE are the principle reaction products of linoleic acid oxygenation.

In the case of the 5-LOX enzyme, HPETE is further metabolized to form the unstable epoxide leukotriene A4 (LTA4) in the presence of 5-lipoxygenase-activating protein. LTA4 is subsequently converted to 5(S)-hydroxy-6-trans8, 11,14-cis-eicosatetranoic acid (5-HETE), or hydrolyzed into LTB4 by leukotriene A4 hydrolase, or the cysteinyl leukotrienes, LTC4, LTD4 and LTE4, following enzymatic conjugation with GSH (Fabre et al. 2002).

Platelet-type 12-LOX exclusively uses AA released from glycerol-phospholipid pools to synthesize 12(S)-HPETE and 12(S)-HETE, whereas leukocyte-type 12-LOX can also synthesize 15(S)-HETE and 12(S)-HETE. In addition to leukocytes and platelets, the expression of 12-LOX isozymes has been detected in various types of cells, such as smooth muscle cells, keratinocytes, endothelial cells and tumor cells. Therefore the expression of these enzymes may potentially regulate various steps in tumor angiogenesis. 15-lipoxygenases (15-LOX) can be subdivided into two isoforms, named 15-LOX-1 and 15-LOX-2 and the intracellular activity of these enzymes is regulated at transcriptional, translational and post-translational levels (Kuhn et al. 2002). 15-LOX-1 is mainly expressed in reticulocytes, eosinophils and airway epithelial cells, as well as in macrophages and atherosclerotic lesions (Conrad 1999). The enzyme plays a role in cell differentiation and maturation, inflammation, asthma, carcinogenesis and atherogenesis (Kuhn et al. 2002). In terms of enzymatic characteristics, 15-LOX-1 preferentially metabolizes linoleic acid primarily to 13-(S)-HODE, but also metabolizes arachidonic acid to 15-(S)-HETE. 15-LOX-2, on the other hand, converts arachidonic acid to 15-(S)-HETE and metabolizes linoleic acid poorly (Brash et al. 1997). While many lines of experimental evidence suggested that 5- and 12-LOX metabolites promote angiogenesis and carcinogenesis, in contrast 15-LOX may play an inhibitory role in tumor angiogenesis and thus, may slow carcinogenesis (van Leyen et al. 1998; Shureiqi and Lippman 2001). The products of LOX metabolism represent either intermediary products such as HPETE, which are transformed enzymatically into secondary products including leukotrienes, hepoxilins, lipoxins and HETEs, which can act as signaling molecules in their own right, or give rise to the production of reactive oxygen species (ROS). Signaling of LOX-derived products can occur through either G protein coupled cell surface receptors, in the case of lipoxins and leukotrienes (Norel and Brink 2004), or through activation of nuclear receptors such as peroxisome proliferator activated receptors (PPAR) in the case of HETEs and HODEs (Michalik et al. 2004).

3   COX and LOX Pathways in Cancer

3.1   Cyclooxygenases and Cancer

COX-derived prostanoids are biologically active lipid mediators involved in a wide range of physiological processes such as modulation of vascular tone, the inflammatory response and gastric cytoprotection. Prostanoids have also been implicated in various disease states such as arthritis, heart disease, pulmonary hypertension and cancer (Smyth et al. 2009). The development of cancer in both humans and experimental animals is consistently linked to an imbalance in COX signaling (Wang and Dubois 2010; Wu et al. 2010). There has been a significant interest in COX-2 and its role in the development and progression of cancer over the past decade, with a number of phase II studies examining celecoxib as a potential chemopreventative agent (Dragovich et al. 2008; Jakobsen et al. 2008; Papadimitrakopoulou et al. 2008; Antonarakis et al. 2009). COX-2 expression has been associated with a poor prognosis in a variety of cancer states (Achiwa et al. 1999; Khuri et al. 2001; Brabender et al. 2002). A number of clinical trials have been carried out to examine the role of COX-2 inhibition in cancer chemoprevention (O'Reilly et al. 1994; Bresalier et al. 2005; Solomon et al. 2005; Heath et al. 2007; Edelman et al. 2008; Fidler et al. 2008; Mutter et al. 2009), with further human trials ongoing. However, conflicting studies aimed at examining the role of non-specific COX or selective COX-2 inhibition will lead to difficulty interpreting these results. For example, an inhibition in mouse lung tumorigenesis has been observed in studies using non-selective COX inhibitors (Rioux and Castonguay 1998; Moody et al. 2001). However, selective COX-2 inhibition with celecoxib resulted in reduced pulmonary inflammation, but no differences in tumor multiplicity and an increase in tumor size in an initiator promoter lung tumor mouse model (Kisley et al. 2002). In addition to these observations, chronic administration of selective COX-2 inhibitors at high concentrations has been associated with an increased risk of cardiovascular events, such as thrombosis, stroke, and myocardial infarction (Bombardier et al. 2000; Bresalier et al. 2005; Solomon et al. 2005). The mechanism whereby these drugs contribute to cardiac complications is thought to be through a disruption in the fine balance between PGI_2 and TXA_2, which modulate platelet activation and aggregation, thereby regulating blood clotting. The role of these respective prostanoids in the regulation of coagulation was demonstrated using a rat model of pulmonary hypertension (Pidgeon et al. 2004), and subsequently confirmed in a COX-gene disrupted mouse model (Cathcart et al. 2008). This model demonstrated that COX-2 gene deletion exacerbated the pulmonary hypertensive response to hypoxia, an effect mediated at least in part through enhanced sensitivity to TXA_2, and a resulting increase in intravascular thrombosis. These findings are further supported by seminal studies of the cardiovascular effects of selective COX-2 inhibitors and the pharmacological roles of downstream receptors (Narumiya and FitzGerald 2001; Funk and FitzGerald 2007).

The effects of COX expression in cancer are thought to be related to the expression profile of down-stream COX-derived prostanoids. Evidence for a role for the prostanoids in carcinogenesis emerges from the numerous epidemiological studies demonstrating that chronic intake of non-steroidal anti-inflammatory drugs (NSAIDs), particularly aspirin, prevents development of the disease (Thun et al. 2002). However, the relationship of the prostanoid profile to cancer growth is not yet completely understood (Yang et al. 1998). Early studies have proposed that the prostaglandin biosynthesis profile of malignant cells may differ from that of normal tissue (McLemore et al. 1988; Hubbard et al. 1989). Increased COX-2 expression has been associated with increased levels of downstream enzymes required for prostanoid synthesis, suggesting that the tumor-promoting effects of COX-2 overexpression may be attributable to specific downstream products of arachidonic acid metabolism (Ermert et al. 2003). In support of this hypothesis, recent observations in our laboratory have demonstrated increased tissue levels of both COX-2 and downstream TXS in the progressive sequence from normal colon to colorectal cancer (Fig. 2). The past number of years has seen considerable interest in targeting downstream effectors of the cyclooxygenase signaling pathway in cancer. Selective targeting of these downstream effectors has the potential advantage of avoiding the cardiovascular effects associated with selective COX-2 inhibition, while maintaining anti-cancer properties.

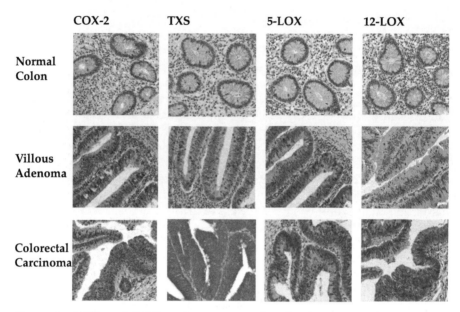

Figure 2 **COX and LOX pathway expression in the progressive sequence from normal colorectal tissue to colorectal cancer.** A panel of colorectal tissue microarrays were immunohistochemically stained for COX-2, thromboxane synthase, 5-lipoxygenase and 12-lipoxygenase across the progressive sequence to colorectal cancer (normal colorectum, villous adenoma, colorectal cancer). COX-2, TXS, and 5-LOX expression were increased across the sequence, with relatively higher levels observed in villous adenoma compared to matched normal mucosae, and increased further in colon carcinoma. Conversely, 12-LOX expression did not appear to be up-regulated in adenoma, but was moderately over-expressed in carcinoma indicating this to be a later event in the progressive sequence (×20 magnification).

Color image of this figure appears in the color plate section at the end of the book.

3.2 Lipoxygenases and Cancer

The role of LOX in the development and progression of cancer is complex due to the variety of LOX genes that have been identified in humans, in addition to different profiles of LOX observed between studies on human tumor biopsies and experimentally induced animal tumor models (Shureiqi and Lippman 2001; Catalano and Procopio 2005). Examining the expression and activity of LOX in matched normal and cancer epithelial tissues from both humans and mice, revealed that human 15-LOX-1 and -2 (or the corresponding mouse orthologs leukocyte 12-LOX and 8-LOX) are usually preferentially expressed in normal tissues and benign lesions, but not in carcinomas of the bladder, breast, colon, lung or prostate (Shappell et al. 1999; Shureiqi et al. 1999; Gonzalez et al. 2004; Subbarayan et al. 2005; Tang et al. 2007). In contrast, 5-LOX and platelet type 12-LOX are generally absent in normal epithelia, can be induced by pro-inflammatory stimuli, and are often constitutively expressed in various epithelial cancers including colon, esophageal, lung prostate and breast cancer (Gao et al. 1995; Gupta et al. 2001; Jiang et al. 2003; Ohd et al. 2003; Chen et al. 2004). The literature emerging on the role of lipoxygenases in tumor growth, for the most part, suggests that distinct LOX isoforms, whose expression are lost during the progression of cancer, may exhibit anti-tumor and anti-angiogenic activity, while other isoforms may exert pro-tumor and pro-angiogenic effects. This view is supported by experimental and human studies in each of the four most prevalent cancers worldwide; breast, lung, colon and prostate cancer, where their development has been shown to be differentially effected by LOX isoform expression. In the case of breast cancer, both 15-LOX-1 and -2 were found to be poorly expressed compared to normal epithelium (Jiang et al. 2006), while elevated levels of both 5-LOX and 12-LOX were identified and found to be associated with higher TNM staging (Jiang et al. 2006). In another study, elevated 12-LOX expression was shown to have prognostic value, when taken in combination with decreased levels of 15-LOX, in patients with breast cancer (Jiang et al. 2003). Similarly, in non-small cell lung cancer (NSCLC), better differentiated tumors expressed greater 15-LOX-2 expression and that this expression was inversely correlated with tumor grade (Gonzalez et al. 2004). In a small study examining colorectal adenomas, 15-LOX-1 expression was downregulated in 87% of tumors relative to normal epithelial mucosa, while no difference in either 5-LOX or 15-LOX-2 were observed (Shureiqi et al. 2005).

The role of various LOX isoforms in prostate cancer has been extensively investigated. In a study involving over 130 prostate cancer patients, Gao et al., found that the level of 12- LOX mRNA expression is correlated with tumor stage (Gao et al. 1995). In this study, the expression of 12-LOX and tumor stage, grade, positive surgical margins and lymph node positivity were evaluated. Overall, 38% of 122 patients demonstrated elevated levels of 12-LOX mRNA in prostate cancer tissue compared with their matching normal tissues. A statistically significantly greater number of cases were found to have an elevated level of 12-LOX among T3, high grade and surgical margin positive than T2, intermediate, and low grade and surgical margin negative prostatic adenocarcinomas. These data suggest that elevation of 12-LOX mRNA expression occurs more frequently in advanced-stage, high-grade prostate cancer (Gao et al. 1995). In addition, urinary levels of 12-HETE, the metabolite produced by 12-LOX, have been reported to be significantly elevated in prostate cancer patients compared to normal individuals, and following radical prostatectomy these levels are significantly reduced, while the procedure has little effect on the levels of other LOX products (Nithipatikom et al. 2006). Each of these studies implicates the involvement of lipoxygenases in tumor differentiation and progression. In our laboratory, we have examined both 5-LOX and 12-LOX expression across the sequence from normal to colorectal cancer in a cohort of 20 patients using full sections and tissue micro arrays. In agreement with previous studies (Melstrom et al. 2008), we found increased expression of 5-LOX in villous adenoma and colorectal adenocarcinoma relative to normal colon mucosae in the same patient (Fig. 2). We also observed a modestly increased expression of 12-LOX in many colorectal cancers, though both villous and tubulovillous adenomas expressed levels of 12-LOX similar to normal colon. While our study is small, it suggests that 5-LOX expression is an early event in the sequence to colon cancer, with increased expression in adenoma frequent, while 12-LOX expression would appear to be a later event, possibly mediating invasion and metastasis. While both 5- and 12-LOX are implicated in tumor progression, the overall picture is controversial with regards to some isoforms. Like platelet-type 12-LOX, 15-LOX-1 over-expression has been reported in prostate tumors where its levels of expression correlate with the Gleason score of the cancer (Kelavkar et al. 2000). Higher Gleason scores correlated with expression of 15-LOX-1, while others have reported that 15-LOX-2 is expressed in normal prostate tissue, but poorly expressed in prostate tumors. The reduced 15-LOX-2 expression is inversely correlated with the Gleason score of the tumor. This report would suggest that 15-LOX-1 is acting very differently in the prostate, compared to its tumor suppressive action in both colon and breast cancer (Shureiqi et al. 2005; Jiang et al. 2006; McCabe et al. 2006; Nie et al. 2006). Therefore, it is likely that the role of LOX expression on cancer growth depends on both the isoforms expressed and the site of the particular tumor. To critically examine the role of each isoform in tumor angiogenesis, it is important to investigate the effect of lipoxygenase metabolism on the various stages of tumor angiogenesis, from endothelial cell survival and growth, to invasion and angiogenesis.

3.3   Cylooxygenases, Lipoxygenases and Tumor Angiogenesis

The induction of angiogenesis is necessary for the supply of oxygen and nutrients to tumors >2 mm in diameter, and is therefore essential for successful tumor growth (Hanahan and Folkman 1996). In order to grow and metastasize, solid tumors secrete a range of pro-angiogenic factors, which tip the delicate balance in favour of angiogenesis (Herbst et al. 2005). Tipping of this balance towards a pro-angiogenic state is known as the 'angiogenic switch' and commonly occurs during tissue hypoxia, in inflammation, and in neoplasia. Cyclooxygenase and lipoxygenase signaling pathways can contribute, at least in part, to tumor development via their role in the regulation of tumor-associated angiogenesis.

Accumulating evidence has linked COX-2 with angiogenesis, with the production of COX-2 derived prostanoids shown to promote the expression of pro-angiogenic factors (Murohara et al. 1998). In addition, COX-2 inhibition by NSAID's leads to restricted angiogenesis and a down-regulation in the production of pro-angiogenic factors. Reports have shown that several drugs that demonstrate anti-angiogenic activity interfere with the expression of the gene encoding COX-2. Thalidomide, a drug which has anti-angiogenic properties, interferes with COX-2 expression at the post-transcriptional level

(D'Amato et al. 1994; Fujita et al. 2001). It has been suggested that COX-enzymes could be important therapeutic targets in the treatment of pathologic angiogenesis, such as that associated with cancer. Accumulating evidence links COX-2 with angiogenesis, suggesting that drugs which selectively target COX-2 and related signaling cascades could be used as anti-angiogenic agents. The major role of COX-2 in angiogenesis is thought to be through the induction of prostanoid synthesis, which then stimulates the expression of pro-angiogenic factors. However, little is known about how the level of individual prostanoids change, or how the expression and activity of the synthases involved in their production are regulated. The effects of COX-2 induced prostanoids on angiogenesis are probably amplified via a positive feedback loop, as the angiogenic factor, VEGF (vascular endothelial growth factor), activates both phospholipase A_2-mediated release of arachidonic acid and COX-2 expression, thereby enhancing PGI_2 and PGE_2 production (Iniguez et al. 2003). However, this mechanism does not necessarily operate in all cell types that participate in pathological angiogenesis (Wheeler-Jones et al. 1997; Hernandez et al. 2001). In addition to physiological angiogenesis, a role for COX-2 in tumor angiogenesis has also been demonstrated. Tumor COX-2 expression was associated with microvessel density (MVD) in a range of cancer types (Tsuji et al. 1998; Daniel et al. 1999; Nie et al. 2000; Cianchi et al. 2001; Pradono et al. 2002).

While some studies suggest that COX-1 plays a role in angiogenesis, most studies indicate that the anti-angiogenic function of NSAID's is mainly due to COX-2 inhibition, particularly in the light of the inability of COX-1 inhibitors to block neovascularisation (Leahy et al. 2000; Masferrer et al. 2000; Dormond et al. 2001). Both selective COX-2 inhibitors and non-selective COX-2 inhibitors attenuate tumor angiogenesis and non-neoplastic angiogenesis in animal models, with selective COX-1 inhibitors generally having little effect (Leahy et al. 2000; Ma et al. 2002; Amano et al. 2003; Chang et al. 2004). More recent studies have demonstrated that COX-1 activation in addition to PGI_2 and subsequent PPARδ activation is an important mechanism underlying the pro-angiogenic function of endothelial progenitor cells (He et al. 2008). While a role for COX-2 in angiogenesis has been clearly demonstrated by molecular, pharmacological, and genetic methods, further studies are required to clearly establish the importance of COX-1.

While the COX-derived prostanoids are thought to be crucial for tumor angiogenesis and progression, the relative contribution of the different prostanoids to angiogenesis is not well known. The role of PGD_2 in this process has not been well investigated, although a recent study demonstrated that the DP suppresses tumor hyperpermeability and angiogenesis *in vivo*, suggesting that DP agonism may be a novel therapeutic strategy for cancer treatment (Murata et al. 2008). PGE_2 appears to be an important modulator of COX-2-regulated angiogenesis, mediating its effects, at least in part, via VEGF and HIF-1α (Huang et al. 2005). This has been further supported by the observation of similar phenotypes of EP2 and COX-2 deficient mice: a reduction in the size of intestinal polyps was observed, and the vascularisation and content of pro-angiogenic factors in these lesions was reduced (Sonoshita et al., 2001). The recent discovery of the PPARδ nuclear receptor for PGI_2 suggests a significant new role for this prostanoid, with PPARδ activation implicated in the control of endothelial cell functions (Bishop-Bailey 2000). Prostacyclin generation by PGIS promotes tumor angiogenesis *via* PPARδ activation (Gupta et al. 2000). Thromboxane A_2 has been shown to both induce and inhibit angiogenesis in a number of studies (Gao et al. 1997; Ashton et al. 1999; Daniel et al. 1999; Nie et al. 2000; Ashton and Ware 2004), indicating that the precise role of this prostanoid in angiogenesis is uncertain. Gene transfer of TXS and PGIS has been shown to alter tumor angiogenesis and tumor growth in a murine colon-cancer model. While tumors from TXS transformants demonstrated an increased growth rate and more abundant vasculature, tumors from PGIS transformants presented with the opposite effects (Pradono et al. 2002). However, this study failed to demonstrate whether the effects of TXS and PGIS overexpression on angiogenesis were due to direct endothelial cell effects or an autocrine effect on tumor cells. While the COX-derived prostanoids all display some sort of bioactivity, only PGE_2, PGI_2 and TXA_2 have demonstrated pro-angiogenic effects, which will be detailed later on in this chapter.

4 Cyclooxygenases, Lipoxygenases and Tumor Angiogenesis

The angiogenic process is a complex one, which can be broken down into a number of individual processes. Angiogenesis is initiated by the production of pro-angiogenic factors from endothelial cells. These secreted factors result in invasion and degradation of the basement membrane surrounding the existing blood vessel. Circulating endothelial cells then migrate to the site of vessel formation, where they proliferate to form a vessel sprout. Further proliferation and differentiation of the endothelial cells then occur to elongate the sprout, forming the lumen of the new vessel (tube formation). Growth factor secretion by endothelial cells attract supporting cells such as pericytes and smooth muscle cells to stabilize the vessel, while a new basement membrane is also formed (Jain 2003). Tumor angiogenesis subsequently plays a major role in invasion and metastasis, which is essential for tumor spread to distant organs (Fig. 3). While COX and LOX pathways do not play a role in all of these processes, they have been strongly implicated in many steps of the angiogenic cascade and subsequently in angiogenesis dependant metastasis. Deregulation of the fine balances controlling angiogenesis is one of the hallmarks of cancer (Hanahan and Weinberg 2011). Altered regulation of COX/LOX signaling pathways has been implicated in the many processes of tumor angiogenesis and progression. The pro- and anti-angiogenic mechanisms of these

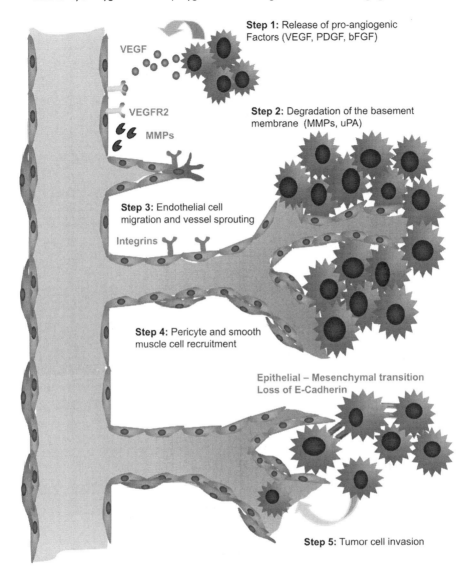

Figure 3 Steps of tumor angiogenesis and progression. Tumor angiogenesis begins with the secretion of angiogenic factors from endothelial cells (VEGF, PDFG, bFGF; step 1), followed by local degradation of the basement membrane surrounding the capillaries (MMPs, uPA; step 2). Endothelial cell migration is then accompanied by the proliferation of cells at the leading edge of the migrating column and sprout formation. As they move, the endothelial cells begin to organize into three-dimensional structures to form new capillary tubes (step 4). The secretion of growth factors from endothelial cells attract supporting cells (smooth muscle cells, pericytes) to stabilise the vessel, with the subsequent formation of a new basement membrane (step 4). Tumor vascularization allows the tumor cells to invade and metastasize to distant organs and is crucial for tumor progression (step 5).

Color image of this figure appears in the color plate section at the end of the book.

pathways are illustrated in Fig. 4. The precise role of these pathways in the aforementioned 'steps' of angiogenesis will be detailed in the following sections.

4.1 COX and LOX Mediated Generation of Angiogenic Factors

It has been suggested that COX-2 induces angiogenesis by stimulating the generation of angiogenic growth factors. In human tumor tissue, COX-2 expression was significantly associated with both VEGF and platelet-derived growth factor (PDGF) expression, while co-expression of COX-2 and VEGF were also independent prognostic factors for overall survival (Tatsuguchi et al. 2004). Endothelial COX-2, but not COX-1 expression is induced by factors that contribute to angiogenesis-dependant disease, such as hypoxia, interleukin-1 (IL-1), VEGF, bFGF, and TNFα. The same stimuli can also induce COX-2 expression in tumor cells, stromal fibroblasts, and macrophages, all of which are additional participants in pathological angiogenesis (Hla and Neilson 1992; Schmedtje et al. 1997; Hernandez et al. 2001). Finetti et al demonstrated that activation of FGFR-1 by PGE_2

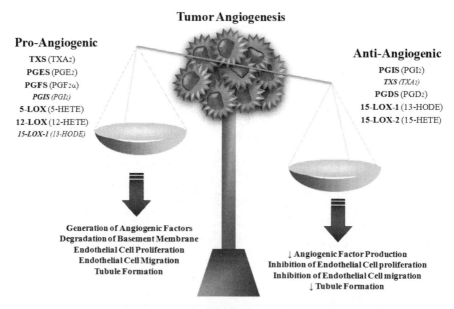

Tumor Angiogenesis

Pro-Angiogenic
TXS (TXA₂)
PGES (PGE₂)
PGFS (PGF₂α)
PGIS (PGI₂)
5-LOX (5-HETE)
12-LOX (12-HETE)
15-LOX-1 (13-HODE)

Anti-Angiogenic
PGIS (PGI₂)
TXS (TXA₂)
PGDS (PGD₂)
15-LOX-1 (13-HODE)
15-LOX-2 (15-HETE)

Generation of Angiogenic Factors
Degradation of Basement Membrane
Endothelial Cell Proliferation
Endothelial Cell Migration
Tubule Formation

↓ Angiogenic Factor Production
Inhibition of Endothelial Cell proliferation
Inhibition of Endothelial Cell migration
↓ Tubule Formation

Figure 4   Tumor angiogenesis may be modulated by both pro- and anti-angiogenic COX and LOX pathways. The expression pattern of tissue-specific synthases (downstream of COX-signaling) and LOX isozymes, and the biological activity of their corresponding products, indicates a critical balance of pro- versus anti-angiogenesis enzymes, which may influence tumor angiogenesis and progression in cancer. The angiogenic role of mediators of COX and LOX signaling is well known in most cases. However, some contradictory studies for a small number of these eicosanoids (PGIS, TXS, 15-LOX-1; shown in italics) suggest that their role in tumor angiogenesis is complex.

and provided evidence for the involvement of FGFR-1 through FGF-2 in mediating PGE_2 angiogenic responses (Finetti et al. 2008). The same lab subsequently demonstrated that synergism between PGE_2 and FGF-2 resulted in robust angiogenesis *in vivo*, with the angiogenic response directly proportional to FGF-2 availability, which determined FGFR-1 activation (Finetti et al. 2009). In non-small cell lung cancer, VEGF was shown to be at least partially COX-2 dependent in cell lines expressing COX-2 (Zhu et al. 2008). Selective COX-2 inhibition reduced expression of angiogenic proteins angiopoietin-1 and VEGF and increased expression of angiopoietin-2 *in vivo* (Kang et al. 2007). Expression of both COX-1 and COX-2 were linked to the angiogenic factor, VEGF and the lympangiogenic factor VEGF-C in esophageal adenocarcinoma, suggesting a role for the COX-isoforms in carcinogenesis (von Rahden et al. 2005). While a role for COX-2 in VEGF-induced angiogenesis is well established, the signaling pathways through which VEGF regulates COX-2 gene expression are unclear. Wu et al demonstrated that COX-2 plays a role in VEGF-induced angiogenesis *via* activation of both p38 and JNK pathways in HUVECs (Wu et al. 2006). A role for COX-2 in angiogenesis was further supported by the observation that activation of the adhesion receptor, $\alpha V\beta 3$, induced both VEGF and endothelial cell proliferation, an effect which was attenuated by both COX-1 and COX-2 inhibition (Murphy et al. 2003). PGE_2 also upregulated VEGF levels in gastric cancer cells, an effect which was mediated via transactivation of the EGFR-MAPK signaling pathways (Ding et al. 2005). These pathways were thus proposed to be mechanisms underlying the contribution of COX-2 to tumor angiogenesis. COX-2 enhanced the expression of angiogenic CXC chemokines, CXCL5 and CXCL8 *in vitro*, while neutralization of these chemokines inhibited the growth of COX-2 overexpressing tumors *in vivo* (Pold et al. 2004). In human patient studies, both COX-2 and VEGF expression were strongly correlated with tumor grade and angiogenesis in prostate cancer (Gyftopoulos et al. 2011), while VEGF expression was associated with COX-2 and microvessel density and overall survival in gastric carcinoma (Kolev et al. 2007). A positive correlation between VEGF and COX-2 was also observed for esophageal, colon, gastric and lung cancer (Williams et al. 2000; Cianchi et al. 2001; Kim et al. 2003; Huang et al. 2005; von Rahden et al. 2005).

The selective COX-2 inhibitor, celecoxib, significantly reduced PGE_2 levels in colorectal tissue homogenates from an implantation model of CRC. This was closely linked with a decrease in VEGF and CD34 (+) expression, as well as microvessel density (MVD) (Wang et al. 2008). PGE_2 induced VEGF expression in colon cancer cells, an effect which was mediated through HIF-1α. PGE_2 also activated ERK and Akt, while inhibition of ERK phosphorylation prevented the induction of VEGF and subsequent HIF-1α expression (Fukuda et al. 2003). PGE_2 has been shown to induce VEGF secretion in endothelial cells through ERK2 activation (Pai et al. 2001), and also strongly induced VEGF in human mast cells, which accumulate in large numbers at angiogenic sites (Abdel-Majid and Marshall 2004). PGE_2 was shown to induce both VEGF and bFGF expression via cAMP signaling (Hoper et al. 1997; Cheng et al. 1998; Pueyo et al. 1998). Notably, COX-2 synthesis is dependent on the presence of EP2 (Sonoshita et al. 2001). PGE_2 therefore not only causes up-regulation of VEGF and bFGF, but also induces COX-2 expression and further PGE_2 synthesis in a positive feedback mechanism. PGE_2, VEGF and bFGF also induce expression of the angiogenic chemokine CXCR4, leading to increased tubular formation *in vitro* (Salcedo et al. 2003). Furthermore, PGE_2 induced expression of the angiogenic chemokine, CXCL1 to induce angiogenesis in human colorectal cancer cells (Wang et al. 2006). Both COX-2 and EP3 expression are required for VEGF expression in stromal fibroblasts, a major source of tumor VEGF (Williams et al. 2000; Amano et al. 2003).

15d-PGJ_2 has been shown to up-regulate VEGF expression in endothelial cells, an effect mediated through PPARγ activation (Inoue et al. 2001). While 15d-PGJ_2 induced VEGF expression in both prostate and bladder cancer cell lines (Haslmayer et al. 2002), this dehydration product of PGD_2 inhibited the expression of pro-angiogenic genes, VEGF and COX-2 in colon cancer cells (Grau et al. 2004), suggesting that its role in tumor angiogenesis is unclear. A more recent study demonstrated that 15d-PGJ_2 upregulated VEGF expression, but inhibited angiogenesis by inhibiting all VEGF-induced angiogenic processes in human umbilical vein endothelial cells (Funovics et al. 2006). PGI_2 has been shown to induce pro-angiogenic VEGF expression in rat intestinal epithelial cells. Furthermore, PGI_2 expression, as well as PGIS expression was induced by Ha-Ras (V-12), suggesting a signaling mechanism for PGI_2 and subsequent VEGF production (Buchanan et al. 2004). Pola et al demonstrated that both VEGF production and angiogenesis may be modulated by PGI_2 through specific activation of the PPAR signaling pathway *in vivo* (Pola et al. 2004), while Biscetti et al subsequently demonstrated that PPARα expression is crucial for iloprost induced VEGF up-regulation and angiogenesis *in vivo* (Biscetti et al. 2009). Downstream of PGE_2, disruption of the gene encoding the EP2 receptor reduced the multiplicity and size of intestinal polyps, VEGF expression and extent of angiogenesis in mice genetically susceptible to intestinal polyp development (Sonoshita et al. 2001; Seno et al. 2002). *EP3* knockout mice also developed less tumor-associated vessels, due to the reduction in VEGF expression (Amano et al. 2003). Investigation of the signaling pathways activated by the $PGF_{2\alpha}$ receptor, FP, showed that this prostanoid could increase angiogenesis in endometrial cancer *in vivo* by EGFR transactivation and induction of VEGF mRNA expression (Sales et al. 2005). TXA_2 has been shown to act as a potent stimulator of angiogenesis, both directly, and by inducing VEGF and PDGF secretion from platelets following aggregation (Arisato et al. 2003; Rhee et al. 2004). VEGF formation was also stimulated following treatment of tumor cells with TXA_2 or TPα stimulation (Wei et al. 2010). Peroxisome proliferator-activated receptor activation has also been shown to play a role in tumor angiogenesis. Treatment with the PPARγ ligand 15-deoxy-delta-PGJ$_2$ downregulated expression of angiopoietin-1 (Ang-1) in gastric cancer cells (Fu et al. 2006).

In addition to cyclooxygenases, a variety of lipoxygenases have been implicated in pathological angiogenesis through their induced or altered expression of various pro-angiogenic factors. A large body of literature links 12-LOX expression with angiogenesis and angiogenic factor expression in prostate cancer. The potential of 12(S)-HETE as a significant stimulator of pathological angiogenesis may lie in its ability to induce the expression of VEGF expression at both protein and promoter levels. This induction has been reported by two independent groups in prostate cancer cells (McCabe et al. 2006; Nie et al. 2006). Both NDGA and baicalein reduced VEGF expression, confirming that the enzymatic activity of 12-LOX plays a significant role in the regulation of VEGF gene expression. In the study by Nie et al., transient transfection with a 12-LOX expression construct enhanced VEGF promoter activity in the prostate cancer cell line, PC-3 (Nie et al. 2006). Additionally the same study reported a potential mechanism for this phenomenon, with 12(S)-HETE capable of inducing signaling via the extracellular-signal related kinase1/2 (ERK1/2) mitogen-activated protein (MAP) kinase pathway in prostate cancer cell lines. In a recent study by the same group, it was demonstrated that 12-LOX increased the expression and functionality of hypoxia inducible factor-1 alpha (HIF-1α) under hypoxic conditions, suggesting a potential mechanism whereby 12-LOX upregulates VEGF expression in prostate cancer cells (Krishnamoorthy et al. 2010). Another potential mechanism of 12(S)-HETE induction of VEGF expression is via direct activation of the BLT2 receptor on endothelial cells. A study by Kim et al demonstrated that blocking the receptor (through which 12(S)-HETE signals) by genetic knockdown or pharmacological inhibition almost completely inhibited VEGF-induced vessel formation in Matrigel-plus assays (Kim et al. 2009). In the same study, the group also reported that VEGF upregulated 12(S)-HETE expression in endothelial cells. Similarly, they reported that following 12-LOX knockdown, VEGF mediated angiogenesis was in HUVECs was inhibited, an effect which was reversed by add-back of 12(S)-HETE.

In another report by Ye et al., cigarette smoke extract was shown to enhance cell proliferation and the expression of 5-lipoxygenase (5-LOX), vascular endothelium growth factor (VEGF), matrix metalloproteinases (MMPs) 2 and 9 in colon cancer cells (Ye et al. 2005). Notably, inhibition of 5-LOX decreased cell proliferation and expressions of VEGF, MMP-2 and MMP-9 induced by cigarette smoke extract. In addition, cigarette smoke extract indirectly stimulated HUVEC proliferation, an effect that was again blocked by the 5-LOX inhibitor. The regulation of VEGF expression by 5-LOX has been reported previously in other cancers. Romano et al reported that 5-LOX is a key regulator of malignant mesothelioma proliferation and survival through VEGF-mediated mechanisms (Romano et al. 2001). The group reported that 5(S)-HETE stimulates the activity of the VEGF promoter in mesothelioma cell lines. In another study, rats treated with 7,12-dimethylbenz(a)anthracene (DMBA) resulted in inflammation associated carcinogenesis in mammary glands (Chatterjee et al. 2011). 5-LOX expression was increased in these cancers, and was associated with both VEGF and MMP-2 expression. Notably, in the same study, the specific 5-LOX inhibitor zileuton significantly reduced cell proliferation and the expression of VEGF and MMP2, and resulted in substantial repair of hyperplastic lesions.

While numerous isoforms of LOX have been shown to stimulate the production of pro-angiogenic factors, there are reports indicating certain LOX isoforms to be anti-angiogenic by decreasing the expression of certain angiogenic factors. Studies have indicated that increased expression of 15-LOX-1 decreases VEGF, nitric oxide and VEGF receptor expression in rabbit skeletal muscles (Viita et al. 2008). In a different study of pathological angiogenesis in ocular disease, intravitreal adenovirus-mediated gene transfer of 15-LOX-1 prevented VEGF induced angiogenesis and consequent pathology in rabbit eyes (Viita et al. 2009). Similar reports of anti-angiogenic activity of 15-LOX in tumor specific models have been reported with inhibitors of 12/15-LOX resulting in 'normalized' vessel morphology and upregulated/restored expression of VEGF in tumor spheroids in mice (Schneider et al. 2010). The anti-angiogenic role of 15-LOX-2 in tumor models is well published, with expression of VEGF suppressed and tumors remaining more dormant in mouse prostate cancer models when 15-LOX-2 was overexpressed in the cells (Tang et al. 2009).

4.2 The Role of COX and LOX Signaling in Invasion and Degradation of the Basement Membrane

Little is known of the role of cyclooxygenase signaling in invasion across and degradation of the basement membrane in angiogenesis. COX-2 overexpression leads to matrix metalloproteinase (MMP) generation, which have been implicated in invasion (Takahashi et al. 1999). COX-2 over-expressing breast tumor cells demonstrated increased invasion across an artificial matrigel basement membrane relative to clones with low COX-2 expression, following treatment with recombinant epidermal growth factor (Prosperi et al. 2004). Urokinase-type plasminogen activator (uPA) and its receptor (uPAR) play a role in degradation of the basement membrane during angiogenesis. The COX-2/PGE$_2$ pathway is thought to be involved in *Helicobacter pylori*-associated activation of the uPA system, with COX-2 inhibition or EP2 antagonism proposed to inhibit angiogenesis and invasion in gastric cancer cells *via* suppression of the uPA system (Iwamoto et al. 2008). In a separate study in gastric cancer, blockade of 5-LOX signaling with a 5-LOX activating protein inhibitor, MK886, resulted in downregulation of matrix metalloproteinase (MMP-7, -9), uPA and its receptor uPAR (Shin et al. 2010). Another recent study by Kummer et al demonstrates a correlation between 5-LOX and MMP-9 expression in resected papillary thyroid carcinoma (PTC) by immunohistochemical analysis (Kummer et al. 2012). The same group also demonstrated that the end product of 5-LOX, 5(S)-HETE, increased MMP-9 protein expression in a dose dependent manner in PTC cell lines. The mechanism whereby 5-LOX induces MMP expression is thought to be via activation of the ERK pathway, as recent studies in both macrophages and tumor cells have indicated that inhibitors of the ERK pathway, and not JNK or p38 MAPK pathways could block 5-LOX induced MMP-2 and MMP-9 expression (Kim et al. 2010; Shin et al. 2010). Interestingly, in a separate study focused on smooth muscle cells, inhibitors of both the ERK and p38 MAPK pathway could block 5-LOX induced expression of MMP-2 production, suggesting a potential different mechanism of regulation in vascular smooth muscle cells (Seo et al. 2010).

The link between lipoxygenase enzymes and MMP expression has also been suggested with 12-LOX. An early study by Mariani et al. indicated that macrophage-derived 12-lipoxygenase metabolites induced the expression of MMP in lung fibroblast cells and the group reported this as a potential mechanism for the induction of resident cell MMP gene expression during the inflammatory lung process (Mariani et al. 1998). These reports suggest that COX, and more in particular LOX enzymes may regulate mechanisms responsible for the induction of matrix-degrading proteases in many cancers.

4.3 Endothelial Cell Migration and Vessel Sprouting via COX and LOX Pathways

Capillary sprout formation ("sprouting angiogenesis") from pre-existing mature blood vessel occurs following degradation of the basement membrane. This creates a break to allow movement (migration) of differentiated endothelial cells towards adjacent tumor cells and stimuli produced by these cells. Angiogenic migration of endothelial cells from the parental venule towards the angiogenic stimulus emanating from the tumor mass is followed by endothelial cell division which, in combination with migration, acts to lengthen the stalk of the endothelial cell sprout. While each of these processes is biologically distinct, numerous reports on the effect of eicosanoids examine endothelial proliferation, migration and tubule formation together in the overall angiogenic process.

COX-2 expression promotes endothelial cell migratory responses through integrin signaling (Dormond et al. 2001). COX-2 over-expression in colon cancer cells stimulated tube formation and extension of co-cultured endothelial cells. Furthermore, selective COX-2 inhibition blocked endothelial cell migration, tubule formation and reduced VEGF expression in the same model, with similar effects observed for COX-1 inhibition (Tsujii et al. 1998). This may explain why in some cases, non-selective COX-2 inhibitors are more effective in reducing tumor growth via angiogenesis inhibition.

Treatment of endothelial cells with exogenous VEGF upregulated COX-2 with a corresponding increased in PGI_2, proliferation and tubule formation (Murphy and Fitzgerald 2001; Murphy et al. 2005). PGE_2 resulted in upregulation of effectors of the fibroblast growth factor-2 (FGF-2) pathway. PGE_2-mediated FGF-2 activation resulted in endothelial cell proliferation and strong angiogenesis *in vivo*, with the magnitude of angiogenic response directly related to the abundance of FGF-2 (Finetti et al. 2009). The angiogenic effects of PGE_2 are thought to be mediated by up-regulation of the angiogenic CXC chemokine, CXCR4 on human microvascular endothelial cells. The ability of PGE_2 to increase tubule formation was completely inhibited by CXCR4 inhibition. The authors demonstrated that PGE_2 is a mediator of VEGF- and bFGF-induced CXCR4-dependent angiogenesis and that the angiogenic effects of PGE_2 require CXCR4 expression (Salcedo et al. 2003). PGE_2 also induced expression of the pro-angiogenic chemokine CXCL1 (growth-activated oncogene α) in human colorectal cancer cells, resulting in an induction in cell migration and tube formation (Wang et al. 2006). A similar effect of PGE_2 on tumor cell migration was observed *in vivo* (Young 2004).

15d-PGJ_2 has been found to inhibit mitogen-induced endothelial cell proliferation and tube formation in an *in vitro* angiogenesis model. These effects are thought to be mediated through down-regulation of VEGF receptors and uPA (Xin et al., 1999). 15d-PGJ_2 treatment also inhibited migration and tube formation in human vascular endothelial cells *in vitro* (Fu et al. 2006; Funovics et al. 2006) as well as well as gastric tumor cell angiogenesis *in vivo*; effects which were restored by addition of Ang-1 (Fu et al. 2006). Furthermore, pharmacological inhibition of COX-2 inhibited TXA_2 generation, endothelial cell migration, and subsequent FGF-induced corneal angiogenesis. Both of these parameters were also inhibited by treatment with a TP antagonist. U-46619, a TP agonist restored both migration and angiogenesis responses invoked by COX-2 inhibition (Daniel et al. 1999). Thromboxane synthase inhibition resulted in a significant reduction (60% reduction in migration compared to vehicle controls) in endothelial cell migration in HUVEC *in vitro* (Jantke et al. 2004). The thromboxane A_2 mimetic U-46619 stimulated endothelial cell migration. Conversely, inhibition of TXA_2 synthesis, which was stimulated by basic fibroblast growth factor (bFGF) or VEGF, reduced endothelial cell migration, implicating TXA_2 in angiogenesis (Nie et al. 2000). The TXA_2 receptor antagonist SQ-29548 inhibited COX-2 dependent microvascular endothelial cell migration and corneal angiogenesis (Daniel et al. 1999). Dual TXS/TP inhibitors dose-dependently inhibited endothelial cell migration in chemotaxis assays. In addition, pre-treatment of endothelial cells with these original dual inhibitors significantly attenuated TP agonist-induced intracellular Ca^{2+} pool mobilization, suggesting a mechanistic link between TXS/TP inhibition and reduced endothelial cell migration (de Leval et al. 2006). Thromboxane synthase inhibitors strongly inhibited capillary tube formation in human vascular endothelial cells (Jantke et al. 2004). Both the number of branches from nodal areas and the length of tube-like structures decreased following treatment with the selective TXS inhibitor, in a dose dependent manner. However, anti-angiogenic roles for thromboxane have also been described, including induction of endothelial cell apoptosis and vascular tube formation (Ashton et al. 1999; Gao et al. 2000; Ashton and Ware 2004). TP stimulation reduced spontaneous endothelial cell migration by 58% (Ashton et al. 1999).

Endothelial cells have also been reported to synthesis various lipoxygenases, which can regulate endothelial cell angiogenic responses. An early study by the Honn laboratory demonstrated that endothelia cells expressed 12-LOX, and that a specific 12-LOX inhibitor BHPP blocked bFGF and VEGF induced endothelial cell proliferation and migration (Nie et al. 2000). In other reports, the broad spectrum lipoxygenase inhibitor NDGA has also been reported to inhibit FGF-2 induced migration of bovine corneal endothelial cells *in vivo* (Rieck et al. 2001). The mechanism whereby 12-LOX mediates endothelia cell migration has been further investigated by other groups in recent years. Hsieh et al., reported that 12-LOX mediated endothelial cell migration is mediated through decreased integrin and vinculin expression, as the 12-LOX inhibitor bacalein up-regulated $\alpha 5\beta 1$ and $\alpha v\beta 3$ integrins resulting in increased actin reorganisation and focal adhesion contact formation in rat endothelia cells (Hsieh et al. 2007). 12-LOX is not the only lipoxygenase isozyme that has been linked to endothelial cell migration and vessel sprouting. 15(S)-HETE, the end product of 15-LOX, has been reported to mediate hypoxia-induced pulmonary vascular remodelling through a direct effect on intimal endothelial cell migration via the Rho-kinase pathway (Ma et al. 2011). Other groups have reported that the HMG-CoA reductase pathway plays a key role in 12/15-LOX induced angiogenesis, and that 15(S)-HETE stimulates Rac1 in human dermal microvascular endothelial cells (Singh et al. 2011). In the same study, the group in Memphis demonstrated that the HMG-CoA reductase inhibitor, simvastatin, blocked 15(S)-HETE-mediated Rac1 expression and thereby blocked 15(S)-HETE mediated endothelial cell migration and tubule formation.

In another study examining 5-LOX expression and angiogenesis, cigarette smoke extract indirectly stimulated HUVEC proliferation, and this effect could be blocked by selective 5-LOX inhibition (Ye et al. 2005). Furthering these observations, the same group also reported that inhibition of COX-2 or 5-LOX in the nude mouse xenograft model reduced colon tumor size induced by cigarette smoke. Perhaps one of the most novel studies linking eicasanoids to endothelial cell migration is the discovery of hemiketal eicosanoids formed by the cross-over of the 5-LOX and COX-2 pathway. The Schneider group at Vanderbilt University Medical School recently identified two cyclical hemiketal eicosanoids, HKD(2) and HKE(2), by LC-mass spectrometry and UV and NMR spectroscopy (Griesser et al. 2011). HKD (2) and HKE (2) stimulated migration and tubulogenesis of microvascular endothelial cells, indicating a novel mechanism of pro-angiogenic activity of 5-LOX and COX-2 co-expression in certain tissues.

4.4   Pericyte and Smooth Muscle Cell Recruitment for Formation of New Basement Membrane

Capillary tube or lumen formation subsequently results in the completion of capillary sprouts and loops and the envelopment of capillaries with new basement membrane structures along with recruitment of perivascular support cells, such as pericytes. Tip cells are crucial to completing this process as they fuse with other tip cells to create a fused (linked) network of capillaries. Pericytes (a single layer of periendothelial smooth muscle cells) modulate endothelial cell function, and are therefore crucial for the development of a mature vascular network. Pericytes regulate vascular function, including vessel diameter (and thus blood flow) and vascular permeability (Gerhardt and Betsholtz 2003). They also provide mechanical support and stability to the vascular wall and maintain endothelial cell survival through direct cell-cell contact and paracrine circuits (Reinmuth et al. 2001; Gerhardt and Betsholtz 2003).

While the role of COX and LOX signaling in this angiogenic process has not been well investigated, COX-2 has been shown to regulate a number of molecules involved in the participation of pericytes and vascular mural cells (VMC) in developing vessels. These include platelet-derived growth factor (a key mediator of endothelial-VMC interaction) and CXCR4, which functions in homing of vascular progenitor cells to sites of active angiogenesis (Nagatsuka et al. 2002; Salcedo et al. 2003).

4.5   Tumor Cell Invasion and Metastasis

Angiogenesis is known to play a major role in tumor progression through processes of tumor invasion and metastasis. Tumor invasion is initiated by receptor-mediated adhesion of tumor cells to matrix proteins, followed by a second phase of matrix-breakdown by tumor-secreted proteases. This process creates an intra-cellular space into which invading cells can migrate (Giese et al. 1995). Tumor spread from the primary site to distant organs is the most clinically important property of malignant tumors and is known as metastasis. Metastasis allows the cancer to survive surgical excision of the primary tumor, and is responsible for increases in tumor burden and increasing difficulty in its clinical management (Fidler and Ellis 1994). Tumor cell invasion is essential for the dissemination of metastatic cells across extracellular matrices and spread to distant organ sites. Studies suggest that the prostanoids and their corresponding synthases can play distinct roles in tumor progression and metastasis (Wang et al. 2007).

PGD_2 expression levels were significantly lower in primary cancer tissue from patients who exhibited hepatic metastasis, relative to those that did not, suggesting that this prostanoid protects against tumor progression (Yoshida et al. 1998). The observation of significantly higher PGE_2 levels in metastatic colorectal tumors, relative to non-metastatic provided evidence for a role for this prostanoid in tumor metastasis (Cianchi et al. 2001).

EP3 may also contribute to metastasis by increasing cell migration via up-regulation of VEGFR-1 signaling. EP4 has also been implicated in metastasis, with a significant decrease in liver metastasis of colon cancer cells observed following EP4 antagonism (Yang et al. 2006). Treatment of colorectal cancer (CRC) cells with PGE_2 up-regulated VEGF receptor-1 expression and increased migration by a mechanism involving EP3 receptor-mediated activation of phosphatidylinositol-3-kinase and extracellular signal-related kinases (Fujino et al. 2011). The effect of PGE_2 on invasion and migration is thought to occur via rapid transactivation and subsequent phosphorylation of the epidermal growth factor receptor (EGFR), leading to induction of Akt. The activation of EGFR occurred via an intracellular Src-mediated event, suggesting a further signaling mechanism underlying the PGE_2-mediated effects (Buchanan et al. 2003). The authors subsequently determined that PGE_2-mediated induction of an EP4-β-arrestin 1-Src complex was required for EGFR transactivation, leading to induction of Akt signaling to stimulate CRC cell migration *in vitro* as well as metastatic spread to the liver *in vivo* (Buchanan et al. 2006). Evidence has also been provided to show that PGE_2 promotes colorectal tumor cell invasiveness by transactivation of c-Met-R (dependant on functional EGFR), thereby increasing phosphorylation and accumulation of β-catenin, and inducing urokinase-type plasminogen activator receptor (uPAR) expression (Pai et al. 2003). $PGF_{2\alpha}$ expression induced cell motility with equivalent potency to PGE_2 in both colorectal adenoma and carcinoma-derived cell lines and also increased the invasiveness of colorectal cancer cells (Qualtrough et al. 2007).

PGI_2 is also well known to have anti-metastatic effects in cancer (Honn et al. 1981). Analogues of PGI_2 are thought to protect against metastasis in colorectal cancer by inhibiting cancer cell-endothelial cell interactions (Yoshida et al. 1999). Specifically, the effects of prostacyclins may be partly mediated through inhibition of CAM-mediated adherence of colorectal tumor cells to endothelial cells of target organs (Daneker et al. 1996). Prostacyclin significantly reduced metastasis formation in a lewis lung carcinoma injected murine model. Treatment with PGI_2 significantly reduced initial lewis lung carcinoma cell density. In addition, lung weight was reduced by over 50% and the number of visible metastatic nodes by over 90% (p<0.05) (Cuneo et al. 2007). PGI_2 has also been shown to reduce the growth of lung micrometastases (Schirner and Schneider 1997).

An *in vivo* study has reported that selective TXS inhibition blocks colorectal liver metastasis (Yokoyama et al. 1995). Specific TXS inhibition has been shown to suppress growth and reduce invasion and migration of bladder cancer cells (Moussa et al. 2005). Glioma cell migration was blocked following treatment with specific TXS inhibitors. A concomitant reduction in TXB_2 generation was also reported here, implicating the TXS pathway as an important regulator of glioma motility (Giese et al. 1999). The role of thromboxane synthase in tumor metastases has been well studied (Giese et al. 1999; Yoshizato et al. 2002; Kim et al. 2003; Jantke et al. 2004;

Moussa et al. 2005; Watkins et al. 2005). TXS expression was associated with increased micro-vessel density (a prognostic factor, predictive of metastasis and poor survival) and metastasis in patients presenting with NSCLC (Yoshimoto et al. 2005). A TXS inhibitor has been shown to block colorectal carcinoma metastasis in an *in vivo* model of the disease (Yokoyama et al. 1995). TXS has also been shown to be involved with renal cell carcinoma metastasis. Early *in vivo* studies with TXS inhibitors have failed to report any beneficial effects on metastasis or spread to the lymph nodes (Stamford et al. 1986). However, when used in combination with a TP antagonist, TXS inhibitors have been found to inhibit metastasis formation from tail vein injected B16a cells, as well as spontaneous metastasis formation from subcutaneous B16a and Lewis lung carcinoma tumors (Honn 1983). Several other studies have subsequently demonstrated a potential role for TXS in promoting tumor invasion and metastasis (Pradono et al. 2002; Nie et al. 2004; Watkins et al. 2005; Sakai et al. 2006). Increased TXS expression was associated with an increase in tumor cell invasion in an astrocytoma cell line (McDonough et al. 1998). Furthermore, other studies have shown that inhibition of TXA_2 generation can inhibit tumor cell migration as well as trefoil peptide-stimulated tumor cell invasion (Rodrigues et al. 2001; Nie et al. 2004). A potential role for the thromboxane receptor in tumor cell invasion and metastasis has also recently been examined. In prostate cancer cells, migration was significantly inhibited by either sustained activation of TP or by inhibition of TP activation, suggesting that TP activation is tightly controlled during cell migration (Nie et al. 2008). More recently again, it was observed that TP-β receptor expression, but not TP-α increased the migratory and invasive capacity of bladder cancer cells (Moussa et al. 2008). The observations of these studies suggest that the contribution of this receptor to tumor cell invasion and metastasis is unclear. Further studies are therefore required to clarify the contribution of TP to this cell survival pathway.

The regulation of cell-cell and cell-matrix adhesion depends largely on the cellular profile of adhesion molecules and matrix metalloproteases and on the signaling mechanisms that regulate adhesive behavior and cellular responses to environmental stimuli. Thus, elucidating the effects of environmental components, such as fatty acids, on tumor cell adhesion is important for the understanding of metastasis. Previous studies have indicated a role for AA metabolism in cell-matrix interactions and integrin signaling. For example, inhibition of cyclooxygenase-2 by non-steroidal anti-inflammatory drugs blocks both platelet aggregation (by suppressing activation of integrin αIIbβ3) (Dominguez-Jimenez et al. 1999) and endothelial cell migration (by suppressing activation of αvβ3 integrin) (Dormond et al. 2002). Other studies have reported the role of LOXs, in particular, 12-LOX, in the regulation of surface integrin expression. For example, adhesion of B16 murine melanoma cells to microvascular endothelial cells was enhanced by pre-treatment of the endothelial cells with the 12-LOX product, 12(S)-HETE, via up-regulation of αvβ3 integrin expression (Tang et al. 1993). In the same cell line, ligation of αIIbβ3 integrin induced 12(S)-HETE production (Raso et al. 2001), implying co regulation of integrin expression and LOX activity in these cells. Similarly, 12(S)-HETE treatment of human endothelial cells enhanced monocyte adhesion through increased very late-acting antigen-4 integrin expression (Patricia et al. 1999). In addition, the interaction of 12-LOX with a number of cellular proteins, including the integrin β4 subunit, has previously been reported by yeast two-hybrid screening (Tang et al. 2000). In a study by Nie et al., 12-LOX transfected prostate cancer cells were reported to be more adhesive toward vitronectin and type I and IV collagen, and were more invasive through matrigel than cells transfected with the control vector (Nie et al. 2003). In the same report, when the cells were subcutaneously grown in nude mice, invasion to the surrounding tissue was more frequently observed in the 12-LOX transfected cells. Similarly, when injected by tail vein into SCID mice with human bone fragment implants, increased metastasis to bone was observed in the cells over-expressing 12-LOX.

In a recent study in colorectal cell lines, over-expression of 12-LOX resulted in increased anchorage-independent growth and a greater capacity for the cells to metastasize *in vivo* (Klampfl et al. 2011). These effects were observed to be due to downregulated E-cadherin and integrin-beta1 expression in the colorectal cells in a 12-LOX dependent manner, disturbing cell-cell interactions. Comprehensive reviews from others and our unit also details the role of 12-LOX and other LOX isoforms in colorectal cancer progression, detailing specific pathways regulating invasion in CRC (Cathcart et al. 2011; Schneider and Pozzi 2011). Among the many mechanisms of 12-LOX mediated tumor invasion is ligand binding to a novel orphan G-coupled receptor GPR31 by 12(S)-HETE and downstream activation of ERK1/2, MEK and NFκB signaling (Guo et al. 2011). Recently, 12(S)-HETE has also been shown to mediate lymphatic invasion in human mammary carcinoma xenografts in mice (Kerjaschki et al. 2011). In the same study, expression of lipoxygenase in sentinel lymph node metastases correlated inversely with metastasis-free survival.

There is less evidence for the role of 5-LOX in cell adhesion, although the generation of MMP-2 is reported to be mediated by activation of phospholipase A_2 and 5-LOX activity in breast cancer cells (Taylor et al. 2002). The authors reported that breast cancer cells in culture release apparently full length soluble EMMPRIN that promoted the release of pro-MMP-2 from fibroblasts. A more recent study by Faronato et al. examined the expression of 5-LOX in clear cell renal carcinoma, and while no association with venous invasion was identified, 5-LOX expression correlated with larger tumor size (Faronato et al. 2007).

5 Controversial Data/Concepts

It has been suggested that that existence of two human TP isoforms may impact on the role of TXA_2 tumor-associated angiogenesis. It has been demonstrated that endothelial expression of TPβ in TP null mice inhibits migration and differentiation *in vitro*, while mice with endothelial overexpression of TPβ also display reduced angiogenesis (Ashton et al. 2004; Ashton and Ware 2004). However, as only humans have TPβ, the animal models on which these observations are flawed as human comparisons are difficult. While it has been suggested that TPα and TPβ have divergent pathological roles in disease, their individual contributions to tumor angiogenesis remain to be elucidated.

There is some evidence that 5-LOX and acetylated COX-2-derived eicosanoids regulate leukocyte-endothelial adherence in response to aspirin (Fiorucci et al. 2003). This mechanism itself could be potentially related to cancer biology as increasingly more aspects regulating normal physiological properties in cells are being adapted by cancer cells. On this note, cell mimicry of cancer cells has been reported, which is implicated in epithelial-mesenchymal transition and in the phenomenon of vascular mimicry. Cancer cells can acquire a geno-phenotype closely resembling platelets and express several megakaryocytic genes including the adhesion receptor αIIbβ3, thrombin receptor and PECAM/CD31, and platelet-type 12-LOX (Timar et al. 2005). Platelet mimicry of cancer cells is typical of pancreatic, breast, prostate, colorectal and urogenital cancers and melanoma. In these cancer types, increased tumor expression of 12-LOX may regulate platelet mimicry and aid them in their haematogenous dissemination and invasion.

6 Anti-angiogenic Targeting of COX/LOX Pathways: Pharmacological Inhibitors, Natural Agents and Dual Targeting Approaches

The last number of years has seen several angiogenesis inhibitors being used as treatments for cancer, which are usually combined with traditional chemotherapeutic approaches to extend progression-free survival. Tumor endothelial cells (ECs) are particularly targets for cancer therapy as they play a major role in angiogenesis. ECs are easily accessible for intravenously administered drugs, are more genetically stable than cancer cells and less-susceptible to drug-induced resistance, demonstrating the suitability of angiogenesis inhibition for cancer treatment (Griffioen and Molema 2000). The majority of anti-angiogenic drugs approved to date involve the VEGF pathway as the single or primary target. However, drug resistance always develops, probably due to targeting a single pathway alone. In support of this, extended blockade of VEGF alone leads to tumor revascularization, which is dependent on the action of other angiogenic factors, such as FGF (Casanovas et al. 2005). Despite their successes, the limitations of anti-VEGF drugs highlight the need to uncover alternative or complementary approaches to blocking tumor angiogenesis and improving the effects of existing treatment regimens.

6.1 Pharmacological Inhibition of COX and LOX Pathways as Anti-angiogenic Therapy

One of the first links between COX-2 and angiogenesis was established during studies of the anti-tumor effects of existing COX inhibitors. In 1997, Seed et al observed that non-selective COX inhibition with diclofenac suppressed the growth of COX-2 positive colon tumor cells *in vivo* by blocking angiogenesis (Seed et al. 1997). Later *in vivo* studies revealed that selective COX-2 inhibitors block angiogenesis, an effect which was reversed by addition of a TP agonist (Daniel et al. 1999). Treatment with NSAIDs such as sulindac inhibited angiogenesis in tumor stroma *in vivo*. This was characterised by an inhibition in cell growth, as well as PGE_2 and VEGF production (Segawa et al. 2009). Treatment with both NSAIDs and selective COX-2 inhibitors significantly reduced microvessel density *in vivo*, while expression of both VEGF and FGF-1 was also reduced. Microvessel density (MVD) was also positively correlated with tumor volume, lending support to this hypothesis (Wu et al. 2005). This reduction in MVD was associated increased tumor endothelial cell apoptosis (Raut et al. 2004; Wu et al. 2005). Furthermore, the selective COX-2 inhibitor, celecoxib, also significantly inhibited both PGE_2 and TXB_2 production *in vivo* (Leahy et al. 2002). In an implantation tumor model of colon cancer, treatment with the selective COX-2 inhibitor, celecoxib, reduced PGE_2 levels and angiogenesis (CD-34 staining for microvessels). Treatment with high concentrations of celecoxib was also associated with significantly reduced levels of VEGF and MMP-2, while microvessel density in tumor tissue was strongly associated with PGE_2, VEGF and MMP-2 levels (Wang et al. 2008). The selective COX-2 inhibitor, rofecoxib, was also shown to negatively regulate angiogenesis, although no effects on apoptosis or metastasis were observed (Fenwick et al. 2003).

Selective COX-1 inhibition has also been investigated as an anti-angiogenic approach for cancer treatment, although studies are limited. In a mouse model of ovarian cancer, the effects of non-selective COX inhibition with ibuprofen on angiogenic parameters were similar to those observed following selective COX-1 inhibition (Li et al. 2009).

A number of *in vitro* and in *vivo* reports indicate anti-angiogenic effects following the inhibition of COX-derived prostanoids (Table 1). Selective inhibition of COX-2, or non-selective inhibition of cyclooxygenases or EP 1/2, significantly reduced the migratory and invasive potential of breast cancer cells (Rozic et al. 2001). Treatment of human intestinal microvascular endothelial cells (HIMEC) with the HDAC inhibitor sodium butyrate, inhibited angiogenic parameters as well as levels of COX-2, PGE_2 and PGI_2 (Ogawa et al. 2003). However, an angiogenic role for PGI_2 has also been identified using a murine corneal model. PGI_2 analogs such as iloprost and carboxyprostacyclin were able to act on nuclear PPARs to induce angiogenesis *in vivo*, an effect which was associated with increased VEGF expression (Pola et al. 2004). Interestingly,

Table 1: Pharmacological and natural inhibitors of COX (A) and LOX (B) signaling pathways that possess anti-angiogenic activity in vitro or in vivo. While the role of both NSAIDs and selective COX-2 inhibitors has been well investigated as anti-angiogenic and anti-cancer agents, less is known about their downstream prostanoid effectors. Furthermore, studies examining the role of LOX as an anti-angiogenic treatment for cancer are limited. While the majority of studies of these agents in tumor angiogenesis have been single agent studies, dual anti-angiogenic targeting has also been examined, either in combination with chemotherapy or other novel inhibitors. ↑, increased; ↓, inhibited.

A

Eicosanoid target	Agent	Molecular target	Type of study	References
PGD$_2$	15-dPGJ$_2$	PPARγ ↑	Cell line	Xin et al., 1999
	15-dPGJ$_2$	PPARγ ↑	Cell line	Fu et al., 2006
	15-dPGJ$_2$	PPARγ ↑	Cell line	Ho et al., 2008
PGE$_2$	Olive oil	PGES↓	Xenograft model	Terzuoli et al., 2010
	Sodium butyrate	PGE$_2$↓	Cell line	Ogawa et al., 2003
	Curcumin	PGE$_2$↓	Cell line	Binion et al., 2008
	AH6809	EP1/EP2↓	Cell line	Rozic et al., 2001
	Ciglitazone & chemotherapy	PPARγ	Xenograft model	Yokoyama et al.
	COX/EP & HDAC inhibitors	HDAC/PGE$_2$	Cell line	Wang et al.
PGF$_{2\alpha}$	PGF$_{2\alpha}$	PGF$_{2\alpha}$ ↑	Cell line	Qualtrough et al., 2007
PGI$_2$	Tranylcypromine	PGIS↓	Xenograft model	Pradono et al., 2002
	Iloprost	PGI$_2$↑	Murine model	Pola et al., 2004
	Carbaprostacyclin	PGI$_2$↑	Murine model	Pola et al., 2004
	Sodium butyrate	PGI$_2$↓	Cell Line	Ogawa et al., 2003
	RO1138452	IP↓	Cell line	Turner et al., 2011
TXA$_2$	Furegrelate	TXS↓	Cell line	Moussa et al., 2005/08
	Ozagrel	TXS↓	Cell line	Moussa et al., 2005/08
	Ozagrel	TXS↓	Murine model	Yokayama et al., 1995
	BM-567	TXA$_2$↓	Cell line	de Leval et al., 2006
	BM-573	TXA$_2$↓	Cell line	de Leval et al., 2006
	PTXA$_2$	TP↓	Cell line	Moussa et al., 2005
	SQ29548	TP↓	Cell line	Moussa et al., 2005
	Seratrodast	TP↓	Xenograft model	Pradono et al., 2002

B

5-LOX	Unknown	5-LOX ↓	Murine model	Jeong et al., 2010
12-LOX	Baicalein	12-LOX ↓	CAM assay/cell line	Liu et al., 2003
	Baicalein	12-LOX ↓	Cell line/Xenograft	Chiu et al., 2011
	Baicalein	12-LOX ↓	Cell line	Wu et al., 2011
	Baicalein	12-LOX ↓	Cell line	Ling et al., 2011
5-LOX /COX-2	Resveratrol	COX-2 and 5-LOX ↓	Cell line/co-culture	Trapp et al., 2010
		COX-2 and 5-LOX ↓	Xenograft model	Yu et al., 2010
COX-2 & 5-LOX	Celecoxib & Zileuton & chemotherapy	COX-2 and 5-LOX ↓	Phase II trial	Edelman et al., 2008

cicaprost, a PGI_2 analog that acts on IP only demonstrated no anti-angiogenic activity *in vivo*, although a more recent study demonstrated that it increased endothelial cell migration and tube formation *in vitro* (Turner et al. 2011). Angiogenic endothelial cells within tumors strongly express the PGD_2 receptor, DP. The synthetic DP agonist, BW245C, significantly reduces tumor growth and hyperpermeability in wild-type mice. In an *in vitro* assay, BW245C strongly improved endothelial barrier function via increased cAMP production, suggesting that this pathway may be a novel regulator of vascular permeability in cancer (Murata et al. 2008). PPARγ activation with 15-d-PGJ_2 inhibits tube formation, with an associated reduction in levels of VEGFR1, VEGFR2 and uPA and increase in PAI-1 *in vitro* (Xin et al. 1999). It has been demonstrated that 15d-PGJ_2 induces vascular endothelial cell apoptosis *in vitro* and *in vivo*, supporting the potential of this PPARγ ligand as an anti-angiogenesis agent (Ho et al. 2008). Modulators of thromboxane activity have also been investigated as anti-angiogenic agents (de Leval et al. 2006). Treatment with TXA_2 inhibitors dose-dependently inhibited endothelial cell migration and VEGF, but had no effect on cell adhesion and viability *in vitro*. Angiogenic parameters were significantly inhibited following treatment with TXS inhibitors or TP antagonists, both *in vitro* and *in vivo* (Nie et al. 2000). Both TXS and TP inhibition inhibited migration and invasion of bladder cancer cells, while TP agonists demonstrated opposing effects (Moussa et al. 2005).

Isoforms of liopxygenases share the same type of protein folding however their molecular interactions vary from isoform to isoform due to differences in size, shape and mode of interaction of the catalytic entity of the substrate binding channels. With respect to pharmacological inhibitors of 5-LOX, the mechanism of action is normally via redox mechanisms, iron-chelating effects or non-redox-related actions. The only FDA approved LOX inhibitor on the market is zileuton which targets 5-LOX and a growing body of evidence supports its chemopreventative roles in cancer. Zileuton has been shown to reduce cell viability and induce apoptosis in a concentration- and time-dependent manner in hepatocellular carcinoma (Xu et al. 2011). 5-LOX is up-regulated in adenomatous colon polyps and cancer compared with normal colonic mucosa and Rev-5901 an inhibitor of 5-LOX inhibits colon cancer cell proliferation both *in vitro* and *in vivo* indicating its potential as a chemopreventive therapy in colon cancer (Melstrom et al. 2008). Interestingly, a more recent study demonstrated that delivery of zileuton by inhalation reduced pulmonary adenomas in the A/J mouse model (Myrdal et al. 2007; Karlage et al. 2010). Another drug with proven selectivity for 12-LOX, baicalein, has its origins in Chinese herbal medicine and has been shown to directly inhibit the proliferation, invasiveness/migration, and induce apoptosis of a variety of tumor cells (Pidgeon et al. 2002; Chiu et al. 2011; Takahashi et al. 2011; Wu et al. 2011). Baicalein has been shown to exert its action in an anti-angiogenic manner, by down-regulating the expression of VEGF and altering integrin expression on both tumor and endothelial cells. Another compound recognized for its pan-LOX inhibitory activity is nordihydroguaiaretic acid (NDGA), which has been used in multiple intervention studies but, like baicalein, has not been licensed for application in humans.

6.2 The Role of Natural Agents in COX and LOX Modulation and Tumor Angiogenesis

A number of natural agents have also demonstrated anti-angiogenic potential in experimental tumor models (Table 1). Curcumin is a natural agent, derived from the spice turmeric. It exhibits anti-inflammatory and anti-carcinogenic properties and also inhibits expression of COX-2 PGES, PGE_2 and 5-LOX (Hong et al. 2004; Koeberle et al. 2009). Curcumin treatment significantly reduced tumor neocapillary density *in vivo* as well as tumor-induced over-expression of COX-2 and serum VEGF in hepatocellular carcinoma cell-implanted nude mice (Yoysungnoen et al. 2006). It was subsequently demonstrated that curcumin inhibits VEGF-mediated angiogenesis in HIMECs (human microvascular endothelial cells) through inhibition of COX-2, PGE_2 and MAP kinase (Binion et al. 2008). These studies suggest that the anti-angiogenic effects of curcumin may be mediated through reduced levels of angiogenic factors COX-2 and VEGF. A product of olive oil, dihydroxyphenylethanol (DPE), has also been shown to have anti-angiogenic effects *in vitro*. DPE also prevented the IL-1β-mediated increase of mPGES-1 expression and PGE_2 generation, as well as the mPGES-1 dependent expression of VEGF, suggesting that its effects are mediated via modulation of the mPGES-1/VEGF axis (Terzuoli et al. 2010).

In addition to the array of pharmacological agents under investigation for their effect on LOX metabolism, a number of naturally occurring agents that may interact with lipoxygenase activity are constantly being discovered, such as dietary carotenoids including lycophyll (present in tomatoes) that may modify the enzymatic function of 5-LOX through binding to the catalytic site with high affinity (Hazai et al. 2006). There are also numerous plant phenols including baicalein, kaempferol, quercetin, nordihydroguaretic acid and resveratrol which are known for their chemopreventative and anti-angiogenic as well as LOX-inhibitory activity in different models (Varinska et al. 2010).

The 12-LOX inhibitor, baicalein, a flavonoid isolated from an important medicinal plant Scutellariae Radix (the root of *Scutellaria baicalensis Georgi*) has been shown to induce apoptosis of various cancer cells, but also has a direct anti-angiogenic effects by inhibiting endothelial cell proliferation, differentiation and migration (Liu et al. 2003). A recent study by Ling et al indicates that baicalein potently suppresses angiogenesis in human umbilical vein endothelial cells induced by VEGF through the p53/Rb signaling pathway resulting in a G1/S cell cycle arrest (Ling et al. 2011.

Resveratrol, trans- 3,5,4'-trihydroxystilbene, was first isolated in 1940 as a constituent of the roots of white hellebore (*Veratrum grandiflorum O. Loes*), but has since been found in various plants, including grapes, berries and peanuts. Resveratrol has been shown to suppress the expression of 5-LOX and COX-2, and to inhibit the proliferation of a wide variety of tumor cells, including lymphoid and myeloid cancers; multiple myeloma; cancers of the breast, prostate, stomach, colon, pancreas and thyroid; melanoma; head and neck squamous cell carcinoma; ovarian carcinoma; and cervical carcinoma (Aggarwal et al. 2004). The growth inhibitory effects of resveratrol are mediated through cell cycle arrest; up-regulation of p21Cip1/WAF1, p53 and Bax; down-regulation of survivin, cyclin D1, cyclin E, Bcl-2, BclxL and cIAPs; and activation of caspases. It has been shown to have direct anti-angiogenic activity through the suppression of activation of several transcription factors, including NF-κB (Yu et al. 2010), AP-1 and Egr-1, to inhibit protein kinases including IkappaBalpha kinase, JNK, MAPK, Akt, PKC, PKD and casein kinase II; and to down-regulate products of genes other than 5-LOX including COX-2, VEGF, IL-1, IL-6, IL-8, AR and PSA. A recent study by Trapp et al indicates that resveratrol stimulated proliferation of isolated vascular endothelial cells (VECs). However, it caused growth inhibition of VECs grown with melanoma cells in three-dimensional co-culture. This effect was associated with increased melanoma cell expression of tumor suppressor protein 53 and matrix protein TSP1, as well as decreased hypoxia-driven expression of hypoxia inducible factor-1α and inhibition of VEGF production (Trapp et al. 2010). This study highlights the importance of further assessing the anti-angiogenic potential of compounds in tissue explants, co-culture or *in vivo* experimental conditions as opposed to endothelial cell cultures alone.

6.3   Dual Targeting of COX and LOX Pathways as Novel Anti-angiogenic Therapeutic Approaches

A concern in the early days of anti-angiogenic drug development was that these drugs would not be useful for combining with traditional treatments of chemotherapy and/or radiation therapy. By compromising blood flow/perfusion anti-angiogenic drugs would deprive tumors of oxygen, thus increasing levels of tumor hypoxia, a resistance factor to radiation and chemotherapy. However, a number of years ago, Teicher et al. reported the first of a series of studies showing that the anti-tumor effects of chemotherapy on transplantable mouse tumors were actually augmented when combined with a drug known to have anti-angiogenic properties (Teicher et al. 1992).

The inhibitory effects of rofecoxib on endothelial cell functions (cell proliferation, migration, tube formation) were increased two-fold when combined with ionizing radiation *in vitro* (Dicker et al. 2001). Combined treatment of glioblastoma cells with celecoxib and radiation significantly reduced tumor MVD *in vivo*. A reduction in angiopoietin-1 and VEGF and an increase in angiopoietin-2 protein expression were also observed, implicating these proteins in the angiogenic response (Kang et al. 2007). Non-selective COX inhibition with ibuprofen decreased cell proliferation and tube formation *in vitro* and also enhanced the anti-tumor effects of chemotherapy with irinotecan *in vivo* (Yao et al. 2005). More recently, combined treatment with chemotherapy and PPARγ ligands has been investigated in an ovarian cancer mouse model. Combined treatment reduced MVD and VEGF as well as expression of COX-2, microsomal PGE synthase and EP3 *in vivo*, suggesting a novel anti-angiogenic strategy for the treatment of ovarian cancer (Yokoyama et al. 2011). Combined treatment of the traditional chemotherapeutic agent, Paclitaxel, with selective COX-2 inhibition resulted in increased anti-endothelial cell effects relative to each agent alone (Merchan et al. 2005). In mice transplanted with bladder cancer cells, treatment with a TP antagonist alone, or in combination with cisplatin, significantly delayed tumor development and increased survival when compared with vehicle and cisplatin alone (Moussa et al. 2008). This suggests that this combined therapy is an attractive anti-angiogenic and anti-tumor strategy and warrants further clinical investigation.

The anti-angiogenic effects of cyclooxygenase inhibition in lung cancer cells were also enhanced when combined with histone deacetylase inhibitors. This was characterized by increased endothelial cell proliferation and capillary formation (Wang et al. 2011). Both 5-LOX and COX-2 inhibitors reduced platelet activating factor-induced angiogenesis in an *in vivo* mouse model, with a concomitant reduction in pulmonary expression of MMP-2 and MMP-9 (Jeong et al. 2010). Combined treatment with a selective COX-2 inhibitor and a 5-LOX inhibitor demonstrated additive effects in reducing tumor growth in mouse models of colon, oesophageal, breast and skin cancer (Chen et al. 2004; Ye et al. 2005; Barry et al. 2009; Fegn and Wang 2009), while dual inhibition of these pathways also reduced liver metastases in an animal model of pancreatic cancer (Wenger et al. 2002). Both COX and LOX pathway inhibitors have been found to enhance the efficacy of traditional chemotherapeutic agents *in vitro* (Moussa et al. 2008). However, clinical trials of combined selective COX-2 and 5-LOX inhibition have proved disappointing. In a phase II trial of advanced NSCLC, no benefit was observed with COX-2 and 5-LOX inhibition alone or combined in addition to chemotherapy. However, subset analysis did suggest some advantage for celecoxib in combination with chemotherapy in patients with moderate to high COX-2 expression (Edelman et al. 2008). A study of VEGF levels on the same trial revealed no correlation with COX-2 or 5-LOX levels. VEGF levels were also unchanged following treatment with COX-2 and/ or 5-LOX inhibitors, suggesting no effect on tumor angiogenesis in this trial (Edelman et al. 2011). Results from this trial continue to be updated. Licofelone is a dual COX and 5-LOX inhibitor, which is in the most advanced phase of clinical development among this group of compounds. While the anti-angiogenic potential of this drug has not yet been tested, it has demonstrated a slightly better efficacy than naproxen with a much better gastrointestinal safety in osteoarthritis studies (Celotti and Durand 2003). A number of other clinically-studied dual COX/LOX inhibitors exist, which have not been assessed as anti-angiogenics, but have demonstrated anti-inflammatory activity (Inagaki et al. 2000; Le Filliatre et al. 2001; Leval et al. 2002; Moon et al. 2005). As both of these pathways have been strongly implicated in tumor angiogenesis, it will be of interest to determine if dual inhibition induces an additive, anti-angiogenic effect.

7   Novel Methodological Approaches for Anti-angiogenic Drug Discovery/Development

The most well-known approach developed to target tumor angiogenesis includes drugs that neutralize pro-angiogenic growth factors such as VEGF, or block signaling through VEGFRs. Currently, all approved anti-angiogenic drugs either target VEGF or VEGFR tyrosine kinase receptors. The concept of 'accidental' angiogenic inhibitors for the treatment of cancer refers to the idea that many anticancer drugs, both old and new, not developed with the intention of inhibiting angiogenesis, may in fact do so, thus contributing to their overall anti-tumor effects (Kerbel et al. 2000). An anti-angiogenic effect was even observed with traditional chemotherapeutic agents (Miller et al. 2001).

Many *in vitro* and *in vivo* methods currently exist to assess angiogenesis. *In vitro* assays for measuring angiogenesis mainly use isolated endothelial cells and allow reproducible measurement of a portion of the angiogenic process. There are also assays which incorporate multiple cell types, such as the aortic ring assay (Nicosia et al. 1982). However, these assays focus only on early steps in angiogenesis do not take into account the remodeling and stabilization which occur once flow begins in a vessel. While cell lines are well utilized in angiogenesis research, results from cell lines rarely correlate with the anti-angiogenic potential of these drugs *in vivo*. There are a number of other *in vivo* assays to measure angiogenesis. The oldest of these use transparent chambers to allow growing blood vessels to be visualized microscopically (Sandison 1924). More recent assays measuring vessel in-growth into subcutaneous implants such as matrigel, modulating vessel formation in normal tissues such as mouse ear and chicken chorioallantoic membrane (CAM), or inducing vessel growth in normally avascular tissues such as the cornea (Jain et al. 1997). The latter assay has an advantage over other *in vivo* assays in that the cornea is initially avascular. This model was originally developed in the rabbit eye, but has recently evolved by changing species to the smaller, less expensive, and genetically tractable mouse (Kenyon et al. 1996). The limitations of this assay, however, include the technically demanding nature of the surgery. Current models of *in vivo* evaluation of tumor angiogenesis generally consist of xenograft implantation of human cancer cell lines into immunodeficient mice. However, these tumor cells represent an extreme deviation from advanced cancers *in vivo*, and are not associated with the tumor stroma, which is a crucial component in tumor angiogenesis and subsequent metastasis (Bhowmick et al. 2004). Generally, xenograft models have limited ability to predict clinical efficacy of anti-angiogenic anti-cancer agents.

Tumor explants and three-dimensional models allow cell:cell and cell: matrix interactions to be examined in live tissue and cells, with endpoints including the measurement of angiogenic gene and protein expression as well as image analysis. More recently orthotopic xenograft of human tumors into nude mice has provided the ability to reproduce both the histology as well as the angiogenic and metastatic pattern of most human cancers at advanced stage (Cespedes et al. 2006). While not useful in examining the contribution of the immune system in this process, this model is more promising than commonly used subcutaneous (SC) xenografts in preclinical drug screening and development. This model allows the evaluation of therapies in individual human tumors derived from different genetic backgrounds, as opposed to the use of inbred animals with a homogeneous genetic background (Cespedes et al. 2006). However, the ability of this model to predict clinical therapeutic response remains to be established.

It is widely accepted that *in vivo* models are critical for defining the mechanism of drug activity and for testing therapeutic regimes. However, only a few models are ideal for this purpose. Zebrafish have become one of the most powerful and versatile models used for biomedical research to identify novel cancer markers/molecular signatures of disease (Amatruda et al. 2002). They are advantageous for drug screening as they account for the complex metabolism that affects drug efficacy or causes toxicity. Since chemicals can be directly delivered into the fish water and proteins can be injected, assessment of cytotoxic, apoptotic or anti-angiogenic effects of potential drug candidates singly, or in combination, is easy and straightforward (Parng et al. 2002; Kari et al. 2007). The transparency of zebrafish also allows direct imaging of mechanisms of cancer progression including cell invasion, intra-/extravasation, and angiogenesis (Stoletov et al. 2007; White et al. 2008). The development of xenograft zebrafish models has allowed the propagation and visualization of human cancer cells engrafted into optically transparent zebrafish (Marques et al. 2009). To date, zebrafish are the only animal organism to date with the potential for large scale, yet cost effective pharmacological screens to identify potential therapies to alter tumor dissemination and/or angiogenesis (Parng et al. 2002). Our group is using a zebrafish model to assess the anti-angiogenic properties of inhibitors of the TXS signaling pathway *in vivo*. While the effect of TXS pathway inhibitors in tumorigenesis has been investigated by our group and others (Jantke et al. 2004; Moussa et al. 2005; Sakai et al. 2006; Cathcart et al. 2011), the anti-angiogenic properties of these inhibitors are not fully understood. Using a transgenic line of zebrafish Tg (fli1:EGFP), which specifically expresses green fluorescent protein in vessels; we examined the effect of TXS-pathway inhibition on the morphology of intersegmental vessels (Fig. 5A). Selective TXS inhibition with ozagrel (10 µM) demonstrated clear effects on intersegmental vessel morphology following only 48 hour incubation (Fig. 5B), suggesting an anti-angiogenic mechanism for this drug. Similar anti-angiogenic effects on intersegmental vessel formation were observed following treatment with both a dual TXS/ EP4 inhibitor and a dual TXS/5-LOX inhibitor (Fig. 5B). These observations support the hypothesis that a dual targeting approach may further enhance the anti-angiogenic potential of single agent targeting alone. These fish can also be used to study the biology and pharmacology of human tumor *in*

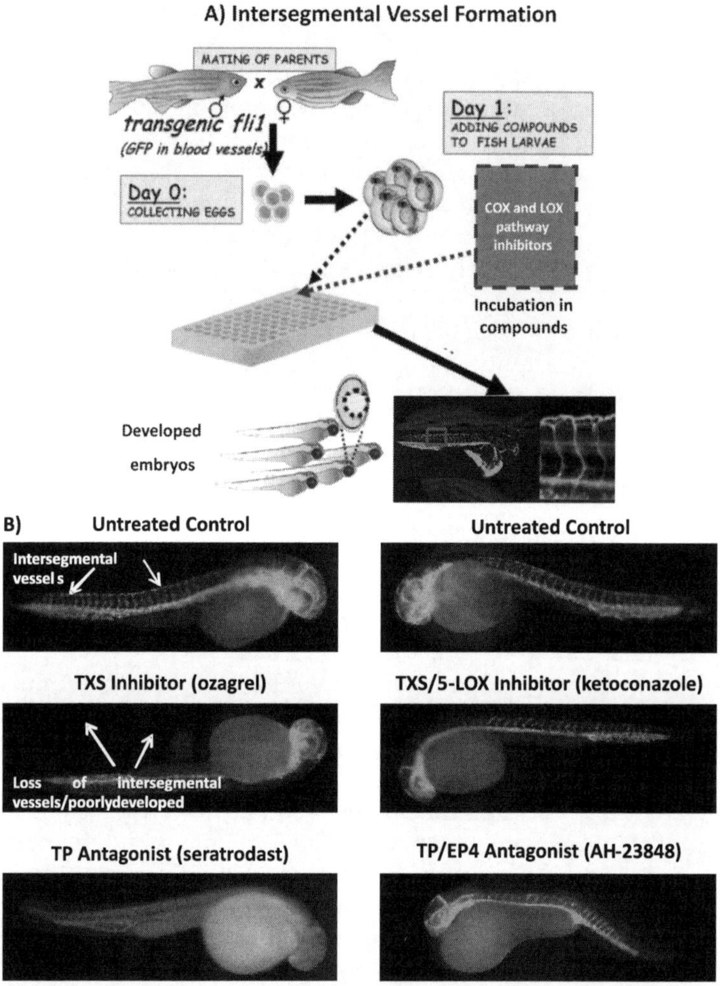

Figure 5 **Zebrafish is a pharmacological screening tool to identify novel anti-cancer therapies.** A transgenic line of zebrafish Tg (fli1:EGFP), which specifically expresses green fluorescent protein in vessels may be used to examine the effect of potential anti-cancer compounds on intersegmental vessel morphology. Zebrafish are mated and eggs collected for screening. Approximately 6 h post-fertilization, developing larvae are added to culture plates (5 per well in duplicate). Screening compounds may be added directly to the zebrafish water for up to 4-days. Imaging and quantification of vessels is carried out by fluorescent microscopy (A). Selective TXS inhibition with ozagrel (10 μM) demonstrated clear effects on intersegmental vessel morphology following only 48 h incubation (B), suggesting an anti-angiogenic mechanism for this drug. Similar effects on intersegmental vessel formation were observed following TP antagonism, dual TXS/EP4 inhibition, and dual TXS/5-LOX inhibition. The morphological differences observed with dual TXS/EP4 inhibitor suggest potential developmental abnormalities and would suggest a titration of the drug is warranted.

Color image of this figure appears in the color plate section at the end of the book.

vivo, by growing human tumor cells or tissue in zebrafish larvae (Marques et al. 2009). These human-to-fish tumor xenografts enable the tumor, and its response to drugs to be analyzed *in vivo*.

8 Conclusions and Perspectives

A complex array of cells and stimuli participate in the regulation of tumor angiogenesis. While COX-2 appears to be the main isoform responsible for the promotion of angiogenesis, the action of downstream prostanoids can also regulate this process, either positively or negatively. The angiogenic process may be dependent on the balance in the generation of these COX-derived prostanoids in any particular scenario. COX-2, 5-LOX and 12-LOX are clearly over-expressed in many cancers and associated with a poor prognosis. While COX-2 has also been associated with tumor angiogenesis, the exact mechanisms underlying this effect are still unclear, although downstream prostanoids such as PGE_2, PGI_2, and TXA_2 have been strongly implicated in the pro-angiogenic effects of COX-2. Similarly both 5 and 12-LOX have been associated with the production of pro-angiogenic factors including VEGF and have been shown to promote various stages of angiogenesis. While COX-2 and its downstream prostanoids have been associated with VEGF, it is likely that multiple pathways are involved; suggesting that single enzyme inhibition would be insufficient. This is supported by the promising data from pharmaceutical agents targeting both the COX and LOX pathways.

Novel *in vivo* angiogenesis assays, such as and zebrafish models (intersegmental vessel screening, xenotransplantation) and explant culturing, may facilitate rapid anti-angiogenic drug discovery for COX/LOX pathway inhibitors with the promise of effective translation into human clinical studies. With respect to new strategies, numerous possibilities exist; among them is the combining of anti-angiogenic drugs with therapeutic modalities other than chemotherapy. Many other types of combination treatment involving anti-angiogenic drugs are currently under preclinical and clinical investigation (Kottke et al. 2010). Finally, identifying patients likely to respond to current anti-angiogenic therapies should enhance personalized cancer treatments and ultimately improve outcomes in this patient cohort, while enabling other patients to be stratified for alternative molecular targeted anti-neoplastic approaches.

REFERENCES

Abdel-Majid, R. M. and J. S. Marshall (2004). "Prostaglandin E2 induces degranulation-independent production of vascular endothelial growth factor by human mast cells." J Immunol 172(2): 1227-1236.

Achiwa, H., Y. Yatabe, T. Hida, T. Kuroishi, K. Kozaki, S. Nakamura, M. Ogawa, T. Sugiura, T. Mitsudomi and T. Takahashi (1999). "Prognostic significance of elevated cyclooxygenase 2 expression in primary, resected lung adenocarcinomas." Clin Cancer Res 5(5): 1001-1005.

Aggarwal, B. B., A. Bhardwaj, R. S. Aggarwal, N. P. Seeram, S. Shishodia and Y. Takada (2004). "Role of resveratrol in prevention and therapy of cancer: preclinical and clinical studies." Anticancer Res 24(5A): 2783-2840.

Alfranca, A., M. A. Iniguez, M. Fresno and J. M. Redondo (2006). "Prostanoid signal transduction and gene expression in the endothelium: role in cardiovascular diseases." Cardiovasc Res 70(3): 446-456.

Amano, H., I. Hayashi, H. Endo, H. Kitasato, S. Yamashina, T. Maruyama, M. Kobayashi, K. Satoh, M. Narita, Y. Sugimoto, T. Murata, H. Yoshimura, S. Narumiya and M. Majima (2003). "Host prostaglandin E(2)-EP3 signaling regulates tumor-associated angiogenesis and tumor growth." J Exp Med 197(2): 221-232.

Amatruda, J. F., J. L. Shepard, H. M. Stern and L. I. Zon (2002). "Zebrafish as a cancer model system." Cancer Cell 1(3): 229-231.

Antonarakis, E. S., E. I. Heath, J. R. Walczak, W. G. Nelson, H. Fedor, A. M. De Marzo, M. L. Zahurak, S. Piantadosi, A. J. Dannenberg, R. T. Gurganus, S. D. Baker, H. L. Parnes, T. L. DeWeese, A. W. Partin and M. A. Carducci (2009). "Phase II, randomized, placebo-controlled trial of neoadjuvant celecoxib in men with clinically localized prostate cancer: evaluation of drug-specific biomarkers." J Clin Oncol 27(30): 4986-4993.

Arisato, T., T. Hashiguchi, K. P. Sarker, K. Arimura, M. Asano, K. Matsuo, M. Osame and I. Maruyama (2003). "Highly accumulated platelet vascular endothelial growth factor in coagulant thrombotic region." J Thromb Haemost 1(12): 2589-2593.

Asano, K., H. Nakamura, C. M. Lilly, M. Klagsbrun and J. M. Drazen (1997). "Interferon gamma induces prostaglandin G/H synthase-2 through an autocrine loop via the epidermal growth factor receptor in human bronchial epithelial cells." J Clin Invest 99(5): 1057-1063.

Ashton, A. W., Y. Cheng, A. Helisch and J. A. Ware (2004). "Thromboxane A_2 receptor agonists antagonize the proangiogenic effects of fibroblast growth factor-2: role of receptor internalization, thrombospondin-1, and alpha(v)beta3." Circ Res 94(6): 735-742.

Ashton, A. W. and J. A. Ware (2004). "Thromboxane A_2 receptor signaling inhibits vascular endothelial growth factor-induced endothelial cell differentiation and migration." Circ Res 95(4): 372-379.

Ashton, A. W., R. Yokota, G. John, S. Zhao, S. O. Suadicani, D. C. Spray and J. A. Ware (1999). "Inhibition of endothelial cell migration, intercellular communication, and vascular tube formation by thromboxane A(2)." J Biol Chem 274(50): 35562-35570.

Barry, M., R. A. Cahill, G. Roche-Nagle, T. G. Neilan, A. Treumann, J. H. Harmey and D. J. Bouchier-Hayes (2009). "Neoplasms escape selective COX-2 inhibition in an animal model of breast cancer." Ir J Med Sci 178(2): 201-208.

Bhowmick, N. A., E. G. Neilson and H. L. Moses (2004). "Stromal fibroblasts in cancer initiation and progression." Nature 432(7015): 332-337.

Binion, D. G., M. F. Otterson and P. Rafiee (2008). "Curcumin inhibits VEGF-mediated angiogenesis in human intestinal microvascular endothelial cells through COX-2 and MAPK inhibition." Gut 57(11): 1509-1517.

Biscetti, F., E. Gaetani, A. Flex, G. Straface, G. Pecorini, F. Angelini, E. Stigliano, T. Aprahamian, R. C. Smith, J. J. Castellot and R. Pola (2009). "Peroxisome proliferator-activated receptor alpha is crucial for iloprost-induced *in vivo* angiogenesis and vascular endothelial growth factor upregulation." J Vasc Res 46(2): 103-108.

Bishop-Bailey, D. (2000). "Peroxisome proliferator-activated receptors in the cardiovascular system." Br J Pharmacol 129(5): 823-834.

Bogatcheva, N. V., M. G. Sergeeva, S. M. Dudek and A. D. Verin (2005). "Arachidonic acid cascade in endothelial pathobiology." Microvasc Res 69(3): 107-127.

Bombardier, C., L. Laine, A. Reicin, D. Shapiro, R. Burgos-Vargas, B. Davis, R. Day, M. B. Ferraz, C. J. Hawkey, M. C. Hochberg, T. K. Kvien and T. J. Schnitzer (2000). "Comparison of upper gastrointestinal toxicity of rofecoxib and naproxen in patients with rheumatoid arthritis. VIGOR Study Group." N Engl J Med 343(21): 1520-1528, 1522 p following 1528.

Brabender, J., J. Park, R. Metzger, P. M. Schneider, R. V. Lord, A. H. Holscher, K. D. Danenberg and P. V. Danenberg (2002). "Prognostic significance of cyclooxygenase 2 mRNA expression in non-small cell lung cancer." Ann Surg 235(3): 440-443.

Brash, A. R., W. E. Boeglin and M. S. Chang (1997). "Discovery of a second 15S-lipoxygenase in humans." Proc Natl Acad Sci USA 94(12): 6148-6152.

Bresalier, R. S., R. S. Sandler, H. Quan, J. A. Bolognese, B. Oxenius, K. Horgan, C. Lines, R. Riddell, D. Morton, A. Lanas, M. A. Konstam and J. A. Baron (2005). "Cardiovascular events associated with rofecoxib in a colorectal adenoma chemoprevention trial." N Engl J Med 352(11): 1092-1102.

Buchanan, F. G., W. Chang, H. Sheng, J. Shao, J. D. Morrow and R. N. DuBois (2004). "Up-regulation of the enzymes involved in prostacyclin synthesis via Ras induces vascular endothelial growth factor." Gastroenterology 127(5): 1391-1400.

Buchanan, F. G., D. L. Gorden, P. Matta, Q. Shi, L. M. Matrisian and R. N. DuBois (2006). "Role of beta-arrestin 1 in the metastatic progression of colorectal cancer." Proc Natl Acad Sci USA 103(5): 1492-1497.

Buchanan, F. G., D. Wang, F. Bargiacchi and R. N. DuBois (2003). "Prostaglandin E2 regulates cell migration via the intracellular activation of the epidermal growth factor receptor." J Biol Chem 278(37): 35451-35457.

Casanovas, O., D. J. Hicklin, G. Bergers and D. Hanahan (2005). "Drug resistance by evasion of antiangiogenic targeting of VEGF signaling in late-stage pancreatic islet tumors." Cancer Cell 8(4): 299-309.

Catalano, A. and A. Procopio (2005). "New aspects on the role of lipoxygenases in cancer progression." Histol Histopathol 20(3): 969-975.

Cathcart, M. C., K. Gately, R. Cummins, E. Kay, O. B. KJ and G. P. Pidgeon (2011). "Examination of thromboxane synthase as a prognostic factor and therapeutic target in non-small cell lung cancer." Mol Cancer 10(1): 25.

Cathcart, M. C., J. Lysaght and G. P. Pidgeon (2011). "Eicosanoid signalling pathways in the development and progression of colorectal cancer: novel approaches for prevention/intervention." Cancer Metastasis Rev 30(3-4): 363-385.

Cathcart, M. C., R. Tamosiuniene, G. Chen, T. G. Neilan, A. Bradford, K. J. O'Byrne, D. J. Fitzgerald and G. P. Pidgeon (2008). "Cyclooxygenase-2-linked attenuation of hypoxia-induced pulmonary hypertension and intravascular thrombosis." J Pharmacol Exp Ther 326(1): 51-58.

Celotti, F. and T. Durand (2003). "The metabolic effects of inhibitors of 5-lipoxygenase and of cyclooxygenase 1 and 2 are an advancement in the efficacy and safety of anti-inflammatory therapy." Prostaglandins Other Lipid Mediat 71(3-4): 147-162.

Cespedes, M. V., I. Casanova, M. Parreno and R. Mangues (2006). "Mouse models in oncogenesis and cancer therapy." Clin Transl Oncol 8(5): 318-329.

Chang, S. H., C. H. Liu, R. Conway, D. K. Han, K. Nithipatikom, O. C. Trifan, T. F. Lane and T. Hla (2004). "Role of prostaglandin E2-dependent angiogenic switch in cyclooxygenase 2-induced breast cancer progression." Proc Natl Acad Sci USA 101(2): 591-596.

Chatterjee, M., S. Das and K. Roy (2011). "Overexpression of 5-lipoxygenase and its relation with cell proliferation and angiogenesis in 7,12-dimethylbenz(alpha) anthracene-induced rat mammary carcinogenesis." Mol Carcinog Dec. 28 [Epub ahead of print].

Chen, X., S. Wang, N. Wu, S. Sood, P. Wang, Z. Jin, D. G. Beer, T. J. Giordano, Y. Lin, W. C. Shih, R. A. Lubet and C. S. Yang (2004). "Overexpression of 5-lipoxygenase in rat and human esophageal adenocarcinoma and inhibitory effects of zileuton and celecoxib on carcinogenesis." Clin Cancer Res 10(19): 6703-6709.

Cheng, T., W. Cao, R. Wen, R. H. Steinberg and M. M. LaVail (1998). "Prostaglandin E2 induces vascular endothelial growth factor and basic fibroblast growth factor mRNA expression in cultured rat Muller cells." Invest Ophthalmol Vis Sci 39(3): 581-591.

Chiu, Y. W., T. H. Lin, W. S. Huang, C. Y. Teng, Y. S. Liou, W. H. Kuo, W. L. Lin, H. I. Huang, J. N. Tung, C. Y. Huang, J. Y. Liu, W. H. Wang, J. M. Hwang and H. C. Kuo (2011). "Baicalein inhibits the migration and invasive properties of human hepatoma cells." Toxicol Appl Pharmacol 255(3): 316-326.

Cianchi, F., C. Cortesini, P. Bechi, O. Fantappie, L. Messerini, A. Vannacci, I. Sardi, G. Baroni, V. Boddi, R. Mazzanti and E. Masini (2001). "Up-regulation of cyclooxygenase 2 gene expression correlates with tumor angiogenesis in human colorectal cancer." Gastroenterology 121(6): 1339-1347.

Conrad, D. J. (1999). "The arachidonate 12/15 lipoxygenases. A review of tissue expression and biologic function." Clin Rev Allergy Immunol 17(1-2): 71-89.

Cuendet, M. and J. M. Pezzuto (2000). "The role of cyclooxygenase and lipoxygenase in cancer chemoprevention." Drug Metabol Drug Interact 17(1-4): 109-157.

Cuneo, K. C., A. Fu, K. L. Osusky and L. Geng (2007). "Effects of vascular endothelial growth factor receptor inhibitor SU5416 and prostacyclin on murine lung metastasis." Anticancer Drugs 18(3): 349-355.

D'Amato, R. J., M. S. Loughnan, E. Flynn and J. Folkman (1994). "Thalidomide is an inhibitor of angiogenesis." Proc Natl Acad Sci USA 91(9): 4082-4085.

Daneker, G. W., S. A. Lund, S. W. Caughman, C. A. Staley and W. C. Wood (1996). "Anti-metastatic prostacyclins inhibit the adhesion of colon carcinoma to endothelial cells by blocking E-selectin expression." Clin Exp Metastasis 14(3): 230-238.

Daniel, T. O., H. Liu, J. D. Morrow, B. C. Crews and L. J. Marnett (1999). "Thromboxane A_2 is a mediator of cyclooxygenase-2-dependent endothelial migration and angiogenesis." Cancer Res 59(18): 4574-4577.

de Leval, X., T. Dassesse, J. M. Dogne, D. Waltregny, A. Bellahcene, V. Benoit, B. Pirotte and V. Castronovo (2006). "Evaluation of original dual thromboxane A_2 modulators as antiangiogenic agents." J Pharmacol Exp Ther 318(3): 1057-1067.

Di Popolo, A., A. Memoli, A. Apicella, C. Tuccillo, A. di Palma, P. Ricchi, A. M. Acquaviva and R. Zarrilli (2000). "IGF-II/IGF-I receptor pathway up-regulates COX-2 mRNA expression and PGE_2 synthesis in Caco-2 human colon carcinoma cells." Oncogene 19(48): 5517-5524.

Dicker, A. P., T. L. Williams and D. S. Grant (2001). "Targeting angiogenic processes by combination rofecoxib and ionizing radiation." Am J Clin Oncol 24(5): 438-442.

Ding, Y. B., R. H. Shi, J. D. Tong, X. Y. Li, G. X. Zhang, W. M. Xiao, J. G. Yang, Y. Bao, J. Wu, Z. G. Yan and X. H. Wang (2005). "PGE_2 up-regulates vascular endothelial growth factor expression in MKN28 gastric cancer cells via epidermal growth factor receptor signaling system." Exp Oncol 27(2): 108-113.

Dirig, D. M., P. C. Isakson and T. L. Yaksh (1998). "Effect of COX-1 and COX-2 inhibition on induction and maintenance of carrageenan-evoked thermal hyperalgesia in rats." J Pharmacol Exp Ther 285(3): 1031-1038.

Dominguez-Jimenez, C., F. Diaz-Gonzalez, I. Gonzalez-Alvaro, J. M. Cesar and F. Sanchez-Madrid (1999). "Prevention of alphaII(b)beta3 activation by non-steroidal antiinflammatory drugs." FEBS Lett 446(2-3): 318-322.

Dormond, O., M. Bezzi, A. Mariotti and C. Ruegg (2002). "Prostaglandin E2 promotes integrin alpha Vbeta 3-dependent endothelial cell adhesion, rac-activation, and spreading through cAMP/PKA-dependent signaling." J Biol Chem 277(48): 45838-45846.

Dormond, O., A. Foletti, C. Paroz and C. Ruegg (2001). "NSAIDs inhibit alpha V beta 3 integrin-mediated and Cdc42/Rac-dependent endothelial-cell spreading, migration and angiogenesis." Nat Med 7(9): 1041-1047.

Dragovich, T., H. Burris, 3rd, P. Loehrer, D. D. Von Hoff, S. Chow, S. Stratton, S. Green, Y. Obregon, I. Alvarez and M. Gordon (2008). "Gemcitabine plus celecoxib in patients with advanced or metastatic pancreatic adenocarcinoma: results of a phase II trial." Am J Clin Oncol 31(2): 157-162.

Edelman, M. J., L. Hodgson, X. Wang, R. Christenson, S. Jewell, E. Vokes and R. Kratzke (2011). "Serum vascular endothelial growth factor and COX-2/5-LOX inhibition in advanced non-small cell lung cancer: Cancer and Leukemia Group B 150304." J Thorac Oncol 6(11): 1902-1906.

Edelman, M. J., D. Watson, X. Wang, C. Morrison, R. A. Kratzke, S. Jewell, L. Hodgson, A. M. Mauer, A. Gajra, G. A. Masters, M. Bedor, E. E. Vokes and M. J. Green (2008). "Eicosanoid modulation in advanced lung cancer: cyclo-oxygenase-2 expression is a positive predictive factor for celecoxib + chemotherapy – Cancer and Leukemia Group B Trial 30203." J Clin Oncol 26(6): 848-855.

Ermert, L., C. Dierkes and M. Ermert (2003). "Immunohistochemical expression of cyclooxygenase isoenzymes and downstream enzymes in human lung tumors." Clin Cancer Res 9(5): 1604-1610.

Fabre, J. E., J. L. Goulet, E. Riche, M. Nguyen, K. Coggins, S. Offenbacher and B. H. Koller (2002). "Transcellular biosynthesis contributes to the production of leukotrienes during inflammatory responses *in vivo*." J Clin Invest 109(10): 1373-1380.

Faronato, M., G. Muzzonigro, G. Milanese, C. Menna, A. R. Bonfigli, A. Catalano and A. Procopio (2007). "Increased expression of 5-lipoxygenase is common in clear cell renal cell carcinoma." Histol Histopathol 22(10): 1109-1118.

Fegn, L. and Z. Wang (2009). "Topical chemoprevention of skin cancer in mice, using combined inhibitors of 5-lipoxygenase and cyclo-oxygenase-2." J Laryngol Otol 123(8): 880-884.

Fenwick, S. W., G. J. Toogood, J. P. Lodge and M. A. Hull (2003). "The effect of the selective cyclooxygenase-2 inhibitor rofecoxib on human colorectal cancer liver metastases." Gastroenterology 125(3): 716-729.

Fidler, I. J. and L. M. Ellis (1994). "The implications of angiogenesis for the biology and therapy of cancer metastasis." Cell 79(2): 185-188.

Fidler, M. J., A. Argiris, J. D. Patel, D. H. Johnson, A. Sandler, V. M. Villaflor, J. T. Coon, L. Buckingham, K. Kaiser, S. Basu and P. Bonomi (2008). "The potential predictive value of cyclooxygenase-2 expression and increased risk of gastrointestinal hemorrhage in advanced non-small cell lung cancer patients treated with erlotinib and celecoxib." Clin Cancer Res 14(7): 2088-2094.

Finetti, F., S. Donnini, A. Giachetti, L. Morbidelli and M. Ziche (2009). "Prostaglandin E(2) primes the angiogenic switch via a synergic interaction with the fibroblast growth factor-2 pathway." Circ Res 105(7): 657-666.

Finetti, F., R. Solito, L. Morbidelli, A. Giachetti, M. Ziche and S. Donnini (2008). "Prostaglandin E2 regulates angiogenesis via activation of fibroblast growth factor receptor-1." J Biol Chem 283(4): 2139-2146.

Fiorucci, S., E. Distrutti, A. Mencarelli, A. Morelli, S. A. Laufor, G. Cirino and J. L. Wallace (2003). "Evidence that 5-lipoxygenase and acetylated cyclooxygenase 2-derived eicosanoids regulate leukocyte-endothelial adherence in response to aspirin." Br J Pharmacol 139(7): 1351-1359.

Fu, Y. G., J. J. Sung, K. C. Wu, A. H. Bai, M. C. Chan, J. Yu, D. M. Fan and W. K. Leung (2006). "Inhibition of gastric cancer cells associated angiogenesis by 15d-prostaglandin J2 through the downregulation of angiopoietin-1." Cancer Lett 243(2): 246-254.

Fujino, H., K. Toyomura, X. B. Chen, J. W. Regan and T. Murayama (2011). "Prostaglandin E regulates cellular migration via induction of vascular endothelial growth factor receptor-1 in HCA-7 human colon cancer cells." Biochem Pharmacol 81(3): 379-387.

Fujita, J., J. R. Mestre, J. B. Zeldis, K. Subbaramaiah and A. J. Dannenberg (2001). "Thalidomide and its analogues inhibit lipopolysaccharide-mediated induction of cyclooxygenase-2." Clin Cancer Res 7(11): 3349-3355.

Fukuda, R., B. Kelly and G. L. Semenza (2003). "Vascular endothelial growth factor gene expression in colon cancer cells exposed to prostaglandin E2 is mediated by hypoxia-inducible factor 1." Cancer Res 63(9): 2330-2334.

Funk, C. D. (1993). "Molecular biology in the eicosanoid field." Prog Nucleic Acid Res Mol Biol 45: 67-98.

Funk, C. D. (2001). "Prostaglandins and leukotrienes: advances in eicosanoid biology." Science 294(5548): 1871-1875.

Funk, C. D. and G. A. FitzGerald (2007). "COX-2 inhibitors and cardiovascular risk." J Cardiovasc Pharmacol 50(5): 470-479.

Funovics, P., C. Brostjan, A. Nigisch, A. Fila, A. Grochot, K. Mleczko, H. Was, G. Weigel, J. Dulak and A. Jozkowicz (2006). "Effects of 15d-PGJ(2) on VEGF-induced angiogenic activities and expression of VEGF receptors in endothelial cells." Prostaglandins Other Lipid Mediat 79(3-4): 230-244.

Furstenberger, G., P. Krieg, K. Muller-Decker and A. J. Habenicht (2006). "What are cyclooxygenases and lipoxygenases doing in the driver's seat of carcinogenesis?" Int J Cancer 119(10): 2247-2254.

Gao, H., W. J. Welch, G. F. DiBona and C. S. Wilcox (1997). "Sympathetic nervous system and hypertension during prolonged TxA_2/PGH_2 receptor activation in rats." Am J Physiol 273(2 Pt 2): H734-739.

Gao, X., D. J. Grignon, T. Chbihi, A. Zacharek, Y. Q. Chen, W. Sakr, A. T. Porter, J. D. Crissman, J. E. Pontes, I. J. Powell and et al. (1995). "Elevated 12-lipoxygenase mRNA expression correlates with advanced stage and poor differentiation of human prostate cancer." Urology 46(2): 227-237.

Gao, Y., R. Yokota, S. Tang, A. W. Ashton and J. A. Ware (2000). "Reversal of angiogenesis *in vitro*, induction of apoptosis, and inhibition of AKT phosphorylation in endothelial cells by thromboxane A(2)." Circ Res 87(9): 739-745.

Gerhardt, H. and C. Betsholtz (2003). "Endothelial–pericyte interactions in angiogenesis." Cell Tissue Res 314(1): 15-23.

Giese, A., C. Hagel, E. L. Kim, S. Zapf, J. Djawaheri, M. E. Berens and M. Westphal (1999). "Thromboxane synthase regulates the migratory phenotype of human glioma cells." Neuro Oncol 1(1): 3-13.

Giese, A., M. A. Loo, M. D. Rief, N. Tran and M. E. Berens (1995). "Substrates for astrocytoma invasion." Neurosurgery 37(2): 294-301; discussion 301-292.

Gilroy, D. W., A. Tomlinson and D. A. Willoughby (1998). "Differential effects of inhibitors of cyclooxygenase (cyclooxygenase 1 and cyclooxygenase 2) in acute inflammation." Eur J Pharmacol 355(2-3): 211-217.

Gonzalez, A. L., R. L. Roberts, P. P. Massion, S. J. Olson, Y. Shyr and S. B. Shappell (2004). "15-Lipoxygenase-2 expression in benign and neoplastic lung: an immunohistochemical study and correlation with tumor grade and proliferation." Hum Pathol 35(7): 840-849.

Grau, R., M. A. Iniguez and M. Fresno (2004). "Inhibition of activator protein 1 activation, vascular endothelial growth factor, and cyclooxygenase-2 expression by 15-deoxy-Delta12,14-prostaglandin J2 in colon carcinoma cells: evidence for a redox-sensitive peroxisome proliferator-activated receptor-gamma-independent mechanism." Cancer Res 64(15): 5162-5171.

Greene, E. R., S. Huang, C. N. Serhan and D. Panigrahy (2011). "Regulation of inflammation in cancer by eicosanoids." Prostaglandins Other Lipid Mediat: Epub.

Griesser, M., T. Suzuki, N. Tejera, S. Mont, W. E. Boeglin, A. Pozzi and C. Schneider (2011). "Biosynthesis of hemiketal eicosanoids by cross-over of the 5-lipoxygenase and cyclooxygenase-2 pathways." Proc Natl Acad Sci USA 108(17): 6945-6950.

Griffioen, A. W. and G. Molema (2000). "Angiogenesis: potentials for pharmacologic intervention in the treatment of cancer, cardiovascular diseases, and chronic inflammation." Pharmacol Rev 52(2): 237-268.

Guo, Y., W. Zhang, C. Giroux, Y. Cai, P. Ekambaram, A. K. Dilly, A. Hsu, S. Zhou, K. R. Maddipati, J. Liu, S. Joshi, S. C. Tucker, M. J. Lee and K. V. Honn (2011). "Identification of the orphan G protein-coupled receptor GPR31 as a receptor for 12-(S)-hydroxyeicosatetraenoic acid." J Biol Chem 286(39): 33832-33840.

Gupta, R. A., T. J., W. F. Krause, M. Geraci, T. M. Willson and S. K. Dey (2000). "Prostacyclin mediated activation of peroxisome proliferator-activated receptor delta in colorectal cancer." Proc Natl Acad Sci USA 97: 13275-13280.

Gupta, S., M. Srivastava, N. Ahmad, K. Sakamoto, D. G. Bostwick and H. Mukhtar (2001). "Lipoxygenase-5 is overexpressed in prostate adenocarcinoma." Cancer 91(4): 737-743.

Gyftopoulos, K., K. Vourda, G. Sakellaropoulos, P. Perimenis, A. Athanasopoulos and E. Papadaki (2011). "The Angiogenic Switch for Vascular Endothelial Growth Factor-A and Cyclooxygenase-2 in Prostate Carcinoma: Correlation with Microvessel Density, Androgen Receptor Content and Gleson Grade." Urol Int 87(4): 464-9.

Hanahan, D. and J. Folkman (1996). "Patterns and emerging mechanisms of the angiogenic switch during tumorigenesis." Cell 86(3): 353-364.

Hanahan, D. and R. A. Weinberg (2011). "Hallmarks of cancer: the next generation." Cell 144(5): 646-674.

Harris, R. C., J. A. McKanna, Y. Akai, H. R. Jacobson, R. N. Dubois and M. D. Breyer (1994). "Cyclooxygenase-2 is associated with the macula densa of rat kidney and increases with salt restriction." J Clin Invest 94(6): 2504-2510.

Haslmayer, P., T. Thalhammer, W. Jager, S. Aust, G. Steiner, C. Ensinger and P. Obrist (2002). "The peroxisome proliferator-activated receptor gamma ligand 15-deoxy-Delta12,14-prostaglandin J2 induces vascular endothelial growth factor in the hormone-independent prostate cancer cell line PC 3 and the urinary bladder carcinoma cell line 5637." Int J Oncol 21(4): 915-920.

Hata, A. N. and R. M. Breyer (2004). "Pharmacology and signaling of prostaglandin receptors: multiple roles in inflammation and immune modulation." Pharmacol Ther 103(2): 147-166.

Hazai, E., Z. Bikadi, F. Zsila and S. F. Lockwood (2006). "Molecular modeling of the non-covalent binding of the dietary tomato carotenoids lycopene and lycophyll, and selected oxidative metabolites with 5-lipoxygenase." Bioorg Med Chem 14(20): 6859-6867.

He, T., T. Lu, L. V. d'Uscio, C. F. Lam, H. C. Lee and Z. S. Katusic (2008). "Angiogenic function of prostacyclin biosynthesis in human endothelial progenitor cells." Circ Res 103(1): 80-88.

Heath, E. I., M. I. Canto, S. Piantadosi, E. Montgomery, W. M. Weinstein, J. G. Herman, A. J. Dannenberg, V. W. Yang, A. O. Shar, E. Hawk and A. A. Forastiere (2007). "Secondary chemoprevention of Barrett's esophagus with celecoxib: results of a randomized trial." J Natl Cancer Inst 99(7): 545-557.

Herbst, R. S., A. Onn and A. Sandler (2005). "Angiogenesis and lung cancer: prognostic and therapeutic implications." J Clin Oncol 23(14): 3243-3256.

Hernandez, G. L., O. V. Volpert, M. A. Iniguez, E. Lorenzo, S. Martinez-Martinez, R. Grau, M. Fresno and J. M. Redondo (2001). "Selective inhibition of vascular endothelial growth factor-mediated angiogenesis by cyclosporin A: roles of the nuclear factor of activated T cells and cyclooxygenase 2." J Exp Med 193(5): 607-620.

Hla, T. and K. Neilson (1992). "Human cyclooxygenase-2 cDNA." Proc Natl Acad Sci USA 89(16): 7384-7388.

Ho, T. C., S. L. Chen, Y. C. Yang, C. Y. Chen, F. P. Feng, J. W. Hsieh, H. C. Cheng and Y. P. Tsao (2008). "15-deoxy-Delta(12,14)-prostaglandin J2 induces vascular endothelial cell apoptosis through the sequential activation of MAPKS and p53." J Biol Chem 283(44): 30273-30288.

Hoffmann, C. (2000). "COX-2 in brain and spinal cord implications for therapeutic use." Curr Med Chem 7(11): 1113-1120.

Hong, J., M. Bose, J. Ju, J. H. Ryu, X. Chen, S. Sang, M. J. Lee and C. S. Yang (2004). "Modulation of arachidonic acid metabolism by curcumin and related beta-diketone derivatives: effects on cytosolic phospholipase A(2), cyclooxygenases and 5-lipoxygenase." Carcinogenesis 25(9): 1671-1679.

Honn, K. V. (1983). "Inhibition of tumor cell metastasis by modulation of the vascular prostacyclin/thromboxane A_2 system." Clin Exp Metastasis 1(2): 103-114.

Honn, K. V., B. Cicone and A. Skoff (1981). "Prostacyclin: a potent antimetastatic agent." Science 212(4500): 1270-1272.

Hoper, M. M., N. F. Voelkel, T. O. Bates, J. D. Allard, M. Horan, D. Shepherd and R. M. Tuder (1997). "Prostaglandins induce vascular endothelial growth factor in a human monocytic cell line and rat lungs via cAMP." Am J Respir Cell Mol Biol 17(6): 748-756.

Hsieh, Y. C., S. J. Hsieh, Y. S. Chang, C. M. Hsueh and S. L. Hsu (2007). "The lipoxygenase inhibitor, baicalein, modulates cell adhesion and migration by up-regulation of integrins and vinculin in rat heart endothelial cells." Br J Pharmacol 151(8): 1235-1245.

Huang, S. P., M. S. Wu, C. T. Shun, H. P. Wang, C. Y. Hsieh, M. L. Kuo and J. T. Lin (2005). "Cyclooxygenase-2 increases hypoxia-inducible factor-1 and vascular endothelial growth factor to promote angiogenesis in gastric carcinoma." J Biomed Sci 12(1): 229-241.

Hubbard, W. C., M. C. Alley, G. N. Gray, K. C. Green, T. L. McLemore and M. R. Boyd (1989). "Evidence for prostanoid biosynthesis as a biochemical feature of certain subclasses of non-small cell carcinomas of the lung as determined in established cell lines derived from human lung tumors." Cancer Res 49(4): 826-832.

Inagaki, M., T. Tsuri, H. Jyoyama, T. Ono, K. Yamada, M. Kobayashi, Y. Hori, A. Arimura, K. Yasui, K. Ohno, S. Kakudo, K. Koizumi, R. Suzuki, S. Kawai, M. Kato and S. Matsumoto (2000). "Novel antiarthritic agents with 1,2-isothiaz-olidine-1,1-dioxide (gamma-sultam) skeleton: cytokine suppressive dual inhibitors of cyclooxygenase-2 and 5-lipoxygenase." J Med Chem 43(10): 2040-2048.

Iniguez, M. A., A. Rodriguez, O. V. Volpert, M. Fresno and J. M. Redondo (2003). "Cyclooxygenase-2: a therapeutic target in angiogenesis." Trends Mol Med 9(2): 73-78.

Inoue, M., H. Itoh, T. Tanaka, T. H. Chun, K. Doi, Y. Fukunaga, N. Sawada, J. Yamshita, K. Masatsugu, T. Saito, S. Sakaguchi, M. Sone, K. Yamahara, T. Yurugi and K. Nakao (2001). "Oxidized LDL regulates vascular endothelial growth factor expression in human macrophages and endothelial cells through activation of peroxisome proliferator-activated receptor-gamma." Arterioscler Thromb Vasc Biol 21(4): 560-566.

Iwamoto, J., Y. Mizokami, K. Takahashi, T. Matsuoka and Y. Matsuzaki (2008). "The effects of cyclooxygenase2-prostaglandinE2 pathway on *Helicobacter pylori*-induced urokinase-type plasminogen activator system in the gastric cancer cells." Helicobacter 13(3): 174-182.

Jain, R. K. (2003). "Molecular regulation of vessel maturation." Nat Med 9(6): 685-693.

Jain, R. K., K. Schlenger, M. Hockel and F. Yuan (1997). "Quantitative angiogenesis assays: progress and problems." Nat Med 3(11): 1203-1208.

Jakobsen, A., J. P. Mortensen, C. Bisgaard, J. Lindebjerg, S. R. Rafaelsen and V. O. Bendtsen (2008). "A COX-2 inhibitor combined with chemoradiation of locally advanced rectal cancer: a phase II trial." Int J Colorectal Dis 23(3): 251-255.

Jantke, J., M. Ladehoff, F. Kurzel, S. Zapf, E. Kim and A. Giese (2004). "Inhibition of the arachidonic acid metabolism blocks endothelial cell migration and induces apoptosis." Acta Neurochir (Wien) 146(5): 483-494.

Jeong, W. C., K. J. Kim, H. W. Ju, H. K. Back, H. K. Kim, S. Y. Im and H. K. Lee (2010). "Cytoplasmic phospholipase A_2 metabolites play a critical role in pulmonary tumor metastasis in mice." Anticancer Res 30(9): 3421-3427.

Jiang, W. G., A. Douglas-Jones and R. E. Mansel (2003). "Levels of expression of lipoxygenases and cyclooxygenase-2 in human breast cancer." Prostaglandins Leukot Essent Fatty Acids 69(4): 275-281.

Jiang, W. G., A. G. Douglas-Jones and R. E. Mansel (2006). "Aberrant expression of 5-lipoxygenase-activating protein (5-LOXAP) has prognostic and survival significance in patients with breast cancer." Prostaglandins Leukot Essent Fatty Acids 74(2): 125-134.

Jiang, W. G., G. Watkins, A. Douglas-Jones and R. E. Mansel (2006). "Reduction of isoforms of 15-lipoxygenase (15-LOX)-1 and 15-LOX-2 in human breast cancer." Prostaglandins Leukot Essent Fatty Acids 74(4): 235-245.

Jones, D. A., D. P. Carlton, T. M. McIntyre, G. A. Zimmerman and S. M. Prescott (1993). "Molecular cloning of human prostaglandin endoperoxide synthase type II and demonstration of expression in response to cytokines." J Biol Chem 268(12): 9049-9054.

Jones, M. K., H. Wang, B. M. Peskar, E. Levin, R. M. Itani, I. J. Sarfeh and A. S. Tarnawski (1999). "Inhibition of angiogenesis by nonsteroidal anti-inflammatory drugs: insight into mechanisms and implications for cancer growth and ulcer healing." Nat Med 5(12): 1418-1423.

Kang, K. B., T. T. Wang, C. T. Woon, E. S. Cheah, X. L. Moore, C. Zhu and M. C. Wong (2007). "Enhancement of glioblastoma radioresponse by a selective COX-2 inhibitor celecoxib: inhibition of tumor angiogenesis with extensive tumor necrosis." Int J Radiat Oncol Biol Phys 67(3): 888-896.

Kari, G., U. Rodeck and A. P. Dicker (2007). "Zebrafish: an emerging model system for human disease and drug discovery." Clin Pharmacol Ther 82(1): 70-80.

Karlage, K. L., E. Mogalian, A. Jensen and P. B. Myrdal (2010). "Inhalation of an ethanol-based zileuton formulation provides a reduction of pulmonary adenomas in the A/J mouse model." AAPS PharmSciTech 11(1): 168-173.

Kelavkar, U. P., C. Cohen, H. Kamitani, T. E. Eling and K. F. Badr (2000). "Concordant induction of 15-lipoxygenase-1 and mutant p53 expression in human prostate adenocarcinoma: correlation with Gleason staging." Carcinogenesis 21(10): 1777-1787.

Kenyon, B. M., E. E. Voest, C. C. Chen, E. Flynn, J. Folkman and R. J. D'Amato (1996). "A model of angiogenesis in the mouse cornea." Invest Ophthalmol Vis Sci 37(8): 1625-1632.

Kerbel, R. S., A. Viloria-Petit, G. Klement and J. Rak (2000). "'Accidental' anti-angiogenic drugs. anti-oncogene directed signal transduction inhibitors and conventional chemotherapeutic agents as examples." Eur J Cancer 36(10): 1248-1257.

Kerjaschki, D., Z. Bago-Horvath, M. Rudas, V. Sexl, C. Schneckenleithner, S. Wolbank, G. Bartel, S. Krieger, R. Kalt, B. Hantusch, T. Keller, K. Nagy-Bojarszky, N. Huttary, I. Raab, K. Lackner, K. Krautgasser, H. Schachner, K. Kaserer, S. Rezar, S. Madlener, C. Vonach, A. Davidovits, H. Nosaka, M. Hammerle, K. Viola, H. Dolznig, M. Schreiber, A. Nader, W. Mikulits, M. Gnant, S. Hirakawa, M. Detmar, K. Alitalo, S. Nijman, F. Offner, T. J. Maier, D. Steinhilber and G. Krupitza (2011). "Lipoxygenase mediates invasion of intrametastatic lymphatic vessels and propagates lymph node metastasis of human mammary carcinoma xenografts in mouse." J Clin Invest 121(5): 2000-2012.

Khuri, F. R., H. Wu, J. J. Lee, B. L. Kemp, R. Lotan, S. M. Lippman, L. Feng, W. K. Hong and X. C. Xu (2001). "Cyclooxygenase-2 overexpression is a marker of poor prognosis in stage I non-small cell lung cancer." Clin Cancer Res 7(4): 861-867.

Kim, C. E., S. J. Lee, K. W. Seo, H. M. Park, J. W. Yun, J. U. Bae, S. S. Bae and C. D. Kim (2010). "Acrolein increases 5-lipoxygenase expression in murine macrophages through activation of ERK pathway." Toxicol Appl Pharmacol 245(1): 76-82.

Kim, G. Y., J. W. Lee, S. H. Cho, J. M. Seo and J. H. Kim (2009). "Role of the low-affinity leukotriene B4 receptor BLT2 in VEGF-induced angiogenesis." Arterioscler Thromb Vasc Biol 29(6): 915-920.

Kim, H. S., H. R. Youm, J. S. Lee, K. W. Min, J. H. Chung and C. S. Park (2003). "Correlation between cyclooxygenase-2 and tumor angiogenesis in non-small cell lung cancer." Lung Cancer 42(2): 163-170.

Kisley, L. R., B. S. Barrett, L. D. Dwyer-Nield, A. K. Bauer, D. C. Thompson and A. M. Malkinson (2002). "Celecoxib reduces pulmonary inflammation but not lung tumorigenesis in mice." Carcinogenesis 23(10): 1653-1660.

Klampfl, T., E. Bogner, W. Bednar, L. Mager, D. Massudom, I. Kalny, C. Heinzle, W. Berger, S. Stattner, J. Karner, M. Klimpfinger, G. Furstenberger, P. Krieg and B. Marian (2011). "Up-regulation of 12(S)-lipoxygenase induces a migratory phenotype in colorectal cancer cells." Exp Cell Res 318(6): 768-78.

Klurfeld, D. M. and A. W. Bull (1997). "Fatty acids and colon cancer in experimental models." Am J Clin Nutr 66(6 Suppl): 1530S-1538S.

Koeberle, A., H. Northoff and O. Werz (2009). "Curcumin blocks prostaglandin E2 biosynthesis through direct inhibition of the microsomal prostaglandin E2 synthase-1." Mol Cancer Ther 8(8): 2348-2355.

Kolev, Y., H. Uetake, S. Iida, T. Ishikawa, T. Kawano and K. Sugihara (2007). "Prognostic significance of VEGF expression in correlation with COX-2, microvessel density, and clinicopathological characteristics in human gastric carcinoma." Ann Surg Oncol 14(10): 2738-2747.

Kottke, T., G. Hall, J. Pulido, R. M. Diaz, J. Thompson, H. Chong, P. Selby, M. Coffey, H. Pandha, J. Chester, A. Melcher, K. Harrington and R. Vile (2010). "Antiangiogenic cancer therapy combined with oncolytic virotherapy leads to regression of established tumors in mice." J Clin Invest 120(5): 1551-1560.

Krieg, P., M. Heidt, M. Siebert, A. Kinzig, F. Marks and G. Furstenberger (2002). "Epidermis-type lipoxygenases." Adv Exp Med Biol 507: 165-170.

Krishnamoorthy, S., R. Jin, Y. Cai, K. R. Maddipati, D. Nie, G. Pages, S. C. Tucker and K. V. Honn (2010). "12-Lipoxygenase and the regulation of hypoxia-inducible factor in prostate cancer cells." Exp Cell Res 316(10): 1706-1715.

Krysan, K., H. Dalwadi, S. Sharma, M. Pold and S. Dubinett (2004). "Cyclooxygenase 2-dependent expression of survivin is critical for apoptosis resistance in non-small cell lung cancer." Cancer Res 64(18): 6359-6362.

Kuhn, H., M. Walther and R. J. Kuban (2002). "Mammalian arachidonate 15-lipoxygenases structure, function, and biological implications." Prostaglandins Other Lipid Mediat 68-69: 263-290.

Kummer, N. T., T. S. Nowicki, J. P. Azzi, I. Reyes, C. Iacob, S. Xie, I. Swati, N. Suslina, S. Schantz, R. K. Tiwari and J. Geliebter (2012). "Arachidonate 5 lipoxygenase expression in papillary thyroid carcinoma promotes invasion via MMP-9 induction." J Cell Biochem 113(6): 1998-2008.

Le Filliatre, G., S. Sayah, V. Latournerie, J. F. Renaud, M. Finet and R. Hanf (2001). "Cyclo-oxygenase and lipoxygenase pathways in mast cell dependent-neurogenic inflammation induced by electrical stimulation of the rat saphenous nerve." Br J Pharmacol 132(7): 1581-1589.

Leahy, K. M., A. T. Koki and J. L. Masferrer (2000). "Role of cyclooxygenases in angiogenesis." Curr Med Chem 7(11): 1163-1170.

Leahy, K. M., R. L. Ornberg, Y. Wang, B. S. Zweifel, A. T. Koki and J. L. Masferrer (2002). "Cyclooxygenase-2 inhibition by celecoxib reduces proliferation and induces apoptosis in angiogenic endothelial cells *in vivo.*" Cancer Res 62(3): 625-631.

Leval, X., F. Julemont, J. Delarge, B. Pirotte and J. M. Dogne (2002). "New trends in dual 5-LOX/COX inhibition." Curr Med Chem 9(9): 941-962.

Li, W., R. J. Xu, Z. Y. Lin, G. C. Zhuo and H. H. Zhang (2009). "Effects of a cyclooxygenase-1-selective inhibitor in a mouse model of ovarian cancer, administered alone or in combination with ibuprofen, a nonselective cyclooxygenase inhibitor." Med Oncol 26(2): 170-177.

Ling, Y., Y. Chen, P. Chen, H. Hui, X. Song, Z. Lu, C. Li, N. Lu and Q. Guo (2011). "Baicalein potently suppresses angiogenesis induced by vascular endothelial growth factor through the p53/Rb signaling pathway leading to G1/S cell cycle arrest." Exp Biol Med (Maywood) 236(7): 851-858.

Liu, J. J., T. S. Huang, W. F. Cheng and F. J. Lu (2003). "Baicalein and baicalin are potent inhibitors of angiogenesis: Inhibition of endothelial cell proliferation, migration and differentiation." Int J Cancer 106(4): 559-565.

Lukiw, W. J., A. Ottlecz, G. Lambrou, M. Grueninger, J. Finley, H. W. Thompson and N. G. Bazan (2003). "Coordinate activation of HIF-1 and NF-kappaB DNA binding and COX-2 and VEGF expression in retinal cells by hypoxia." Invest Ophthalmol Vis Sci 44(10): 4163-4170.

Ma, C., Y. Li, J. Ma, Y. Liu, Q. Li, S. Niu, Z. Shen, L. Zhang, Z. Pan and D. Zhu (2011). "Key role of 15-lipoxygenase/15-hydroxyeicosatetraenoic acid in pulmonary vascular remodeling and vascular angiogenesis associated with hypoxic pulmonary hypertension." Hypertension 58(4): 679-688.

Ma, L., P. del Soldato and J. L. Wallace (2002). "Divergent effects of new cyclooxygenase inhibitors on gastric ulcer healing: Shifting the angiogenic balance." Proc Natl Acad Sci USA 99(20): 13243-13247.

Mariani, T. J., S. Sandefur, J. D. Roby and R. A. Pierce (1998). "Collagenase-3 induction in rat lung fibroblasts requires the combined effects of tumor necrosis factor-alpha and 12-lipoxygenase metabolites: a model of macrophage-induced, fibroblast-driven extracellular matrix remodeling during inflammatory lung injury." Mol Biol Cell 9(6): 1411-1424.

Marques, I. J., F. U. Weiss, D. H. Vlecken, C. Nitsche, J. Bakkers, A. K. Lagendijk, L. I. Partecke, C. D. Heidecke, M. M. Lerch and C. P. Bagowski (2009). "Metastatic behaviour of primary human tumours in a zebrafish xenotransplantation model." BMC Cancer 9: 128.

Masferrer, J. L., K. M. Leahy, A. T. Koki, B. S. Zweifel, S. L. Settle, B. M. Woerner, D. A. Edwards, A. G. Flickinger, R. J. Moore and K. Seibert (2000). "Antiangiogenic and antitumor activities of cyclooxygenase-2 inhibitors." Cancer Res 60(5): 1306-1311.

McCabe, N. P., S. H. Selman and J. Jankun (2006). "Vascular endothelial growth factor production in human prostate cancer cells is stimulated by overexpression of platelet 12-lipoxygenase." Prostate 66(7): 779-787.

McDonough, W., N. Tran, A. Giese, S. A. Norman and M. E. Berens (1998). "Altered gene expression in human astrocytoma cells selected for migration: I. Thromboxane synthase." J Neuropathol Exp Neurol 57(5): 449-455.

McLemore, T. L., W. C. Hubbard, C. L. Litterst, M. C. Liu, S. Miller, N. A. McMahon, J. C. Eggleston and M. R. Boyd (1988). "Profiles of prostaglandin biosynthesis in normal lung and tumor tissue from lung cancer patients." Cancer Res 48(11): 3140-3147.

Melstrom, L. G., D. J. Bentrem, M. R. Salabat, T. J. Kennedy, X. Z. Ding, M. Strouch, S. M. Rao, R. C. Witt, C. A. Ternent, M. S. Talamonti, R. H. Bell and T. A. Adrian (2008). "Overexpression of 5-lipoxygenase in colon polyps and cancer and the effect of 5-LOX inhibitors *in vitro* and in a murine model." Clin Cancer Res 14(20): 6525-6530.

Merchan, J. R., D. R. Jayaram, J. G. Supko, X. He, G. J. Bubley and V. P. Sukhatme (2005). "Increased endothelial uptake of paclitaxel as a potential mechanism for its antiangiogenic effects: potentiation by Cox-2 inhibition." Int J Cancer 113(3): 490-498.

Michalik, L., B. Desvergne and W. Wahli (2004). "Peroxisome-proliferator-activated receptors and cancers: complex stories." Nat Rev Cancer 4(1): 61-70.

Miller, K. D., C. J. Sweeney and G. W. Sledge, Jr. (2001). "Redefining the target: chemotherapeutics as antiangiogenics." J Clin Oncol 19(4): 1195-1206.

Moody, T. W., J. Leyton, H. Zakowicz, T. Hida, Y. Kang, S. Jakowlew, L. You, L. Ozbun, H. Zia, J. Youngberg and A. Malkinson (2001). "Indomethacin reduces lung adenoma number in A/J mice." Anticancer Res 21(3B): 1749-1755.

Moon, C., M. Ahn, M. B. Wie, H. M. Kim, C. S. Koh, S. C. Hong, M. D. Kim, N. Tanuma, Y. Matsumoto and T. Shin (2005). "Phenidone, a dual inhibitor of cyclooxygenases and lipoxygenases, ameliorates rat paralysis in experimental autoimmune encephalomyelitis by suppressing its target enzymes." Brain Res 1035(2): 206-210.

Morteau, O., S. G. Morham, R. Sellon, L. A. Dieleman, R. Langenbach, O. Smithies and R. B. Sartor (2000). "Impaired mucosal defense to acute colonic injury in mice lacking cyclooxygenase-1 or cyclooxygenase-2." J Clin Invest 105(4): 469-478.

Moussa, O., A. W. Ashton, M. Fraig, E. Garrett-Mayer, M. A. Ghoneim, P. V. Halushka and D. K. Watson (2008). "Novel role of thromboxane receptors beta isoform in bladder cancer pathogenesis." Cancer Res 68(11): 4097-4104.

Moussa, O., J. S. Yordy, H. Abol-Enein, D. Sinha, N. K. Bissada, P. V. Halushka, M. A. Ghoneim and D. K. Watson (2005). "Prognostic and functional significance of thromboxane synthase gene overexpression in invasive bladder cancer." Cancer Res 65(24): 11581-11587.

Murata, T., M. I. Lin, K. Aritake, S. Matsumoto, S. Narumiya, H. Ozaki, Y. Urade, M. Hori and W. C. Sessa (2008). "Role of prostaglandin D2 receptor DP as a suppressor of tumor hyperpermeability and angiogenesis *in vivo*." Proc Natl Acad Sci USA 105(50): 20009-20014.

Murohara, T., J. R. Horowitz, M. Silver, Y. Tsurumi, D. Chen, A. Sullivan and J. M. Isner (1998). "Vascular endothelial growth factor/vascular permeability factor enhances vascular permeability via nitric oxide and prostacyclin." Circulation 97(1): 99-107.

Murphy, J. F. and D. J. Fitzgerald (2001). "Vascular endothelial growth factor induces cyclooxygenase-dependent proliferation of endothelial cells via the VEGF-2 receptor." Faseb J 15(9): 1667-1669.

Murphy, J. F., F. Lennon, C. Steele, D. Kelleher, D. Fitzgerald and A. C. Long (2005). "Engagement of CD44 modulates cyclooxygenase induction, VEGF generation, and proliferation in human vascular endothelial cells." Faseb J 19(3): 446-448.

Murphy, J. F., C. Steele, O. Belton and D. J. Fitzgerald (2003). "Induction of cyclooxygenase-1 and -2 modulates angiogenic responses to engagement of alphavbeta3." Br J Haematol 121(1): 157-164.

Mutter, R., B. Lu, D. P. Carbone, I. Csiki, L. Moretti, D. H. Johnson, J. D. Morrow, A. B. Sandler, Y. Shyr, F. Ye and H. Choy (2009). "A phase II study of celecoxib in combination with paclitaxel, carboplatin, and radiotherapy for patients with inoperable stage IIIA/B non-small cell lung cancer." Clin Cancer Res 15(6): 2158-2165.

Myrdal, P. B., K. Karlage, P. J. Kuehl, B. S. Angersbach, B. A. Merrill and P. D. Wightman (2007). "Effects of novel 5-lipoxygenase inhibitors on the incidence of pulmonary adenomas in the A/J murine model when administered via nose-only inhalation." Carcinogenesis 28(5): 957-961.

Nagatsuka, I., N. Yamada, S. Shimizu, M. Ohira, H. Nishino, S. Seki and K. Hirakawa (2002). "Inhibitory effect of a selective cyclooxygenase-2 inhibitor on liver metastasis of colon cancer." Int J Cancer 100(5): 515-519.

Narumiya, S. and G. A. FitzGerald (2001). "Genetic and pharmacological analysis of prostanoid receptor function." J Clin Invest 108(1): 25-30.

Nicosia, R. F., R. Tchao and J. Leighton (1982). "Histotypic angiogenesis *in vitro*: light microscopic, ultrastructural, and radioautographic studies." *In vitro* 18(6): 538-549.

Nie, D. (2007). "Cyclooxygenases and lipoxygenases in prostate and breast cancers." Front Biosci 12: 1574-1585.

Nie, D., M. Che, A. Zacharek, Y. Qiao, L. Li, X. Li, M. Lamberti, K. Tang, Y. Cai, Y. Guo, D. Grignon and K. V. Honn (2004). "Differential expression of thromboxane synthase in prostate carcinoma: role in tumor cell motility." Am J Pathol 164(2): 429-439.

Nie, D., Y. Guo, D. Yang, Y. Tang, Y. Chen, M. T. Wang, A. Zacharek, Y. Qiao, M. Che and K. V. Honn (2008). "Thromboxane A_2 receptors in prostate carcinoma: expression and its role in regulating cell motility via small GTPase Rho." Cancer Res 68(1): 115-121.

Nie, D., S. Krishnamoorthy, R. Jin, K. Tang, Y. Chen, Y. Qiao, A. Zacharek, Y. Guo, J. Milanini, G. Pages and K. V. Honn (2006). "Mechanisms regulating tumor angiogenesis by 12-lipoxygenase in prostate cancer cells." J Biol Chem 281(27): 18601-18609.

Nie, D., M. Lamberti, A. Zacharek, L. Li, K. Szekeres, K. Tang, Y. Chen and K. V. Honn (2000). "Thromboxane A(2) regulation of endothelial cell migration, angiogenesis, and tumor metastasis." Biochem Biophys Res Commun 267(1): 245-251.

Nie, D., J. Nemeth, Y. Qiao, A. Zacharek, L. Li, K. Hanna, K. Tang, G. G. Hillman, M. L. Cher, D. J. Grignon and K. V. Honn (2003). "Increased metastatic potential in human prostate carcinoma cells by overexpression of arachidonate 12-lipoxygenase." Clin Exp Metastasis 20(7): 657-663.

Nie, D., K. Tang, C. Diglio and K. V. Honn (2000). "Eicosanoid regulation of angiogenesis: role of endothelial arachidonate 12-lipoxygenase." Blood 95(7): 2304-2311.

Nithipatikom, K., M. A. Isbell, W. A. See and W. B. Campbell (2006). "Elevated 12- and 20-hydroxyeicosatetraenoic acid in urine of patients with prostatic diseases." Cancer Lett 233(2): 219-225.

Nolte, R. T., G. B. Wisely, S. Westin, J. E. Cobb, M. H. Lambert, R. Kurokawa, M. G. Rosenfeld, T. M. Willson, C. K. Glass and M. V. Milburn (1998). "Ligand binding and co-activator assembly of the peroxisome proliferator-activated receptor-gamma." Nature 395(6698): 137-143.

Norel, X. and C. Brink (2004). "The quest for new cysteinyl-leukotriene and lipoxin receptors: recent clues." Pharmacol Ther 103(1): 81-94.

O'Reilly, M. S., L. Holmgren, Y. Shing, C. Chen, R. A. Rosenthal, M. Moses, W. S. Lane, Y. Cao, E. H. Sage and J. Folkman (1994). "Angiostatin: a novel angiogenesis inhibitor that mediates the suppression of metastases by a Lewis lung carcinoma." Cell 79(2): 315-328.

Ogawa, H., P. Rafiee, P. J. Fisher, N. A. Johnson, M. F. Otterson and D. G. Binion (2003). "Sodium butyrate inhibits angiogenesis of human intestinal microvascular endothelial cells through COX-2 inhibition." FEBS Lett 554(1-2): 88-94.

Ohd, J. F., C. K. Nielsen, J. Campbell, G. Landberg, H. Lofberg and A. Sjolander (2003). "Expression of the leukotriene D4 receptor CysLT1, COX-2, and other cell survival factors in colorectal adenocarcinomas." Gastroenterology 124(1): 57-70.

Okahara, K., B. Sun and J. Kambayashi (1998). "Upregulation of prostacyclin synthesis-related gene expression by shear stress in vascular endothelial cells." Arterioscler Thromb Vasc Biol 18(12): 1922-1926.

Pai, R., T. Nakamura, W. S. Moon and A. S. Tarnawski (2003). "Prostaglandins promote colon cancer cell invasion; signaling by cross-talk between two distinct growth factor receptors." Faseb J 17(12): 1640-1647.

Pai, R., I. L. Szabo, B. A. Soreghan, S. Atay, H. Kawanaka and A. S. Tarnawski (2001). "PGE(2) stimulates VEGF expression in endothelial cells via ERK2/JNK1 signaling pathways." Biochem Biophys Res Commun 286(5): 923-928.

Papadimitrakopoulou, V. A., W. N. William, Jr., A. J. Dannenberg, S. M. Lippman, J. J. Lee, F. G. Ondrey, D. E. Peterson, L. Feng, A. Atwell, A. K. El-Naggar, C. O. Nathan, J. I. Helman, B. Du, B. Yueh and J. O. Boyle (2008). "Pilot randomized phase II study of celecoxib in oral premalignant lesions." Clin Cancer Res 14(7): 2095-2101.

Parng, C., W. L. Seng, C. Semino and P. McGrath (2002). "Zebrafish: a preclinical model for drug screening." Assay Drug Dev Technol 1(1 Pt 1): 41-48.

Patricia, M. K., J. A. Kim, C. M. Harper, P. T. Shih, J. A. Berliner, R. Natarajan, J. L. Nadler and C. C. Hedrick (1999). "Lipoxygenase products increase monocyte adhesion to human aortic endothelial cells." Arterioscler Thromb Vasc Biol 19(11): 2615-2622.

Patrignani, P., M. R. Panara, A. Greco, O. Fusco, C. Natoli, S. Iacobelli, F. Cipollone, A. Ganci, C. Creminon, J. Maclouf and et al. (1994). "Biochemical and pharmacological characterization of the cyclooxygenase activity of human blood prostaglandin endoperoxide synthases." J Pharmacol Exp Ther 271(3): 1705-1712.

Pichiule, P., J. C. Chavez and J. C. LaManna (2004). "Hypoxic regulation of angiopoietin-2 expression in endothelial cells." J Biol Chem 279(13): 12171-12180.

Pidgeon, G. P., M. Kandouz, A. Meram and K. V. Honn (2002). "Mechanisms controlling cell cycle arrest and induction of apoptosis after 12-lipoxygenase inhibition in prostate cancer cells." Cancer Res 62(9): 2721-2727.

Pidgeon, G. P., R. Tamosiuniene, G. Chen, I. Leonard, O. Belton, A. Bradford and D. J. Fitzgerald (2004). "Intravascular thrombosis after hypoxia-induced pulmonary hypertension: regulation by cyclooxygenase-2." Circulation 110(17): 2701-2707.

Pola, R., E. Gaetani, A. Flex, T. R. Aprahamian, M. Bosch-Marce, D. W. Losordo, R. C. Smith and P. Pola (2004). "Comparative analysis of the *in vivo* angiogenic properties of stable prostacyclin analogs: a possible role for peroxisome proliferator-activated receptors." J Mol Cell Cardiol 36(3): 363-370.

Pold, M., L. X. Zhu, S. Sharma, M. D. Burdick, Y. Lin, P. P. Lee, A. Pold, J. Luo, K. Krysan, M. Dohadwala, J. T. Mao, R. K. Batra, R. M. Strieter and S. M. Dubinett (2004). "Cyclooxygenase-2-dependent expression of angiogenic CXC chemokines ENA-78/CXC Ligand (CXCL) 5 and interleukin-8/CXCL8 in human non-small cell lung cancer." Cancer Res 64(5): 1853-1860.

Pradono, P., R. Tazawa, M. Maemondo, M. Tanaka, K. Usui, Y. Saijo, K. Hagiwara and T. Nukiwa (2002). "Gene transfer of thromboxane A(2) synthase and prostaglandin I(2) synthase antithetically altered tumor angiogenesis and tumor growth." Cancer Res 62(1): 63-66.

Prosperi, J. R., S. R. Mallery, K. A. Kigerl, A. A. Erfurt and F. M. Robertson (2004). "Invasive and angiogenic phenotype of MCF-7 human breast tumor cells expressing human cyclooxygenase-2." Prostaglandins Other Lipid Mediat 73 (3-4): 249-264.

Pueyo, M. E., Y. Chen, G. D'Angelo and J. B. Michel (1998). "Regulation of vascular endothelial growth factor expression by cAMP in rat aortic smooth muscle cells." Exp Cell Res 238(2): 354-358.

Qualtrough, D., A. Kaidi, S. Chell, H. N. Jabbour, A. C. Williams and C. Paraskeva (2007). "Prostaglandin F(2alpha) stimulates motility and invasion in colorectal tumor cells." Int J Cancer 121(4): 734-740.

Raso, E., J. Tovari, K. Toth, S. Paku, M. Trikha, K. V. Honn and J. Timar (2001). "Ectopic alphaIIbbeta3 integrin signaling involves 12-lipoxygenase- and PKC-mediated serine phosphorylation events in melanoma cells." Thromb Haemost 85(6): 1037-1042.

Raut, C. P., S. Nawrocki, L. M. Lashinger, D. W. Davis, S. Khanbolooki, H. Xiong, L. M. Ellis and D. J. McConkey (2004). "Celecoxib inhibits angiogenesis by inducing endothelial cell apoptosis in human pancreatic tumor xenografts." Cancer Biol Ther 3(12): 1217-1224.

Reinmuth, N., W. Liu, Y. D. Jung, S. A. Ahmad, R. M. Shaheen, F. Fan, C. D. Bucana, G. McMahon, G. E. Gallick and L. M. Ellis (2001). "Induction of VEGF in perivascular cells defines a potential paracrine mechanism for endothelial cell survival." Faseb J 15(7): 1239-1241.

Rhee, J. S., M. Black, U. Schubert, S. Fischer, E. Morgenstern, H. P. Hammes and K. T. Preissner (2004). "The functional role of blood platelet components in angiogenesis." Thromb Haemost 92(2): 394-402.

Ricciotti, E. and G. A. FitzGerald (2011). "Prostaglandins and inflammation." Arterioscler Thromb Vasc Biol 31(5): 986-1000.

Rieck, P. W., S. Cholidis and C. Hartmann (2001). "Intracellular signaling pathway of FGF-2-modulated corneal endothelial cell migration during wound healing *in vitro*." Exp Eye Res 73(5): 639-650.

Rioux, N. and A. Castonguay (1998). "Prevention of NNK-induced lung tumorigenesis in A/J mice by acetylsalicylic acid and NS-398." Cancer Res 58(23): 5354-5360.

Rodrigues, S., Q. D. Nguyen, S. Faivre, E. Bruyneel, L. Thim, B. Westley, F. May, G. Flatau, M. Mareel, C. Gespach and S. Emami (2001). "Activation of cellular invasion by trefoil peptides and src is mediated by cyclooxygenase- and thromboxane A_2 receptor-dependent signaling pathways." Faseb J 15(9): 1517-1528.

Romano, M., A. Catalano, M. Nutini, E. D'Urbano, C. Crescenzi, J. Claria, R. Libner, G. Davi and A. Procopio (2001). "5-lipoxygenase regulates malignant mesothelial cell survival: involvement of vascular endothelial growth factor." Faseb J 15(13): 2326-2336.

Rozic, J. G., C. Chakraborty and P. K. Lala (2001). "Cyclooxygenase inhibitors retard murine mammary tumor progression by reducing tumor cell migration, invasiveness and angiogenesis." Int J Cancer 93(4): 497-506.

Sakai, H., T. Suzuki, Y. Takahashi, M. Ukai, K. Tauchi, T. Fujii, N. Horikawa, T. Minamimura, Y. Tabuchi, M. Morii, K. Tsukada and N. Takeguchi (2006). "Upregulation of thromboxane synthase in human colorectal carcinoma and the cancer cell proliferation by thromboxane A_2." FEBS Lett 580(14): 3368-3374.

Salcedo, R., X. Zhang, H. A. Young, N. Michael, K. Wasserman, W. H. Ma, M. Martins-Green, W. J. Murphy and J. J. Oppenheim (2003). "Angiogenic effects of prostaglandin E2 are mediated by up-regulation of CXCR4 on human microvascular endothelial cells." Blood 102(6): 1966-1977.

Sales, K. J., T. List, S. C. Boddy, A. R. Williams, R. A. Anderson, Z. Naor and H. N. Jabbour (2005). "A novel angiogenic role for prostaglandin F2alpha-FP receptor interaction in human endometrial adenocarcinomas." Cancer Res 65(17): 7707-7716.

Sandison, J. C. (1924). "A new method for the microscopic study of living growing tissues by the introduction of a transparent chamber in the rabbits ear. " Anat. Rec. 28: 281-287.

Schirner, M. and M. R. Schneider (1997). "Inhibition of metastasis by cicaprost in rats with established SMT2A mammary carcinoma growth." Cancer Detect Prev 21(1): 44-50.

Schmedtje, J. F., Jr., Y. S. Ji, W. L. Liu, R. N. DuBois and M. S. Runge (1997). "Hypoxia induces cyclooxygenase-2 via the NF-kappaB p65 transcription factor in human vascular endothelial cells." J Biol Chem 272(1): 601-608.

Schneider, C. and A. Pozzi (2011). "Cyclooxygenases and lipoxygenases in cancer." Cancer Metastasis Rev 30(3-4): 277-294.

Schneider, M., M. Wortmann, P. K. Mandal, W. Arpornchayanon, K. Jannasch, F. Alves, S. Strieth, M. Conrad and H. Beck (2010). "Absence of glutathione peroxidase 4 affects tumor angiogenesis through increased 12/15-lipoxygenase activity." Neoplasia 12(3): 254-263.

Schwab, J. M., R. Beschorner, R. Meyermann, F. Gozalan and H. J. Schluesener (2002). "Persistent accumulation of cyclooxygenase-1-expressing microglial cells and macrophages and transient upregulation by endothelium in human brain injury." J Neurosurg 96(5): 892-899.

Seed, M. P., J. R. Brown, C. N. Freemantle, J. L. Papworth, P. R. Colville-Nash, D. Willis, K. W. Somerville, S. Asculai and D. A. Willoughby (1997). "The inhibition of colon-26 adenocarcinoma development and angiogenesis by topical diclofenac in 2.5% hyaluronan." Cancer Res 57(9): 1625-1629.

Segawa, E., S. Hashitani, Y. Toyohara, H. Kishimoto, K. Noguchi, K. Takaoka and M. Urade (2009). "Inhibitory effect of sulindac on DMBA-induced hamster cheek pouch carcinogenesis and its derived cell line." Oncol Rep 21(4): 869-874.

Seno, H., M. Oshima, T. O. Ishikawa, H. Oshima, K. Takaku, T. Chiba, S. Narumiya and M. M. Taketo (2002). "Cyclooxygenase 2- and prostaglandin E(2) receptor EP(2)-dependent angiogenesis in Apc(Delta716) mouse intestinal polyps." Cancer Res 62(2): 506-511.

Seo, K. W., S. J. Lee, C. E. Kim, M. R. Yun, H. M. Park, J. W. Yun, S. S. Bae and C. D. Kim (2010). "Participation of 5-lipoxygenase-derived LTB(4) in 4-hydroxynonenal-enhanced MMP-2 production in vascular smooth muscle cells." Atherosclerosis 208(1): 56-61.

Shappell, S. B., W. E. Boeglin, S. J. Olson, S. Kasper and A. R. Brash (1999). "15-lipoxygenase-2 (15-LOX-2) is expressed in benign prostatic epithelium and reduced in prostate adenocarcinoma." Am J Pathol 155(1): 235-245.

Shemesh, M., M. Gurevich, D. Mizrachi, L. Dombrovski, Y. Stram, M. J. Fields and L. S. Shore (1997). "Expression of functional luteinizing hormone (LH) receptor and its messenger ribonucleic acid in bovine uterine veins: LH induction of cyclooxygenase and augmentation of prostaglandin production in bovine uterine veins." Endocrinology 138(11): 4844-4851.

Shin, V. Y., H. C. Jin, E. K. Ng, J. J. Sung, K. M. Chu and C. H. Cho (2010). "Activation of 5-lipoxygenase is required for nicotine mediated epithelial-mesenchymal transition and tumor cell growth." Cancer Lett 292(2): 237-245.

Shureiqi, I. and S. M. Lippman (2001). "Lipoxygenase modulation to reverse carcinogenesis." Cancer Res 61(17): 6307-6312.

Shureiqi, I., K. J. Wojno, J. A. Poore, R. G. Reddy, M. J. Moussalli, S. A. Spindler, J. K. Greenson, D. Normolle, A. A. Hasan, T. S. Lawrence and D. E. Brenner (1999). "Decreased 13-S-hydroxyoctadecadienoic acid levels and 15-lipoxygenase-1 expression in human colon cancers." Carcinogenesis 20(10): 1985-1995.

Shureiqi, I., Y. Wu, D. Chen, X. L. Yang, B. Guan, J. S. Morris, P. Yang, R. A. Newman, R. Broaddus, S. R. Hamilton, P. Lynch, B. Levin, S. M. Fischer and S. M. Lippman (2005). "The critical role of 15-lipoxygenase-1 in colorectal epithelial cell terminal differentiation and tumorigenesis." Cancer Res 65(24): 11486-11492.

Singh, N. K., V. Kundumani-Sridharan and G. N. Rao (2011). "12/15-Lipoxygenase gene knockout severely impairs ischemia-induced angiogenesis due to lack of Rac1 farnesylation." Blood 118(20): 5701-5712.

Smyth, E. M., T. Grosser, M. Wang, Y. Yu and G. A. FitzGerald (2009). "Prostanoids in health and disease." J Lipid Res 50 Suppl: S423-428.

Solomon, S. D., J. J. McMurray, M. A. Pfeffer, J. Wittes, R. Fowler, P. Finn, W. F. Anderson, A. Zauber, E. Hawk and M. Bertagnolli (2005). "Cardiovascular risk associated with celecoxib in a clinical trial for colorectal adenoma prevention." N Engl J Med 352(11): 1071-1080.

Sonoshita, M., K. Takaku, N. Sasaki, Y. Sugimoto, F. Ushikubi, S. Narumiya, M. Oshima and M. M. Taketo (2001). "Acceleration of intestinal polyposis through prostaglandin receptor EP2 in Apc(Delta 716) knockout mice." Nat Med 7(9): 1048-1051.

Stamford, I. F., P. B. Melhuish, M. A. Carroll, C. J. Corrigan, S. Patel and A. Bennett (1986). "Survival of mice with NC carcinoma is unchanged by drugs that are thought to inhibit thromboxane synthesis or increase prostacyclin formation." Br J Cancer 54(2): 257-263.

Stoletov, K., V. Montel, R. D. Lester, S. L. Gonias and R. Klemke (2007). "High-resolution imaging of the dynamic tumor cell vascular interface in transparent zebrafish." Proc Natl Acad Sci USA 104(44): 17406-17411.

Subbarayan, V., X. C. Xu, J. Kim, P. Yang, A. Hoque, A. L. Sabichi, N. Llansa, G. Mendoza, C. J. Logothetis, R. A. Newman, S. M. Lippman and D. G. Menter (2005). "Inverse relationship between 15-lipoxygenase-2 and PPAR-gamma gene expression in normal epithelia compared with tumor epithelia." Neoplasia 7(3): 280-293.

Takahashi, H., M. C. Chen, H. Pham, E. Angst, J. C. King, J. Park, E. Y. Brovman, H. Ishiguro, D. M. Harris, H. A. Reber, O. J. Hines, A. S. Gukovskaya, V. L. Go and G. Eibl (2011). "Baicalein, a component of Scutellaria baicalensis, induces apoptosis by Mcl-1 down-regulation in human pancreatic cancer cells." Biochim Biophys Acta 1813(8): 1465-1474.

Takahashi, Y., F. Kawahara, M. Noguchi, K. Miwa, H. Sato, M. Seiki, H. Inoue, T. Tanabe and T. Yoshimoto (1999). "Activation of matrix metalloproteinase-2 in human breast cancer cells overexpressing cyclooxygenase-1 or -2." FEBS Lett 460(1): 145-148.

Tamura, M., S. Sebastian, B. Gurates, S. Yang, Z. Fang and S. E. Bulun (2002). "Vascular endothelial growth factor up-regulates cyclooxygenase-2 expression in human endothelial cells." J Clin Endocrinol Metab 87(7): 3504-3507.

Tang, D. G., B. Bhatia, S. Tang and R. Schneider-Broussard (2007). "15-lipoxygenase 2 (15-LOX2) is a functional tumor suppressor that regulates human prostate epithelial cell differentiation, senescence, and growth (size)." Prostaglandins Other Lipid Mediat 82(1-4): 135-146.

Tang, D. G., I. M. Grossi, Y. Q. Chen, C. A. Diglio and K. V. Honn (1993). "12(S)-HETE promotes tumor-cell adhesion by increasing surface expression of alpha V beta 3 integrins on endothelial cells." Int J Cancer 54(1): 102-111.

Tang, K., R. L. Finley, Jr., D. Nie and K. V. Honn (2000). "Identification of 12-lipoxygenase interaction with cellular proteins by yeast two-hybrid screening." Biochemistry 39(12): 3185-3191.

Tang, Y., M. T. Wang, Y. Chen, D. Yang, M. Che, K. V. Honn, G. D. Akers, S. R. Johnson and D. Nie (2009). "Downregulation of vascular endothelial growth factor and induction of tumor dormancy by 15-lipoxygenase-2 in prostate cancer." Int J Cancer 124(7): 1545-1551.

Tatsuguchi, A., K. Matsui, Y. Shinji, K. Gudis, T. Tsukui, T. Kishida, Y. Fukuda, Y. Sugisaki, A. Tokunaga, T. Tajiri and C. Sakamoto (2004). "Cyclooxygenase-2 expression correlates with angiogenesis and apoptosis in gastric cancer tissue." Hum Pathol 35(4): 488-495.

Taylor, P. M., R. J. Woodfield, M. N. Hodgkin, T. R. Pettitt, A. Martin, D. J. Kerr and M. J. Wakelam (2002). "Breast cancer cell-derived EMMPRIN stimulates fibroblast MMP2 release through a phospholipase A(2) and 5-lipoxygenase catalyzed pathway." Oncogene 21(37): 5765-5772.

Teicher, B. A., E. A. Sotomayor and Z. D. Huang (1992). "Antiangiogenic agents potentiate cytotoxic cancer therapies against primary and metastatic disease." Cancer Res 52(23): 6702-6704.

Terzuoli, E., S. Donnini, A. Giachetti, M. A. Iniguez, M. Fresno, G. Melillo and M. Ziche (2010). "Inhibition of hypoxia inducible factor-1alpha by dihydroxyphenylethanol, a product from olive oil, blocks microsomal prostaglandin-E synthase-1/vascular endothelial growth factor expression and reduces tumor angiogenesis." Clin Cancer Res 16(16): 4207-4216.

Thun, M. J., S. J. Henley and C. Patrono (2002). "Nonsteroidal anti-inflammatory drugs as anticancer agents: mechanistic, pharmacologic, and clinical issues." J Natl Cancer Inst 94(4): 252-266.

Timar, J., J. Tovari, E. Raso, L. Meszaros, B. Bereczky and K. Lapis (2005). "Platelet-mimicry of cancer cells: epiphenomenon with clinical significance." Oncology 69(3): 185-201.

Trapp, V., B. Parmakhtiar, V. Papazian, L. Willmott and J. P. Fruehauf (2010). "Anti-angiogenic effects of resveratrol mediated by decreased VEGF and increased TSP1 expression in melanoma-endothelial cell co-culture." Angiogenesis 13(4): 305-315.

Tsujii, M., S. Kawano, S. Tsuji, H. Sawaoka, M. Hori and R. N. DuBois (1998). "Cyclooxygenase regulates angiogenesis induced by colon cancer cells." Cell 93(5): 705-716.

Turner, E. C., E. P. Mulvaney, H. M. Reid and B. T. Kinsella (2011). "Interaction of the human prostacyclin receptor with the PDZ adapter protein PDZK1: role in endothelial cell migration and angiogenesis." Mol Biol Cell 22(15): 2664-2679.

van Leyen, K., R. M. Duvoisin, H. Engelhardt and M. Wiedmann (1998). "A function for lipoxygenase in programmed organelle degradation." Nature 395(6700): 392-395.

Varinska, L., L. Mirossay, G. Mojzisova and J. Mojzis (2010). "Antiangogenic effect of selected phytochemicals." Pharmazie 65(1): 57-63.

Viita, H., K. Kinnunen, E. Eriksson, J. Lahteenvuo, M. Babu, G. Kalesnykas, T. Heikura, S. Laidinen, T. Takalo and S. Yla-Herttuala (2009). "Intravitreal adenoviral 15-lipoxygenase-1 gene transfer prevents vascular endothelial growth factor A-induced neovascularization in rabbit eyes." Hum Gene Ther 20(12): 1679-1686.

Viita, H., J. Markkanen, E. Eriksson, M. Nurminen, K. Kinnunen, M. Babu, T. Heikura, S. Turpeinen, S. Laidinen, T. Takalo and S. Yla-Herttuala (2008). "15-lipoxygenase-1 prevents vascular endothelial growth factor A- and placental growth factor-induced angiogenic effects in rabbit skeletal muscles via reduction in growth factor mRNA levels, NO bioactivity, and downregulation of VEGF receptor 2 expression." Circ Res 102(2): 177-184.

von Rahden, B. H., H. J. Stein, F. Puhringer, I. Koch, R. Langer, G. Piontek, J. R. Siewert, H. Hofler and M. Sarbia (2005). "Coexpression of cyclooxygenases (COX-1, COX-2) and vascular endothelial growth factors (VEGF-A, VEGF-C) in esophageal adenocarcinoma." Cancer Res 65(12): 5038-5044.

Wang, D. and R. N. DuBois (2007). "Measurement of eicosanoids in cancer tissues." Methods Enzymol 433: 27-50.

Wang, D. and R. N. Dubois (2010). "Eicosanoids and cancer." Nat Rev Cancer 10(3): 181-193.

Wang, D., H. Wang, J. Brown, T. Daikoku, W. Ning, Q. Shi, A. Richmond, R. Strieter, S. K. Dey and R. N. DuBois (2006). "CXCL1 induced by prostaglandin E2 promotes angiogenesis in colorectal cancer." J Exp Med 203(4): 941-951.

Wang, L., W. Chen, X. Xie, Y. He and X. Bai (2008). "Celecoxib inhibits tumor growth and angiogenesis in an orthotopic implantation tumor model of human colon cancer." Exp Oncol 30(1): 42-51.

Wang, M. T., K. V. Honn and D. Nie (2007). "Cyclooxygenases, prostanoids, and tumor progression." Cancer Metastasis Rev 26(3-4): 525-34.

Wang, X., G. Li, A. Wang, Z. Zhang, J. R. Merchan and B. Halmos (2011). "Combined histone deacetylase and cyclooxygenase inhibition achieves enhanced antiangiogenic effects in lung cancer cells." Mol Carcinog Nov 28 [Epub ahead of print].

Watkins, G., A. Douglas-Jones, R. E. Mansel and W. G. Jiang (2005). "Expression of thromboxane synthase, TBXAS1 and the thromboxane A_2 receptor, $TBXA_2R$, in human breast cancer." Int Semin Surg Oncol 2: 23.

Wei, J., W. Yan, X. Li, Y. Ding and H. H. Tai (2010). "Thromboxane receptor alpha mediates tumor growth and angiogenesis via induction of vascular endothelial growth factor expression in human lung cancer cells." Lung Cancer 69(1): 26-32.

Wenger, F. A., M. Kilian, M. Bisevac, C. Khodadayan, M. von Seebach, I. Schimke, H. Guski and J. M. Muller (2002). "Effects of Celebrex and Zyflo on liver metastasis and lipidperoxidation in pancreatic cancer in Syrian hamsters." Clin Exp Metastasis 19(8): 681-687.

Wheeler-Jones, C., R. Abu-Ghazaleh, R. Cospedal, R. A. Houliston, J. Martin and I. Zachary (1997). "Vascular endothelial growth factor stimulates prostacyclin production and activation of cytosolic phospholipase A_2 in endothelial cells via p42/p44 mitogen-activated protein kinase." FEBS Lett 420(1): 28-32.

White, R. M., A. Sessa, C. Burke, T. Bowman, J. LeBlanc, C. Ceol, C. Bourque, M. Dovey, W. Goessling, C. E. Burns and L. I. Zon (2008). "Transparent adult zebrafish as a tool for *in vivo* transplantation analysis." Cell Stem Cell 2(2): 183-189.

Williams, C. S., M. Tsujii, J. Reese, S. K. Dey and R. N. DuBois (2000). "Host cyclooxygenase-2 modulates carcinoma growth." J Clin Invest 105(11): 1589-1594.

Willson, T. M., P. J. Brown, D. D. Sternbach and B. R. Henke (2000). "The PPARs: from orphan receptors to drug discovery." J Med Chem 43(4): 527-550.

Wu, B., J. Li, D. Huang, W. Wang, Y. Chen, Y. Liao, X. Tang, H. Xie and F. Tang (2011). "Baicalein mediates inhibition of migration and invasiveness of skin carcinoma through Ezrin in A431 cells." BMC Cancer 11: 527.

Wu, G., J. Luo, J. S. Rana, R. Laham, F. W. Sellke and J. Li (2006). "Involvement of COX-2 in VEGF-induced angiogenesis via P38 and JNK pathways in vascular endothelial cells." Cardiovasc Res 69(2): 512-519.

Wu, W. K., J. J. Sung, C. W. Lee, J. Yu and C. H. Cho (2010). "Cyclooxygenase-2 in tumorigenesis of gastrointestinal cancers: an update on the molecular mechanisms." Cancer Lett 295(1): 7-16.

Wu, Y. L., S. L. Fu, Y. P. Zhang, M. M. Qiao and Y. Chen (2005). "Cyclooxygenase-2 inhibitors suppress angiogenesis and growth of gastric cancer xenografts." Biomed Pharmacother 59 Suppl 2: S289-292.

Xin, X., S. Yang, J. Kowalski and M. E. Gerritsen (1999). "Peroxisome proliferator-activated receptor gamma ligands are potent inhibitors of angiogenesis *in vitro* and *in vivo*." J Biol Chem 274(13): 9116-9121.

Xu, X. M., J. J. Deng, G. J. Yuan, F. Yang, H. T. Guo, M. Xiang, W. Ge and Y. G. Wu (2011). "5-Lipoxygenase contributes to the progression of hepatocellular carcinoma." Mol Med Report 4(6): 1195-1200.

Yang, L., Y. Huang, R. Porta, K. Yanagisawa, A. Gonzalez, E. Segi, D. H. Johnson, S. Narumiya and D. P. Carbone (2006). "Host and direct antitumor effects and profound reduction in tumor metastasis with selective EP4 receptor antagonism." Cancer Res 66(19): 9665-9672.

Yang, V. W., J. M. Shields, S. R. Hamilton, E. W. Spannhake, W. C. Hubbard, L. M. Hylind, C. R. Robinson and F. M. Giardiello (1998). "Size-dependent increase in prostanoid levels in adenomas of patients with familial adenomatous polyposis." Cancer Res 58(8): 1750-1753.

Yao, M., W. Zhou, S. Sangha, A. Albert, A. J. Chang, T. C. Liu and M. M. Wolfe (2005). "Effects of nonselective cyclooxygenase inhibition with low-dose ibuprofen on tumor growth, angiogenesis, metastasis, and survival in a mouse model of colorectal cancer." Clin Cancer Res 11(4): 1618-1628.

Ye, Y. N., W. K. Wu, V. Y. Shin, I. C. Bruce, B. C. Wong and C. H. Cho (2005). "Dual inhibition of 5-LOX and COX-2 suppresses colon cancer formation promoted by cigarette smoke." Carcinogenesis 26(4): 827-834.

Ye, Y. N., W. K. Wu, V. Y. Shin and C. H. Cho (2005). "A mechanistic study of colon cancer growth promoted by cigarette smoke extract." Eur J Pharmacol 519(1-2): 52-57.

Yokoyama, I., S. Hayashi, T. Kobayashi, M. Negita, M. Yasutomi, K. Uchida and H. Takagi (1995). "Prevention of experimental hepatic metastasis with thromboxane synthase inhibitor." Res Exp Med (Berl) 195(4): 209-215.

Yokoyama, Y., B. Xin, T. Shigeto and H. Mizunuma (2011). "Combination of ciglitazone, a peroxisome proliferator-activated receptor gamma ligand, and cisplatin enhances the inhibition of growth of human ovarian cancers." J Cancer Res Clin Oncol 137(8): 1219-1228.

Yoshida, N., T. Yoshikawa, S. Nakagawa, K. Sakamoto, Y. Nakamura, Y. Naito and M. Kondo (1999). "Effect of shear stress and a stable prostaglandin I2 analogue on adhesive interactions of colon cancer cells and endothelial cells." Clin Exp Immunol 117(3): 430-434.

Yoshida, T., S. Ohki, M. Kanazawa, H. Mizunuma, Y. Kikuchi, H. Satoh, Y. Andoh, A. Tsuchiya and R. Abe (1998). "Inhibitory effects of prostaglandin D2 against the proliferation of human colon cancer cell lines and hepatic metastasis from colorectal cancer." Surg Today 28(7): 740-745.

Yoshimoto, A., K. Kasahara, A. Kawashima, M. Fujimura and S. Nakao (2005). "Characterization of the prostaglandin biosynthetic pathway in non-small cell lung cancer: a comparison with small cell lung cancer and correlation with angiogenesis, angiogenic factors and metastases." Oncol Rep 13(6): 1049-1057.

Yoshizato, K., S. Zapf, M. Westphal, M. E. Berens and A. Giese (2002). "Thromboxane synthase inhibitors induce apoptosis in migration-arrested glioma cells." Neurosurgery 50(2): 343-354.

Young, M. R. (2004). "Tumor-derived prostaglandin E2 and transforming growth factor-beta stimulate endothelial cell motility through inhibition of protein phosphatase-2A and involvement of PTEN and phosphatidylinositide 3-kinase." Angiogenesis 7(2): 123-131.

Yoysungnoen, P., P. Wirachwong, P. Bhattarakosol, H. Niimi and S. Patumraj (2006). "Effects of curcumin on tumor angiogenesis and biomarkers, COX-2 and VEGF, in hepatocellular carcinoma cell-implanted nude mice." Clin Hemorheol Microcirc 34(1-2): 109-115.

Yu, H. B., H. F. Zhang, X. Zhang, D. Y. Li, H. Z. Xue, C. E. Pan and S. H. Zhao (2010). "Resveratrol inhibits VEGF expression of human hepatocellular carcinoma cells through a NF-kappa B-mediated mechanism." Hepatogastroenterology 57(102-103): 1241-1246.

Yu, Z., C. Schneider, W. E. Boeglin, L. J. Marnett and A. R. Brash (2003). "The lipoxygenase gene ALOXE3 implicated in skin differentiation encodes a hydroperoxide isomerase." Proc Natl Acad Sci USA 100(16): 9162-9167.

Zhu, Y. M., N. S. Azahri, D. C. Yu and P. J. Woll (2008). "Effects of COX-2 inhibition on expression of vascular endothelial growth factor and interleukin-8 in lung cancer cells." BMC Cancer 8: 218.

Chapter 8

Cathepsins and Other Proteases in Tumor Angiogenesis

Manu O. Platt* and Jerald E. Dumas

Wallace H. Coulter Department of Biomedical Engineering, 315 Ferst Dr, Suite 1308, Georgia Institute of Technology and Emory University, Atlanta, GA 30332 USA

CHAPTER OUTLINE

ABSTRACT

Cathepsins and other proteases play a role in tumor angiogenesis mainly through degradation and remodeling of the extracellular matrix proteins, which promotes vessel ingrowth. Cathepsins are cysteine proteases usually resident in lysosomal compartments, but are released to the extracellular environment by diseased cells. Tight regulation of cathepsin activity prevents aberrant cathepsins from functioning extracellularly, but microenvironmental changes in pH, oxidation state, and overexpression of enzymes overcome these regulatory mechanisms to degrade the collagens, laminins, fibronectins, and other extracellular matrix proteins between the

Correspondence Author Wallace H. Coulter, Department of Biomedical Engineering, 315 Ferst Dr. Suite 1308, Georgia Institute of Technology. Atlanta, GA 30332, phone: +1 404 385 8531. Fax: +1 404 385 8109. Email: manu.platt@bme.gatech.edu

growing tumor and the vasculature. These enzymes are the most potent collagenases and elastases and are therefore able to greatly assist in carving a way for the tumor to metastasize as well as for vessels to penetrate. Endothelial cells, tumor cells, tumor associated macrophages and other cell types have been shown to upregulate these enzymes and play different roles in the angiogenic process. This chapter will discuss cathepsin structure and function, and the regulation of their activity in normal and tumor environments by cells, extracellular matrix, glycosaminoglycans, and other biological molecules. A particular focus will be on cathepsins B, K, L, S, and V as well as cystatin C, their endogenous protein inhibitor as these are most implicated in this disease, particularly breast, lung, and prostate cancer as well as their metastasis to bone. Traditional and novel methods to detect cathepsin activity and the benefits or trade-offs of each will be discussed. Finally, therapeutic strategies that target inhibition of proteases will be presented with a discussion of their controversy and difficulty passing clinical trials due to severe side effects.

Key words: cathepsins, extracellular matrix, angiogenesis, cancer, proteases

1   Introduction

Cysteine proteases, matrix metalloproteinases, serine proteases, and aspartic proteases, all play roles in tumor angiogenesis primarily through degradation and remodeling of the extracellular matrix proteins to promote vessel in-growth. This chapter will focus on the papain family of cysteine proteases, commonly called cathepsins, as there has been a recent surge in studies and important findings that implicate them heavily in angiogenesis, tumor progression, and metastasis. Cathepsins are cysteine proteases that usually resident in lysosomal compartments, but they are released to the extracellular environment by diseased cells. Tight regulation of cathepsin activity prevents aberrant cathepsins from functioning extracellularly, but microenvironmental changes in pH, oxidation state, and overexpression of enzymes overcome these regulatory mechanisms to degrade the collagens, laminins, fibronectins, and other extracellular matrix proteins between the growing tumor and the vasculature. These enzymes are the most potent collagenases and elastases and are therefore able to significantly facilitate degradation of the ECM barrier, which aids the tumor to metastasize as well as for vessels to penetrate. Endothelial cells, tumor cells, tumor associated macrophages and other cell types have been shown to upregulate these enzymes as each plays a different role in the angiogenic process. There are 11 members of the cathepsin family (Table 1), each with different preferences for substrate cleavage, and each family member may play a unique or redundant role in the tissue remodeling occurring around the growing tumor. Parsing the contribution of each enzyme has presented a challenge as the sequence homology among family members has obfuscated specific identification with synthetic substrates, matrices, or probes. It has also confounded pharmacological inhibitor development and proper dosing, yielding side effects that have ended some clinical trials for use as chemotherapeutics.

Table 1 Cathepsin family members and properties

Preferred names	Other names	Tissue expression	pH	Gene symbol
Cathepsin B	Cathepsin B1, APP secretase	Ubiquitous	pH 4-6 (optimal), stable to pH 7	CTSB
Cathepsin C	Dipeptidyl peptidase I, cathepsin J, dipeptidyl-transferase	Ubiquitous	pH 6 (optimal), pH 4-7.5 (stable)	CTSC
Cathepsin F	SmCF	heart, skeletal muscle, brain, testis ovary	pH 5.2-6.8 (optimal), pH 4.5-7.2 (stable)	CTSF
Cathepsin H	Cathepsin B/3, BANA hydrolase	brain, kidney, liver, inflamed tonsil	pH 6.8 (optimal), pH 5-8 (stable)	CTSH
Cathepsin L	Spase, cathepsin L1	Ubiquitous	pH 6 (optimal), pH 4-7 (stable) varying	CTSL
Cathepsin K	cathepsin O(1), cathepsin O2, cathepsin X	predominantly in bone (osteoclasts), present in most epithelial tissues	pH 6 (optimal), pH 4-8 (stable)	CTSK
Cathepsin O		widely expressed		CTSO
Cathepsin S		alveolar macrophages, spleen, testis, epithelial cells; CD4+ T-cells	pH 6 (optimal), pH 4.5-8 (stable)	CTSS
Cathepsin V	cathepsin L2, cathepsin U	thymus, testis, brain, corneal epithelium, skin	pH 5.7 (optimal), pH < 4 to > 7.2 (stable)	CTSL2
Cathepsin W	Lymphopain	spleen, natural killer, and cytotoxic T-cells		CTSW
Cathepsin X	cathepsin B2, cathepsin P, Y, cathepsin Z	widely expressed, ubiquitously in primary tumors and cancer cell lines		CTSZ

1.1 Cathepsin Structure and Functions

Cysteine cathepsins catalyze peptide bond cleavage of other proteins using their SH- groups at the active site. "Cathepsin" is the main term used for this superfamily of cysteine proteases, and their activity is highly regulated. Having first been identified in lysosomes of different cell types, they are primarily involved in the intracellular breakdown of proteins in lysosomes, where up to 50% of proteins are degraded. They prefer acidic pH for optimal activity and most lose this activity quickly in neutral to basic pH; cathepsin S is the exception to this rule as it maintains its activity at neutral pH (Chapman et al. 1997). At the cathepsin active site, a cysteine and histidine residue pair serves as a nucleophile to hydrolyze the amide bond in the substrate. Appropriate positioning of the protons of these residues depends on the pH of the solution (Drenth et al. 1968). Thirdly, like many other protease families (trypsin from the serine protease family and MMP-9 from the matrix metalloproteinase) they are synthesized as zymogens, inactive precursors that require proteolytic cleavage of a propeptide region to be active.

In addition to all of these inherent regulatory mechanisms, there are several inhibitors constantly present in high ratios to the cathepsins to prevent their activity if they are unexpectedly released from cells (Topol and Nissen 1995; McGrath 1999; Turk et al. 2000; Jormsjo et al. 2002; Turk 2006). Cystatins are the extracellular family of inhibitors, and the stefins are generally found intracellularly to prevent unwarranted cathepsin activity. Ultimately, expression of a cathepsin does not mean it will play a physiological role; it must be moved to an acidic environment, activated, and be free from inhibition, but once this happens, protein turnover can be rapid. Among the 11 family members, all tissue matrix protein degradation is possible, although there is selectivity and preference, within the family, for certain substrates over others, and even differences with cell and tissue expression under healthy, physiological conditions. See Table 1 for a descriptive list.

1.2   Regulation of Cathepsin Activity in Normal and Tumor Environments

Because of their destructive potential, cathepsins are highly regulated, but overexpression of cathepsins in tumor environments shifts homeostatic balances, and the disease can progress (Fig. 1). Cathepsins are produced by many of the cell types comprising the tumor, and have been linked to various pathophysiological consequences including tumor growth, extracellular matrix degradation, metastasis, and important for this summary, angiogenesis. That overexpression, in combination with local acidification of the tumor microenvironment due to increased glucose used by tumor cells, provides an environment perfect for cathepsin degradation of extracellular matrix. Regulation of cathepsin activity occurs at several levels from transcription to translation, but also includes the cleavage of propeptide to the mature, active cathepsin, and targeting either to the lysosome or for secretion. Mechanisms of cathepsin secretion are not entirely understood although there is some evidence that the overloading of the lysosomal pathway leads to aberrant trafficking of cathepsin loaded vesicles (Collette et al. 2004). More specifically, secretory lysosomes (Griffiths 2002) and alternative targeting pathways suggest that overexpression of cathepsins B and L overloads the lysosomal targeting (the traditional mannose-6-phosphate receptor mediated path) and causes them to be secreted instead (Linebaugh et al. 1999; Ahn et al. 2002). Procathepsin L aggregation into multivesicular bodies occurs when it is overexpressed and may lead to mannose-6-phosphate receptor independent secretion (Yeyeodu et al. 2000). Cathepsin K is known to be secreted at the ruffled border in osteoclasts by still unidentified mechanisms (Littlewood-Evans et al. 1997;

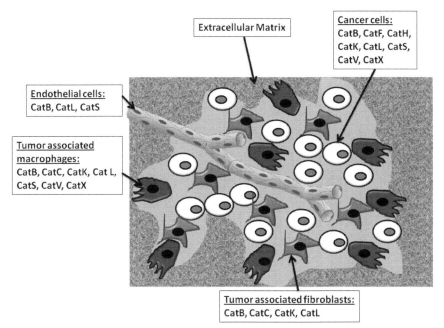

Figure 1 The tumor microenvironment comprised of heterogeneous cell types with each type known to express different cathepsin family members.

Color image of this figure appears in the color plate section at the end of the book.

Dodds et al. 2001), and a number of other family members have been secreted by a host of other cell types.

After secretion, cathepsins must be activated. Procathepsins are activated into the mature enzyme by cleavage of the propeptide. This can occur if other active proteases cleave the propeptide off of each other eliciting a positive feedback of cathepsin activation (Chapman et al. 1997; McQueney et al. 1997). They are also capable of autocatalytically cleaving their own propeptide under acidic conditions to activate themselves (McQueney et al. 1997; Menard et al. 1998; Vasiljeva et al. 2005). Changes in conformation changed due to pH reposition the propeptide into the active site cleft and it becomes hydrolyzed; in this vein, many reports also discuss the inhibitory ability of the free propeptide on cathepsin activity (Carmona et al. 1996; Maubach et al. 1997; Burden et al. 2007).

Cells create pericellular spaces to optimize cathepsin activity due to their size and enclosure capabilities. Lacunae are one such space created by osteoclasts. These specialized pits are created pericellularly by osteoclasts where they serve to resorb bone. Vacuolar type H+-ATPases pump protons into the space to acidify it, and cathepsin K is secreted where it degrades the collagen I components of bone to release the mineral (Dodds et al. 2001). This pericellular space constructed by the cell activates cathepsin K and other enzymes, and controls its volume and environment such that it will perform the necessary tasks (collagenolytic activity to release calcium

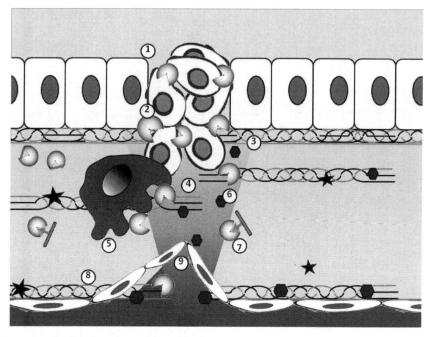

Figure 2   Multiple roles proteases play in tumor angiogenesis. 1) Epithelial to mesenchymal transition and initiation of cancer. 2) Cancer cells begin to overexpress and secrete cathepsins which 3) assist in basement membrane degradation and invasion to become malignant. 4) Local acidification of the peritumor space occurs and 5) macrophages that have infiltrated into the region also secrete cathepsins and mobilize proton pumps to assist extracellular acidification. 6) pro-angiogenic factors such as VEGF are cleaved by proteases from extracellular matrix and are secreted by cancerous cells and tumor associated macrophages. 7) Any endogenous inhibitors present bind to free cathepsins and prevent their proteolytic activity. 8) Endothelial cells receiving pro-angiogenic signals begin to degrade their basement membrane of initial sprouting and invasion, and **9)** endothelial cells begin to migrate and form neovessels towards the recruiting cytokines and hypoxic regions of the growing tumor.

Color image of this figure appears in the color plate section at the end of the book.

mineral from bone in this scenario) but will lose activity if it escapes. Outside of that, the factors required by cathepsin K for high catalytic activity are minimal. Macrophages have also been shown to create these microenvironmental spaces when apposed to insoluble elastin fragments *in vitro* in such a way that ruthenium red or cationized ferritin captured inside of this space could be imaged. This study further showed the mobilization of proton pumps to acidify that extracellular space (Punturieri et al. 2000). In a study of the aggressive breast cancer cell line MDA-MB-231, a group measured the local drops in pH both proximal to the cells' apical surface, and then even greater when measuring the pH below the cell between it and the tissue culture plate (Montcourrier et al. 1997). The conclusions from these data is that cells are capable of producing the necessary components and microenvironmental conditions for cathepsin activity and will even restrict those activating components to fence in the powerful proteases it is releasing extracellularly. Many tumors as a whole have locally acidic environments, and one study of breast cancer significantly reduced cathepsin activity by injecting bicarbonate at the tumor site to locally alkalinize the space (Robey et al. 2009) again verifying the importance of a slightly acidic pH on this enzymatic activity.

1.3 Glycosaminoglycans and Other Matrix Factors

Glycosaminoglycans (GAGs) are bound to proteoglycans in matrix and even on cell surface as a part of the glycocalyx (Jean et al. 2008; Jimenez-Vergara et al. 2010). These charged molecules have been shown to complex with cathepsins to modify their activity towards different ECM substrates, sometimes helping it and sometimes hindering it (Li et al. 2004). In particular, chondroitin sulfate forms complexes with cathepsin K to increase its collagenolytic activity but to suppress its elastase activity (Li et al. 2002; Li et al. 2004). Keratin sulfate behaved similarly, but dermatan sulfate inhibits cathepsin K's collagenolytic activity, suggesting that these complex interactions in cancer tissue, *in vivo*, is far more difficult to describe as more contributors are identified in these processes.

Another mechanism by which proteases regulate angiogenesis is through cleavage of bioactive factors from extracellular matrices. A few of note are cathepsin L release of endostatin from collagen XVIII; arresten, canstatin, tumstatin, and metastatin cleaved from collagen IV, collagen VIII, XV, laminin, fibronectin, are other ECM proteins that release either pro- or anti-angiogenic fragments after proteolytic cleavage (Daly et al. 2003; Mott and Werb 2004; Nyberg et al. 2005; van Hinsbergh et al. 2006).

2 Cathepsins in Vascular Cells

Angiogenesis is a process that begins with degradation of endothelial basement membrane, endothelial cell sprouting and generating new tubes and branches from pre-existing blood vessels. It is important to summarize studies describing the regulation of cathepsins in endothelial cells and smooth muscle cells, the major cells comprising the blood vessel wall, even though they were studied in the context of atherosclerotic plaques and local matrix degradation that assists plaque progression. However, this can be considered a parallel environment to the tumor; both are comprised of many different cell types, macrophage infiltration, and tissue remodeling that alters the properties of that tissue. Additionally, the larger number of studies can provide helpful insight into the regulatory mechanisms that may be occurring in tumor environments. As an example, cathepsins B, K, L, and S are elastinolytic and collagenolytic, proving capable of degrading the main structural components of the blood vessel wall (Topol and Nissen

1995; Gacko and Glowinski 1998; Chen et al. 2002), and depending on the location of the primary tumor, that tissue as well. The acidic environment created by macrophage secretions in the plaque-forming, intimal region can be analogized to the environment created by tumor associated macrophages, and these activities can be extrapolated to known and potential roles for cathepsins in tumor angiogenesis. We will focus on some of the unique properties of specific family members and discuss their identified roles in blood vessels, then extrapolate that to new vessel formation that is essential for angiogenesis.

2.1   Cathepsin B

Cathepsin B is one of the more abundant cathepsins with lysosomal concentrations as high as one millimolar (Turk et al. 2000). However, on endothelial cells, cathepsin B is relocated to the cell surface where it binds Annexin II in caveolae and assists in extracellular matrix proteolysis for angiogenesis and tube formation (Mai et al. 2000; Cavallo-Medved et al. 2009). It also has anti-angiogenic functions as it too generates endostatin after cleaving collagen XVIII (Ferreras et al. 2000). With these dual roles, cathepsin B is considered to be the "angiogenic switch" (Im et al. 2005) by proteolyzing extracellular matrix to initiate cascades that promote angiogenesis (Kobayashi et al. 1991). Collagen IV, laminin, and fibronectin, three major constituents of basement membrane, are all degradable by cathepsin B and this proteolysis was shown at both acid and neutral pH (Buck et al. 1992), implicating its putative involvement in cell invasion and, of course, angiogenesis (Calkins et al. 1998). Further, pancreatic tumor angiogenesis was reduced in the RIP1-Tag2 mouse model in a cathepsin B null mouse, implicating cathepsin B specific involvement in an animal model (Gocheva et al. 2006).

2.2   Cathepsin K

Cathepsin K is the most potent collagenase, capable of cleaving mature types I and II collagen in the native triple helix and in the telopeptide regions (Garnero et al. 1998) while other collagenases can only cleave at one site or the other. Although this protease was first characterized with osteoclast bone resorption, it has also been identified in endothelial cells (Platt et al. 2007), smooth muscle cells (Sukhova et al. 1998), and atherosclerotic plaques (Sukhova et al. 2003; Lutgens et al. 2006). ApoE null mice that were also null for Cathepsin K had smaller plaques, increased collagen content, and less elastic lamina breaks (Lutgens et al. 2006). Interferon-γ and interleukins 6 and 13 have been shown to stimulate cathepsin K in vascular cells and macrophages, while TGFβ and IL-10 have inhibited cathepsin K expression and activity in a number of different cell types (Ebisui et al. 1995; Ling et al. 1995; Lugering et al. 1998; Sukhova et al. 1998; Wang et al. 2000; Zheng et al. 2000; Kamolmatyakul et al. 2004; van den Brule et al. 2005). More importantly, however, cathepsin K has been identified in a number of human cancers: skin, thyroid, cervical, lung, breast, and prostate (Littlewood-Evans et al. 1997; Gaumann et al. 2001; Brubaker et al. 2003; Lindeman et al. 2004; Rapa et al. 2006).

2.3   Cathepsins L and V

Cathepsin L has also been identified in endothelial cells and smooth muscle cells of both mice and humans (Liu et al. 2006; Platt et al. 2006), but it has more direct links to angiogenesis as having been implicated as necessary for neovascularization by endothelial progenitor cells (Urbich et al. 2005). Alternatively, cleavage of collagen XVIII by cathepsin L releases the anti-angiogenic endostatin fragment and was first shown in hemangioendothelioma (EOMA) cells, a tumor cell line of endothelial origin (Felbor et al. 2000). It has also uniquely been shown to have important nuclear functions in processing histones for cell differentiation (Duncan et al. 2008). Cathepsin L cleaves a variety of ECM substrates: fibronectin, laminin, and type I, IV, and XVIII collagen, even at neutral pH (Maciewicz and Etherington 1988). Though it was formerly considered as an elastase, but it has been confused with cathepsin V, a close family member sharing 80% sequence homology (Reiser et al. 2010), and now it is thought that much of that activity should have been attributed to cathepsin V, which has also been recently shown to be the most potent elastase even above cathepsins K and S (Yasuda et al. 2004) although it is generally believed to be intracellular (Tolosa et al. 2003). To confuse things further, mouse cathepsin L is the ortholog to human cathepsin V and that there is no mouse equivalent to human cathepsin L (Yasuda et al. 2004; Reiser et al. 2010); this is especially important when comparing mouse cathepsin L knockout studies and parallels drawn to human disease. It only functions at acidic pH and loses its activity quicker than the others as the pH rises (Barrett and Kirschke 1981). Disease pathologies include tumor metastasis, arthritis, bone disorders, aortic aneurysms (Gacko and Glowinski 1998; McGrath 1999; Liu et al. 2006; Jean et al. 2008; Reiser et al. 2010).

2.4   Cathepsin S

Cathepsin S is a potent elastase and has the special characteristic that at neutral pH, it is still active, making it unique among the papain family of cysteine proteases (Chapman et al. 1997). Cathepsin S has been identified in endothelial cells, smooth muscle cells, and macrophages (Sukhova et al. 1998). In *cathepsin S* null mice, there was decreased elastic lamina breakdown and reduced plaque development when crossed with LDLR null mice to produce the atherosclerotic mouse model (Sukhova et al. 2003), indicating the importance of cathepsin S in extracellular matrix remodeling in diseased conditions. In another study, the importance of cathepsin S in microvessel formation was illustrated by treating endothelial cells with

TNFα, IFNγ, VEGF, and bFGF and observing increased expression of cathepsin S at the mRNA and protein levels. This was associated with microtubule formation and invasion through Matrigel and collagen matrix, but could be blocked with the broad cathepsin protein inhibitor cystatin C, or the small molecule analog E-64. TIMP-1 did not have a significant effect in this study (Shi et al. 2003). Cathepsin S cleavage of elastin has particular relevance for tumor angiogenesis in that previous studies have shown that degraded elastin contributed to angiogenesis (Nackman et al. 1997). and that tumor progression was greater when cancer cells interacted with elastin (Lapis and Timar 2002).

2.5   Cystatin C

Cystatins are the group of inhibitors of lysosomal cysteine proteases, of which, cystatin C is the most prominent and tightly binding (Chapman et al. 1997). This 13 kD protein is translated with a 26 amino acid signal peptide that targets it for secretion approximately one hour after post-translational processing (Merz et al. 1997). At this point its three-dimensional conformation forms a wedge shape that will block the substrate binding cleft of the cathepsin in a tight, reversible fashion (Bode and Huber 2000) with an inhibition constant on the subnanomolar range (Hall et al. 1998).

3   Major Cancers with Cathepsin Proteolytic Activity

Breast, lung, and prostate cancer are the most common forms of the human disease. Two major forms of these cancers are adenocarcinoma and squamous cell carcinoma. Adenocarcinomas originate from the glands and typically have a higher vessel count (Yuan et al. 1995; Sikora et al. 1997) than squamous cell carcinoma, which originates from the epithelium. Therefore, adenocarcinomas are more likely to metastasize. Angiogenesis is an important prognostic indicator of these cancers, and can be indirectly measured by counting microvessels, which have been correlated with tumor size (Cox et al. 2000). We will specifically discuss these cancers from the vantage point of the cells and proteases they express, and their implications for angiogenesis.

3.1   Breast Cancer

Breast cancer is one of the most common cancers among American women affecting one in eight women. The adiposity, extracellular matrix, and microenvironmental regulatory mechanisms of breast tumors promote cathepsin activity contributing to invasion and angiogenesis. There have been a number of studies implicating cathepsins as either biomarkers for breast cancer, or as agents of breast cancer progression indicative of poor prognosis. Cathepsin B and H levels has been shown to be elevated in both tissue and serum samples (Gabrijelcic et al. 1992). Furthermore, cathepsin

B, along with cathepsin L, has been reported as a predictor of relapse and prognostic indicator or shorter survival (Lah et al. 1995; Thomssen et al. 1995; Saleh et al. 2003). In a study that used a novel multiplex zymography assay on frozen breast tissue samples, cathepsin K activity was found to be increased by more than 50-fold in tumor specimens compared to normal, and this was measured in only 10 µg of protein from the tumor homogenate (Chen and Platt 2011). Additionally, this study found that cathepsin L was 9-fold increased in the tumors, but MMP-2 and MMP-9 were only up by 2-3 fold by comparison in the same samples (Chen and Platt 2011).

Ductal carcinoma forms in the lining of the milk ducts, which is surrounded by adipose tissue, lobes, and stromata, but even normal human breast tissue (i.e., lobules) in close proximity support tumor angiogenesis as shown in pre-clinical models (Jensen et al. 1982; Schmeisser et al. 2006). Adipocytes are a prominent cell type in the tumor microenvironment and they also contribute to the types of matrices that are deposited that provide cues for the tumor cells as well as infiltrating macrophages and fibroblasts. Cathepsins B, K, L, S, V, H, and D expression are increased in the tumor environment stimulating adipogenesis (Mohamed and Sloane 2006; Taleb et al. 2006; Xiao et al. 2006; Han et al. 2009; Masson et al. 2011), and interestingly, in avascular areas, tumor associated macrophages, which can comprise up to 50% of cell mass in breast carcinoma (Lewis et al. 1995), are recruited to increase local VEGF levels (Fox et al. 2007; Chen and Platt 2011) that are cleaved from the ECM by cysteine cathepsins. Taken together, one can see the putative links between angiogenesis and adipogenesis in breast cancer.

For optimal cathepsin activity, as described earlier, an acidic environment is preferred. Breast cancer cells have a high capacity to acidify the ECM both apically and basally (Montcourrier et al. 1997), providing an optimal condition for cathepsin proteolytic activity. Mechanisms behind this local acidification in the breast tumor environment have implicated proton pump mobilization and hypoxia to the plasma membrane. MDA-MB-231, one of the more aggressive lines of breast cancer cells, have the ability to acidify their extracellular pH to as low as 5.5 while the MCF-7 cell line is capable of lowering pH to around 6 (Montcourrier et al. 1997). Similarly to macrophages and osteoclast (Sundquist et al. 1990; Reddy et al. 1995), these breast cancer cells use V-ATPase proton pumps to lower pH, and this has been verified by immunocytochemical labeling of the V-ATPase proton pump to the cell surface in highly metastatic yet was absent in less aggressive cells which preferred the Na+/H+ exchangers and bicarbonate transporters (Sennoune et al. 2004). V-ATPase inhibitors significantly reduced both migration and invasion by these cells showing the importance of local acidification for tumor progression (Sennoune et al. 2004). In mice, oral delivery of bicarbonate increased the extracellular pH around the tumor, significantly reduced rate of lymph node metastasis, and pericellular cathepsin B activity (Robey et al. 2009) again verifying the importance of local acidification in tumor advancement.

As a secondary mechanism, hypoxia triggers glycolytic pathways, lactic acid production, and acidification is accomplished via lactic acid production. Hypoxia also triggers increased HIF-1α, a key transcription factor that binds to VEGF promoter stimulating its gene and protein expression in endothelial cells (Ferrara et al. 2003). HIF1α has been linked to angiogenesis in breast cancer (Schneider and Miller 2005), and plays roles in acidification by turning on carbonic anhydrase IX, a pH regulator that converts water and carbon dioxide to carbonic acid (Sundquist et al. 1990). Increased expression and activity of carbonic anhydrase IX has been shown in human patients with invasive breast cancer tumors as HIF-1α dependent and indicator of poor prognosis (Jensen et al. 1982; Chia et al. 2001).

3.2   Lung Cancer

Throughout the world, lung cancer is the leading cause of cancer deaths in men and women, and like other cancers, angiogenesis is associated with a poor prognosis (Cox et al. 2000; Greenlee et al. 2000). During normal granuloma in emphysema, typically caused by smoking or excessive air pollutants, alveolar macrophages defend the lung tissue against pathogens, actively releasing cathepsin B, D, K, L, and S (Chapman et al. 1984; Lesser et al. 1989; Samokhin et al. 2011) with increased elastinolytic activity destroying elastic tissue of the alveoli (Padilla et al. 1988; Golovatch et al. 2009). They are also upregulated in tumors.

Non-small cell lung cancer (NSCLC) is the most common type of lung cancer and is further categorized into two groups: adenocarcinoma and squamous cell. Neovascularization is more prevalent in lung adenocarcinoma, the most common type of NSCLC which originates in the peripheral lung tissue (Cox et al. 2000); squamous cell carcinoma originates near the central bronchus, rich in collagen and elastin (Pories et al. 2008). Cathepsins B and L protein levels (5.6 and 2.2-fold increase for cathepsins B and L, respectively) and activity were upregulated in NSCLC in human tissue samples as determined by enzyme-linked immunosorbent assay (ELISA) and activity-based substrates (Ledakis et al. 1996; Werle et al. 2000). In contrast, cathepsin D, an aspartic protease, showed no protein level or activity correlation when comparing normal and tumor lung tissues (Ledakis et al. 1996). These observations suggest that cysteine cathepsins may be better suited for lung cancer prognosis determination.

Ambiguity exists in studies to define the roles of several cathepsins in lung cancers. Cathepsin S protein expression has been shown to increase twice as high in lung tumors of NSCLC patients as determined by ELISA; interestingly, it was found that patients with a higher level of cathepsin S had a significantly higher survival rate than patients with lower cathepsin S values (Kos et al. 2001). This observation illustrates that protein levels are not always correlated to activity as well as a possible dual role of cathepsin S in lung cancer (Kos et al. 2001). Multiplex cathepsin zymography has indicated at least 3-fold increases for cathepsins K, L, and S activities, respectively, in human lung tumor tissue over normal tissue (Chen and Platt 2011). Cathepsin H protein levels has been shown to decrease in human lung tumors to about 64% of normal lung tissue; however, decreased cathepsin H protein levels in serum from human lung cancer patients could possibly be indicate enhanced secretion of the enzyme (Delebecq et al. 2000; Schweiger et al. 2000). Patients with recurrent intrapulmonary metastasis expressed higher expression of cathepsin B in biopsied tissue (Fujise et al. 2000). Furthermore, cathepsin B has been found to be an independent prognostic factor of patients suffering from NSCLC in immunochemical studies of human tumor tissue (Schweiger et al. 2000). In contrast, MMP-9, a collagenase, exhibited no significant prognostic capability (Fujise et al. 2000).

3.3 Prostate Cancer

Prostate cancer is the third most common cause of death in men of all ages worldwide (Ferlay et al. 2010) and is marked by increased expression of cathepsins B, H, K, L, S and X that play multiple roles from matrikine release after proteolytic cleavage, matrix degradation for migration, and roles in angiogenesis (Friedrich et al. 1999; Brubaker et al. 2003; Berdowska 2004; Nagler et al. 2004; Coulson-Thomas et al. 2010). Prostate cancer is unique among neoplasms because it grows slowly so when diagnosed, more aggressive treatment options are less often pursued. This is important to mention, because, according to the National Cancer Institute, if prostate cancer is found early, when still localized to the prostate or regional vicinity, there is a 100% 5-year survival rate, with precipitous declines in survival after that (De Becker et al. 2007). Like other cancers, angiogenesis is a major contributor to the metastasis of prostate cancer, and prostate cancer cells even have been shown to upregulate VEGF (Ferrer et al. 1998).

Dramatic changes in ECM molecule content, such as decreased laminin and collagen VII, in advanced prostate tumor samples facilitated tumor migration and metastasis (Takeichi 1991; Xue et al. 1998; Brar et al. 2003). Cathepsins and other proteases are upregulated in these environments and are responsible for this ECM remodeling that assists the tumor progression and angiogenesis. Cathepsins B and S have been found to be co-expressed in the early development of primary human prostatic adenocarcinomas in comparison to normal tissue using immunohistochemistry (Fernandez et al. 2001). Of a total of 38 adenocarcinomas cathepsin B and S were expressed in 74% and 84% of the tissues, respectively, with macrophages being the major sources of expression (Fernandez et al. 2001). Interestingly, the activities, measured using fluorometric methods, of cathepsin B, H and L decline in human prostate tumor tissue samples as compared to normal tissue (Friedrich et al. 1999). Although activity does not always correlate with protein expression, this discrepancy between the two studies is surprising. The decreased activity could be a result of a decrease in the epithelial component and an increase of the stromal component during

Table 2 Cathepsin levels in three human cancers: prostate, breast, and lung

Cancer type	Cysteine cathepsin	Cancer vs. normal	Sample	Observation	References
	B	↑	tissue	protein	(Sinha et al. 1995; Sinha et al. 1998; Friedrich et al. 1999; Fernandez et al. 2001)
		→	tissue	activity	(Friedrich et al. 1999)
		↑	cell culture	activity	(Friedrich et al. 1999)
	L	→	tissue	activity	(Friedrich et al. 1999)
		↑	cell culture	activity	(Friedrich et al. 1999)
Prostate	S	↑	tissue	protein	(Fernandez et al. 2001)
	K	↑	tissue	protein	(Brubaker et al. 2003)
	X	↑	tissue	protein	(Nagler et al. 2004)
	H	↑	tissue	protein	(Waghray et al. 2002)
		→	tissue	activity	(Friedrich et al. 1999)
		↑	cell culture	activity	(Friedrich et al. 1999)
		↑	tissue	protein	(Gabrijelcic et al. 1992; Lah et al. 1997; Lah et al. 2000; Saleh et al. 2003)
		↑	tissue	poor prognosis	(Lah et al. 1992; Lah et al. 1995; Lah et al. 1997; Maguire et al. 1998; Levicar et al. 2002)
	B	↑	tissue	predictor of relapse	((Lah et al. 1992; Thomssen et al. 1995; Lah et al. 1997; Foekens et al. 1998; Lah et al. 2000)
Breast		↑	tissue	protein	(Lah et al. 1992; Lah et al. 1992; Yano et al. 2001)
	L	↑	tissue	protein	(Lah et al. 1992; Lah et al. 2000; Levicar et al. 2002)
		↑	tissue	poor prognosis	(Thomssen et al. 1995; Foekens et al. 1998)
		↑	tissue	predictor of relapse	
	H	↑	tissue	protein	(Gabrijelcic et al. 1992; Berdowska 2004)
		↑	serum, tissue	activity/protein	(Berdowska 2004)
	K	↑	tissue	protein	(Littlewood-Evans et al. 1997)

B	↑	tissue	protein	(Krepela et al. 1998; Werle et al. 1999; Fujise et al. 2000)
	↑	tissue	poor prognosis	(Inoue et al. 1994; Kayser et al. 2003)
	↑	tissue	activity	(Ebert et al. 1994; Werle et al. 1994; Krepela et al. 1998; Werle et al. 1999; Werle et al. 2000)
C	↑	tissue	poor prognosis	(Ebert et al. 1994; Werle et al. 1999; Werle et al. 2000)
	↔	tissue	activity	(Krepela et al. 1997)
L	↑	tissue	protein	(Werle et al. 1995; Ledakis et al. 1996)
	↑	tissue	activity	(Werle et al. 1995; Ledakis et al. 1996)
	↑	tissue	poor prognosis	(Kos et al. 2001)
	↔	serum	poor prognosis	(Kos et al. 2001)
H	↑	serum	protein	(Delebecq et al. 2000)
	→	tissue	protein	(Schweiger et al. 2000)
	↑	serum	protein	(Schweiger et al. 2000)

Lung

benign prostatic hyperplasia, and the change of cathepsin activity could be attributed to this change as stromal cells can express cathepsins.

Cathepsin K activity and protein expression has been observed in human prostate cancer tissues and cell lines (Brubaker et al. 2003; Stewart et al. 2004). Cathepsin K expression, measured by RT-PCR, is observed in prostate cancer cell lines: DU-145, PC-3, LNCaP, C4, and C4-2 (Brubaker et al. 2003). And, *in situ* hybridization of human prostate tissue detected cathepsin K mRNA for 5 out of 7 samples, while normal samples had minimum cathepsin K mRNA (Brubaker et al. 2003). Prostate cancer cells have increased expression of the receptor activator nuclear factor kB ligand (RANKL), the key cytokine that binds to its cognate receptor RANK on the cell membrane of osteoclast progenitors, facilitating osteoclastogenesis (Mundy 2002; Odero-Marah et al. 2008). These findings highlight the relationship between prostate cancer cells, osteoclastogenesis, and increased cathepsin K, which is expressed by osteoclasts.

Prostate cancer may be androgen-sensitive (WK Chung), and androgen deprivation strategies have shown some efficacy in limiting angiogenesis in prostate cancer (Marshall and Narayan 1993; Woodward et al. 2005). Hence, hormonal therapy is a key treatment for prostate cancer. However, not all prostate cancers are responsive to hormonal therapy making them castration resistant prostate cancer; the absence of sufficient medications for these types of malignancies enhances the need for on anti-angiogenic therapies as alternative treatments to control prostate cancer.

3.4   Bone Metastasis

Once the primary tumor induces angiogenesis, tumor cells invade the blood vessels and migrate into the capillary bed in bone. Patients with advanced cancers will most likely experience the bulk of the tumor burden in the bones (Mundy 2002). Tumors metastasize to bone and this occurs in 65-80% of patients with metastatic breast and prostate cancers (Mundy 2002; Mundy 2006). The bone tumor microenvironment provides an ideal condition for tumor growth making it highly susceptible to metastasis. Subsequently, the tumor cells extravasate from the blood vessel and enter what will become the "vicious cycle" of bone (Mundy 2002). Tumor cells release growth factors such as bone morphogenetic proteins (BMPs), fibroblast growth factors (FGFs), and platelet-derived growth factor (PDGF) that enhance the differentiation of pre-osteoblastic cells as well as RANKL that to promote osteoclastogenesis in osteolytic lesions. As osteoblasts and osteoclasts are coupled by RANK receptor/ligand, an increase in osteoblast population increases the differentiation of monocyte/macrophages in the bone marrow to osteoclasts. Osteoclasts degrade the bone ECM via mineral dissolution and degradation of type I collagen by cathepsin K, an essential protease in this cycle (Teitelbaum 2000). Growth factors released from the bone ECM support the proliferation of tumor cells, perpetuating the cycle.

4   Assays to Detect Cathepsin Activity

Detection of mature cathepsins and quantification of specific proteolytic activity has proven difficult. Procathepsins are stable at neutral pH but are not proteolytically active. After the cleavage of the propeptide, the active and mature cathepsin is generally unstable in neutral environments (of course with the exception of cathepsin S), and susceptible to pH-induced denaturation and degradation which complicates measurements of its proteolytic activity in cell extracts and the extracellular environment. Of course this is important to identify cathepsin expression levels, their activity, and their roles in tumor progression and angiogenesis. Here, we will discuss many different methods used to measure and identify cathepsins in cancer tissues along with the pros and cons of each.

4.1   Traditional Methods

Commonly used methods of cell and molecular biology have all been used to characterize, identify, and quantify cathepsin expression at the transcriptional and translational levels. Immunohistochemistry, immunoblotting (Western blots), enzyme linked immunosorbent assays (ELISA), and real-time polymerase chain reaction (RT-PCR) have been the most popular methods. However, these methods do not measure enzymatic activity, which is important when considering their functions as proteases and their synthesis and maturation processes. The zymogen or procathepsin form is distinctly separated from the mature form in Western blots due to the propeptide cleavage reducing the molecular weight by around 7–10 kDa. However, when performed immunohistochemical labeling of tissues, most antibodies do not distinguish between the pro-form and the mature form leading to the majority of the labeled proteins generally being the inactive procathepsin forms, which are more stable at neutral pH. This is, perhaps, one of the causes behind the overestimation of cathepsin levels that have led to improper dosing of cathepsin pharmaceutical inhibitors, and the resultant side effects that have ended the phase II clinical trials early loss (Desmarais et al. 2008; Bromme and Lecaille 2009).

4.2   Novel Methods

4.2.1   Quenched-fluorophore synthetic substrates

To quantify the activity of cathepsins in culture or tissues, a combination of synthetic quenched fluorescent substrates has been employed with the quencher molecule benzyloxycarbonyl linked to the fluorophore, 7-amino-4-methylcoumarin (AMC) by short peptide sequences that provide selectivity for certain cathepsins over others. As examples, Z-RR-AMC (arginine-arginine linker) is selective for cathepsin B; Z-FR-AMC (phenylalanine-arginine linker) for cathepsin L, and Z-GPR-AMC (glycine-proline-arginine) for cathepsin K (Barrett and Kirschke 1981;

Aibe et al. 1996). These peptides had initially been perceived to be specific for a certain cathepsin, but promiscuity and cross-reactivity have confounded these studies. These efforts at measuring cathepsin activity must be interpreted carefully as both limits of detection and cross-reactivity with other cathepsin family members have been problematic. For example, although Z-GPR-AMC has an order of magnitude higher affinity and susceptibility to hydrolysis for cathepsin K ($k_{cat}/K_m = 1.2 \times 10^5$ $M^{-1}s^{-1}$) than cathepsin B ($k_{cat}/K_m = 9.5 \times 10^3$) (Hagar and Vichinsky 2008), cathepsin B is the most abundant cathepsin (Turk et al. 2000) and able to hydrolyze a significant amount of substrate in cellular assays. CA-074, the cathepsin B inhibitor, was added to the cellular extracts to prevent this cross-reactivity, but the significant reduction in AMC release due to cathepsin B inhibition distorted the results suggesting minimal contribution of cathepsin K even though immune-based approaches have shown otherwise (Brubaker et al. 2003). Methods are schematized in Fig. 3.

4.2.2   Activity-based Probes

Labeling of mature cathepsins with active site probes designed to bind to mature, properly folded cathepsins capable of substrate hydrolysis has proven beneficial in a number of contexts, but the probes are usually promiscuous among all cathepsin members. Radioactive, fluorescent, or biotinylated activity-based probes have been coupled with blotting and histological protocols (Greenbaum et al. 2002), and have increased sensitivity to visualize the mature form in a blot. Dr. Matthew Bogyo has been a pioneer in the development of these probes and modified their reporter moieties for use with near infrared and *in vivo* labeling (Greenbaum et al. 2002; Joyce et al. 2004; Yasuda et al. 2004; Cavallo-Medved et al. 2009).

4.2.3   Gelatin Zymography

Gelatin zymography is a technique that has been used widely to detect matrix metalloproteinase (MMP) activity (Snoek-van Beurden and Von den Hoff 2005) and less so for cathepsins (Hashimoto et al. 2006; Klose et al. 2006; Chen and Platt 2011). This technique has many benefits: 1) it does not require antibodies, 2) separation of proteins by molecular mass visually confirms enzyme identity, 3) densitometry can be used for quantitative analysis, and 4) versatility of the assay enables inhibition with small molecules to corroborate the identity of the enzyme of interest. Native PAGE and non-reducing methods have been reported for cathepsins B and L, respectively (Hashimoto et al. 2006; Klose et al. 2006; Chen and Platt 2011), but a protocol for cathepsin K zymography has never been described. Multiplex zymography's utility as a supplemental screening tool of pathological specimens was effectively shown to profile cathepsin K, L, and S activities in breast, lung, and cervical tissue at three different stages of tumor progression (Chen and Platt 2011). This matrix of information was captured by this assay after pathological grading of the biopsied tissue indicating that quantitative comparisons with cathepsin zymography can supplement the gold standard histological methods of determining whether biopsied tissue is cancerous or not.

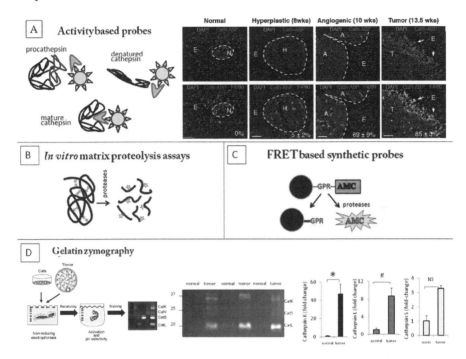

Figure 3 Assays to detect cathepsin activity. A) Activity based probes bind to cathepsins in their active conformation at the active site and provide fluorogenic, radioactive, or chromogenic signals of the number of active cathepsins. Image shown is reprinted with permission from Gocheva et al. (Gocheva et al. 2006). Cathepsins are highly activated in infiltrating macrophages during tumor progression at the angiogenic islet and tumor stages of RT2 tumorigenesis. Pancreatic tissue from RT2 pancreatic cancer model mice were injected with Cath-ABP and pancreatic tissue was stained with an -F4/80 antibody to indicate macrophage content. Normal (N), hyperplastic (H), angiogenic (A), and tumor (T) islets are shown. Arrows are used to indicate the invasive tumor front. B) *In vivo* matrix proteolysis assays. Fluorogenically labeled matrix proteins are cleaved by proteases to generate increased fluorescence that can be assayed spectrophotometrically or by fluorescence microscopy. C) FRET based synthetic peptides follow similar methodology. Quencher molecule tethered to fluorescent moiety by amino acid sequence susceptible to protease cleavage will release fluorescence after being sufficiently separated by the hydrolysis of the linker. D) Gelatin zymography schematic describes the method and representative zymogram shown with permission from Chen et al. (Chen and Platt 2011) illustrating increased activity of mature cathepsins K and L in breast cancer tissue compared to normal by 50-fold and 9-fold, respectively. Cathepsin S increased activity was not statistically significant.

Color image of this figure appears in the color plate section at the end of the book.

This also can significantly reduce cost compared to immunobased methods such as ELISA, Western blotting, and immunohistochemistry, and remove concerns of nonspecific antibody binding and pro-cathepsin detection interference. Procathepsins K, S, and L are detectable by immunoblotting, but do not give detectable zymography signals (Wagener et al. 2001 and data not shown). Cathepsins B and X are not active by this assay presumably due to the improper refolding of their occluding loops in the native conformations (Maynard et al. 2003; Nagler et al. 2004).

4.2.4 *In vitro* Proteolysis Assays

The most common means of examining the amount of active cathepsin within a system is by analyzing the kinetics of the proteolytic reactions with natural or synthetic substrates tagged with chromogenic, fluorogenic, or radioactive reporters. Fluorogenic elastin, gelatin, and collagen IV have been used with good effect to image live cell proteolysis, determine enzyme activity in cell extracts, and even to image activity *in situ* (Galis et al. 1994; Galis et al. 1995; Friedrich et al. 1999). Matrix proteins are heavily labeled with BODIPY, fluorescein, or rhodamine such that the fluorophores quench each other due to their close proximity. Upon proteolytic cleavage, the fluorophores are sufficiently separated by distance that the fluorescent is able to be measured and an increase in fluorescence indicates increased proteolytic activity. Appropriate combinations of inhibitors must be co-incubated to determine enzyme specificity since many of these enzymes cleave the native substrates.

5 Therapeutic Strategies

The inhibition of angiogenesis is a primary strategy in precluding the growth and metastasis of cancerous tumors. As a result of this strategy, cysteine cathepsins along with other proteases have become targets of inhibition. There are two major approaches in inhibiting cathepsins: proteins and small molecules. These will be discussed regarding their mechanisms, efficacy, and progression towards clinical use.

5.1 Proteins as Inhibitors

Protein therapies can be difficult and expensive to produce, but protein inhibitors can reproduce body's natural responses to turn up or down specific responses. The biotechnology company Amgen has been leading the way with the largest number of protein products on the market. Their product Anakinra is an IL-1 receptor antagonist that reduces rheumatoid arthritis and has long term safety and efficacy (Fleischmann et al. 2006). Specific delivery to the target site is challenging as diffusive limitations, innate proteases and defense mechanisms, all present challenges that small molecules easily circumvent, but the efficacy can be huge if it works. The leaky vasculature of tumors due to the neovessels that are forming as well as increased blood flow to that area have allowed for many protein products to escape the vessels and localize to the tumor have also supported more direct delivery to tumor regions, but proteases degrade proteins; increased extracellular proteases at tumor sites can be inhibitory to effective delivery of therapeutic concentrations of protein drugs.

Antibodies are highly selective, tightly binding proteins, naturally produced by B cells to fight infections, and they have been employed for chemotherapeutic purposes. Perhaps, the most famous example of this is trastuzumab, marketed as Herceptin, the recombinant antibody against the epidermal growth factor receptor family member HER2 that is upregulated in a subset of breast cancers. Trastuzumab effectively inhibits tumor cell proliferation in breast cancer cells overexpressing HER2 (Baselga et al. 1998). Antibodies are available for all cysteine cathepsins and have been studied in several *in vivo* models to inhibit cathepsins s and l to block angiogenesis and tumor progression (Frade et al. 2008; Ward et al. 2010). Anti-cathepsin S antibodies were delivered in 6-8 week old female nude mice on either flank, synergistically with anti-VEGF therapy, and inhibited mean vessel area from 1400 mm^2 to 400 mm^2 in human xenograft models (Aase et al. 2001; Ward et al. 2010). Anti-cathepsin L therapy inhibited the growth of tumors induced by the injection A375SM human melanoma cells into male athymic nude mice (Frade et al. 2008). Tumor volume was reduced to below 100 mm^3 with anti-cathepsin L therapy compared to tumor volumes surpassing 800 mm^3 for controls (Frade et al. 2008).

There are two families of endogenous inhibitors of cysteine cathepsins: the extracellular cystatins and the intracellular stefins (Turk and Bode 1991), and the concentration of these inhibitors are typically reduced at tumor sites (Watson and Kreuzaler 2009). These are tight binding inhibitors that can be used for pharmaceutical means to mimic the natural response to suppress overactive proteases with controlled release. Intracellular delivery of cystatin via poly (lactide-co-glycolide) (PLGA) nanoparticles effectively inhibited cathepsin B in MCF-10A neo T cells (Cegnar et al. 2006). Furthermore, palmitoylated (covalent attachment of a fatty acid chain) cystatin inhibited intracellular cathepsin B at it was internalized by MCF-10A neo T cells, which impaired both intracellular and extracellular cathepsin B activity that contribute to invasion (Kocevar et al. 2007). It can be conjectured that these technologies would be applicable in the control of angiogenesis.

Serpins have the ability to inhibit a broad range of cathepsin activity despite being a serine protease inhibitor (Jayakumar et al. 2003; Watson and Kreuzaler 2009; Higgins et al. 2010). Serpin research has produced positive results in the treatment of cancer as it is a promising anti-angiogenic agent, and interestingly, most of these serpins had a non-inhibitory role, acting more as a substrate as it reduced the migration rate of both MDA-MB-435 (from 1.42 to 0.23 μm/hr) and MDA-MB-231 (from 4.48 to 0.19 μm/hour) breast cancer cells by binding to the cell surface as determined by immunostaining (Sheng et al. 1996; O'Reilly et al. 1999; Celerier et al. 2002; Daly et al. 2003; Dawson et al. 2009). Thus, suggesting that serpins could have a dual role to inhibit cathepsins released by the cell, and affect cancer cell migration rate. Delivering maspin, a different serine protease inhibitor, to LnCaP tumors established in dorsal black flaps of male athymic mice blocked tumor growth and reduced the tumor volume from 116.3 mm^3 to 33.9 mm^3 (Zhang et al. 2000). Suppression of breast cancer growth in nude mice has been accomplished via transfection with myoepithelium-derived serine inhibitor (MEPI) (Xiao et al. 1999). Incorporation of MEPI, significantly inhibited tumor growth in mammary fat pads in nude mice, reducing it from 1600 mm^3 for the controls to 435 mm^3 (Xiao et al. 1999). The results from these studies illustrate the importance of cathepsin activity in the growth of tumors as tumor growth was suppressed by the inhibition effects of maspin.

5.2   Small Molecule Inhibitors

Small molecule inhibitors target both intracellular and extracellular cathepsins and are an attractive options to bind to proteases and inhibit their contributions to angiogenic programs, as they have high lipophilicity and low molecular weight (i.e., less than 500) (Bumcrot et al. 2006), and can be designed to be reversible or irreversible. E-64 (trans-epoxy succinyl-l-leucylamido-(4-guanidino) butane), which was isolated from *aspergillus japonicas* (mold) (Barrett et al. 1982), is a potent, irreversible cysteine protease inhibitor that makes a covalent change to the active site. Its structure is very similar to cystatin C, the endogenous protein inhibitor of the cysteine cathepsins that binds at subnanomalar K$_i$ values by inserting a wedge shape to fill the cleft at the active site (Hall et al. 1998). E-64's similarities to cystatin C and tight binding affinity have led to modifications of its structure to produce derivatives such as E-64c to improve cell permeability (Drahl et al. 2005) and E-64d which is also cell permeable, but must be activated by esterases to inhibit intracellular cathepsins (Claveau et al. 2000). E-64c phase III clinical trials in Japan for muscular dystrophy produced negative results since the drug covalently modified other proteins beyond the cathepsins it was targeting (Drahl et al. 2005). An *in vitro* invasion assay using MCF-10AT breast cancer cells indicated a 12% reduction of resorption in the presence of E-64c (Bervar et al. 2003). However, these epoxy-based inhibitors will need to be further studied as they can also cause immune responses and other side effects (Wun-Shaing W. Chang 2007; Chen et al. 2010).

Reversible inhibitors may be preferred because they are less toxic, but a consequence is that the inhibition is less effective (Chen et al. 2010), and does not effectively block angiogenesis in animal models of tumor progression. Leupeptin is a reversible inhibitor of serine and cysteine proteases isolated from *Streptomyces* (soil bacteria) (Kim et al. 1998). Leupeptin has been shown to have inconclusive results in the reduction of angiogenesis in pancreatic and prostate cancer models (Briozzo et al. 1991; Lafleur et al. 2001; Jankun et al. 2006).

Cancer cells have developed mechanisms to block apoptotic pathways, but cell death can still be achieved by sensitization to the lysosomal death pathway. Since cathepsins, particularly cathepsins B and L, are known to facilitate cellular apoptosis through the lysosomal death pathway (Fehrenbacher and Jaattela 2005), a second approach to prevent cancer and angiogenesis is to exploit their abilities to drive programmed cell death instead of inhibiting them. Strategies are being developed to use lysosomal enzymes, such as cathepsins, to trigger the apoptotic pathway (Bareford and Swaan 2007).

5.3 Clinical Trials of Cathepsin Inhibitors

There are a multitude of small molecules developed to target cathepsins K and S as these two have been specifically implicated in bone metastasis and angiogenesis, respectively. Both of these cathepsins have been implicated in several orthopedic diseases such as osteoporosis and arthritis. Therefore, there are several inhibitors that are in various stages of clinical trials with a compilation in Table 3. Such treatments involve small molecules that contain nucleophilic "warheads" that bind to the active site of the enzymes (Wun-Shaing W. Chang 2007).

5.3.1 Cathepsin K Inhibition

Cathepsin K is upregulated in the three major forms of human cancer: breast, lung, and prostate. As there are several oral therapeutics that target cathepsin K for orthopedic disease purposes, including bone metastases, a discussion of these molecules is relevant here, as they may be applicable to tumor progression and anti-angiogenic therapies. From 2004-2010, there were more than 50 patent applications of potential cathepsin K inhibitors (Wijkmans and Gossen 2011), yet to date, no cathepsin K inhibitors have been approved for clinical use. It was once suspected that side effects from cathepsin K inhibitors would have been minimal as the enzyme's predominant role was for bone resorption. Lysosomotropic drugs were developed such that they could target cathepsins in the lysosomes, but these classes of drugs facilitated serious side effects. Although they were successful in reducing target goals, off-targeting effects diminished other cathepsins, and led to lysosomal accumulation of protein, causing skin rashes, lung fibrosis, and cartilage loss (Desmarais et al. 2008; Bromme and Lecaille 2009).

To minimize side effects, non-lysosomotropic drugs that target secreted cathepsin K such as odanacatib, have been developed and exhibit reduced skin and lung related side effects during clinical trials (Bromme and Lecaille 2009; Costa et al. 2011). Non-lysosomotropic drugs target cathepsins that are released from the cell, and they do not enter the lysosome, reducing the chance of affecting other off-target cathepsins. Interactions with other drugs

Table 3 Cathepsin inhibitors and their progression through clinical trials.

Drug	Common name	Target	Chemical nature	Indication	Phase
MK-0822	Odanacatib	Cathepsin K	nitrile-based, non-basic, nonlysosomotropic	breast cancer; metastatic bone disease	III
AAE-581	Balicatib	Cathepsin K	nitrile-based, basic, lysosomotropic	osteoporosis	Discontinued at II
CRA-013783		Cathepsin K		osteoporosis	I
MIV-701		Cathepsin K	not disclosed	bone degradation diseases	Discontinued
MIV-710/711		Cathepsin K		bone degradation diseases	I
GSK462795	Relacatib	Cathepsin K	a-heteroatom cyclic ketone, monobasic	bone metastatis, osteoporosis, osteo-arthritis	Discontinued at II
AFG-495		Cathepsin K	nitrile-based, basic, lysosomotropic	osteoporosis	Discontinued at II
OST-4077		Cathepsin K	cyclic ketone	bone degradation diseases	Preclinical
SB-553484		Cathepsin K		osteoporosis, bone metastasis	Preclinical
RWJ-445380		Cathepsin S		rheumatoid arthritis	II
CRA-028129		Cathepsin S		psoriasis	I
Undisclosed		Cathepsin S		autoimmune diseases	Preclinical
AM-3840		Cathepsin S	bicyclic ketone	rheumatoid arthritis, neuropathic pain	Preclinical
AM-3701		Cathepsin K	bicyclic ketone	osteoporosis, bone metastasis, osteo-arthritis	Preclinical
AM-3876		Cathepsins S, K	bicyclic ketone	osteoarthritis, rheumatoid arthritis	Preclinical
ONO-5334		Cathepsin K		osteoporosis and osteopenia	II
Bevacizumab	Avastin	VEGF	IgG1 antibody	colorectal cancer	appproved
Sorafenib	Nexavar	VEGF	tosylate salt	renal cell carcinoma	appproved
Sunitinib	Sutent	VEGF	receptor protein	renal cell carcinoma	appproved

have also negatively affected the development of cathepsin K inhibitors. Relacatib, a cyclic ketone inhibitor of cathepsin K, was discontinued after possible drug interactions with acetaminophen and ibuprofen, over the counter medicines for pain, and atorvastatin, commonly used statin originally marketed as Lipitor (Costa et al. 2011).

5.3.2   Cathepsin S Inhibition

Research of cathepsin S inhibition remains a promising strategy to control angiogenesis in cancer. Johnson & Johnson Pharmaceutical Research and Development currently has a cathepsin S inhibitor, RWJ-445380, in Phase II clinical trials, with rheumatoid arthritis as a primary target (Palermo and Joyce 2008). Cathepsin S antibodies, such as Fsn0503, have been shown to inhibit tumor cell invasion and block capillary tube formation(Ward et al. 2010). Companies such as Amura (UK) are currently developing synergistic cathepsin S and K inhibitors for osteoarthritis (Vasiljeva et al. 2007). The promising success of these new cathepsin S inhibitors is exciting as these drugs possess potential for dual applications in both orthopedics and angiogenesis. VEGF initiates angiogenesis when it is secreted from the tumor and has been conjectured to foster a balance between cathepsins (B, S, L) and cathepsin inhibitors (Chang et al. 2009). Treatment of human umbilical vein endothelial cells (HUVEC) with both cathepsin S inhibitor and anti-VEGF has already been shown to have a synergistic effect on the inhibition of microvascular development (Ward et al. 2010).

5.4   Novel Technologies

With the success of numerous cathepsin inhibitory technologies, there has been a demand for novel drug delivery systems. Several polymeric biomaterial/nanotechnology platforms are being designed to target tumor angiogenesis with the goal of improving the pharmacological efficacy via protection and controlled release. Therapeutic agents can either be conjugated within the polymer backbone or encapsulated in nanoparticles. Nanoconjugate materials include N-(2-hydroxyproplyl) methacrylamide (HPMA), poly(lactic co-glycolic acid) (PLGA), polysaccharides, dendrimers, and other natural/synthetic polymers (Banerjee et al. 2011). Selection criteria of the polymer system include conjugation potential, degradation properties, FDA-approval, and mechanical integrity. Conjugated polymers require the cleavage of covalent bonds for release, while encapsulated nanoparticles provide controlled release via surface or bulk erosion (Banerjee et al. 2011).

A wide variety of such platforms are currently being developed for protease inhibition. Cystatin released from PLGA nanoparticles inhibits cathepsin B activity *in vitro* studies (Cegnar et al. 2004), protecting the cystatin against general proteolytic degradation. With the advent of anti-VEGF platforms, conjugate polymers have become useful tools. A polymer-peptide conjugate incorporating the soluble fragment of VEGF receptor Flt-1 has been shown to reduce tumor growth in a human xenograft model

(Line et al. 2005), while a polyethyleneimine conjugated system has shown similar effects *in vitro* (Kim et al. 2005). Such biomaterials platforms could also provide effective delivery of other small molecule and antibody/ protein inhibitors such as those discussed in the previous section.

Smart biomaterials that incorporate protease degradable cross-links are currently under development to facilitate greater interaction between the material and the microenvironment. A majority of such materials are sensitive towards MMP developments (Lutolf et al. 2003; Seliktar et al. 2004). However, with the importance of cathepsins in biology moving to the forefront, biomaterials based on cathepsin degradation are being studied. When delivered with an ethylene vinylacetate copolymer pellet system, which produced a sustained releases of bFGF, leupeptin, and SJA6017 (a synthesized peptide cysteine cathepsin inhibitor) into guinea pig and rat corneas for the treatment of corneal angiogenesis, reduction was observed (Tamada et al. 2000). The delivery of bFGF with 10 nanomole of SJA6017 significantly reduced the number of blood vessels, while the same dose of leupeptin produced insignificant results (Tamada et al. 2000) suggesting that the delivery vehicle can impart inhibitory efficacy.

With the advent of RNAi technologies, the silencing of the cathepsin genes has been studied as anti-angiogenic strategies. Silencing of proteases such as cathepsin B in mice inhibits invasion as well as apoptosis of prostate cancer cells (Nalla et al. 2010), when uPAR (urokinase-type plasminogen receptor) and cathepsin B were silenced with RNAi, glioma cell invasion and angiogenesis was reduced in a nude mouse model (Gondi et al. 2004). These studies showed great promise for the use of RNAi to silence proteases that facilitate angiogenesis, but like other modalities, side effects will need to be addressed. Complications resulting from RNAi therapies can result in off-target effects as well as the saturation and silencing of endogenous pathways of essential RNA, causing toxicity (Lopez-Fraga et al. 2008). For example, off-target knockdown effects of cell adhesion molecules could compromise the migration of monocytes, compromising the immune system.

6 Matrix Metalloproteinases (MMPs) vs Cathepsins

Although this focus is on cysteine cathepsins, MMPs and aspartic cathepsins (such as cathepsin D) have been widely studied in cancer biology and deserve mention.

6.1 MMP Structure and Function

MMPs are the class of zinc and calcium dependent endopeptidases that have been highly investigated in vascular biology and matrix remodeling. Currently, 14 MMPs have been studied in vascular cells. Like cathepsins, they are synthesized as zymogens, with a propeptide that must be cleaved to activate the latent enzyme. MMPs and members of other protease families are capable of this cleavage putting into motion a cascade of proteolytic activity to increase their combined activity. Together, as a family, MMPs are capable of degrading all of the components of the vascular extracellular matrix and are classified according to their substrate preference: collagenases, gelatinases, stromelysins, membrane-type MMPs, and others (Newby 2005).

MMPs, like cathepsins play dual roles in angiogenesis. Pro-angiogenic factors such as fibroblast growth factor (bFGF), vascular endothelial growth factor (VEGF), and transforming growth factor beta (TGF-β are released from matrix tethers as cleaved by MMPs. Conversely, MMPs are able to cleave the precursors of endogenous inhibitors of angiogenesis, angiostatin, endostatin, and tumstatin. For example, MMP-9 has been widely studied as a major contributor in angiogenesis (Bergers et al. 2000; Bergers and Benjamin 2003). However, tumstatin is generated by MMP-9 proteolysis of the basement membrane of collagen IV (Hamano et al. 2003). Thus, the inhibition of MMP-9 disturbs the regulation of the angiogenesis process. Increased angiogenesis and increased lung tumor growth, transplanted subcutaneously, was associated in MMP-9 null mice (Hamano et al. 2003). Of note, increased levels of MMP-9 and angiostatin were observed when MMP-9-overexpressing cells were transplanted into wild-type mice (Pozzi et al. 2002). Subsequently, a smaller tumor volume and decreased vascularization was observed.

TIMPs are the endogenous, potent inhibitors of MMPs and are also studied in vascular physiology to observe any shift in the balance from inhibition to proteolysis by this family. There is a 1:1 stoichiometric binding of TIMP to MMP that involves the N-terminal cysteine displacement of the water molecule at the active site from the zinc ion (Gomis-Ruth et al. 1997).

6.1.1 Proteolytic Activity Comparisons of Cathepsins and MMPs

Cysteine cathepsins function best at acidic pH, not the neutral pH preferred by MMPs, and are often overlooked because of their preference for acidic milieus, even though 60% of elastinolytic activity of macrophages is cathepsin dependent while only 25–30% is attributed to MMPs (Yasuda et al. 2004). Cathepsin V has been shown to be the most potent mammalian elastase with activity 3–4x higher than other cathepsins and 2–6 times higher than pancreatic elastases MMP-2 and MMP-12 (Yasuda et al. 2004). Cathepsin K solubilizes twice as much insoluble type I collagen than MMP-9 over the same time periods at equimolar concentrations (Garnero et al. 1998). In recent times, cathepsin overexpression is implicated in aggressive, malignant cancers including lung, colorectal, skin, bone, brain, and pancreatic cancers as well as delivering a poor prognosis for patient survival.

6.2 MMP Inhibition

MMP inhibitors (MMPIs) have mostly been developed to treat cancer by suppressing tumor growth, and they have been evaluated over a wide range of animal models and *in vivo* pharmacological experiments. One initial challenge was poor oral bioavailability, which led to a higher inhibitor dosage requirement, and secondly, targeting issues abound as MMPs also share similar structure at active sites, making their inhibitors non-selective. This limitation has further complicated the ability to achieve adequate inhibitory concentrations.

Hydroxamate, a chelating agent that binds Zn2+ ion inhibiting MMP activity (Jani et al. 2005), has been shown to increase metastasis in certain animal models (Deryugina and Quigley 2006). Lung metastasis was not significantly reduced when treated with batimastat, a metalloproteinase inhibitor, in immunodeficient mice injected with human carcinoma cells. Of note, batimastat induced liver metastasis, and unusual occurrence (Deryugina and Quigley 2006). Clinical trials involving MMPIs failed in the late 1990s due to broad inhibition of other MMPs causing severe musculoskeletal diseases (Turk 2006; Palermo and Joyce 2008). Interestingly, negative correlations between MMP expression and cancer progression have also been observed in several cancers such as pancreatic, colorectal, breast, cervical, and melanoma (Mook et al. 2004; Garcea et al. 2005; Hofmann et al. 2005; Deryugina and Quigley 2006). Just as with cathepsins, it appears that turning MMPs off can elicit as many side effects as their upregulation indicating the important balance of their activity in healthy and disease environments.

7   Conclusions and Perspectives

Targeting proteases to inhibit tumorigenesis and angiogenesis has been a challenging enterprise due to the many different factors discussed above, but also continues to be a promising new arena with potential for heavy impact in the field. As the study of proteases expands beyond just being biomarkers and advances toward cathepsin-targeted therapeutics, a better understanding of the contributions of different cell types that produce, secrete, and activate them will be necessary, and how these levels are differentially regulated across cancer tumors originating in a number of organs. Clinical considerations of supplementing chemotherapeutic regimens with protease inhibitors will also require careful balance of the side effects and benefits after inhibiting these enzymes as they are important for homeostasis, but also critical in advancing cancer angiogenesis, metastasis, and progression.

REFERENCES

Aase, K., G. von Euler, X. Li, A. Ponten, P. Thoren, R. Cao, Y. Cao, B. Olofsson, S. Gebre-Medhin, M. Pekny, K. Alitalo, C. Betsholtz and U. Eriksson (2001). "Vascular endothelial growth factor-B-deficient mice display an atrial conduction defect." Circulation 104: 358-364.

Ahn, K., S. Yeyeodu, J. Collette, V. Madden, J. Arthur, L. Li and A. H. Erickson (2002). "An alternate targeting pathway for procathepsin L in mouse fibroblasts." Traffic 3(2): 147-159.

Aibe, K., H. Yazawa, K. Abe, K. Teramura, M. Kumegawa, H. Kawashima and K. Honda (1996). "Substrate specificity of recombinant osteoclast-specific cathepsin K from rabbits." Biol Pharm Bull 19: 1026-1031.

Banerjee, D., R. Harfouche and S. Sengupta (2011). "Nanotechnology-mediated targeting of tumor angiogenesis." Vasc Cell 3(1): 3.

Bareford, L. M. and P. W. Swaan (2007). "Endocytic mechanisms for targeted drug delivery." Adv Drug Deliv Rev 59: 748-758.

Barrett, A. J., A. A. Kembhavi, M. A. Brown, H. Kirschke, C. G. Knight, M. Tamai and K. Hanada (1982). "L-trans-Epoxysuccinyl-leucylamido(4-guanidino)butane (E-64) and its analogues as inhibitors of cysteine proteinases including cathepsins B, H and L." Biochem J 201: 189-198.

Barrett, A. J. and H. Kirschke (1981). "Cathepsin B, Cathepsin H, and cathepsin L." Methods Enzymol 80 Pt C: 535-561.

Bartsch, G., J. Frick, I. Ruegg, M. Bucher, O. Holliger, M. Oberholzer and H. P. Rohr (1979). "Electron microscopic stereological analysis of the normal human prostate and of benign prostatic hyperplasia." J Urol 122: 481-486.

Bartsch, G., H. R. Muller, M. Oberholzer and H. P. Rohr (1979). "Light microscopic stereological analysis of the normal human prostate and of benign prostatic hyperplasia." J Urol 122: 487-491.

Baselga, J., L. Norton, J. Albanell, Y. M. Kim and J. Mendelsohn (1998). "Recombinant humanized anti-HER2 antibody (Herceptin) enhances the antitumor activity of paclitaxel and doxorubicin against HER2/neu overexpressing human breast cancer xenografts." Cancer Res 58: 2825-2831.

Berdowska, I. (2004). "Cysteine proteases as disease markers." Clin Chim Acta 342: 41-69.

Bergers, G. and L. E. Benjamin (2003). "Tumorigenesis and the angiogenic switch." Nat Rev Cancer 3: 401-410.

Bergers, G., R. Brekken, G. McMahon, T. H. Vu, T. Itoh, K. Tamaki, K. Tanzawa, P. Thorpe, S. Itohara, Z. Werb and D. Hanahan (2000). "Matrix metalloproteinase-9 triggers the angiogenic switch during carcinogenesis." Nat Cell Biol 2: 737-744.

Bervar, A., I. Zajc, N. Sever, N. Katunuma, B. F. Sloane and T. T. Lah (2003). "Invasiveness of transformed human breast epithelial cell lines is related to cathepsin B and inhibited by cysteine proteinase inhibitors." Biol Chem 384: 447-455.

Bode, W. and R. Huber (2000). "Structural basis of the endoproteinase-protein inhibitor interaction." Biochim Biophys Acta 1477: 241-252.

Brar, P. K., B. L. Dalkin, C. Weyer, K. Sallam, I. Virtanen and R. B. Nagle (2003). "Laminin alpha-1, alpha-3, and alpha-5 chain expression in human prepubertal [correction of prepubetal] benign prostate glands and adult benign and malignant prostate glands." Prostate 55: 65-70.

Briozzo, P., J. Badet, F. Capony, I. Pieri, P. Montcourrier, D. Barritault and H. Rochefort (1991). "MCF7 mammary cancer cells respond to bFGF and internalize it following its release from extracellular matrix: a permissive role of cathepsin D." Exp Cell Res 194: 252-259.

Bromme, D. and F. Lecaille (2009). "Cathepsin K inhibitors for osteoporosis and potential off-target effects." Expert Opin Investig Drugs 18: 585-600.

Brubaker, K. D., R. L. Vessella, L. D. True, R. Thomas and E. Corey (2003). "Cathepsin K mRNA and protein expression in prostate cancer progression." J Bone Miner Res 18: 222-230.

Buck, M. R., D. G. Karustis, N. A. Day, K. V. Honn and B. F. Sloane (1992). "Degradation of extracellular-matrix proteins by human cathepsin B from normal and tumour tissues." Biochem J 282 (Pt 1): 273-278.

Bumcrot, D., M. Manoharan, V. Koteliansky and D. W. Y. Sah (2006). "RNAi therapeutics: a potential new class of pharmaceutical drugs." Nat Chem Biol 2: 711-719.

Burden, R. E., P. Snoddy, C. A. Jefferies, B. Walker and C. J. Scott (2007). "Inhibition of cathepsin L-like proteases by cathepsin V propeptide." Biol Chem 388: 541-545.

Calkins, C. C., M. Sameni, J. Koblinski, B. F. Sloane and K. Moin (1998). "Differential localization of cysteine protease inhibitors and a target cysteine protease, cathepsin B, by immuno-confocal microscopy." J Histochem Cytochem 46 : 745-751.

Carmona, E., E. Dufour, C. Plouffe, S. Takebe, P. Mason, J. S. Mort and R. Menard (1996). "Potency and selectivity of the cathepsin L propeptide as an inhibitor of cysteine proteases." Biochemistry 35: 8149-8157.

Cavallo-Medved, D., D. Rudy, G. Blum, M. Bogyo, D. Caglic and B. F. Sloane (2009). "Live-cell imaging demonstrates extracellular matrix degradation in association with active cathepsin B in caveolae of endothelial cells during tube formation." Exp Cell Res 315: 1234-1246.

Cegnar, M., J. Kos and J. Kristl (2006). "Intracellular delivery of cysteine protease inhibitor cystatin by polymeric nanoparticles." J Nanosci Nanotechnol 6: 3087-3094.

Cegnar, M., A. Premzl, V. Zavasnik-Bergant, J. Kristl and J. Kos (2004). "Poly(lactide-co-glycolide) nanoparticles as a carrier system for delivering cysteine protease inhibitor cystatin into tumor cells." Exp Cell Res 301: 223-231.

Celerier, J., A. Cruz, N. Lamande, J.-M. Gasc and P. Corvol (2002). "Angiotensinogen and its cleaved derivatives inhibit angiogenesis." Hypertension 39: 224-228.

Chang, S.-H., K. Kanasaki, V. Gocheva, G. Blum, J. Harper, M. A. Moses, S.-C. Shih, J. A. Nagy, J. Joyce, M. Bogyo, R. Kalluri and H. F. Dvorak (2009). "VEGF-A induces angiogenesis by perturbing the cathepsin-cysteine protease inhibitor balance in venules, causing basement membrane degradation and mother vessel formation." Cancer Res 69: 4537-4544.

Chapman, H. A., R. J. Riese and G. P. Shi (1997). "Emerging roles for cysteine proteases in human biology." Annu Rev Physiol 59: 63-88.

Chapman, H. A., O. L. Stone and Z. Vavrin (1984). "Degradation of fibrin and elastin by intact human alveolar macrophages in vitro. Characterization of a plasminogen activator and its role in matrix degradation." J Clin Invest 73: 806-815.

Chen, B. and M. O. Platt (2011). "Multiplex zymography captures stage-specific activity profiles of cathepsins K, L, and S in human breast, lung, and cervical cancer." J Transl Med 9: 109-109.

Chen, J.-C., B.-J. Uang, P.-C. Lyu, J.-Y. Chang, K.-J. Liu, C.-C. Kuo, H.-P. Hsieh, H.-C. Wang, C.-S. Cheng, Y.-H. Chang, M. D.-T. Chang, W.-S. W. Chang and C.-C. Lin (2010). "Design and synthesis of alpha-ketoamides as cathepsin S inhibitors with potential applications against tumor invasion and angiogenesis." J Med Chem 53: 4545-4549.

Chen, J., C.-H. Tung, U. Mahmood, V. Ntziachristos, R. Gyurko, M. C. Fishman, P. L. Huang and R. Weissleder (2002). "*In vivo* imaging of proteolytic activity in atherosclerosis." Circulation 105: 2766-2771.

Chia, S. K., C. C. Wykoff, P. H. Watson, C. Han, R. D. Leek, J. Pastorek, K. C. Gatter, P. Ratcliffe and A. L. Harris (2001). "Prognostic significance of a novel hypoxia-regulated marker, carbonic anhydrase IX, in invasive breast carcinoma." J Clin Oncol 19(16): 3660-3668.

Claveau, D., D. Riendeau and J. A. Mancini (2000). "Expression, maturation, and rhodamine-based fluorescence assay of human cathepsin K expressed in CHO cells." Biochem Pharmacol 60: 759-769.

Collette, J., J. P. Bocock, K. Ahn, R. L. Chapman, G. Godbold, S. Yeyeodu and A. H. Erickson (2004). "Biosynthesis and alternate targeting of the lysosomal cysteine protease cathepsin L." Int Rev Cytol 241: 1-51.

Costa, A. G., N. E. Cusano, B. C. Silva, S. Cremers and J. P. Bilezikian (2011). "Cathepsin K: its skeletal actions and role as a therapeutic target in osteoporosis." Nat Rev Rheumatol 7: 447-456.

Coulson-Thomas, V. J., T. F. Gesteira, Y. M. Coulson-Thomas, C. M. Vicente, I. L. S. Tersariol, H. B. Nader and L. Toma (2010). "Fibroblast and prostate tumor cell cross-talk: fibroblast differentiation, TGF-β, and extracellular matrix down-regulation." Exp Cell Res 316: 3207-3226.

Cox, G., J. L. Jones, R. A. Walker, W. P. Steward and K. J. O'Byrne (2000). "Angiogenesis and non-small cell lung cancer." Lung Cancer 27: 81-8100.

Daly, M. E., A. Makris, M. Reed and C. E. Lewis (2003). "Hemostatic regulators of tumor angiogenesis: a source of antiangiogenic agents for cancer treatment?" J Natl Cancer Inst 95: 1660-1673.

Dawson, M. R., D. G. Duda, D. Fukumura and R. K. Jain (2009). "VEGFR1-activity-independent metastasis formation." Nature 461.

De Becker, A., P. Van Hummelen, M. Bakkus, I. Vande Broek, J. De Wever, M. De Waele and I. Van Riet (2007). "Migration of culture-expanded human mesenchymal stem cells through bone marrow endothelium is regulated by matrix metalloproteinase-2 and tissue inhibitor of metalloproteinase-3." Haematologica 92(4): 440-449.

Delebecq, T. J., H. Porte, F. Zerimech, M. C. Copin, V. Gouyer, E. Dacquembronne, M. Balduyck, A. Wurtz and G. Huet (2000). "Overexpression level of stromelysin 3 is related to the lymph node involvement in non-small cell lung cancer." Clin Cancer Res 6: 1086-1092.

Deryugina, E. I. and J. P. Quigley (2006). "Matrix metalloproteinases and tumor metastasis." Cancer Metastasis Rev 25: 9-34.

Desmarais, S., W. C. Black, R. Oballa, S. Lamontagne, D. Riendeau, P. Tawa, L. T. Duong, M. Pickarski and M. D. Percival (2008). "Effect of cathepsin K inhibitor basicity on in vivo off-target activities." Mol Pharmacol 73: 147-156.

Dodds, R. A., I. E. James, D. Rieman, R. Ahern, S. M. Hwang, J. R. Connor, S. D. Thompson, D. F. Veber, F. H. Drake, S. Holmes, M. W. Lark and M. Gowen (2001). "Human osteoclast cathepsin K is processed intracellularly prior to attachment and bone resorption." J Bone Miner Res 16: 478-486.

Drahl, C., B. F. Cravatt and E. J. Sorensen (2005). "Protein-reactive natural products." Angew Chem Int Ed Engl 44(36): 5788-5809.

Drenth, J., J. N. Jansonius, R. Koekoek, H. M. Swen and B. G. Wolthers (1968). "Structure of papain." Nature 218: 929-932.

Duncan, E. M., T. L. Muratore-Schroeder, R. G. Cook, B. A. Garcia, J. Shabanowitz, D. F. Hunt and C. D. Allis (2008). "Cathepsin L proteolytically processes histone H3 during mouse embryonic stem cell differentiation." Cell 135: 284-294.

Ebert, W., H. Knoch, B. Werle, G. Trefz, T. Muley and E. Spiess (1994). "Prognostic value of increased lung tumor tissue cathepsin B." Anticancer Res 14: 895-899.

Ebisui, C., T. Tsujinaka, T. Morimoto, K. Kan, S. Iijima, M. Yano, E. Kominami, K. Tanaka and M. Monden (1995). "Interleukin-6 induces proteolysis by activating intracellular proteases (cathepsins B and L, proteasome) in C2C12 myotubes." Clin Sci (Lond) 89: 431-439.

Fehrenbacher, N. and M. Jaattela (2005). "Lysosomes as targets for cancer therapy." Cancer Res 65(8): 2993-2995.

Felbor, U., L. Dreier, R. A. Bryant, H. L. Ploegh, B. R. Olsen and W. Mothes (2000). "Secreted cathepsin L generates endostatin from collagen XVIII." EMBO J 19: 1187-1194.

Ferlay, J., D. M. Parkin and E. Steliarova-Foucher (2010). "Estimates of cancer incidence and mortality in Europe in 2008." Eur J Cancer 46: 765-781.

Fernandez, P. L., X. Farre, A. Nadal, E. Fernandez, N. Peiro, B. F. Sloane, G. P. Shi, H. A. Chapman, E. Campo and A. Cardesa (2001). "Expression of cathepsins B and S in the progression of prostate carcinoma." Int J Cancer 95: 51-55.

Ferrara, N., H.-P. Gerber and J. LeCouter (2003). "The biology of VEGF and its receptors." Nat Med 9: 669-676.

Ferrer, F. A., L. J. Miller, R. I. Andrawis, S. H. Kurtzman, P. C. Albertsen, V. P. Laudone and D. L. Kreutzer (1998). "Angiogenesis and prostate cancer: in vivo and in vitro expression of angiogenesis factors by prostate cancer cells." Urology 51: 161-167.

Ferreras, M., U. Felbor, T. Lenhard, B. R. Olsen and J. Delaisse (2000). "Generation and degradation of human endostatin proteins by various proteinases." FEBS Lett 486: 247-251.

Fleischmann, R. M., J. Tesser, M. H. Schiff, J. Schechtman, G. R. Burmester, R. Bennett, D. Modafferi, L. Zhou, D. Bell and B. Appleton (2006). "Safety of extended treatment with anakinra in patients with rheumatoid arthritis." Ann Rheum Dis 65: 1006-1012.

Foekens, J. A., J. Kos, H. A. Peters, M. Krasovec, M. P. Look, N. Cimerman, M. E. Meijer-van Gelder, S. C. Henzen-Logmans, W. L. van Putten and J. G. Klijn (1998). "Prognostic significance of cathepsins B and L in primary human breast cancer." J Clin Oncol 16(3): 1013-1021.

Fox, S. B., D. G. Generali and A. L. Harris (2007). "Breast tumour angiogenesis." Breast Cancer Res 9: 216-216.

Frade, R., N. Rousselet and D. Jean (2008). "Intratumoral gene delivery of anti-cathepsin L single-chain variable fragment by lentiviral vector inhibits tumor progression induced by human melanoma cells." Cancer Gene Ther 15: 591-604.

Friedrich, B., K. Jung, M. Lein, I. Turk, B. Rudolph, G. Hampel, D. Schnorr and S. A. Loening (1999). "Cathepsins B, H, L and cysteine protease inhibitors in malignant prostate cell lines, primary cultured prostatic cells and prostatic tissue." Eur J Cancer 35: 138-144.

Fujise, N., A. Nanashim, Y. Taniguchi, S. Matsuo, K. Hatano, Y. Matsumoto, Y. Tagawa and H. Ayabe (2000). "Prognostic impact of cathepsin B and matrix metalloproteinase-9 in pulmonary adenocarcinomas by immunohistochemical study." Lung Cancer 27: 19-26.

Gabrijelcic, D., B. Svetic, D. Spaic, J. Skrk, M. Budihna, I. Dolenc, T. Popovic, V. Cotic and V. Turk (1992). "Cathepsins B, H and L in human breast carcinoma." Eur J Clin Chem Clin Biochem 30: 69-74.

Gacko, M. and S. Glowinski (1998). "Cathepsin D and cathepsin L activities in aortic aneurysm wall and parietal thrombus." Clin Chem Lab Med 36: 449-452.

Galis, Z. S., G. K. Sukhova, M. W. Lark and P. Libby (1994). "Increased expression of matrix metalloproteinases and matrix degrading activity in vulnerable regions of human atherosclerotic plaques." J Clin Invest 94: 2493-2503.

Galis, Z. S., G. K. Sukhova and P. Libby (1995). "Microscopic localization of active proteases by in situ zymography: detection of matrix metalloproteinase activity in vascular tissue." FASEB J 9: 974-980.

Garcea, G., C. P. Neal, C. J. Pattenden, W. P. Steward and D. P. Berry (2005). "Molecular prognostic markers in pancreatic cancer: a systematic review." Eur J Cancer 41: 2213-2236.

Garnero, P., O. Borel, I. Byrjalsen, M. Ferreras, F. H. Drake, M. S. McQueney, N. T. Foged, P. D. Delmas and J. M. Delaisse (1998). "The collagenolytic activity of cathepsin K is unique among mammalian proteinases." J Biol Chem 273: 32347-32352.

Gaumann, A., T. Hansen, H. H. Kohler, F. Kommoss, W. Mann, J. Maurer, C. J. Kirkpatrick and J. Kriegsmann (2001). "The expression of cathepsins in osteoclast-like giant cells of an anaplastic thyroid carcinoma with tracheal perforation." Pathol Res Pract 197: 257-262.

Gocheva, V., W. Zeng, D. Ke, D. Klimstra, T. Reinheckel, C. Peters, D. Hanahan and J. A. Joyce (2006). "Distinct roles for cysteine cathepsin genes in multistage tumorigenesis." Genes Dev 20: 543-556.

Golovatch, P., B. A. Mercer, V. Lemaitre, A. Wallace, R. F. Foronjy and J. D'Armiento (2009). "Role for cathepsin K in emphysema in smoke-exposed guinea pigs." Exp Lung Res 35: 631-645.

Gomis-Ruth, F. X., K. Maskos, M. Betz, A. Bergner, R. Huber, K. Suzuki, N. Yoshida, H. Nagase, K. Brew, G. P. Bourenkov, H. Bartunik and W. Bode (1997). "Mechanism of inhibition of the human matrix metalloproteinase stromelysin-1 by TIMP-1." Nature 389: 77-81.

Gondi, C. S., S. S. Lakka, D. H. Dinh, W. C. Olivero, M. Gujrati and J. S. Rao (2004). "RNAi-mediated inhibition of cathepsin B and uPAR leads to decreased cell invasion, angiogenesis and tumor growth in gliomas." Oncogene 23: 8486-8496.

Greenbaum, D., A. Baruch, L. Hayrapetian, Z. Darula, A. Burlingame, K. F. Medzihradszky and M. Bogyo (2002). "Chemical approaches for functionally probing the proteome." Mol Cell Proteomics 1: 60-68.

Greenlee, R. T., T. Murray, S. Bolden and P. A. Wingo (2000). "Cancer statistics, 2000." CA Cancer J Clin 50: 7-33.

Griffiths, G. (2002). "What's special about secretory lysosomes?" Semin Cell Dev Biol 13: 279-284.

Hagar, W. and E. Vichinsky (2008). "Advances in clinical research in sickle cell disease." Br J Haematol 141: 346-356.

Hall, A., I. Ekiel, R. W. Mason, F. Kasprzykowski, A. Grubb and M. Abrahamson (1998). "Structural basis for different inhibitory specificities of human cystatins C and D." Biochemistry 37: 4071-4079.

Hamano, Y., M. Zeisberg, H. Sugimoto, J. C. Lively, Y. Maeshima, C. Yang, R. O. Hynes, Z. Werb, A. Sudhakar and R. Kalluri (2003). "Physiological levels of tumstatin, a fragment of collagen IV alpha3 chain, are generated by MMP-9 proteolysis and suppress angiogenesis via alphaV beta3 integrin." Cancer Cell 3: 589-601.

Han, J., T. Luo, Y. Gu, G. Li, W. Jia and M. Luo (2009). "Cathepsin K regulates adipocyte differentiation: possible involvement of type I collagen degradation." Endocr J 56: 55-63.

Hashimoto, Y., C. Kondo, T. Kojima, H. Nagata, A. Moriyama, T. Hayakawa and N. Katunuma (2006). "Significance of 32-kDa cathepsin L secreted from cancer cells." Cancer Biother Radiopharm 21: 217-224.

Higgins, W. J., D. M. Fox, P. S. Kowalski, J. E. Nielsen and D. M. Worrall (2010). "Heparin enhances serpin inhibition of the cysteine protease cathepsin L." J Biol Chem 285: 3722-3729.

Hofmann, U. B., R. Houben, E.-B. Brocker and J. C. Becker (2005). "Role of matrix metalloproteinases in melanoma cell invasion." Biochimie 87: 307-314.

Im, E., A. Venkatakrishnan and A. Kazlauskas (2005). "Cathepsin B regulates the intrinsic angiogenic threshold of endothelial cells." Mol Biol Cell 16: 3488-3500.

Inoue, T., T. Ishida, K. Sugio and K. Sugimachi (1994). "Cathepsin B expression and laminin degradation as factors influencing prognosis of surgically treated patients with lung adenocarcinoma." Cancer Res 54: 6133-6136.

Jani, M., H. Tordai, M. Trexler, L. Banyai and L. Patthy (2005). "Hydroxamate-based peptide inhibitors of matrix metalloprotease 2." Biochimie 87: 385-392.

Jankun, J., S. H. Selman, J. Aniola and E. Skrzypczak-Jankun (2006). "Nutraceutical inhibitors of urokinase: potential applications in prostate cancer prevention and treatment." Oncol Rep 16: 341-346.

Jayakumar, A., Y. Kang, M. J. Frederick, S. C. Pak, Y. Henderson, P. R. Holton, K. Mitsudo, G. A. Silverman, E. L.-N. AK, D. Bromme and G. L. Clayman (2003). "Inhibition of the cysteine proteinases cathepsins K and L by the serpin headpin (SERPINB13): a kinetic analysis." Arch Biochem Biophys 409(2): 367-374.

Jean, D., N. Rousselet and R. Frade (2008). "Cathepsin L expression is up-regulated by hypoxia in human melanoma cells: role of its 5'-untranslated region." Biochem J 413: 125-134.

Jensen, H. M., I. Chen, M. R. DeVault and A. E. Lewis (1982). "Angiogenesis induced by "normal" human breast tissue: a probable marker for precancer." Science 218: 293-295.

Jimenez-Vergara, A. C., V. Guiza-Arguello, S. Becerra-Bayona, D. J. Munoz-Pinto, R. E. McMahon, A. Morales, L. Cubero-Ponce and M. S. Hahn (2010). "Approach for fabricating tissue engineered vascular grafts with stable endothelialization." Ann Biomed Eng 38: 2885-2895.

Jormsjo, S., D. M. Wuttge, A. Sirsjo, C. Whatling, A. Hamsten, S. Stemme and P. Eriksson (2002). "Differential expression of cysteine and aspartic proteases during progression of atherosclerosis in apolipoprotein E-deficient mice." Am J Pathol 161: 939-945.

Joyce, J. A., A. Baruch, K. Chehade, N. Meyer-Morse, E. Giraudo, F.-Y. Tsai, D. C. Greenbaum, J. H. Hager, M. Bogyo and D. Hanahan (2004). "Cathepsin cysteine proteases are effectors of invasive growth and angiogenesis during multistage tumorigenesis." Cancer Cell 5: 443-453.

Kamolmatyakul, S., W. Chen, S. Yang, Y. Abe, R. Moroi, A. M. Ashique and Y. P. Li (2004). "IL-1alpha stimulates cathepsin K expression in osteoclasts via the tyrosine kinase-NF-kappaB pathway." J Dent Res 83: 791-796.

Kayser, K., N. Richter, P. Hufnagl, G. Kayser, J. Kos and B. Werle (2003). "Expression, proliferation activity and clinical significance of cathepsin B and cathepsin L in operated lung cancer." Anticancer Res 23: 2767-2772.

Kim, I. S., Y. B. Kim and K. J. Lee (1998). "Characterization of the leupeptin-inactivating enzyme from Streptomyces exfoliatus SMF13 which produces leupeptin." Biochem J 331 (Pt 2): 539-545.

Kim, W. J., J. W. Yockman, M. Lee, J. H. Jeong, Y.-H. Kim and S. W. Kim (2005). "Soluble Flt-1 gene delivery using PEI-g-PEG-RGD conjugate for anti-angiogenesis." J Control Release 106: 224-234.

Klose, A., P. Zigrino, R. Dennhofer, C. Mauch and N. Hunzelmann (2006). "Identification and discrimination of extracellularly active cathepsins B and L in high-invasive melanoma cells." Anal Biochem 353: 57-62.

Kobayashi, H., M. Schmitt, L. Goretzki, N. Chucholowski, J. Calvete, M. Kramer, W. A. Gunzler, F. Janicke and H. Graeff (1991). "Cathepsin B efficiently activates the soluble and the tumor cell receptor-bound form of the proenzyme urokinase-type plasminogen activator (Pro-uPA)." J Biol Chem 266: 5147-5152.

Kocevar, N., N. Obermajer, B. Strukelj, J. Kos and S. Kreft (2007). "Improved acylation method enables efficient delivery of functional palmitoylated cystatin into epithelial cells." Chem Biol Drug Des 69: 124-131.

Kos, J., A. Sekirnik, G. Kopitar, N. Cimerman, K. Kayser, A. Stremmer, W. Fiehn and B. Werle (2001). "Cathepsin S in tumours, regional lymph nodes and sera of patients with lung cancer: relation to prognosis." Br J Cancer 85: 1193-1200.

Krepela, E., J. Prochazka, B. Karova, J. Cermak and H. Roubkova (1997). "Cathepsin B, thiols and cysteine protease inhibitors in squamous-cell lung cancer." Neoplasma 44: 219-239.

Krepela, E., J. Prochazka, B. Karova, J. Cermak and H. Roubkova (1998). "Cysteine proteases and cysteine protease inhibitors in non-small cell lung cancer." Neoplasma 45: 318-331.

Lafleur, M. A., M. D. Hollenberg, S. J. Atkinson, V. Knauper, G. Murphy and D. R. Edwards (2001). "Activation of pro-(matrix metalloproteinase-2) (pro-MMP-2) by thrombin is membrane-type-MMP-dependent in human umbilical vein endothelial cells and generates a distinct 63 kDa active species." Biochem J 357: 107-115.

Lah, T. T., G. Calaf, E. Kalman, B. G. Shinde, J. Russo, D. Jarosz, J. Zabrecky, R. Somers and I. Daskal (1995). "Cathepsins D, B and L in breast carcinoma and in transformed human breast epithelial cells (HBEC)." Biol Chem Hoppe Seyler 376: 357-363.

Lah, T. T., E. Kalman, D. Najjar, E. Gorodetsky, P. Brennan, R. Somers and I. Daskal (2000). "Cells producing cathepsins D, B, and L in human breast carcinoma and their association with prognosis." Hum Pathol 31: 149-160.

Lah, T. T., M. Kokalj-Kunovar, M. Drobnic-Kosorok, J. Babnik, R. Golouh, I. Vrhovec and V. Turk (1992). "Cystatins and cathepsins in breast carcinoma." Biol Chem Hoppe Seyler 373: 595-604.

Lah, T. T., M. Kokalj-Kunovar, B. Strukelj, J. Pungercar, D. Barlic-Maganja, M. Drobnic-Kosorok, L. Kastelic, J. Babnik, R. Golouh and V. Turk (1992). "Stefins and lysosomal cathepsins B, L and D in human breast carcinoma." Int J Cancer 50: 36-44.

Lah, T. T., J. Kos, A. Blejec, S. Frkovic-Georgio, R. Golouh, I. I. Vrhovec and V. V. Turk (1997). "The Expression of Lysosomal Proteinases and Their Inhibitors in Breast Cancer: Possible Relationship to Prognosis of the Disease." Pathol Oncol Res 3(2): 89-99.

Lapis, K. and J. Timar (2002). "Role of elastin-matrix interactions in tumor progression." Semin Cancer Biol 12: 209-217.

Ledakis, P., W. T. Tester, N. Rosenberg, D. Romero-Fischmann, I. Daskal and T. T. Lah (1996). "Cathepsins D, B, and L in malignant human lung tissue." Clin Cancer Res 2: 561-568.

Lesser, M., J. C. Chang, N. I. Galicki, J. Edelman and C. Cardozo (1989). "Cathepsin B and D activity in alveolar macrophages from rats with pulmonary granulomatous inflammation or acute lung injury." Agents Actions 28: 264-271.

Levicar, N., J. Kos, A. Blejec, R. Golouh, I. Vrhovec, S. Frkovic-Grazio and T. T. Lah (2002). "Comparison of potential biological markers cathepsin B, cathepsin L, stefin A and stefin B with urokinase and plasminogen activator inhibitor-1 and clinicopathological data of breast carcinoma patients." Cancer Detect Prev 26: 42-49.

Lewis, C. E., R. Leek, A. Harris and J. O. McGee (1995). "Cytokine regulation of angiogenesis in breast cancer: the role of tumor-associated macrophages." J Leukoc Biol 57: 747-751.

Li, Z., W.-S. Hou, C. R. Escalante-Torres, B. D. Gelb and D. Bromme (2002). "Collagenase activity of cathepsin K depends on complex formation with chondroitin sulfate." J Biol Chem 277: 28669-28676.

Li, Z., Y. Yasuda, W. Li, M. Bogyo, N. Katz, R. E. Gordon, G. B. Fields and D. Bromme (2004). "Regulation of collagenase activities of human cathepsins by glycosaminoglycans." J Biol Chem 279: 5470-5479.

Lindeman, J. H. N., R. Hanemaaijer, A. Mulder, P. D. S. Dijkstra, K. Szuhai, D. Bromme, J. H. Verheijen and P. C. W. Hogendoorn (2004). "Cathepsin K is the principal protease in giant cell tumor of bone." Am J Pathol 165: 593-600.

Line, B. R., A. Mitra, A. Nan and H. Ghandehari (2005). "Targeting tumor angiogenesis: comparison of peptide and polymer-peptide conjugates." J Nucl Med 46: 1552-1560.

Linebaugh, B. E., M. Sameni, N. A. Day, B. F. Sloane and D. Keppler (1999). "Exocytosis of active cathepsin B enzyme activity at pH 7.0, inhibition and molecular mass." Eur J Biochem 264: 100-109.

Ling, H., S. Vamvakas, G. Busch, J. Dammrich, L. Schramm, F. Lang and A. Heidland (1995). "Suppressing role of transforming growth factor-beta 1 on cathepsin activity in cultured kidney tubule cells." Am J Physiol 269: 911-917.

Littlewood-Evans, A., T. Kokubo, O. Ishibashi, T. Inaoka, B. Wlodarski, J. A. Gallagher and G. Bilbe (1997). "Localization of cathepsin K in human osteoclasts by in situ hybridization and immunohistochemistry." Bone 20: 81-86.

Littlewood-Evans, A. J., G. Bilbe, W. B. Bowler, D. Farley, B. Wlodarski, T. Kokubo, T. Inaoka, J. Sloane, D. B. Evans and J. A. Gallagher (1997). "The osteoclast-associated protease cathepsin K is expressed in human breast carcinoma." Cancer Res 57: 5386-5390.

Liu, J., G. K. Sukhova, J.-T. Yang, J. Sun, L. Ma, A. Ren, W.-H. Xu, H. Fu, G. M. Dolganov, C. Hu, P. Libby and G.-P. Shi (2006). "Cathepsin L expression and regulation in human abdominal aortic aneurysm, atherosclerosis, and vascular cells." Atherosclerosis 184: 302-311.

Lopez-Fraga, M., N. Wright and A. Jimenez (2008). "RNA interference-based therapeutics: new strategies to fight infectious disease." Infect Disord Drug Targets 8: 262-273.

Lugering, N., T. Kucharzik, H. Gockel, C. Sorg, R. Stoll and W. Domschke (1998). "Human intestinal epithelial cells down-regulate IL-8 expression in human intestinal microvascular endothelial cells; role of transforming growth factor-beta 1 (TGF-beta1)." Clin Exp Immunol 114: 377-384.

Lutgens, E., S. P. M. Lutgens, B. C. G. Faber, S. Heeneman, M. M. J. Gijbels, M. P. J. de Winther, P. Frederik, I. van der Made, A. Daugherty, A. M. Sijbers, A. Fisher, C. J. Long, P. Saftig, D. Black, M. J. A. P. Daemen and K. B. J. M. Cleutjens (2006). "Disruption of the cathepsin K gene reduces atherosclerosis progression and induces plaque fibrosis but accelerates macrophage foam cell formation." Circulation 113: 98-9107.

Lutolf, M. P., J. L. Lauer-Fields, H. G. Schmoekel, A. T. Metters, F. E. Weber, G. B. Fields and J. A. Hubbell (2003). "Synthetic matrix metalloproteinase-sensitive hydrogels for the conduction of tissue regeneration: engineering cell-invasion characteristics." Proc Natl Acad Sci USA 100: 5413-5418.

Maciewicz, R. A. and D. J. Etherington (1988). "A comparison of four cathepsins (B, L, N and S) with collagenolytic activity from rabbit spleen." Biochem J 256: 433-440.

Maguire, T. M., S. G. Shering, C. M. Duggan, E. W. McDermott, N. J. O'Higgins and M. J. Duffy (1998). "High levels of cathepsin B predict poor outcome in patients with breast cancer." Int J Biol Markers 13: 139-144.

Mai, J., D. M. Waisman and B. F. Sloane (2000). "Cell surface complex of cathepsin B/annexin II tetramer in malignant progression." Biochim Biophys Acta 1477: 215-230.

Marshall, S. and P. Narayan (1993). "Treatment of prostatic bleeding: suppression of angiogenesis by androgen deprivation." J Urol 149: 1553-1554.

Masson, O., C. Prebois, D. Derocq, A. Meulle, C. Dray, D. Daviaud, D. Quilliot, P. Valet, C. Muller and E. Liaudet-Coopman (2011). "Cathepsin-D, a key protease in breast cancer, is up-regulated in obese mouse and human adipose tissue, and controls adipogenesis." PLoS One 6.

Maubach, G., K. Schilling, W. Rommerskirch, I. Wenz, J. E. Schultz, E. Weber and B. Wiederanders (1997). "The inhibition of cathepsin S by its propeptide-specificity and mechanism of action." Eur J Biochem 250: 745-750.

Maynard, S. E., J.-Y. Min, J. Merchan, K.-H. Lim, J. Li, S. Mondal, T. A. Libermann, J. P. Morgan, F. W. Sellke, I. E. Stillman, F. H. Epstein, V. P. Sukhatme and S. A. Karumanchi (2003). "Excess placental soluble fms-like tyrosine kinase 1 (sFlt1) may contribute to endothelial dysfunction, hypertension, and proteinuria in preeclampsia." J Clin Invest 111: 649-658.

McGrath, M. E. (1999). "The lysosomal cysteine proteases." Annu Rev Biophys Biomol Struct 28: 181-204.

McQueney, M. S., B. Y. Amegadzie, K. D'Alessio, C. R. Hanning, M. M. McLaughlin, D. McNulty, S. A. Carr, C. Ijames, J. Kurdyla and C. S. Jones (1997). "Autocatalytic activation of human cathepsin K." J Biol Chem 272(21): 13955-13960.

Menard, R., E. Carmona, S. Takebe, E. Dufour, C. Plouffe, P. Mason and J. S. Mort (1998). "Autocatalytic processing of recombinant human procathepsin L. Contribution of both intermolecular and unimolecular events in the processing of procathepsin L in vitro." J Biol Chem 273: 4478-4484.

Merz, G. S., E. Benedikz, V. Schwenk, T. E. Johansen, L. K. Vogel, J. I. Rushbrook and H. M. Wisniewski (1997). "Human cystatin C forms an inactive dimer during intracellular trafficking in transfected CHO cells." J Cell Physiol 173: 423-432.

Mohamed, M. M. and B. F. Sloane (2006). "Cysteine cathepsins: multifunctional enzymes in cancer." Nat Rev Cancer 6: 764-775.

Montcourrier, P., I. Silver, R. Farnoud, I. Bird and H. Rochefort (1997). "Breast cancer cells have a high capacity to acidify extracellular milieu by a dual mechanism." Clin Exp Metastasis 15: 382-392.

Mook, O. R., W. M. Frederiks and C. J. Van Noorden (2004). "The role of gelatinases in colorectal cancer progression and metastasis." Biochim Biophys Acta 1705(2): 69-89.

Mott, J. D. and Z. Werb (2004). "Regulation of matrix biology by matrix metalloproteinases." Curr Opin Cell Biol 16: 558-564.

Mundy, G. R. (2002). "Metastasis to bone: causes, consequences and therapeutic opportunities." Nat Rev Cancer 2: 584-593.

Mundy, G. R. (2006). "Nutritional modulators of bone remodeling during aging." Am J Clin Nutr 83(2): 427S-430S.

Nackman, G. B., F. J. Karkowski, V. J. Halpern, H. P. Gaetz and M. D. Tilson (1997). "Elastin degradation products induce adventitial angiogenesis in the Anidjar/ Dobrin rat aneurysm model." Surgery 122(1): 39-44.

Nagler, D. K., S. Kruger, A. Kellner, E. Ziomek, R. Menard, P. Buhtz, M. Krams, A. Roessner and U. Kellner (2004). "Up-regulation of cathepsin X in prostate cancer and prostatic intraepithelial neoplasia." Prostate 60: 109-119.

Nalla, A. K., B. Gorantla, C. S. Gondi, S. S. Lakka and J. S. Rao (2010). "Targeting MMP-9, uPAR, and cathepsin B inhibits invasion, migration and activates apoptosis in prostate cancer cells." Cancer Gene Ther 17: 599-613.

Newby, A. C. (2005). "Dual role of matrix metalloproteinases (matrixins) in intimal thickening and atherosclerotic plaque rupture." Physiol Rev 85: 1-31.

Nyberg, P., L. Xie and R. Kalluri (2005). "Endogenous inhibitors of angiogenesis." Cancer Res 65: 3967-3979.

O'Reilly, M. S., S. Pirie-Shepherd, W. S. Lane and J. Folkman (1999). "Antiangiogenic activity of the cleaved conformation of the serpin antithrombin." Science 285: 1926-1928.

Odero-Marah, V. A., R. Wang, G. Chu, M. Zayzafoon, J. Xu, C. Shi, F. F. Marshall, H. E. Zhau and L. W. K. Chung (2008). "Receptor activator of NF-kappaB Ligand (RANKL) expression is associated with epithelial to mesenchymal transition in human prostate cancer cells." Cell Res 18: 858-870.

Padilla, M. L., N. I. Galicki, J. Kleinerman, M. Orlowski and M. Lesser (1988). "High cathepsin B activity in alveolar macrophages occurs with elastase-induced emphysema but not with bleomycin-induced pulmonary fibrosis in hamsters." Am J Pathol 131: 92-9101.

Palermo, C. and J. A. Joyce (2008). "Cysteine cathepsin proteases as pharmacological targets in cancer." Trends Pharmacol Sci 29: 22-28.

Platt, M. O., R. F. Ankeny and H. Jo (2006). "Laminar shear stress inhibits cathepsin L activity in endothelial cells." Arterioscler Thromb Vasc Biol 26: 1784-1790.

Platt, M. O., R. F. Ankeny, G.-P. Shi, D. Weiss, J. D. Vega, W. R. Taylor and H. Jo (2007). "Expression of cathepsin K is regulated by shear stress in cultured endothelial cells and is increased in endothelium in human atherosclerosis." Am J Physiol Heart Circ Physiol 292: 1479-1486.

Pories, S. E., D. Zurakowski, R. Roy, C. C. Lamb, S. Raza, A. Exarhopoulos, R. G. Scheib, S. Schumer, C. Lenahan, V. Borges, G. W. Louis, A. Anand, N. Isakovich, J. Hirshfield-Bartek, U. Wewer, M. M. Lotz and M. A. Moses (2008). "Urinary metalloproteinases: noninvasive biomarkers for breast cancer risk assessment." Cancer Epidemiol Biomarkers Prev 17: 1034-1042.

Pozzi, A., W. F. LeVine and H. A. Gardner (2002). "Low plasma levels of matrix metalloproteinase 9 permit increased tumor angiogenesis." Oncogene 21: 272-281.

Punturieri, A., S. Filippov, E. Allen, I. Caras, R. Murray, V. Reddy and S. J. Weiss (2000). "Regulation of elastinolytic cysteine proteinase activity in normal and cathepsin K-deficient human macrophages." J Exp Med 192: 789-799.

Rapa, I., M. Volante, S. Cappia, R. Rosas, G. V. Scagliotti and M. Papotti (2006). "Cathepsin K is selectively expressed in the stroma of lung adenocarcinoma but not in bronchioloalveolar carcinoma. A useful marker of invasive growth." Am J Clin Pathol 125: 847-854.

Reddy, V. Y., Q. Y. Zhang and S. J. Weiss (1995). "Pericellular mobilization of the tissue-destructive cysteine proteinases, cathepsins B, L, and S, by human monocyte-derived macrophages." Proc Natl Acad Sci USA 92: 3849-3853.

Reiser, J., B. Adair and T. Reinheckel (2010). "Specialized roles for cysteine cathepsins in health and disease." J Clin Invest 120: 3421-3431.

Robey, I. F., B. K. Baggett, N. D. Kirkpatrick, D. J. Roe, J. Dosescu, B. F. Sloane, A. I. Hashim, D. L. Morse, N. Raghunand, R. A. Gatenby and R. J. Gillies (2009). "Bicarbonate increases tumor pH and inhibits spontaneous metastases." Cancer Res 69: 2260-2268.

Saleh, Y., M. Siewinski, T. Sebzda, M. Jelen, P. Ziolkowski, J. Gutowicz, M. Grybos and M. Pawelec (2003). "Inhibition of cathepsin B activity in human breast cancer tissue by cysteine peptidase inhibitor isolated from human placenta: immunohistochemical and biochemical studies." Folia Histochem Cytobiol 41: 161-167.

Samokhin, A. O., J. Y. Gauthier, M. D. Percival and D. Bromme (2011). "Lack of cathepsin activities alter or prevent the development of lung granulomas in a mouse model of sarcoidosis." Respir Res 12: 13-13.

Schmeisser, A., M. Christoph, A. Augstein, R. Marquetant, M. Kasper, R. C. Braun-Dullaeus and R. H. Strasser (2006). "Apoptosis of human macrophages by Flt-4 signaling: implications for atherosclerotic plaque pathology." Cardiovasc Res 71: 774-784.

Schneider, B. P. and K. D. Miller (2005). "Angiogenesis of breast cancer." J Clin Oncol 23(8): 1782-1790.

Schweiger, A., A. Staib, B. Werle, M. Krasovec, T. T. Lah, W. Ebert, V. Turk and J. Kos (2000). "Cysteine proteinase cathepsin H in tumours and sera of lung cancer patients: relation to prognosis and cigarette smoking." Br J Cancer 82: 782-788.

Seliktar, D., A. H. Zisch, M. P. Lutolf, J. L. Wrana and J. A. Hubbell (2004). "MMP-2 sensitive, VEGF-bearing bioactive hydrogels for promotion of vascular healing." J Biomed Mater Res A 68: 704-716.

Sennoune, S. R., K. Bakunts, G. M. Martinez, J. L. Chua-Tuan, Y. Kebir, M. N. Attaya and R. Martinez-Zaguilan (2004). "Vacuolar H+-ATPase in human breast cancer cells with distinct metastatic potential: distribution and functional activity." Am J Physiol Cell Physiol 286: 1443-1452.

Sheng, S., J. Carey, E. A. Seftor, L. Dias, M. J. Hendrix and R. Sager (1996). "Maspin acts at the cell membrane to inhibit invasion and motility of mammary and prostatic cancer cells." Proc Natl Acad Sci USA 93: 11669-11674.

Shi, G. P., G. K. Sukhova, M. Kuzuya, Q. Ye, J. Du, Y. Zhang, J. H. Pan, M. L. Lu, X. W. Cheng, A. Iguchi, S. Perrey, A. M. E. Lee, H. A. Chapman and P. Libby (2003). "Deficiency of the cysteine protease cathepsin S impairs microvessel growth." Circ Res 92: 493-500.

Sikora, J., J. Slodkowska, A. Radomyski, D. Giedronowicz, J. Kobos, W. Kupis and P. Rudzinski (1997). "Immunohistochemical evaluation of tumour angiogenesis in adenocarcinoma and squamous cell carcinoma of lung." Rocz Akad Med Bialymst 42 Suppl 1: 271-279.

Sinha, A. A., B. J. Quast, M. J. Wilson, P. K. Reddy, D. F. Gleason and B. F. Sloane (1998). "Codistribution of procathepsin B and mature cathepsin B forms in human prostate tumors detected by confocal and immunofluorescence microscopy." Anat Rec 252: 281-289.

Sinha, A. A., M. J. Wilson, D. F. Gleason, P. K. Reddy, M. Sameni and B. F. Sloane (1995). "Immunohistochemical localization of cathepsin B in neoplastic human prostate." Prostate 26: 171-178.

Snoek-van Beurden, P. A. and J. W. Von den Hoff (2005). "Zymographic techniques for the analysis of matrix metalloproteinases and their inhibitors." Biotechniques 38(1): 73-83.

Stewart, D. A., C. R. Cooper and R. A. Sikes (2004). "Changes in extracellular matrix (ECM) and ECM-associated proteins in the metastatic progression of prostate cancer." Reprod Biol Endocrinol 2: 2-2.

Sukhova, G. K., G. P. Shi, D. I. Simon, H. A. Chapman and P. Libby (1998). "Expression of the elastolytic cathepsins S and K in human atheroma and regulation of their production in smooth muscle cells." J Clin Invest 102: 576-583.

Sukhova, G. K., Y. Zhang, J.-H. Pan, Y. Wada, T. Yamamoto, M. Naito, T. Kodama, S. Tsimikas, J. L. Witztum, M. L. Lu, Y. Sakara, M. T. Chin, P. Libby and G.-P. Shi (2003). "Deficiency of cathepsin S reduces atherosclerosis in LDL receptor-deficient mice." J Clin Invest 111: 897-906.

Sundquist, K., P. Lakkakorpi, B. Wallmark and K. Vaananen (1990). "Inhibition of osteoclast proton transport by bafilomycin A1 abolishes bone resorption." Biochem Biophys Res Commun 168: 309-313.

Takeichi, M. (1991). "Cadherin cell adhesion receptors as a morphogenetic regulator." Science 251: 1451-1455.

Taleb, S., R. Cancello, K. Clement and D. Lacasa (2006). "Cathepsin s promotes human preadipocyte differentiation: possible involvement of fibronectin degradation." Endocrinology 147: 4950-4959.

Tamada, Y., C. Fukiage, D. L. Boyle, M. Azuma and T. R. Shearer (2000). "Involvement of cysteine proteases in bFGF-induced angiogenesis in guinea pig and rat cornea." J Ocul Pharmacol Ther 16: 271-283.

Teitelbaum, S. L. (2000). "Bone resorption by osteoclasts." Science 289: 1504-1508.

Thomssen, C., M. Schmitt, L. Goretzki, P. Oppelt, L. Pache, P. Dettmar, F. Janicke and H. Graeff (1995). "Prognostic value of the cysteine proteases cathepsins B and cathepsin L in human breast cancer." Clin Cancer Res 1: 741-746.

Tolosa, E., W. Li, Y. Yasuda, W. Wienhold, L. K. Denzin, A. Lautwein, C. Driessen, P. Schnorrer, E. Weber, S. Stevanovic, R. Kurek, A. Melms and D. Bromme (2003). "Cathepsin V is involved in the degradation of invariant chain in human thymus and is overexpressed in myasthenia gravis." J Clin Invest 112: 517-526.

Topol, E. J. and S. E. Nissen (1995). "Our preoccupation with coronary luminology. The dissociation between clinical and angiographic findings in ischemic heart disease." Circulation 92: 2333-2342.

Turk, B. (2006). "Targeting proteases: successes, failures and future prospects." Nat Rev Drug Discov 5: 785-799.

Turk, B., D. Turk and V. Turk (2000). "Lysosomal cysteine proteases: more than scavengers." Biochim Biophys Acta 1477: 98-9111.

Turk, V. and W. Bode (1991). "The cystatins: protein inhibitors of cysteine proteinases." FEBS Lett 285: 213-219.

Urbich, C., C. Heeschen, A. Aicher, K.-i. Sasaki, T. Bruhl, M. R. Farhadi, P. Vajkoczy, W. K. Hofmann, C. Peters, L. A. Pennacchio, N. D. Abolmaali, E. Chavakis, T. Reinheckel, A. M. Zeiher and S. Dimmeler (2005). "Cathepsin L is required for endothelial progenitor cell-induced neovascularization." Nat Med 11: 206-213.

van den Brule, S., P. Misson, F. Buhling, D. Lison and F. Huaux (2005). "Overexpression of cathepsin K during silica-induced lung fibrosis and control by TGF-beta." Respir Res 6: 84-84.

van Hinsbergh, V. W. M., M. A. Engelse and P. H. A. Quax (2006). "Pericellular proteases in angiogenesis and vasculogenesis." Arterioscler Thromb Vasc Biol 26: 716-728.

Vasiljeva, O., M. Dolinar, J. R. Pungercar, V. Turk and B. Turk (2005). "Recombinant human procathepsin S is capable of autocatalytic processing at neutral pH in the presence of glycosaminoglycans." FEBS Lett 579: 1285-1290.

Vasiljeva, O., T. Reinheckel, C. Peters, D. Turk, V. Turk and B. Turk (2007). "Emerging roles of cysteine cathepsins in disease and their potential as drug targets." Curr Pharm Des 13: 387-403.

Wagener, F. A., N. G. Abraham, Y. van Kooyk, T. de Witte and C. G. Figdor (2001). "Heme-induced cell adhesion in the pathogenesis of sickle-cell disease and inflammation." Trends Pharmacol Sci 22: 52-54.

Waghray, A., D. Keppler, B. F. Sloane, L. Schuger and Y. Q. Chen (2002). "Analysis of a truncated form of cathepsin H in human prostate tumor cells." J Biol Chem 277: 11533-11538.

Wang, Z., T. Zheng, Z. Zhu, R. J. Homer, R. J. Riese, H. A. Chapman, S. D. Shapiro and J. A. Elias (2000). "Interferon gamma induction of pulmonary emphysema in the adult murine lung." J Exp Med 192: 1587-1600.

Ward, C., D. Kuehn, R. E. Burden, J. A. Gormley, T. J. Jaquin, M. Gazdoiu, D. Small, R. Bicknell, J. A. Johnston, C. J. Scott and S. A. Olwill (2010). "Antibody targeting of cathepsin S inhibits angiogenesis and synergistically enhances anti-VEGF." PLoS One 5.

Watson, C. J. and P. A. Kreuzaler (2009). "The role of cathepsins in involution and breast cancer." J Mammary Gland Biol Neoplasia 14: 171-179.

Werle, B., W. Ebert, W. Klein and E. Spiess (1994). "Cathepsin B in tumors, normal tissue and isolated cells from the human lung." Anticancer Res 14: 1169-1176.

Werle, B., W. Ebert, W. Klein and E. Spiess (1995). "Assessment of cathepsin L activity by use of the inhibitor CA-074 compared to cathepsin B activity in human lung tumor tissue." Biol Chem Hoppe Seyler 376: 157-164.

Werle, B., C. Kraft, T. T. Lah, J. Kos, U. Schanzenbacher, K. Kayser, W. Ebert and E. Spiess (2000). "Cathepsin B in infiltrated lymph nodes is of prognostic significance for patients with nonsmall cell lung carcinoma." Cancer 89: 2282-2291.

Werle, B., H. Lotterle, U. Schanzenbacher, T. T. Lah, E. Kalman, K. Kayser, H. Bulzebruck, J. Schirren, M. Krasovec, J. Kos and E. Spiess (1999). "Immunochemical analysis of cathepsin B in lung tumours: an independent prognostic factor for squamous cell carcinoma patients." Br J Cancer 81: 510-519.

Wijkmans, J. and J. Gossen (2011). "Inhibitors of cathepsin K: a patent review (2004-2010)." Expert Opin Ther Pat 21: 1611-1629.

Woodward, W. A., P. Wachsberger, R. Burd and A. P. Dicker (2005). "Effects of androgen suppression and radiation on prostate cancer suggest a role for angiogenesis blockade." Prostate Cancer Prostatic Dis 8(2): 127-132.

Wun-Shaing W. Chang, H.-R. W., Chi-Tai Yeh, Cheng-Wen Wu, and Jang-Yang Chang (2007). "Lysosomal Cysteine Proteinase Cathepsin S as a Potential Target for Anti-Cancer Therapy." J Cancer Mol. 3: 5-14.

Xiao, G., Y. E. Liu, R. Gentz, Q. A. Sang, J. Ni, I. D. Goldberg and Y. E. Shi (1999). "Suppression of breast cancer growth and metastasis by a serpin myoepithelium-derived serine proteinase inhibitor expressed in the mammary myoepithelial cells." Proc Natl Acad Sci USA 96: 3700-3705.

Xiao, Y., H. Junfeng, L. Tianhong, W. Lu, C. Shulin, Z. Yu, L. Xiaohua, J. Weixia, Z. Sheng, G. Yanyun, L. Guo and L. Min (2006). "Cathepsin K in adipocyte differentiation and its potential role in the pathogenesis of obesity." J Clin Endocrinol Metab 91: 4520-4527.

Xue, Y., J. Li, M. A. Latijnhouwers, F. Smedts, R. Umbas, T. W. Aalders, F. M. Debruyne, J. J. De La Rosette and J. A. Schalken (1998). "Expression of periglandular tenascin-C and basement membrane laminin in normal prostate, benign prostatic hyperplasia and prostate carcinoma." Br J Urol 81: 844-851.

Yano, M., K. Hirai, Z. Naito, M. Yokoyama, T. Ishiwata, Y. Shiraki, M. Inokuchi and G. Asano (2001). "Expression of cathepsin B and cystatin C in human breast cancer." Surg Today 31: 385-389.

Yasuda, Y., Z. Li, D. Greenbaum, M. Bogyo, E. Weber and D. Bromme (2004). "Cathepsin V, a novel and potent elastolytic activity expressed in activated macrophages." J Biol Chem 279: 36761-36770.

Yeyeodu, S., K. Ahn, V. Madden, R. Chapman, L. Song and A. H. Erickson (2000). "Procathepsin L self-association as a mechanism for selective secretion." Traffic 1: 724-737.

Yuan, A., P. C. Yang, C. J. Yu, Y. C. Lee, Y. T. Yao, C. L. Chen, L. N. Lee, S. H. Kuo and K. T. Luh (1995). "Tumor angiogenesis correlates with histologic type and metastasis in non-small-cell lung cancer." Am J Respir Crit Care Med 152: 2157-2162.

Zhang, M., O. Volpert, Y. H. Shi and N. Bouck (2000). "Maspin is an angiogenesis inhibitor." Nat Med 6: 196-199.

Zheng, T., Z. Zhu, Z. Wang, R. J. Homer, B. Ma, R. J. Riese, Jr., H. A. Chapman, Jr., S. D. Shapiro and J. A. Elias (2000). "Inducible targeting of IL-13 to the adult lung causes matrix metalloproteinase- and cathepsin-dependent emphysema." J Clin Invest 106(9): 1081-1093.

Chapter 9

Krüppel-like Factor 5 (*KLF5*) and Tumor Angiogenesis

Rohinton S. Tarapore and Yizeng Yang*

660 Clinical Research Building, 415 Curie Blvd, Philadelphia, PA 19104

CHAPTER OUTLINE

► Introduction
► Angiogenesis
► Conclusions and perspectives
► References

ABSTRACT

KLF5, a Krüppel-like family transcription factor, plays a potential role in cell proliferation, differentiation, tissue homeostasis and tumorigenesis. *KLF5* is abundantly expressed in embryonic smooth muscle cells and is down-regulated with vascular development, but re-induced in proliferative smooth muscles in response to vascular injury. Homologous deletion of *KLF5* is lethal and mice die before E8.5. Heterozygous mice appear grossly normal; however, these mice show cardiac hypertrophy and diminished angiogenesis and arterial wall thickening. *KLF5* is involved in the regulation of angiogenesis during tumorigenesis and cancer progression by targeting or remodeling tumor microenvironment. *KLF5* activates many genes that can contribute to angiogenesis such as platelet-derived growth factor (PDGF) A/B, Egr-1, plasminogen activator inhibitor-1 (PAI-1), inducible nitric oxide synthase (iNOS), and vascular endothelial growth factor receptors (VEGFR). *KLF5* is regulated by many transcriptional regulators and nuclear receptors such as retinoic acid receptor-alpha (RARα), NF-κB, peroxisome proliferator-activated receptors (PPARγ) and adenovirus E1A-associated cellular p300 transcriptional co-activator protein (p300). RARα antagonist activates *KLF5* and induces angiogenesis.

Correspondence Author   660 Clinical Research Building, 415 Curie Blvd, Philadelphia, PA 19104.
Email: yizeng@mail.med.upenn.edu

KLF5 also plays an important role on the recruitment and activation of inflammatory cells that release cytokines and chemokines to induce hypoxia. This chapter will review recent studies carried out on the biological functions of *KLF5* with an insight on the tumor angiogenesis.

Key words: KLF5; Angiogenesis; Tumorigenesis

1   Introduction

1.1   Krüppel-like Factor Family

The Krüppel-like factor (*KLF*) family of transcription factors is named after the developmental factor, Krüppel, inactivation of which causes a crippled phenotype in fruit flies (*Drosophila melanogaster*). All members of the *KLF* family have a highly conserved C-terminal CH2H2 zinc-finger DNA binding domain. The amino-terminal domain is highly variable and provides functional identity through recruitment and regulation of specific protein-protein interactions (Zhang et al. 2002; Suske et al. 2005). This family of *KLFs* consists of more than 17 members which are designated *KLF*1-17 (Suske et al. 2005; van Vliet et al. 2006) though many have alternative names according to the tissues in which they were originally shown to be enriched. Members of the *KLF* family are divided into 3 subgroups based on structural and functional features acting as either activator or repressor of their target genes depending on the cellular environment. *KLFs* regulate diverse processes including cell growth, differentiation and signal transduction (Dang et al. 2000; Huber et al. 2001).

1.1.1   Krüppel-like factor 5 (*KLF5*)

KLF5 (BTEB2) is also named as intestinal-enriched Krüppel-like factor (IKLF) on account of its high level of expression in intestinal epithelia. The *KLF5* gene is located on chromosome 13q21 that is 18Kb in length with 4 exons (see Figure 1). Its full length cDNA consists of 3,350 base pairs with a 1,052 bp 3-untranslated region (3′ UTR), a 320 bp 5′ UTR and a 1,374bp coding sequence for 457 amino acid. The C-terminal end contains 3 zinc-finger (ZF) domains that aid in DNA binding. A proline rich transactivation domain (TAD) is located upstream of the ZF domain (Sogawa et al. 1993; Kojima et al. 1997).

Gene

Figure 1   **The human *KLF5* gene structure.**   The human *KLF5* genome contains four exons (exon 1, 585 bp; intron 1, 2,272 bp; exon 2,874 bp; intron 2, 1,089 bp; exon 3, 60 bp; intron 3, 11,824 bp; and exon 4, 1,831 bp).

Though first discovered in the intestinal epithelia, *KLF5* is widely expressed in different tissues at variable levels. High levels of *KLF5* mRNA have been observed in mouse and human tissues such as intestine, stomach, pancreas, lung, placenta, prostate, testis and skeletal muscle (Sogawa et al. 1993; Conkright et al. 1999; Shi et al. 1999). Besides epithelial tissues, *KLF5* is also expressed in cardiovascular smooth muscle cells, neurons, lymphoid cells and cornea (Watanabe et al. 1999; Yang et al. 2003; Chiambaretta et al. 2004; Yanagi et al. 2008). *KLF5* shows temporal change in expression during embryogenesis suggesting that it might play an essential role in proliferating cells more so than in differentiated cells (Sogawa et al. 1993; Watanabe et al. 1999; Yang et al. 2003; Yanagi et al. 2008). By analyzing human gut tissues, it has been demonstrated that *KLF5* protein is also exclusively expressed in the proliferating cells at the base of the intestinal crypts but not in the terminally differentiated epithelial cells in the villi (McConnell et al. 2007).

1.1.1.1 Target Genes of *KLF5*

KLF5 is an important transcriptional factor and has been demonstrated to regulate many genes involved in cell cycle, cell survival, proliferation, migration and differentiation (see Table 1) (Dang et al. 2000; Black et al. 2001; Sun et al. 2001; Chen et al. 2002; Shindo et al. 2002; Chen et al. 2003; Bateman et al. 2004; Nandan et al. 2004; Yang et al. 2008). The proximal promoters in most *KLF5* target genes contain one or more GC-rich elements (Sogawa et al. 1993; Kojima et al. 1997; Zhang and Teng 2003). *KLF5* has been shown to bind to Sp1 sites, GC boxes and CACCC boxes (Sogawa et al. 1993; Kojima et al. 1997; Zhang and Teng 2003). However, no strictly conserved consensus core sequences have been identified yet. Microarray studies (Chen et al. 2006) in theTSU-Pr1 bladder cancer cell line have identified at least 58 genes that are differentially expressed upon *KLF5* induction. These genes include fibroblast growth factor (FGF)-binding protein (HBP17/FGF-BP), beta-2-adrenergic receptor (ADBR2), breast cancer anti-estrogen resistance protein 3 (BCAR3), glycosyl phosphatidylinositol-anchored protein (CD24), ER transmembrane protein (Dri42)/dual specificity phosphatase 5 (DUSP5), eukaryotic translation elongation factor 1 alpha 2 (EEF1A2), epithelial membrane protein 1. (EMP1), exostosin-1 (EXT1), Integrin alpha-6 (ITGA6), lipocortin III, meningioma (disrupted in balanced translocation) 1 (MN1), inhibitor of cyclin dependent kinase (p27), microsomal prostaglandin E synthase 1(PIG12), retinoic acid induced-gene G protein coupling receptor (RAIG1), spindle assembly gene cluster (SAS), secreted glycoprotein gene (Slit), transforming growth factor beta (TGFβ), TGF-related protein M2 (TGFM2), tissular matrix metallopeptidase inhibitor 2 (TIMP2), tropomodulin gene (TMOD) and wingless-type MMTV (mouse mammary tumor virus)-integration site family, member 7A (Wnt7a). Wan et al using a lung-specific knockout mouse model demonstrated that *KLF5* regulates the expression of hundreds of genes associated with lipid metabolism, angiogenesis, survival

Table 1   Direct target genes of *KLF5*

Functions	Target genes	References
Angiogenesis	*PDGF, FGF-BP, VEGFα*	(Shindo et al. 2002; Aizawa et al. 2004; Usui et al. 2004; Chen et al. 2006; Wan et al. 2008)
Apoptosis	*Survivin, Pim1*	(Zhu et al. 2006; Zhao et al. 2008)
Cell cycle	*Cyclin B1, cyclin D1, cdc2, p15, p27*	(Bateman et al. 2004; Nandan et al. 2004; Nandan et al. 2005; Chen et al. 2006; Guo et al. 2009; Suzuki et al. 2009)
Differentiation	*iNOS, SMemb/NMHC-B, PAI-1, SM22α, EGR-1, PPARγ*	(Watanabe et al. 1999; Adam et al. 2000; Nagai et al. 2000; Nagai et al. 2001; Nagai et al. 2005; Oishi et al. 2005)
Fatty acid metabolism	*Cpt1b, Ucp2, Usp3, FASN*	(Oishi et al. 2008; Lee et al. 2009)
Inflammation	*NF-κB, MCP-1*	(Chanchevalap et al. 2006; Kumekawa et al. 2008)
Migration	*MMP9, ILK*	(Shinoda et al. 2008; Yang et al. 2008)
Stemness	*Fbxo15, Esrrb, Oct3/4, Nanog, Tcl1*	(Jiang et al. 2008; Parisi et al. 2008)
Others	*EGFR, DAF, Lama1, γ-globin, Lactoferrin, TCR Dβ1, KLF4, MAO/AB*	(Teng et al. 1998; Shi et al. 1999; Wong et al. 2001; Dang et al. 2002; Yang et al. 2003; Piccinni et al. 2004; Shih and Chen 2004; Zhang et al. 2005; Yang et al. 2007; Shao et al. 2008)

and several paracrine signaling including platelet derived growth factor – (PDGF)-FGF, VEGF, bone morphogenetic protein (BMP) and TGFβ (Wan et al. 2008).

1.1.1.2   Role of *KLF5* in Proliferation and Differentiation

In the mouse embryo, *KLF5* is abundant from E7 and is expressed throughout the primitive gut beginning at E10.5 (Ohnishi et al. 2000). At E16.5-E17.5, *KLF5* remains consistently at high levels throughout the gastrointestinal tract and can also be detected in the epithelia of the tongue. Afterward, it becomes progressively confined to the crypts of the small intestine. In adults, *KLF5* is predominantly present in the proliferating compartments of the gastrointestinal epithelia in the basal layer of the esophagus, small intestinal crypts and the lower third of the colonic crypts (Conkright et al. 1999; Ohnishi et al. 2000; Goldstein et al. 2007; Yang et al. 2008).

Functionally, it is well-accepted that *KLF5* is a positive regulator of proliferation in non-transformed epithelial cells (Sun et al. 2001), which increases proliferation in intestinal epithelial cells, IEC-6, IEC-18 and ICME (Bateman et al. 2004; Chanchevalap et al. 2004). Transgenic expression of *KLF5* in murine esophageal epithelia results in increased basal cell proliferation (Goldstein et al. 2007). *KLF5* is also important in maintaining the transit-amplifying cell state in esophageal epithelia by transcriptionally activating EGFR (epidermal factor growth factor receptor) and ILK (integrin-linked kinase) to regulate proliferation and migration respectively (Yang et al. 2007; Yang et al. 2008).

KLF5 also plays an essential role in epithelial differentiation and homeostasis. During epithelial homeostasis, stem cells divide to produce progenitor cells and the progenitor cells further proliferate to generate cell mass needed for mature epithelia (Blanpain et al. 2007). Overexpression of *KLF5* is observed in proliferating epithelial cells (immortal but untransformed epithelial cell lines) that mostly include progenitor cells (Chen et al. 2002; Chen et al. 2003; Bateman et al. 2004; Chanchevalap et al. 2004). *KLF5* also controls the differentiation of smooth muscle cells (SMCs) and adipocytes. The proliferative SMCs express *KLF5*, but significantly reduce following activation of SMCs to modulate differentiation in response to injury, which is a very important mechanism for vascular regeneration (Walsh and Takahashi 2001; Suzuki et al. 2005). Further studies confirmed that *KLF5* could upregulate the expression of genes involved in the differentiation of SMCs like SMemb/NMHC-β (Watanabe et al. 1999) and SM22α (Adam et al. 2000). 3T3-L1 preadipocytes express high levels of *KLF5* at an early stage of differentiation followed by induction of PPARγ2 (Belli and Bosco 1992). Constitutive overexpression of dominant negative *KLF5* inhibits adipocyte differentiation while overexpression of wild-type *KLF5* induces differentiation, even in the absence of hormonal stimulation (Belli and Bosco 1992). It appears that *KLF5* expression is induced by C/EBPβ and δ and in turn *KLF5* (in concert with C/EBP β/δ) activates the expression of PPARγ2 (Belli and Bosco 1992).

1.1.1.3 *KLF5* and Cancer

KLF5 is believed to play a dynamic context-dependent role in tumorigenesis. Since *KLF5* promotes cell survival and proliferation, it was implicated as an oncogenic factor. However, accumulating functional and genetic studies suggest *KLF5* can be a tumor-suppressor or oncogenic factor under certain situations (see Table 2).

In cancers derived from salivary gland, breast and prostate, *KLF5* gene shows frequent genetic alterations. Numerous comparative genomic-hybridization studies have found that *KLF5* gene locus (chromosome 13q21) is the second most frequent deletion in different human cancers (Knuutila et al. 1999; Dong and Chen 2009). The deletion of the 13q21 region is associated with metastasis and higher tumor grade in prostate cancer (Hyytinen et al. 1999; Dong et al. 2000; Chen et al. 2001; Dong et al. 2001; Dong and Chen 2009). In prostate cancer, the majority of *KLF5* alterations are hemizygous deletions (Shindo et al. 2002) that are equivalent to haplosufficiency (i.e. 50% of protein expression). The excessive degradation of *KLF5* protein by the overexpression of E3 ubiquitin ligase WWP1 is common in many breast and prostate cancer cell lines (Chen et al. 2002; Chen et al. 2003). Clinical studies (Kwak et al. 2008) demonstrate that *KLF5* is detected in early tumor stages and is less observed in late stage tumors and cancers. A study which analyzed 247 gastric cancers showed that *KLF5* is expressed in 46% of

Table 2   Role of *KLF5* in various cancers

Cancers	Expression	References
Melanoma	mRNA is downregulated in Ras mutated cancer cell lines by microarray	(Bloethner et al. 2005)
Nasopharyngeal	mRNA is downregulated	(Fang et al. 2008)
Salivary gland	Gain in gene copy number by CGH	(Giefing et al. 2008)
Esophagus	mRNA level is upregulated in stem-like cancer cells by qPCR. Inhibits cell proliferation, survival and invasion.	(Chaib et al. 2001; Chen et al. 2003; Huang et al. 2009)
Stomach	Protein expression is high in early-staged and is negative in lymph node metastasis by IHC	(Kwak et al. 2008)
Breast	Gene copy number loss, mRNA is low in ER+ cancer cell lines, high in ER- tumors	(Chen et al. 2002; Tong et al. 2006; Ben-Porath et al. 2008)
Prostate	Loss in gene copy number, mRNA is lost in some cancer cell lines	(Ohnishi et al. 2003; Chen et al. 2006)

tumors (113/247). *KLF5* was more frequently detected in early stage tumors than in late stage tumors, in cases without lymph node metastasis and the primary tumor was smaller than 5 cm in size (Kwak et al. 2008; Dong and Chen 2009). In breast cancers, increased levels of *KLF5* mRNA are directly associated with a reduced disease free period and overall survival rate (Tong et al. 2006). Similar clinical outcomes related to *KFLF5* mRNA have been achieved in prostate cancers (Chaib et al. 2001). In an esophageal cancer cell line derived from poorly differentiated esophageal squamous carcinoma, stable expression of *KLF5* significantly inhibits cell proliferation and invasion (Yang et al. 2005). A recent study suggested that p53 family protein play critical role for this dynamic context-dependent function of *KLF5* (Yang et al. 2011).

2   Angiogenesis

Angiogenesis occurs through life beginning in the fetus and continuing on through old age. Blood capillaries are essential in all tissues for diffusion, exchange of nutrients and metabolites. Changes in metabolic activity of the tissue and the oxygen content play a pivotal role in the regulation of angiogenesis. Hemodynamic (blood flow) factors are critical for structural adaptation of blood vessel walls and for the survival of vascular network. Stimulation of angiogenesis can be of therapeutic value in the areas of ischemic heart disease, peripheral arterial disease and wound healing. Inhibiting angiogenesis can also be significant in therapies targeting cancer, ophthalmic conditions, rheumatoid arthritis and other ailments. In healthy tissue, capillaries can grow and regress according to functional demands. A lack of exercise can lead to capillary regression while exercise stimulates angiogenesis in skeletal muscle and heart (Kraus et al. 2004). In short, angiogenesis occurs during developmental and vascular remodeling as a series of controlled events that lead to neo-vascularization thereby supporting the changing requirement of tissue (Folkman and Hanahan 1991).

2.1 Tumor Angiogenesis

A tumor consists of a population of rapidly dividing and growing cancer cells that have lost their ability to divide in a controlled fashion. Mutations (alterations in the gene) are frequent which allow the cancer cells (or a sub-population of cells within the tumor) to develop resistance to anti-cancer drugs and escape therapy. Tumors cannot grow beyond a certain size usually 2–3 mm^3 due to the lack of oxygen and other essential nutrients without angiogenesis. Tumor cells induce angiogenesis by secreting growth factors like VEGF and FGF. These factors can induce capillary growth into the tumor resulting in increased tumor size (Weis and Cheresh 2011), which is a necessary event for the transition from a cluster of cells (the size of the tip of a ball point pen) to a large tumor. Angiogenesis also plays an essential role for the spread of a tumor or metastasis – single cells can break away from an established solid tumor, enter the blood vessel and be carried to distant sites where they can implant and begin the growth of a secondary tumor (Bergers and Benjamin 2003).

Tumor blood vessels have perivascular detachment, vessel dilation and irregular shape. Studies show that tumor blood vessels are not smooth like normal tissues and their main role is to feed the tumor oxygen and other nutrients. Endothelial cells are originally from bone marrow and eventually integrated into the growing blood vessels. These endothelial cells differentiate and migrate into the perivascular space to form tumor cells. VEGF, an important growth factor, plays an essential role in the formation of blood vessels that lead to tumor growth, allowing vessels to expand – a process called sprouting angiogenesis (Brown and Giaccia 1998; Rafii et al. 2002; Bergers and Benjamin 2003). A lot of research has been carried out on angiogenesis. Angiogenesis-based tumor therapy relies on the use of natural or synthetic angiogenesis inhibitors like angiostatin, endostatin and tumorstatin. The first FDA-approved therapy targeting angiogenesis is a monoclonal antibody directed against an isoform of VEGF. Avastin is the commercial name of this antibody.

2.2 *KLF5* and Tumor Angiogenesis

Cardiovascular remodeling is a complicated process involving the activation of mesenchymal cells (smooth muscle cells, myofibroblasts), production of extracellular matrix and angiogenesis. External stresses such as physical injury, ischemia or pressure overload activate the mesenchymal cells and trigger early response genes resulting in remodeling (Carmeliet and Collen 2000; Libby 2000; Nagai et al. 2005). *KLF5* is abundantly expressed in embryonic vascular smooth muscles in the cardiovascular system but is downregulated in adult vessels. However, it is strongly re-induced in activated smooth muscle cells and myofibroblasts in athero- and arteriosclerosis (Shiojima et al. 1999). Studies (Weber et al. 1988; Leslie et al. 1991) identify *KLF5* as an important transcriptional factor in cardiovascular remodeling. *In vitro* analysis showed *KLF5* activates many genes promoters like PDGF-A/B, iNOS, PAI-1 and VEGF receptor, which are induced in cardiovascular remodeling (Nagai et al. 2005). Animal models also highlight the significance of *KLF5* in cardiovascular remodeling as evidenced by *KLF5*+/− mice in which inflammatory response and Angiogiotensin II-induced cardiac hypertrophy were attenuated. Numerous studies carried out by Nagai et al (Nagai et al. 2000; Nagai et al. 2001; Nagai et al. 2005) indicate that *KLF5* plays a crucial role in cardiac remodeling involving mesenchymal cell activation, development of interstitial fibrosis and angiogenesis.

Neovascularization and angiogenesis in tumor tissues are important processes for sustained tumor growth and metastasis. The role of epithelial growth factor receptor B (ERBB) family in angiogenesis and the requirement of Erbb2 in cardiac development have been demonstrated (Petit et al. 1997; Russell et al. 1999). Inhibiting oncogene-mediated angiogenesis is a major biologic process for the tumoricidal effect of therapeutics used in the treatment of Erbb-overexpression of breast carcinomas (Izumi et al. 2002; Klos et al. 2003). Studies on *ERBB2*-transfected fibroblasts demonstrate downregulation of angiogenic factors like Serpinf1 and other inhibitors like tissular inhibitor of matrix metalloprotease 3 (TIMP3) and secreted protein acidic and rich in cysteine (SPARC) and an increased expression of *KLF5*. Mice heterozygous for a *KLF5* knockout allele showed impaired angiogenic activity which is consistent with *KLF5* expression in activated endothelial cells (Shindo et al. 2002). In addition, *KLF5* was identified as one of the factors induced upon vascular injury (Shindo et al. 2002). *KLF5* overexpression may contribute to the transformed phenotype of ERBB-transfected cells. *KLF5* transfection in NIH-3T3 fibroblasts positively regulated anchorage-independent proliferation and led to a loss of cell-cell contact inhibition (Sun et al. 2001).

3 Conclusions and Perspectives

The Krüppel-like Factor 5 (*KLF5/IKLF/BTEB2*) plays an important role in carcinogenesis, cell proliferation, differentiation, cell cycle regulation and angiogenesis. The function of *KLF5* is cell-context dependent in various physiological and pathological conditions – wound healing (Ogata et al. 2000), cardiovascular remodeling (Shindo et al. 2002), embryonic development (Ohnishi et al. 2000) and regulation of adipogenesis and energy metabolism (Oishi et al. 2005; Oishi et al. 2008). Thus, *KLF5* is an attractive target for the development of novel therapeutics towards the treatment of cardiovascular and metabolic diseases and cancer since the level of *KLF5* expression correlates with the prognosis of certain cancers (Tong et al. 2006). The knowledge of the complex interplay between *KLF5* and other factors has enabled the identification of target points with the potential for modulation with the use of small molecules. *KLF5* genes and small molecules that modulate the expression and function of *KLF5* may be very useful in our combat against cancer and other metabolic diseases.

REFERENCES

Adam, P. J., C. P. Regan, M. B. Hautmann and G. K. Owens (2000). "Positive- and negative-acting Krüppel-like transcription factors bind a transforming growth factor beta control element required for expression of the smooth muscle cell differentiation marker SM22alpha in vivo." J Biol Chem 275(48): 37798-37806.

Aizawa, K., T. Suzuki, N. Kada, A. Ishihara, K. Kawai-Kowase, T. Matsumura, K. Sasaki, Y. Munemasa, I. Manabe, M. Kurabayashi, T. Collins and R. Nagai (2004). "Regulation of platelet-derived growth factor-A chain by Krüppel-like factor 5: new pathway of cooperative activation with nuclear factor-kappaB." J Biol Chem 279(1): 70-76.

Bateman, N. W., D. Tan, R. G. Pestell, J. D. Black and A. R. Black (2004). "Intestinal tumor progression is associated with altered function of KLF5." J Biol Chem 279(13): 12093-12101.

Belli, A. and C. Bosco (1992). "Influence of stretch-shortening cycle on mechanical behaviour of triceps surae during hopping." Acta Physiol Scand 144(4): 401-408.

Ben-Porath, I., M. W. Thomson, V. J. Carey, R. Ge, G. W. Bell, A. Regev and R. A. Weinberg (2008). "An embryonic stem cell-like gene expression signature in poorly differentiated aggressive human tumors." Nat Genet 40(5): 499-507.

Bergers, G. and L. E. Benjamin (2003). "Tumorigenesis and the angiogenic switch." Nat Rev Cancer 3(6): 401-410.

Black, A. R., J. D. Black and J. Azizkhan-Clifford (2001). "Sp1 and krüppel-like factor family of transcription factors in cell growth regulation and cancer." J Cell Physiol 188(2): 143-160.

Blanpain, C., V. Horsley and E. Fuchs (2007). "Epithelial stem cells: turning over new leaves." Cell 128(3): 445-458.

Bloethner, S., B. Chen, K. Hemminki, J. Muller-Berghaus, S. Ugurel, D. Schadendorf and R. Kumar (2005). "Effect of common B-RAF and N-RAS mutations on global gene expression in melanoma cell lines." Carcinogenesis 26(7): 1224-1232.

Brown, J. M. and A. J. Giaccia (1998). "The unique physiology of solid tumors: opportunities (and problems) for cancer therapy." Cancer Res 58(7): 1408-1416.

Carmeliet, P. and D. Collen (2000). "Transgenic mouse models in angiogenesis and cardiovascular disease." J Pathol 190(3): 387-405.

Chaib, H., E. K. Cockrell, M. A. Rubin and J. A. Macoska (2001). "Profiling and verification of gene expression patterns in normal and malignant human prostate tissues by cDNA microarray analysis." Neoplasia 3(1): 43-52.

Chanchevalap, S., M. O. Nandan, B. B. McConnell, L. Charrier, D. Merlin, J. P. Katz and V. W. Yang (2006). "Krüppel-like factor 5 is an important mediator for lipopolysaccharide-induced proinflammatory response in intestinal epithelial cells." Nucleic Acids Res 34(4): 1216-1223.

Chanchevalap, S., M. O. Nandan, D. Merlin and V. W. Yang (2004). "All-trans retinoic acid inhibits proliferation of intestinal epithelial cells by inhibiting expression of the gene encoding Krüppel-like factor 5." FEBS Lett 578(1-2): 99-105.

Chen, C., M. S. Benjamin, X. Sun, K. B. Otto, P. Guo, X. Y. Dong, Y. Bao, Z. Zhou, X. Cheng, J. W. Simons and J. T. Dong (2006). "KLF5 promotes cell proliferation and tumorigenesis through gene regulation and the TSU-Pr1 human bladder cancer cell line." Int J Cancer 118(6): 1346-1355.

Chen, C., H. V. Bhalala, H. Qiao and J. T. Dong (2002). "A possible tumor suppressor role of the KLF5 transcription factor in human breast cancer." Oncogene 21(43): 6567-6572.

Chen, C., H. V. Bhalala, R. L. Vessella and J. T. Dong (2003). "KLF5 is frequently deleted and down-regulated but rarely mutated in prostate cancer." Prostate 55(2): 81-88.

Chen, C., W. W. Brabham, B. G. Stultz, H. F. Frierson, Jr., J. C. Barrett, C. L. Sawyers, J. T. Isaacs and J. T. Dong (2001). "Defining a common region of deletion at 13q21 in human cancers." Genes Chromosomes Cancer 31(4): 333-344.

Chiambaretta, F., F. De Graeve, G. Turet, G. Marceau, P. Gain, B. Dastugue, D. Rigal and V. Sapin (2004). "Cell and tissue specific expression of human Krüppel-like transcription factors in human ocular surface." Mol Vis 10: 901-909.

Conkright, M. D., M. A. Wani, K. P. Anderson and J. B. Lingrel (1999). "A gene encoding an intestinal-enriched member of the Krüppel-like factor family expressed in intestinal epithelial cells." Nucleic Acids Res 27(5): 1263-1270.

Dang, D. T., J. Pevsner and V. W. Yang (2000). "The biology of the mammalian Krüppel-like family of transcription factors." Int J Biochem Cell Biol 32(11-12): 1103-1121.

Dang, D. T., W. Zhao, C. S. Mahatan, D. E. Geiman and V. W. Yang (2002). "Opposing effects of Krüppel-like factor 4 (gut-enriched Krüppel-like factor) and Krüppel-like factor 5 (intestinal-enriched Krüppel-like factor) on the promoter of the Krüppel-like factor 4 gene." Nucleic Acids Res 30(13): 2736-2741.

Dong, J. T., J. C. Boyd and H. F. Frierson, Jr. (2001). "Loss of heterozygosity at 13q14 and 13q21 in high grade, high stage prostate cancer." Prostate 49(3): 166-171.

Dong, J. T. and C. Chen (2009). "Essential role of *KLF5* transcription factor in cell proliferation and differentiation and its implications for human diseases." Cell Mol Life Sci 66(16): 2691-2706.

Dong, J. T., C. Chen, B. G. Stultz, J. T. Isaacs and H. F. Frierson, Jr. (2000). "Deletion at 13q21 is associated with aggressive prostate cancers." Cancer Res 60(14): 3880-3883.

Fang, W., X. Li, Q. Jiang, Z. Liu, H. Yang, S. Wang, S. Xie, Q. Liu, T. Liu, J. Huang, W. Xie, Z. Li, Y. Zhao, E. Wang, F. M. Marincola and K. Yao (2008). "Transcriptional patterns, biomarkers and pathways characterizing nasopharyngeal carcinoma of Southern China." J Transl Med 6: 32.

Folkman, J. and D. Hanahan (1991). "Switch to the angiogenic phenotype during tumorigenesis." Princess Takamatsu Symp 22: 339-347.

Giefing, M., M. Wierzbicka, M. Rydzanicz, R. Cegla, M. Kujawski and K. Szyfter (2008). "Chromosomal gains and losses indicate oncogene and tumor suppressor gene candidates in salivary gland tumors." Neoplasma 55(1): 55-60.

Goldstein, B. G., H. H. Chao, Y. Yang, Y. A. Yermolina, J. W. Tobias and J. P. Katz (2007). "Overexpression of Krüppel-like factor 5 in esophageal epithelia in vivo leads to increased proliferation in basal but not suprabasal cells." Am J Physiol Gastrointest Liver Physiol 292(6): G1784-1792.

Guo, P., X. Y. Dong, X. Zhang, K. W. Zhao, X. Sun, Q. Li and J. T. Dong (2009). "Proproliferative factor KLF5 becomes anti-proliferative in epithelial homeostasis upon signaling-mediated modification." J Biol Chem 284(10): 6071-6078.

Huang, D., Q. Gao, L. Guo, C. Zhang, W. Jiang, H. Li, J. Wang, X. Han, Y. Shi and S. H. Lu (2009). "Isolation and identification of cancer stem-like cells in esophageal carcinoma cell lines." Stem Cells Dev 18(3): 465-473.

Huber, T. L., A. C. Perkins, A. E. Deconinck, F. Y. Chan, P. E. Mead and L. I. Zon (2001). "neptune, a Krüppel-like transcription factor that participates in primitive erythropoiesis in *Xenopus*." Curr Biol 11(18): 1456-1461.

Hyytinen, E. R., H. F. Frierson, Jr., J. C. Boyd, L. W. Chung and J. T. Dong (1999). "Three distinct regions of allelic loss at 13q14, 13q21-22, and 13q33 in prostate cancer." Genes Chromosomes Cancer 25(2): 108-114.

Izumi, Y., L. Xu, E. di Tomaso, D. Fukumura and R. K. Jain (2002). "Tumour biology: herceptin acts as an anti-angiogenic cocktail." Nature 416(6878): 279-280.

Jiang, J., Y. S. Chan, Y. H. Loh, J. Cai, G. Q. Tong, C. A. Lim, P. Robson, S. Zhong and H. H. Ng (2008). "A core KLF circuitry regulates self-renewal of embryonic stem cells." Nat Cell Biol 10(3): 353-360.

Klos, K. S., X. Zhou, S. Lee, L. Zhang, W. Yang, Y. Nagata and D. Yu (2003). "Combined trastuzumab and paclitaxel treatment better inhibits ErbB-2-mediated angiogenesis in breast carcinoma through a more effective inhibition of Akt than either treatment alone." Cancer 98(7): 1377-1385.

Knuutila, S., Y. Aalto, K. Autio, A. M. Bjorkqvist, W. El-Rifai, S. Hemmer, T. Huhta, E. Kettunen, S. Kiuru-Kuhlefelt, M. L. Larramendy, T. Lushnikova, O. Monni, H. Pere, J. Tapper, M. Tarkkanen, A. Varis, V. M. Wasenius, M. Wolf and Y. Zhu (1999). "DNA copy number losses in human neoplasms." Am J Pathol 155(3): 683-694.

Kojima, S., A. Kobayashi, O. Gotoh, Y. Ohkuma, Y. Fujii-Kuriyama and K. Sogawa (1997). "Transcriptional activation domain of human BTEB2, a GC box-binding factor." J Biochem 121(2): 389-396.

Kraus, R. M., H. W. Stallings, 3rd, R. C. Yeager and T. P. Gavin (2004). "Circulating plasma VEGF response to exercise in sedentary and endurance-trained men." J Appl Physiol 96(4): 1445-1450.

Kumekawa, M., G. Fukuda, S. Shimizu, K. Konno and M. Odawara (2008). "Inhibition of monocyte chemoattractant protein-1 by Krüppel-like factor 5 small interfering RNA in the tumor necrosis factor- alpha-activated human umbilical vein endothelial cells." Biol Pharm Bull 31(8): 1609-1613.

Kwak, M. K., H. J. Lee, K. Hur, J. Park do, H. S. Lee, W. H. Kim, K. U. Lee, K. J. Choe, P. Guilford and H. K. Yang (2008). "Expression of Krüppel-like factor 5 in human gastric carcinomas." J Cancer Res Clin Oncol 134(2): 163-167.

Lee, M. Y., J. S. Moon, S. W. Park, Y. K. Koh, Y. H. Ahn and K. S. Kim (2009). "KLF5 enhances SREBP-1 action in androgen-dependent induction of fatty acid synthase in prostate cancer cells." Biochem J 417(1): 313-322.

Leslie, K. O., D. J. Taatjes, J. Schwarz, M. vonTurkovich and R. B. Low (1991). "Cardiac myofibroblasts express alpha smooth muscle actin during right ventricular pressure overload in the rabbit." Am J Pathol 139(1): 207-216.

Libby, P. (2000). "Changing concepts of atherogenesis." J Intern Med 247(3): 349-358.

McConnell, B. B., A. M. Ghaleb, M. O. Nandan and V. W. Yang (2007). "The diverse functions of Krüppel-like factors 4 and 5 in epithelial biology and pathobiology." Bioessays 29(6): 549-557.

Nagai, R., K. Kowase and M. Kurabayashi (2000). "Transcriptional regulation of smooth muscle phenotypic modulation." Ann N Y Acad Sci 902: 214-222; discussion 222-213.

Nagai, R., T. Suzuki, K. Aizawa, S. Miyamoto, T. Amaki, K. Kawai-Kowase, K. I. Sekiguchi and M. Kurabayashi (2001). "Phenotypic modulation of vascular smooth muscle cells: dissection of transcriptional regulatory mechanisms." Ann N Y Acad Sci 947: 56-66; discussion 66-57.

Nagai, R., T. Suzuki, K. Aizawa, T. Shindo and I. Manabe (2005). "Significance of the transcription factor KLF5 in cardiovascular remodeling." J Thromb Haemost 3(8): 1569-1576.

Nandan, M. O., S. Chanchevalap, W. B. Dalton and V. W. Yang (2005). "Krüppel-like factor 5 promotes mitosis by activating the cyclin B1/Cdc2 complex during oncogenic Ras-mediated transformation." FEBS Lett 579(21): 4757-4762.

Nandan, M. O., H. S. Yoon, W. Zhao, L. A. Ouko, S. Chanchevalap and V. W. Yang (2004). "Krüppel-like factor 5 mediates the transforming activity of oncogenic H-Ras." Oncogene 23(19): 3404-3413.

Ogata, T., M. Kurabayashi, Y. Hoshino, K. Sekiguchi, S. Ishikawa, Y. Morishita and R. Nagai (2000). "Inducible expression of basic transcription element-binding protein 2 in proliferating smooth muscle cells at the vascular anastomotic stricture." J Thorac Cardiovasc Surg 119(5): 983-989.

Ohnishi, S., F. Laub, N. Matsumoto, M. Asaka, F. Ramirez, T. Yoshida and M. Terada (2000). "Developmental expression of the mouse gene coding for the Krüppel-like transcription factor KLF5." Dev Dyn 217(4): 421-429.

Ohnishi, S., S. Ohnami, F. Laub, K. Aoki, K. Suzuki, Y. Kanai, K. Haga, M. Asaka, F. Ramirez and T. Yoshida (2003). "Downregulation and growth inhibitory effect of epithelial-type Krüppel-like transcription factor KLF4, but not KLF5, in bladder cancer." Biochem Biophys Res Commun 308(2): 251-256.

Oishi, Y., I. Manabe, K. Tobe, M. Ohsugi, T. Kubota, K. Fujiu, K. Maemura, N. Kubota, T. Kadowaki and R. Nagai (2008). "SUMOylation of Krüppel-like transcription factor 5 acts as a molecular switch in transcriptional programs of lipid metabolism involving PPAR-delta." Nat Med 14(6): 656-666.

Oishi, Y., I. Manabe, K. Tobe, K. Tsushima, T. Shindo, K. Fujiu, G. Nishimura, K. Maemura, T. Yamauchi, N. Kubota, R. Suzuki, T. Kitamura, S. Akira, T. Kadowaki and R. Nagai (2005). "Krüppel-like transcription factor KLF5 is a key regulator of adipocyte differentiation." Cell Metab 1(1): 27-39.

Parisi, S., F. Passaro, L. Aloia, I. Manabe, R. Nagai, L. Pastore and T. Russo (2008). "KLF5 is involved in self-renewal of mouse embryonic stem cells." J Cell Sci 121(Pt 16): 2629-2634.

Petit, A. M., J. Rak, M. C. Hung, P. Rockwell, N. Goldstein, B. Fendly and R. S. Kerbel (1997). "Neutralizing antibodies against epidermal growth factor and ErbB-2/neu receptor tyrosine kinases down-regulate vascular endothelial growth factor production by tumor cells in vitro and in vivo: angiogenic implications for signal transduction therapy of solid tumors." Am J Pathol 151(6): 1523-1530.

Piccinni, S. A., A. L. Bolcato-Bellemin, A. Klein, V. W. Yang, M. Kedinger, P. Simon-Assmann and O. Lefebvre (2004). "Krüppel-like factors regulate the Lama1 gene encoding the laminin alpha1 chain." J Biol Chem 279(10): 9103-9114.

Rafii, S., B. Heissig and K. Hattori (2002). "Efficient mobilization and recruitment of marrow-derived endothelial and hematopoietic stem cells by adenoviral vectors expressing angiogenic factors." Gene Ther 9(10): 631-641.

Russell, K. S., D. F. Stern, P. J. Polverini and J. R. Bender (1999). "Neuregulin activation of ErbB receptors in vascular endothelium leads to angiogenesis." Am J Physiol 277(6 Pt 2): H2205-2211.

Shao, J., V. W. Yang and H. Sheng (2008). "Prostaglandin E2 and Krüppel-like transcription factors synergistically induce the expression of decay-accelerating factor in intestinal epithelial cells." Immunology 125(3): 397-407.

Shi, H., Z. Zhang, X. Wang, S. Liu and C. T. Teng (1999). "Isolation and characterization of a gene encoding human Krüppel-like factor 5 (IKLF): binding to the CAAT/GT box of the mouse lactoferrin gene promoter." Nucleic Acids Res 27(24): 4807-4815.

Shih, J. C. and K. Chen (2004). "Regulation of MAO-A and MAO-B gene expression." Curr Med Chem 11(15): 1995-2005.

Shindo, T., I. Manabe, Y. Fukushima, K. Tobe, K. Aizawa, S. Miyamoto, K. Kawai-Kowase, N. Moriyama, Y. Imai, H. Kawakami, H. Nishimatsu, T. Ishikawa, T. Suzuki, H. Morita, K. Maemura, M. Sata, Y. Hirata, M. Komukai, H. Kagechika, T. Kadowaki, M. Kurabayashi and R. Nagai (2002). "Krüppel-like zinc-finger transcription factor *KLF5*/BTEB2 is a target for angiotensin II signaling and an essential regulator of cardiovascular remodeling." Nat Med 8(8): 856-863.

Shinoda, Y., N. Ogata, A. Higashikawa, I. Manabe, T. Shindo, T. Yamada, F. Kugimiya, T. Ikeda, N. Kawamura, Y. Kawasaki, K. Tsushima, N. Takeda, R. Nagai, K. Hoshi, K. Nakamura, U. I. Chung and H. Kawaguchi (2008). "Krüppel-like factor 5 causes cartilage degradation through transactivation of matrix metalloproteinase 9." J Biol Chem 283(36): 24682-24689.

Shiojima, I., M. Aikawa, J. Suzuki, Y. Yazaki and R. Nagai (1999). "Embryonic smooth muscle myosin heavy chain SMemb is expressed in pressure-overloaded cardiac fibroblasts." Jpn Heart J 40(6): 803-818.

Sogawa, K., H. Imataka, Y. Yamasaki, H. Kusume, H. Abe and Y. Fujii-Kuriyama (1993). "cDNA cloning and transcriptional properties of a novel GC box-binding protein, BTEB2." Nucleic Acids Res 21(7): 1527-1532.

Sun, R., X. Chen and V. W. Yang (2001). "Intestinal-enriched Krüppel-like factor (Krüppel-like factor 5) is a positive regulator of cellular proliferation." J Biol Chem 276(10): 6897-6900.

Suske, G., E. Bruford and S. Philipsen (2005). "Mammalian SP/KLF transcription factors: bring in the family." Genomics 85(5): 551-556.

Suzuki, T., K. Aizawa, T. Matsumura and R. Nagai (2005). "Vascular implications of the Krüppel-like family of transcription factors." Arterioscler Thromb Vasc Biol 25(6): 1135-1141.

Suzuki, T., D. Sawaki, K. Aizawa, Y. Munemasa, T. Matsumura, J. Ishida and R. Nagai (2009). "Krüppel-like factor 5 shows proliferation-specific roles in vascular remodeling, direct stimulation of cell growth, and inhibition of apoptosis." J Biol Chem 284(14): 9549-9557.

Teng, C., H. Shi, N. Yang and H. Shigeta (1998). "Mouse lactoferrin gene. Promoter-specific regulation by EGF and cDNA cloning of the EGF-response-element binding protein." Adv Exp Med Biol 443: 65-78.

Tong, D., K. Czerwenka, G. Heinze, M. Ryffel, E. Schuster, A. Witt, S. Leodolter and R. Zeillinger (2006). "Expression of KLF5 is a prognostic factor for disease-free survival and overall survival in patients with breast cancer." Clin Cancer Res 12(8): 2442-2448.

Usui, S., N. Sugimoto, N. Takuwa, S. Sakagami, S. Takata, S. Kaneko and Y. Takuwa (2004). "Blood lipid mediator sphingosine 1-phosphate potently stimulates platelet-derived growth factor-A and -B chain expression through S1P1-Gi-Ras-MAPK-dependent induction of Krüppel-like factor 5." J Biol Chem 279(13): 12300-12311.

van Vliet, J., L. A. Crofts, K. G. Quinlan, R. Czolij, A. C. Perkins and M. Crossley (2006). "Human KLF17 is a new member of the Sp/KLF family of transcription factors." Genomics 87(4): 474-482.

Walsh, K. and A. Takahashi (2001). "Transcriptional regulation of vascular smooth muscle cell phenotype." Z Kardiol 90 Suppl 3: 12-16.

Wan, H., F. Luo, S. E. Wert, L. Zhang, Y. Xu, M. Ikegami, Y. Maeda, S. M. Bell and J. A. Whitsett (2008). "Krüppel-like factor 5 is required for perinatal lung morphogenesis and function." Development 135(15): 2563-2572.

Watanabe, N., M. Kurabayashi, Y. Shimomura, K. Kawai-Kowase, Y. Hoshino, I. Manabe, M. Watanabe, M. Aikawa, M. Kuro-o, T. Suzuki, Y. Yazaki and R. Nagai (1999). "BTEB2, a Krüppel-like transcription factor, regulates expression of the SMemb/Nonmuscle myosin heavy chain B (SMemb/NMHC-B) gene." Circ Res 85(2): 182-191.

Weber, K. T., J. S. Janicki, S. G. Shroff, R. Pick, R. M. Chen and R. I. Bashey (1988). "Collagen remodeling of the pressure-overloaded, hypertrophied nonhuman primate myocardium." Circ Res 62(4): 757-765.

Weis, S. M. and D. A. Cheresh (2011). "Tumor angiogenesis: molecular pathways and therapeutic targets." Nat Med 17(11): 1359-1370.

Wong, W. K., K. Chen and J. C. Shih (2001). "Regulation of human monoamine oxidase B gene by Sp1 and Sp3." Mol Pharmacol 59(4): 852-859.

Yanagi, M., T. Hashimoto, N. Kitamura, M. Fukutake, O. Komure, N. Nishiguchi, T. Kawamata, K. Maeda and O. Shirakawa (2008). "Expression of Krüppel-like factor 5 gene in human brain and association of the gene with the susceptibility to schizophrenia." Schizophr Res 100(1-3): 291-301.

Yang, X. O., R. T. Doty, J. S. Hicks and D. M. Willerford (2003). "Regulation of T-cell receptor D beta 1 promoter by KLF5 through reiterated GC-rich motifs." Blood 101(11): 4492-4499.

Yang, Y., B. G. Goldstein, H. H. Chao and J. P. Katz (2005). "KLF4 and KLF5 regulate proliferation, apoptosis and invasion in esophageal cancer cells." Cancer Biol Ther 4(11): 1216-1221.

Yang, Y., B. G. Goldstein, H. Nakagawa and J. P. Katz (2007). "Krüppel-like factor 5 activates MEK/ERK signaling via EGFR in primary squamous epithelial cells." FASEB J 21(2): 543-550.

Yang, Y., H. Nakagawa, M. P. Tetreault, J. Billig, N. Victor, A. Goyal, A. R. Sepulveda and J. P. Katz (2011). "Loss of transcription factor KLF5 in the context of p53 ablation drives invasive progression of human squamous cell cancer." Cancer Res 71(20): 6475-6484.

Yang, Y., M. P. Tetreault, Y. A. Yermolina, B. G. Goldstein and J. P. Katz (2008). "Krüppel-like factor 5 controls keratinocyte migration via the integrin-linked kinase." J Biol Chem 283(27): 18812-18820.

Zhang, D., X. L. Zhang, F. J. Michel, J. L. Blum, F. A. Simmen and R. C. Simmen (2002). "Direct interaction of the Krüppel-like family (KLF) member, BTEB1, and PR mediates progesterone-responsive gene expression in endometrial epithelial cells." Endocrinology 143(1): 62-73.

Zhang, P., P. Basu, L. C. Redmond, P. E. Morris, J. W. Rupon, G. D. Ginder and J. A. Lloyd (2005). "A functional screen for Krüppel-like factors that regulate the human gamma-globin gene through the CACCC promoter element." Blood Cells Mol Dis 35(2): 227-235.

Zhang, Z. and C. T. Teng (2003). "Phosphorylation of Krüppel-like factor 5 (KLF5/IKLF) at the CBP interaction region enhances its transactivation function." Nucleic Acids Res 31(8): 2196-2208.

Zhao, Y., M. S. Hamza, H. S. Leong, C. B. Lim, Y. F. Pan, E. Cheung, K. C. Soo and N. G. Iyer (2008). "Krüppel-like factor 5 modulates p53-independent apoptosis through Pim1 survival kinase in cancer cells." Oncogene 27(1): 1-8.

Zhu, N., L. Gu, H. W. Findley, C. Chen, J. T. Dong, L. Yang and M. Zhou (2006). "KLF5 Interacts with p53 in regulating survivin expression in acute lymphoblastic leukemia." J Biol Chem 281(21): 14711-14718.

Chapter 10

Angiogenesis in Gynecologic Cancers
The Promise of Innovative Therapies

Whitfield Growdon, Rosemary Foster
Leslie Bradford and Bo R. Rueda[*]

Vincent Center for Reproductive Biology
Vincent Department of Obstetrics and Gynecology
Massachusetts General Hospital, Boston, MA, 02116

CHAPTER OUTLINE

- ▶ Introduction
- ▶ Ovarian cancer
- ▶ Endometrial cancer
- ▶ Cervical cancer
- ▶ Conclusions and perspectives
- ▶ References

ABSTRACT

Angiogenesis is a fundamental process that tumors require to develop, grow and allow metastatic cells to successfully implant and form new tumor. While treatable, gynecologic cancers pose many challenges in the upfront and particularly recurrent setting. Scientific insight into the pathogenesis of these tumors is important for

Correspondence Author: Massachusetts General Hospital, Thier 901A, 55 Fruit St, Boston, MA 02114. Email: brueda@partners.org; Website: http://www.VCRB.org

setting. Scientific insight into the pathogenesis of these tumors is important for improving outcomes related to current therapeutic strategies. To date, preclinical investigation has demonstrated that cancers of the ovary, uterus and cervix manifest heightened activation of specific pathways that promote the formation of new vasculature. These findings have provided scientific rationale for the numerous clinical trials that have tested diverse therapies that inhibit signaling pathways contributing to angiogenesis in gynecologic cancers. While it is unclear how anti-angiogenic therapy will be used in the treatment of ovarian, uterine and cervical cancer, recent evidence supports its use in subsets of patients with these malignancies, particularly in the recurrent setting. Defining a molecular signature associated with response to specific anti-angiogenic therapies will be crucial to the success of this approach in the future.

Key words: Gynecologic Cancer, Angiogenesis, Targeted Therapy

1   Introduction

Tumor cells require a robust vascular system to maintain growth and promote invasion (Kerbel 2008). In fact, without appropriate angiogenesis tumors fail to grow beyond 1-2 mm and may remain dormant (Crawford and Ferrara 2009). As early as the 1970's, insightful pioneers envisioned that the metastatic spread of solid tumors was angiogenesis-dependent (Folkman 1971) suggesting that targeting this process of altered neovascularization would hold great promise in the treatment of solid tumors. Decades later, large clinical trials testing anti-angiogenic agents in lung, kidney, colon, brain and breast cancers have been conducted with a spectrum of responses and proposed clinical benefits. In colon, brain, kidney and lung cancer, anti-angiogenic therapy has been demonstrated to offer clear benefits as both first line (lung, colon) (Hurwitz et al. 2004; Sandler et al. 2006; Reck et al. 2009) and second line therapies (brain, kidney, colon) (Yang et al. 2003; Motzer et al. 2006; Escudier et al. 2007; Escudier et al. 2007; Giantonio et al. 2007; Motzer et al. 2007; Vredenburgh et al. 2007; Norden et al. 2008; Cortes-Funes 2009). These compounds have been used as single agents, though most are approved to be used in concert with other drugs, including immunotherapies and cytotoxic chemotherapies (Ranieri et al. 2006). Angiogenesis is a key component of all tumor cell biology and has recently become a promising therapeutic target for women undergoing treatment for gynecologic cancers (Rasila et al. 2005; Spannuth et al. 2008).

Many anti-angiogenic therapeutics target the vascular endothelial growth factor (VEGF) signaling pathway as their primary mechanism of action. The VEGF family includes six related proteins. The most important member is VEGF-A, which was discovered first and was called simply "VEGF" before other variants (VEGF-B, C and D) with more specialized functions involving embryonic and site specific angiogenesis were described. (Li and Eriksson 2001; Kerbel 2008; Fauconnet et al. 2009) VEGF-A will be referenced as simply "VEGF" in this review. Increased VEGF expression has been hypothesized to result from numerous mechanisms including paracrine control of stromal secretion of circulating VEGF and secretion of VEGF from tumor cells which stimulates neovascularization in an autocrine feedback loop (Hopfl et al. 2004; Monk et al. 2010). While the exact mechanism of activation remains imperfectly understood, VEGF has been established as a key factor promoting endothelial cell migration, fenestration and continuous remodeling of neovascularization in solid tumors (Rasila et al. 2005).

Correlative scientific investigations have demonstrated that pro-angiogenic signaling cascades are activated in ovarian, endometrial and cervical cancer. Molecular alterations that upregulate angiogenic factors such as VEGF, platelet derived growth factor (PDGF) and lesser known molecules have been linked to progression of pre-malignant lesions, advanced and refractory disease, and poor survival. These studies have demonstrated that gynecologic malignancies manifest potent signatures associated with active angiogenesis. Given these data, clinical trials in ovarian, endometrial and cervical cancer analyzing the efficacy of an array of novel anti-angiogenic antibodies and small molecules have demonstrated responses in the upfront, recurrent, and consolidation settings. These investigations provide evidence to indicate that a significant subset of gynecologic cancer patients will benefit from therapeutic strategies targeting new blood vessel formation.

This chapter will present the pre-clinical data defining common pro-angiogenic factors in ovarian, endometrial and cervical cancer with a discussion of their underlying biology. These pre-clinical investigations provide rationale for subsequent clinical trials that have employed anti-angiogenic strategies to antagonize cancer cell growth. The agents that will be reviewed by disease site will be bevacizumab, VEGF-Trap, sunitinib, sorafinib and pazopanib (Table 1), although many new agents are in the ever evolving pipeline. This innovative therapeutic approach for ovarian, endometrial and cervical cancer has clear and emerging benefits, as well as significant toxicities and costs that need to be addressed both on a patient, as well as societal level.

Table 1 Anti-angiogenic therapies currently under investigation for the treatment of gynecologic cancers

Agent	VEGFR[a]	VEGF[b]	EGFR[c]	PDGFR[d]	c-KIT	Ras
Bevacizumab		x				
VEGF-Trap	x					
Sunitinib	x		x	x	x	
Sorafinib	x			x	x	x
Pazopanib	x			x	x	

[a]Vascular endothelial growth factor receptor
[b]Vascular endothelial growth factor
[c]Epidermal growth factor receptor
[d]Platelet derived growth factor receptor

2 Ovarian Cancer

Within the gynecologic malignancies, angiogenic signaling has been most characterized in ovarian cancer. Cyclical changes in serum levels of vascular growth factors suggest that angiogenesis plays a key role in basic physiologic ovarian function (Kumaran et al. 2009). Current data suggest that angiogenesis plays a key role in metastatic spread, and increased angiogenic signaling is a poor prognostic factor in ovarian cancer (Hollingsworth et al. 1995; Abulafia et al. 1997; Alvarez et al. 1999; Goodheart et al. 2002; Gadducci et al. 2003; Rasila et al. 2005; Rubatt et al. 2009). Ovarian cancer cell lines and human tumors express higher levels of pro-angiogenic factors, such as hypoxia-inducible factor-1 (HIF-1), VEGF and PDGF as well as lower levels of anti-angiogenic factors, such as endostatins (Papetti and Herman 2002; Duhoux and Machiels 2010). The fundamental role of heightened angiogenesis in ovarian malignancy is supported by numerous clinical trials that have demonstrated efficacy of anti-angiogenic therapies in the upfront and recurrent clinical settings.

2.1 Preclinical

VEGF, formerly known as vascular permeability factor (VPF), is the most widely studied angiogenic factor in ovarian cancer, as well as in other solid tumor types. VEGF is normally expressed in endothelial cells and functions to promote proliferation, migration, stabilization, and cell survival. It also plays a key role in the mobilization of endothelial progenitor cells from bone marrow. VEGF is thought to have a direct effect on tumor cell proliferative and invasive properties (Frumovitz and Sood 2007; Duhoux and Machiels 2010). In advanced ovarian cancer, VEGF induces vascular permeability and subsequent intra-peritoneal hyperosmolarity thereby contributing to malignant ascites (Manenti et al. 2005; Belotti et al. 2008). Within tumors, VEGF serves not only to produce more blood vessels, but also to stabilize new, but poorly formed, vasculature (Saaristo et al. 2000; Frumovitz and Sood 2007; Kobold et al. 2009).

Initial preclinical studies focused on detecting VEGF mRNA expression in primary tissue and cell lines derived from patient ascites using RT-PCR and in situ hybridization analyses (Boocock et al. 1995). VEGF was found to be constitutively expressed in benign and malignant ovarian tissue (Olson et al. 1994). Paley and colleagues analyzed VEGF protein expression by immunohistochemistry (IHC), and found significantly higher levels in malignant tissue compared to normal tissue (42 versus 7%, P = 0.02) (Paley et al. 2000). In addition, the volume of ascites and tumor burden was correlated with tumor and serum VEGF expression in ovarian cancer xenograft models (Yoneda et al. 1998; Manenti et al. 2005). Numerous investigations have confirmed that VEGF is a key mediator of vascular endothelial cell proliferation in ovarian cancer by contributing to tumor invasiveness and metastases (Decaussin et al. 1999; Saaristo et al. 2000; Masood et al. 2001; Spannuth et al. 2009; Duhoux and Machiels 2010).

From a clinical standpoint, VEGF levels also appear to have prognostic value in the setting of ovarian cancer. Tempfer and colleagues investigated serum levels of VEGF in women with all stages of ovarian cancer and found that higher levels correlated closely with worse prognosis (Tempfer et al. 1998) although other investigators did not detect a significant difference in serum VEGF levels between cancer patients and benign controls (Obermair et al. 1998). These data suggest that serum VEGF levels will be most associated with prognosis once invasive disease has been diagnosed. In support of this concept, a comparison of VEGF serum levels among women with invasive cancer, borderline tumors, and benign masses revealed high levels of serum VEGF were associated with worse survival only in the invasive cancer group (Cooper et al. 2002).

Preclinical studies investigating the role of anti-VEGF antibodies as a possible treatment for ovarian cancer began in the late 1990s. Mesiano et al. investigated the effect of a function-blocking anti-VEGF antibody administered to nude mice harboring ovarian cancer. Treatment significantly decreased tumor growth as compared to placebo (Mesiano et al. 1998). Others blocked VEGF signaling in both orthotopic ovarian cancer xenograft and intra-peritoneal carcinomatosis models using numerous anti-angiogenic therapies including VEGF-Trap, VEGF shRNA, sunitinib and bevacizumab, either alone or in combination with platinum based chemotherapy. These investigations found that therapeutic intervention markedly reduced ascites volume and tumor burden in the animals with post-therapy analyses suggesting that this effect was mediated through significant micro-vessel remodeling (Byrne et al. 2003; Downs et al. 2005; Manenti et al. 2005; Mabuchi et al. 2008; Bai et al. 2009; Bauerschlag et al. 2010; Ghosh and Maity 2010; Merritt et al. 2010). These translational studies supported the use of anti-angiogenic agents in clinical trials. Initial clinical studies carried out in the recurrent setting were encouraging leading to investigation of anti-angiogenic agents as upfront therapy. The details of these studies are discussed below.

2.2. Clinical

2.2.1 Recurrent Epithelial Ovarian Cancer

Numerous retrospective reports and clinical trials investigating VEGF inhibitors in ovarian cancer have been reported. As early as 2005, retrospective reports in high and low grade ovarian cancer suggested benefit of utilizing bevacizumab in the recurrent, chemotherapy refractory setting (Monk et al. 2005; Bidus et al. 2006) and early clinical trials similarly focused on this patient population. In 2007, two phase II trials were published showing response rates as high as 15-21%, with a clinical benefit rate (stable disease rate + response rate) of greater than 60% as shown in Table 2 (Burger et al. 2007; Cannistra et al. 2007). Additional trials have tested bevacizumab in combination with other anti-neoplastic agents, including oral cytoxan and erlotinib demonstrating 10-20% response rates, even in the setting of

Table 2 Selected phase II studies of VEGF pathway-specific inhibitors

Study	Patients	Treatment	Response	SD, %	Median PFS	Median OS	Most common grade ≥ 3 AEs
Burger [20]	Persistent/recurrent EOC or PPC after 1-2 CT regimens; N = 62; median age, 57.0 years	Bevacizumab 15 mg/kg IV q3 weeks; median no. of cycles, 7	RR, 21%	52	4.7 months	16.9 months	Hypertension (10%), and GI events (7%)
Cannistra [21]	Platinum-resistant EOC/PSC after 2-3 CT regimens; N = 44; median age, 59.5 years	Bevacizumab 15 mg/kg IVq3 weeks; median no. of cycles, 5	RR, 16%	61	4.4 months	10.7 months[a]	GI perforation[b] (11%), small intestinal obstruction[b] (9%), hypertension (9%), abdominal pain (5%), or digestion (5%), fatigue (5%), and dyspnea (5%)
Garcia [22]	Recurrent EOC or PPC after 1-3 CT regimens; N = 70; median age, 60.0 years	Bevacizumab 10 mg/kg IV q2 weeks and cyclophosphamide 50 mg daily orally; median no. of cycles, 5	RR, 24%	63	6-months PFS, 56%	16.9 months	Lymphopenia (14 episodes), pain (13 episodes), hypertension[c] (11 episodes), fatigue (6 episodes), GI obstruction (5 episodes), hyponatremia (5 episodes), and vomiting (4 episodes)
Tillmanns [23]	Recurrent/platinum-resistant EOC/PPC after ≥ 1 prior regimen; N = 48[d]; median age, 61 years	Bevacizumab 10 mg/kg IV q2 weeks and nab-paclitaxel 100 mg/m²; median no. of cycles, NR	RR, 46%	30.8	8.3 months	16.5 months	Bowel obstruction (4%), nausea (4%), and nose bleed (4%)

Study	Population	Regimen	Response		Survival		Adverse events
Penson [24]	Newly diagnosed, CT-naive, stage ≥ 1C EOC, FTC, PPC, or UPC; N = 62; median age, 58.0 years	Carboplatin AUC 5, paclitaxel 175 mg/m², and bevacizumab 15 mg/kg IV q3 weeks for 6-8 cycles followed by bevacizumab maintenance for 1 year; all patients completed the planned no. of chemotherapy cycles (median no. of consolidation doses, NR)	RR, 76%	21	29.8 months	Not reached	Chemotherapy phase: Neutropenia (n = 14), metabolic (n = 8), hypertension (n = 6), thrombocytopenia (n = 4), neuropathy (n = 4), and allergic reaction (n = 4). Maintenance phase: Hypertension (n = 5), and musculoskeletal pain (n = 3)
Rose [25]	First-line therapy of advanced EOC/ FTC/PPC following cytoreduction; N = 95[d]; median age, 58.0 years	Oxaliplatin 85 mg/m², docetaxel 75 mg/m², and bevacizumab 15 mg/kg IV q3 weeks for 6 cycles followed by bevacizumab maintenance to complete 1 year; median no. of cycles, NR	CR, 33%; PR, 29%	33	1-y PFS, 70%[e]	NR	Neutropenia (39%), leukopenia (11%), hypertension (9%), and fatigue (7%)
Tew [26]	Recurrent EOC after 2-3 CT regimens; N = 162[d]; median age, 58.0 years	VEGF Trap 2 or 4 mg/kg IV q2 weeks; median no. of cycles, NR	RR, 11%	NR	NR	NR	Hypertension (9%), and proteinuria (4%)

Abbreviations: AEs, adverse events; AUC, area under the concentration-time curve; CR, complete response; CT, chemotherapy; EOC, epithelial ovarian carcinoma; FTC, fallopian tube cancer; GI, gastrointestinal; IV, intravenously; NR, not reported; OS, overall survival; PFS, progression-free survival; PPC, primary peritoneal cancer; PR, partial response; PSC, peritoneal serous carcinoma; RR, response rate (complete response + partial response); SD, stable disease; UPC, uterine papillary serous carcinoma; VEGF, vascular endothelial growth factor.

[a]Survival duration at study termination.

[b]Categorized as a serious AE (not as a grade ≥ 3 AE).

[c]Categorized as a bevacizumab-related AE, all of grade 3 severity.

[d]Preliminary data reported in an abstract.

[e]Based on 55 patients with measurable disease.

Adapted from Burger RA et al (Burger 2011).

platinum refractory disease (Garcia et al. 2008; Nimeiri et al. 2008). Although erlotinib appeared to have no obvious synergy with bevacizumab, cytoxan was observed to have a 24% response rate suggesting bevacizumab may potentiate cytotoxic chemotherapy in the recurrent setting. In 2012, the OCEANS trial reported that combining bevacizumab with carboplatinum and gemcitabine significantly prolonged the progression free survival (PFS) by 4 months in women with recurrent platinum sensitive ovarian cancer compared to the combination without bevacizumab (Aghajanian et al. 2011; Aghajanian et al. 2012). Notably, the experimental arm of this trial administered the bevacizumab to patients until progression of disease raising the possibility that extended anti-angiogenic therapy also contributed to the survival benefit observed. These clinical data suggest that bevacizumab may become an important component of recurrent ovarian cancer treatment strategies.

Despite the benefits of bevacizumab observed in women with recurrent ovarian cancer, significant toxicities were observed in this pretreated population. The most common toxicities observed in the trials were elevated blood pressure, protein in the urine and gastro-intestinal symptoms, including spontaneous bowel perforations (Burger 2007; Burger et al. 2007). In two of the bevacizumab studies, the bowel perforation rates were found to be 11–15% of the population, a highly concerning rate leading to marked levels of morbidity (Cannistra et al. 2007; Nimeiri et al. 2008). In-depth analysis of this serious complication revealed the risk of gastro-intestinal perforation (GIP) in these clinical trials was positively associated with the number of previous lines of chemotherapy (Cannistra et al. 2007). While GIP is certainly a great concern, the OCEANS trial data presented in 2011 found an absence of GIP in the 242 women who received bevacizumab with 6 cycles of carboplatinum and gemcitabine (Aghajanian et al. 2011; Aghajanian et al. 2012). Thus, the OCEANS data provide evidence to suggest that the rate of GIP with bevacizumab use is likely to be small in the setting of less than two previous lines of chemotherapy.

In addition to bevacizumab, a growing number of other anti-angiogenic agents have entered clinical trial. A recent phase II trial using VEGF-Trap (aflibercept) in concert with docetaxel demonstrated a 54% overall response rate in recurrent ovarian cancer (Coleman et al. 2011). In addition, numerous novel receptor tyrosine kinase (RTK) inhibitors directed against VEGF receptor have been tested. Agents such as sorafinib, sunitinib and pazopanib interact with multiple additional RTKs in addition to the VEGF receptor including the PDGF receptor, epidermal growth factor receptor (EGFR), and c-Kit. Limited clinical trials in women with recurrent ovarian cancer have been performed using these multiple RTK inhibitors (Friedlander et al. 2010; Biagi et al. 2011; Bodnar et al. 2011; Burger 2011; Matei et al. 2011). These investigations have not revealed responses as robust as those observed with bevacizumab, although little pre-selection based on individualized tumor pathway activation has been performed. Such stratification may be required to enrich for relevant clinical responses (Merritt et al. 2010; Burger 2011).

The mechanism of action for bevacizumab and the family of RTK inhibitors are depicted in Figure 1.

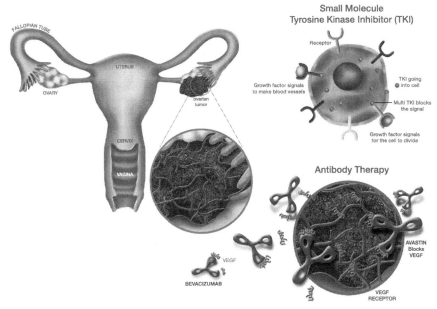

Figure 1 Novel antibody and small tyrosine kinase inhibitor therapies for ovarian cancer. These compounds are directed against circulating vascular endothelial growth factor (VEGF) (a.k.a. bevacizumab) and the VEGF receptor (sorafinib). In clinical trials, these therapies have demonstrated efficacy against ovarian cancer and hold promise as therapeutic options in the recurrent setting.

Color image of this figure appears in the color plate section at the end of the book.

As previously demonstrated in the preclinical studies, anti-angiogenic therapy also holds great promise for those women with debilitating ascites from recurrent ovarian cancer. A growing body of case reports and case series has demonstrated that agents such as bevacizumab are effective at decreasing ascites, even in the platinum refractory setting (Cohn et al. 2006; Numnum et al. 2006; Hamilton et al. 2008; Kobold et al. 2009). These reports confirm the preclinical findings that suggest that the vascular permeability observed in malignant ascites is strongly governed by VEGF, making anti-VEGF directed therapies attractive palliative options in this setting.

2.2.2 Primary Epithelial Ovarian Cancer

Only bevacizumab has been tested in the upfront setting as a therapy for ovarian cancer with the encouraging finding that prolonged bevacizumab treatment results in a significant PFS benefit. Numerous phase II trials demonstrated that the addition of bevacizumab to conventional 6 cycles of intravenous carboplatinum and paclitaxel resulted in a tolerable combination that offered high response rates (>70%) with minimal toxicity

(Micha et al. 2007; Penson et al. 2010), even when the bevacizumab was continued for an entire year (Penson et al. 2010). These trials suggested that this strategy of maintenance bevacizumab was safe, feasible and effective and therefore merited additional study. Two large phase III trials that were reported as presentations in 2010 suggested that administration for 12-15 months of bevacizumab in concert with and after 6 cycles of carboplatinum and paclitaxel chemotherapy resulted in a significant improvement in PFS of 1.7 (ICON7) to 2.8 (GOG218) months compared to arms that did not have a maintenance bevacizumab phase (Burger 2011; Burger et al. 2011; Perren et al. 2011). These large phase III trials encompassed different populations, with ICON7 allowing for high risk stages I and II patients, whereas GOG218 included only those women with stages III and IV epithelial ovarian cancer. Both trials were powered to detect differences in PFS only and no statistical differences were observed in the overall survival (OS) rates between the arms, perhaps because bevacizumab is so effective in the recurrent setting that sequential therapy is as effective as upfront therapy. Interestingly, a subset analysis of poor prognosis patients with more cancer bulk ("suboptimally debulked") in ICON7 did reveal a marked OS benefit possibly suggesting those patients with the greatest tumor volume benefit the most from upfront bevacizumab (Perren et al. 2011).

Toxicities observed in these large phase III trials were not insignificant, though the risks of GIP were less than 2%, confirming the impression that the greatest risks lie with those patients who have received multiple lines of chemotherapy. Only hypertension and proteinuria were shown be statistically over-represented in those patients that had maintenance bevacizumab. All other significant toxicities appeared to be equally distributed amongst the arms suggesting that bevacizumab does not potentiate the hematologic, neurologic or other toxicities commonly experienced by women undergoing first line therapy for ovarian cancer (Burger et al. 2011; Perren et al. 2011).

One potential adverse event not discussed in these trials, which certainly affects the widespread clinical utility of bevacizumab, is the heightened cost of the therapy. Since it is a humanized monoclonal antibody, bevacizumab is expensive to produce and its per dose administered cost is similarly high. A recent cost analysis of GOG218 suggested that bevacizumab increased the cost of one patient's upfront therapy for ovarian cancer more than 30-fold (Cohn et al. 2011). The analysis goes on to question whether or not the observed PFS improvement merits this marked increase in cost. The authors conclude that at the time of publication, treatment with bevacizumab as a first line therapy does not meet cost effective benchmarks and would need to have a price reduction of approximately 25% to justify the 2.8-month PFS benefit observed (Cohn et al. 2011).

2.3 Epithelial Ovarian Cancer Summary

Ovarian cancer manifests potent new blood vessel formation to promote growth and metastatic spread leading many to suggest anti-angiogenic strategies that either target VEGF ligand or receptor will hold much promise to improve therapy for this devastating disease (Kaye 2007). In the clinic, anti-angiogenic therapeutic strategies have merit both in the first line and recurrent settings. Bevacizumab appears to have significant first line activity to prolong PFS, especially in patients with bulky tumors and residual disease and when given as maintenance for at least 12 additional cycles after initial platinum based chemotherapy. While statistically significant, the clinical effect is limited to less than 3 additional months without disease and is still awaiting global approval. Currently, extensive efforts are underway to identify clinical and molecular signatures associated with potent responses to bevacizumab, such as elevated serum VEGF and other angiogenesis mediators, to determine prior to drug administration, which patients would be most likely to benefit. Pre-selection would decrease the number of patients receiving the medication, while theoretically increasing the efficacy thus possibly making this expensive class of medicine more cost effective.

In the recurrent setting, a full range of anti-angiogenic medicines with varied mechanisms of action in addition to VEGF inhibition have been tested with some manifesting better signals than others. Correlative scientific investigation is of paramount importance in this setting as the agents are more diverse and unlikely to be effective in the absence of a more defined molecular signature. Investigators have tested molecular markers, including histologic microvessel density (MVD), serum VEGF and several gene polymorphisms in angiogenesis mediators to determine if any associations exist with response (Schultheis et al. 2008; Han et al. 2010; Smerdel et al. 2010). This limited body of literature suggests that alterations in serum and tumor VEGF, IL-8 and chemokine receptor 2 (CXCR2) may be promising markers that could be used prospectively to identify patients likely to respond to anti-angiogenic therapy. To date, however, no studies have prospectively selected patients based on molecular signature. Currently, pairing anti-angiogenic therapeutics with cytotoxic chemotherapies appears to be the most promising clinical strategy to provide clinical benefit for these women.

3 Endometrial Cancer

3.1 Introduction

The development of epithelial endometrial cancer follows a multistage process of carcinogenesis. Irreversible changes in cellular DNA are sequentially promoted, resulting in cellular clonal proliferation. Clinically, this results in a spectrum of progression from endometrial hyperplasia to hyperplasia with atypia to invasive carcinoma observed in endometrioid histologic subtypes (Dedes et al. 2011). Pro-angiogenic factors such as VEGF play a significant role in malignant progression and act to mediate invasion (Guidi et al. 1996). VEGF, however, is a crucial factor in normal menstruation and has been linked to the physiologic development of the corpus luteum, a process that relies heavily on neovascularization. Mutation of the VEGF receptor in animal models, for instance, results in decreased serum progesterone and interferes with corpus luteum development (Kamat et al. 1995; Ferrara et al. 1998). During the secretory phase of menstruation, endometrial tissues express a significantly higher amount of VEGF RNA than what is expressed during other menstrual phases (Torry et al. 1996). Likewise, serum levels of VEGF vary throughout the menstrual cycle (Heer et al. 1998). VEGF has a complex interaction with female sex steroid hormones suggesting its involvement both in the evolution of endometrial cancer from endometrial hyperplasia and in supporting malignant invasion.

Hyperplasia and eventual carcinoma develop when regulatory mechanisms mediating endometrial proliferation go awry, as in the case of excessive exposure to estrogen. Interestingly, Charnok-Jones and colleagues demonstrated that treating endometrial cancer cell lines with estrogen resulted in a time- and dose-dependent increase in VEGF mRNA (Charnock-Jones et al. 1993) suggesting VEGF as a putative mediator of estrogen induced proliferation. This finding has been confirmed in subsequent studies on cell lines both with estrogen and progesterone. Treatment with progesterone alone, however, does not increase VEGF mRNA levels (Bausero et al. 1998; Perrot-Applanat et al. 2000; Krikun et al. 2004) perhaps explaining, in part, the observed anti-hyperplastic effects observed with progesterone therapy.

3.2   Preclinical

Retrospective reviews of endometrial carcinoma suggest that markers of angiogenesis associate with several prognostic factors. Immunohistochemical staining for MVD counts have also been employed in the preclinical setting to investigate angiogenesis in endometrial hyperplasia and carcinoma. Abulafia et al compared MVD counts in endometrial hyperplasia compared to Stage I endometrial carcinoma (Abulafia et al. 1995). MVD counts were significantly higher in complex versus simple hyperplasia, and these counts were, in turn, lower than those observed in invasive carcinoma (p < 0.05). Within the cohort of Stage I endometrial carcinoma specimens, significantly higher MVD counts were observed in those with myometrial invasion than those without (p < 0.01). The authors concluded that higher tumor grade and depth of invasion were directly correlated with angiogenic intensity(Abulafia et al. 1995). Kaku et al confirmed these findings with their investigation of 85 specimens from patients with Stage I and II endometrial carcinoma (Kaku et al. 1997). MVD was strongly correlated with tumor grade, depth of myometrial invasion, as well as lymphovascular space invasion. There was no correlation between MVD and stage of disease (Kaku et al. 1997). An additional preclinical study supported the finding that higher MVD counts are associated with worse PFS and OS confirming that angiogenesis appears to be a clinically relevant signature in endometrial cancer (Kirschner et al. 1996).

Although a clear etiology of angiogenic stimulation in endometrial cancer is unclear, stromal tissue, including adipose, may directly stimulate the formation of new blood vessels. The idea that stromal cells recruited by tumor cells provide a trophic environment for tumor growth is not new. Mesenchymal stromal cells are thought to be primarily derived from the bone marrow (Nakamizo et al. 2005; Spaeth et al. 2009). More recently, it has been shown that stromal cells derived from the omental adipose also promote tumor cell proliferation *in vitro* (Klopp et al. 2012). More importantly, however, the adipose derived stromal cells stimulated vascularization and promoted growth of endometrial tumors in immune-compromised mice with little centralized tumor necrosis. Collectively, these findings support the concept that intra abdominal visceral adipose tissue may contain a unique population of stromal cells that contributes to the aggressive pathology of intra-abdominal endometrial epithelial cancers (Klopp et al. 2012). This concept could become even more relevant in obese patients in whom excessive adipose tissue could further promote tumor growth.

The exact mechanisms by which adipose cells contribute to endometrial cancer are not completely known. However, it has long been recognized that leptin, a secreted product of endometrial cancer, is primarily derived from adipocytes (Huang and Li 2000). Independent of its role in regulating the body's metabolism, leptin stimulates normal and tumor cell proliferation, promotes tumor cell migration and invasion and enhances angiogenesis and aromatase activity (Rose et al. 2002). Importantly, leptin has been implicated in the pathology of endometrial cancer (Petridou et al. 2002; Yuan et al. 2004; Carino et al. 2008).

Leptin is a well recognized mediator of a number of vascular promoting factors including IL-1β, VEGF, IL-8 and has been shown to upregulate pro-angiogenic/pro-inflammatory factors in benign and endometrial cancer cells (Carino et al. 2008; Catalano et al. 2009). These data suggest leptin supports endometrial cancer proliferation via its contribution to vascular development as has been shown in syngeneic and xenograft mouse models of breast carcinoma (Zheng et al. 2012; Gonzalez et al. 2006).

While leptin appears to promote angiogenesis, progestins may act to inhibit this process. Low grade endometrial cancer is a hormonally sensitive disease that can regress with progestin therapy and this effect may be mediated by impairing tumor angiogenesis (Reed et al. 2010; Gunderson et al. 2012). The effect of treatment with medroxyprogesterone on tumor growth was first demonstrated by Gross et al in 1981 (Gross et al. 1981). The authors suggested that medroxyprogesterone acetate inhibits angiogenesis by interfering with tumor growth and collagenase production which may, in turn, impair neoplastic invasion (Gross et al. 1981). This was further investigated by Jikihara et al, who began by implanting endometrial carcinoma into white rabbit corneas (Jikihara et al. 1992). Angiogenesis was induced in 70.4% of corneas. When the cancer tissue was transplanted with a pellet of medroxyprogesterone acetate, angiogenesis was only induced in 21.5% of corneas (Jikihara et al. 1992). These studies provided evidence to suggest a potential link between progestin therapy and modulation of angiogenic potential in endometrial cancer.

Limited investigations have tested anti-angiogenic therapies against *in vivo* endometrial cancer animal models. Investigators have attempted to both define the degree of response and understand the potential pathways of resistance when novel angiogenic therapies are utilized. One investigation used bevacizumab to treat nude mice with endometrial cancer xenografts and found that xenograft growth was impeded by single agent therapy. This effect was associated with alterations in pathways that stimulate proliferation and anti-apoptosis such as altered expression of matrix metalloprotease family protein. In this model, activation of the c-*jun* oncogene was associated with development of resistance to bevacizumab (Davies et al. 2011).

Collectively, these preclinical data in endometrial cancer suggest that angiogenesis plays an important role in both normal homeostasis of the endometrium andtransformation into an invasive phenotype. Targeting this process may be of great therapeutic benefit thus providing rationale for clinical investigations in endometrial cancer.

3.3 Clinical

The majority of endometrial cancers are low grade, minimally invasive of the myometrium, and largely cured with staging surgery with recurrence rates as low as 2% in stage I disease. Advanced staged endometrial cancers with disease outside of the uterus, however, present a clinical dilemma with few effective treatment options as these patients frequently develop recurrent or persistent cancer refractory to conventional therapies. While the standard of care has been platinum based cytotoxic chemotherapy, the correlative scientific data suggest angiogenesis is a mediator of cancer cell survival and proliferation. Consequently, investigators are actively pursuing anti-angiogenesis therapeutics as a promising avenue in the clinic.

Agents with anti-angiogenic activity have been assessed in clinical trials in recurrent or persistent endometrial cancer on a limited basis and the results have shown modest activity. One phase II trial tested single agent sorafenib, a multi-target tyrosine kinase inhibitor (TKI). The 5% response rate in this study suggests that only a minority of patients will benefit from the use of this therapy alone (Nimeiri et al. 2010). Another phase II trial investigated sunitinib in this same population and reported a 15% response rate with major toxicities limited to fatigue and hypertension (Correa et al. 2010). Bevacizumab was also recently studied in a phase II trial (GOG 229E) resulting in a 13.5% response rate which compares favorably with other single agents tested in the same setting. Commonly observed toxicities (grade 3 or 4) included hemorrhage (2/52), hypertension (4/52) and pain (4/52), and no deaths or bowel perforations were observed. Interestingly, tumor and serum VEGF levels were quantified by immunohistochemistry and ELISA and correlated with response and survival. High tumor VEGF staining correlated with improved survival, while elevated serum VEGF levels correlated with decreased survival and treatment failure (Aghajanian et al. 2011).

These initial studies provide evidence that anti-angiogenic agents in endometrial cancer merit additional study to determine both drug synergy and those tumor factors that may predict response to therapy. In particular, it is unknown how standard platinum based chemotherapy and anti-angiogenesis therapeutics would synergize as first line therapies for advanced stage or high grade endometrial cancers.

4   Cervical Cancer

4.1   Introduction

The majority of research regarding the role of angiogenesis in cervical cancer focuses on carcinoma progression and maintenance as promoted by human papilloma virus (HPV) mediated inactivation of p53 (Nakamura et al. 2009; Monk et al. 2010). Cervical cancer is highly preventable if treated in the pre-invasive period but the prognosis for women with advanced or recurrent disease can be guarded and treatment options are limited. Investigators hypothesize that because of the strong association between HPV infection, heightened neovascularization and progression of cervical carcinoma, strategies antagonizing angiogenesis mediators such as VEGF will offer new therapeutic opportunities (Monk et al. 2010).

4.2   Preclinical

In the 1970's, Stafl and Mattingly first reported on epithelial proliferation and capillary compression in cervical neoplasia suggesting that neovascularization was observed in cases of carcinoma in situ (CIS) that progressed to invasive cancer (Stafl and Mattingly 1975). Subsequent reports built on these observations finding a stepwise activation of angiogenesis as dysplasia transformed into invasive carcinoma. One group found that almost 3% of patients with CIS produced abnormal vessels noted at the time of colposcopy compared to 50% of patients with micro-invasive disease and 100% of patients with frankly invasive cancer (Sillman et al. 1981). An additional study confirmed that dysplastic cervical lesions manifested heightened neovascularization (Smith-McCune and Weidner 1994). In this investigation, no significant differences in vascular structure were noted in cervical intraepithelial neoplasia (CIN) versus normal controls, but a significant difference between CIN I and CIN III was observed. This group then developed an animal model in the K14-HPV16 transgenic mouse, in which expression of the HPV type 16 early gene was targeted to basal squamous epithelial cells by regulatory elements in the human keratin 14 (HK-14) promoter. This multistage squamous carcinogenesis model allowed for imunnohistochemical staining of vessels to demonstrate a parallel upregulation of angiogenesis during the early stages of carcinogenesis and provided evidence that HPV infection activates angiogenesis mechanisms (Smith-McCune 1997; Willmott and Monk 2009).

Subsequent preclinical studies have determined that CIN has greater MVD and VEGF protein expression compared to normal epithelium, and that this difference increases with the grade, or severity, of CIN (Davidson et al. 1997; Lee et al. 2002). More specifically, Guidi et al studied this relationship between angiogenesis and VEGF expression in premalignant versus malignant disease of the cervix. A total of 66 specimens were examined (16 benign, 17 LSIL, 18 HSIL, and 15 invasive carcinoma). Factor VIII-related antigen immunohistochemistry and in situ hybridization assessing VEGF mRNA expression were preformed. The authors concluded that VEGF mRNA and MVD counts were significantly higher in high grade lesions (HSIL) and invasive carcinoma compared to low grade SIL lesions and benign samples. Increased MVD counts were significantly associated with increased VEGF mRNA expression (Guidi et al. 1995).

Building upon these insights, additional studies demonstrated that invasive cervical carcinoma expressed the highest level of VEGF, with the greatest levels found in early stage disease, suggesting that VEGF may play a role in early invasion (Kodama et al. 1999). Conflicting data, however, do exist. An investigation that compared 69 cases of invasive squamous cell carcinoma of the cervix with 48 cone biopsy specimens of CIN III suggested that the histologic feature associated with invasive carcinoma was neovascularization, as assessed by factor VIII-related antigen immunohistochemical staining (56.5 versus 0% in cancer versus benign, respectively) (Leung et al. 1994). A smaller study utilized a similar assay to evaluate 18 cone biopsy specimens of invasive cancer versus 14 cone biopsy specimens of microinvasive disease as well as 22 control specimens with normal cervix. These authors found that carcinoma in situ had decreased neovascularization, compared to microinvasive disease which demonstrated higher MVD counts (Abulafia et al. 1996).

Other investigators have characterized MVD in cervical cancers and correlated the findings with clinical parameters with conflicting results at times. Wiggins et al focused their investigation on malignant versus control specimens and confirmed previous findings that MVD counts in patients with squamous cell carcinoma were significantly different from the benign controls. They did not observe a correlation between MVD counts and node status, parametrial involvement, depth of invasion, or gross disease, though MVD counts were significantly associated with lymphovascular space invasion. Four patients in this cohort recurred in less than a year and were found to have high MVD counts, but had negative nodes and no lymphovascular space invasion at the time of initial diagnosis. The authors hypothesized that angiogenesis may be a predictor of recurrent disease, and thus possibly a prognostic factor (Wiggins et al. 1995). Subsequent studies, however, found no correlation between tumor vascularity and tumor stage or prognosis (Kainz et al. 1995; Rutgers et al. 1995). In contrast, Schlenger et al found tumor vascularity to be the strongest independent prognostic factor (Schlenger et al. 1995). Tumors from 42 patients with cervical cancer, FIGO Stages IB-IVA were assessed using immunohistochemical staining for Factor VIII-related antigen. Tumor vascularity was independent of age, FIGO stage, tumor size, lymph node involvement, or lymphovascular space involvement. Despite the lack of association with known risk factors, these authors observed that patients with higher tumor vascularity had a shorter disease-free interval (P = 0.025) (Schlenger et al. 1995). This observation has been duplicated by a study that found MVD counts to be an independent factor in recurrent disease. Tumors from 22 patients were assessed using immunohistochemical staining for Factor VIII-related antigen. MVD counts were compared to factors such as age, tumor size, grade, and lymphovascular space invasion. High MVD counts and size were identified as independent factors (Dinh et al. 1996). The largest study to date examined 130 specimens by IHC staining for CD31 to determine MVD. MVD counts in conjunction with depth of invasion, regional lymph node involvement, and lymphovascular invasion were found to be independent prognostic indicators for overall survival (Dellas et al. 1997).

VEGF protein expression in the tumor also appears to be correlated with tumor histology. The majority of cervical carcinomas are of squamous histology, arising from the ectocervix. Adenocarcinomas, which generally arise from the glandular epithelium of the cervical canal, express higher levels of VEGF compared to squamous cell cervical cancers (Santin et al. 1999). Clinically, adenocarcinomas may behave in a more aggressive fashion, suggesting that VEGF may be related to prognosis (Tokumo et al. 1998; Santin et al. 1999). Studies that focus specifically on the relationship between tumor VEGF protein expression as assessed by IHC and clinical prognosis have demonstrated that VEGF expression is a prognosis factor for both disease-free and overall survival. Specifically, higher levels of VEGF protein expression appear to be correlated with tumor size, lymphovascular space invasion, and lymph node metastases as well as a shorter disease-free interval (Cheng et al. 1999; Cheng et al. 2000; Loncaster et al. 2000). Investigators have also examined common *VEGF* gene polymorphisms and found that specific isotypes associate with development of cervical cancer as well as worsened survival (Kim et al. 2010). These data identify mediators of heightened angiogenesis as clinically relevant drivers of cervical cancer proliferation and provide rationale for therapeutic intervention targeting angiogenesis.

4.3 Clinical

Like endometrial cancer, therapeutic strategies targeting angiogenesis in cervical cancer have been limited to recurrent cervical cancer. The results have been encouraging and suggest that these therapies can offer benefit in a setting known to be highly refractory to conventional cytotoxic therapy. Early phase I data employing a novel angiogenesis inhibitor TNP-470 showed limited single agent activity in a cohort of heavily pretreated patients demonstrating a 7% response rate, with a 20% disease stabilization rate (Kudelka et al. 1997). Years later, retrospective data utilizing bevacizumab also reported significant activity in this setting with one remarkable report of six patients suggesting one complete response, one partial response and three stable disease when combined with gemcitabine or 5-FU cytotoxic chemotherapies (Wright et al. 2006). Coupled with the clear scientific basis for targeting angiogenesis in cervical cancer, these early data provided evidence that these therapies would offer a beneficial clinical signal.

Bevacizumab was the first anti-angiogenic therapy to be tested in a phase II clinical trial for women with recurrent or persistent cervical cancer. In this trial (GOG227C), Monk and colleagues observed a 10.9% response rate, with a 24% stable disease rate at 6 months in this heavily pretreated and radiated population. The median PFS was 3.4 months, with responders manifesting a 6.2-month duration of response (Monk et al. 2009). An additional clinical trial of over 150 patients has evaluated either pazopanib (targeting VEGFR, PGFR, c-Kit) and lapatinib (targeting ErbB1, ErbB2) in this same population finding that pazopanib manifested a markedly improved response rate (9% vs. 5%) with a significant improvement in OS of 11 weeks (50 vs. 39 weeks) over lapatinib. The toxicity profile was observed to be similar amongst the groups in the trial, with diarrhea and abdominal pain among the most common side effects. An additional phase II trial tested sunitinib in this population finding no objective responses, though an 84% stable disease rate. Because of the lack of response, and an unanticipated 26% intestinal fistula rate, sunitinib was not believed to be merit further investigation (Mackay et al. 2010).

These trials affirm that anti-angiogenic strategies offer a subset of cervical cancer patients benefit that merits further investigation on a larger scale incorporating conventional platinum based therapies in the first line and recurrent settings (Monk et al. 2010). Currently, several phase III trials in cervical cancer are underway testing the synergy of bevacizumab with cisplatinum, topotecan and paclitaxel in order to further refine the most appropriate regimen for advanced stage, recurrent or persistent cervical cancer.

5   Conclusions and Perspectives

In summary, translational and clinical trial investigations strongly suggest that increased angiogenesis is a key promoter in the development and progression of epithelial ovarian, endometrial and cervical cancer. Markers of heightened or altered angiogenesis, such as MVD or VEGF expression have been shown to be of prognostic value in these gynecologic malignancies, and clinical trials targeting mediators of neovascularization have been shown to have clear benefits in significant subsets of patients. Many challenges linger, and these include developing biomarkers that predict response, managing the costs of these expensive therapies that can threaten to bankrupt healthcare systems and being vigilant in assessing whether or not anti-angiogenic therapies have unintended consequences, such as a rebound effect, that could render tumors more resistant to conventional chemotherapies.

REFERENCES

Abulafia, O., W. E. Triest and D. M. Sherer (1996). "Angiogenesis in squamous cell carcinoma in situ and microinvasive carcinoma of the uterine cervix." Obstet Gynecol 88(6): 927-932.

Abulafia, O., W. E. Triest and D. M. Sherer (1997). "Angiogenesis in primary and metastatic epithelial ovarian carcinoma." Am J Obstet Gynecol 177(3): 541-547.

Abulafia, O., W. E. Triest, D. M. Sherer, C. C. Hansen and F. Ghezzi (1995). "Angiogenesis in endometrial hyperplasia and stage I endometrial carcinoma." Obstet Gynecol 86(4 Pt 1): 479-485.

Aghajanian, C., S. V. Blank, B. A. Goff, P. L. Judson, M. G. Teneriello, A. Husain, M. A. Sovak, J. Yi and L. R. Nycum (2012). "OCEANS: A Randomized, Double-Blind, Placebo-Controlled Phase III Trial of Chemotherapy With or Without Bevacizumab in Patients With Platinum-Sensitive Recurrent Epithelial Ovarian, Primary Peritoneal, or Fallopian Tube Cancer." J Clin Oncol. 30(17): 2039-45.

Aghajanian, C., M. W. Sill, K. M. Darcy, B. Greer, D. S. McMeekin, P. G. Rose, J. Rotmensch, M. N. Barnes, P. Hanjani and K. K. Leslie (2011). "Phase II trial of bevacizumab in recurrent or persistent endometrial cancer: a Gynecologic Oncology Group study." J Clin Oncol 29(16): 2259-2265.

Aghajanian, C. A., N. J. Finkler, T. Rutherford, D. A. Smith, J. Yi, H. Parmar, L. R. Nycum and M. A. Sovak (2011). "OCEANS: A randomized, double-blinded, placebo-controlled phase III trial of chemotherapy with or without bevacizumab (BEV) in patients with platinum-sensitive recurrent epithelial ovarian (EOC), primary peritoneal (PPC), or fallopian tube cancer (FTC)." J Clin Oncol 29(Suppl).

Alvarez, A. A., H. R. Krigman, R. S. Whitaker, R. K. Dodge and G. C. Rodriguez (1999). "The prognostic significance of angiogenesis in epithelial ovarian carcinoma." Clin Cancer Res 5(3): 587-591.

Bai, Y., H. Deng, Y. Yang, X. Zhao, Y. Wei, G. Xie, Z. Li, X. Chen, L. Chen, Y. Wang, D. Su, Z. Qian, Q. Zhong, H. Luo and T. Yi (2009). "VEGF-targeted short hairpin RNA inhibits intraperitoneal ovarian cancer growth in nude mice." Oncology 77(6): 385-394.

Bauerschlag, D. O., C. Schem, S. Tiwari, J. H. Egberts, M. T. Weigel, H. Kalthoff, W. Jonat, N. Maass and I. Meinhold-Heerlein (2010). "Sunitinib (SU11248) inhibits growth of human ovarian cancer in xenografted mice." Anticancer Res 30(9): 3355-3360.

Bausero, P., F. Cavaille, G. Meduri, S. Freitas and M. Perrot-Applanat (1998). "Paracrine action of vascular endothelial growth factor in the human endometrium: production and target sites, and hormonal regulation." Angiogenesis 2(2): 167-182.

Belotti, D., C. Calcagno, A. Garofalo, D. Caronia, E. Riccardi, R. Giavazzi and G. Taraboletti (2008). "Vascular endothelial growth factor stimulates organ-specific host matrix metalloproteinase-9 expression and ovarian cancer invasion." Mol Cancer Res 6(4): 525-534.

Biagi, J. J., A. M. Oza, H. I. Chalchal, R. Grimshaw, S. L. Ellard, U. Lee, H. Hirte, J. Sederias, S. P. Ivy and E. A. Eisenhauer (2011). "A phase II study of sunitinib in patients with recurrent epithelial ovarian and primary peritoneal carcinoma: an NCIC Clinical Trials Group Study." Ann Oncol 22(2): 335-340.

Bidus, M. A., J. C. Webb, J. D. Seidman, G. S. Rose, C. R. Boice and J. C. Elkas (2006). "Sustained response to bevacizumab in refractory well-differentiated ovarian neoplasms." Gynecol Oncol 102(1): 5-7.

Bodnar, L., M. Gornas and C. Szczylik (2011). "Sorafenib as a third line therapy in patients with epithelial ovarian cancer or primary peritoneal cancer: a phase II study." Gynecol Oncol 123(1): 33-36.

Boocock, C. A., D. S. Charnock-Jones, A. M. Sharkey, J. McLaren, P. J. Barker, K. A. Wright, P. R. Twentyman and S. K. Smith (1995). "Expression of vascular endothelial growth factor and its receptors flt and KDR in ovarian carcinoma." J Natl Cancer Inst 87(7): 506-516.

Burger, R. A. (2007). "Experience with bevacizumab in the management of epithelial ovarian cancer." J Clin Oncol 25(20): 2902-2908.

Burger, R. A. (2011). "Overview of anti-angiogenic agents in development for ovarian cancer." Gynecol Oncol 121(1): 230-238.

Burger, R. A., M. F. Brady, M. A. Bookman, G. F. Fleming, B. J. Monk, H. Huang, R. S. Mannel, H. D. Homesley, J. Fowler, B. E. Greer, M. Boente, M. J. Birrer and S. X. Liang (2011). "Incorporation of bevacizumab in the primary treatment of ovarian cancer." N Engl J Med 365(26): 2473-2483.

Burger, R. A., M. W. Sill, B. J. Monk, B. E. Greer and J. I. Sorosky (2007). "Phase II trial of bevacizumab in persistent or recurrent epithelial ovarian cancer or primary peritoneal cancer: a Gynecologic Oncology Group Study." J Clin Oncol 25(33): 5165-5171.

Byrne, A. T., L. Ross, J. Holash, M. Nakanishi, L. Hu, J. I. Hofmann, G. D. Yancopoulos and R. B. Jaffe (2003). "Vascular endothelial growth factor-trap decreases tumor burden, inhibits ascites, and causes dramatic vascular remodeling in an ovarian cancer model." Clin Cancer Res 9(15): 5721-5728.

Cannistra, S. A., U. A. Matulonis, R. T. Penson, J. Hambleton, J. Dupont, H. Mackey, J. Douglas, R. A. Burger, D. Armstrong, R. Wenham and W. McGuire (2007). "Phase II study of bevacizumab in patients with platinum-resistant ovarian cancer or peritoneal serous cancer." J Clin Oncol 25(33): 5180-5186.

Carino, C., A. B. Olawaiye, S. Cherfils, T. Serikawa, M. P. Lynch, B. R. Rueda and R. R. Gonzalez (2008). "Leptin regulation of proangiogenic molecules in benign and cancerous endometrial cells." Int J Cancer 123(12): 2782-2790.

Catalano, S., C. Giordano, P. Rizza, G. Gu, I. Barone, D. Bonofiglio, F. Giordano, R. Malivindi, D. Gaccione, M. Lanzino, F. De Amicis and S. Ando (2009). "Evidence that leptin through STAT and CREB signaling enhances cyclin D1 expression and promotes human endometrial cancer proliferation." J Cell Physiol 218(3): 490-500.

Charnock-Jones, D. S., A. M. Sharkey, J. Rajput-Williams, D. Burch, J. P. Schofield, S. A. Fountain, C. A. Boocock and S. K. Smith (1993). "Identification and localization of alternately spliced mRNAs for vascular endothelial growth factor in human uterus and estrogen regulation in endometrial carcinoma cell lines." Biol Reprod 48(5): 1120-1128.

Cheng, W. F., C. A. Chen, C. N. Lee, T. M. Chen, F. J. Hsieh and C. Y. Hsieh (1999). "Vascular endothelial growth factor in cervical carcinoma." Obstet Gynecol 93(5 Pt 1): 761-765.

Cheng, W. F., C. A. Chen, C. N. Lee, L. H. Wei, F. J. Hsieh and C. Y. Hsieh (2000). "Vascular endothelial growth factor and prognosis of cervical carcinoma." Obstet Gynecol 96(5 Pt 1): 721-726.

Cohn, D. E., K. H. Kim, K. E. Resnick, D. M. O'Malley and J. M. Straughn, Jr. (2011). "At what cost does a potential survival advantage of bevacizumab make sense for the primary treatment of ovarian cancer? A cost-effectiveness analysis." J Clin Oncol 29(10): 1247-1251.

Cohn, D. E., S. Valmadre, K. E. Resnick, L. A. Eaton, L. J. Copeland and J. M. Fowler (2006). "Bevacizumab and weekly taxane chemotherapy demonstrates activity in refractory ovarian cancer." Gynecol Oncol 102(2): 134-139.

Coleman, R. L., L. R. Duska, P. T. Ramirez, J. V. Heymach, A. A. Kamat, S. C. Modesitt, K. M. Schmeler, R. B. Iyer, M. E. Garcia, D. L. Miller, E. F. Jackson, C. S. Ng, V. Kundra, R. Jaffe and A. K. Sood (2011). "Phase 1-2 study of docetaxel plus aflibercept in patients with recurrent ovarian, primary peritoneal, or fallopian tube cancer." Lancet Oncol 12(12): 1109-1117.

Cooper, B. C., J. M. Ritchie, C. L. Broghammer, J. Coffin, J. I. Sorosky, R. E. Buller, M. J. Hendrix and A. K. Sood (2002). "Preoperative serum vascular endothelial growth factor levels: significance in ovarian cancer." Clin Cancer Res 8(10): 3193-3197.

Correa, R., H. Mackay, H. W. Hirte, R. Morgan, S. A. Welch, G. F. Fleming, L. Wang, C. Blattle, S. P. Ivy and A. M. Oza (2010). "A phase II study of sunitinib in recurrent or metastatic endometrial carcinoma: A trial of the Princess Margaret Hospital, The University of Chicago, and California Cancer Phase II Consortia." J Clin Oncol 28(15s).

Cortes-Funes, H. (2009). "The role of antiangiogenesis therapy: bevacizumab and beyond." Clin Transl Oncol 11(6): 349-355.

Crawford, Y. and N. Ferrara (2009). "VEGF inhibition: insights from preclinical and clinical studies." Cell Tissue Res 335(1): 261-269.

Davidson, B., I. Goldberg and J. Kopolovic (1997). "Angiogenesis in uterine cervical intraepithelial neoplasia and squamous cell carcinoma: an immunohistochemical study." Int J Gynecol Pathol 16(4): 335-338.

Davies, S., D. Dai, G. Pickett, K. W. Thiel, V. P. Korovkina and K. K. Leslie (2011). "Effects of bevacizumab in mouse model of endometrial cancer: Defining the molecular basis for resistance." Oncol Rep 25(3): 855-862.

Decaussin, M., H. Sartelet, C. Robert, D. Moro, C. Claraz, C. Brambilla and E. Brambilla (1999). "Expression of vascular endothelial growth factor (VEGF) and its two receptors (VEGF-R1-Flt1 and VEGF-R2-Flk1/KDR) in non-small cell lung carcinomas (NSCLCs): correlation with angiogenesis and survival." J Pathol 188(4): 369-377.

Dedes, K. J., D. Wetterskog, A. Ashworth, S. B. Kaye and J. S. Reis-Filho (2011). "Emerging therapeutic targets in endometrial cancer." Nat Rev Clin Oncol 8(5): 261-271.

Dellas, A., H. Moch, E. Schultheiss, G. Feichter, A. C. Almendral, F. Gudat and J. Torhorst (1997). "Angiogenesis in cervical neoplasia: microvessel quantitation in precancerous lesions and invasive carcinomas with clinicopathological correlations." Gynecol Oncol 67(1): 27-33.

Dinh, T. V., E. V. Hannigan, E. R. Smith, M. J. Hove, V. Chopra and T. To (1996). "Tumor angiogenesis as a predictor of recurrence in stage Ib squamous cell carcinoma of the cervix." Obstet Gynecol 87(5 Pt 1): 751-754.

Downs, L. S., Jr., L. M. Rogers, Y. Yokoyama and S. Ramakrishnan (2005). "Thalidomide and angiostatin inhibit tumor growth in a murine xenograft model of human cervical cancer." Gynecol Oncol 98(2): 203-210.

Duhoux, F. P. and J. P. Machiels (2010). "Antivascular therapy for epithelial ovarian cancer." J Oncol 2010: 372547.

Escudier, B., T. Eisen, W. M. Stadler, C. Szczylik, S. Oudard, M. Siebels, S. Negrier, C. Chevreau, E. Solska, A. A. Desai, F. Rolland, T. Demkow, T. E. Hutson, M. Gore, S. Freeman, B. Schwartz, M. Shan, R. Simantov and R. M. Bukowski (2007). "Sorafenib in advanced clear-cell renal-cell carcinoma." N Engl J Med 356(2): 125-134.

Escudier, B., A. Pluzanska, P. Koralewski, A. Ravaud, S. Bracarda, C. Szczylik, C. Chevreau, M. Filipek, B. Melichar, E. Bajetta, V. Gorbunova, J. O. Bay, I. Bodrogi, A. Jagiello-Gruszfeld and N. Moore (2007). "Bevacizumab plus interferon alfa-2a for treatment of metastatic renal cell carcinoma: a randomised, double-blind phase III trial." Lancet 370(9605): 2103-2111.

Fauconnet, S., S. Bernardini, I. Lascombe, G. Boiteux, A. Clairotte, F. Monnien, E. Chabannes and H. Bittard (2009). "Expression analysis of VEGF-A and VEGF-B: relationship with clinicopathological parameters in bladder cancer." Oncol Rep 21(6): 1495-1504.

Ferrara, N., H. Chen, T. Davis-Smyth, H. P. Gerber, T. N. Nguyen, D. Peers, V. Chisholm, K. J. Hillan and R. H. Schwall (1998). "Vascular endothelial growth factor is essential for corpus luteum angiogenesis." Nat Med 4(3): 336-340.

Folkman, J. (1971). "Tumor angiogenesis: therapeutic implications." N Engl J Med 285(21): 1182-1186.

Friedlander, M., K. C. Hancock, D. Rischin, M. J. Messing, C. A. Stringer, G. M. Matthys, B. Ma, J. P. Hodge and J. J. Lager (2010). "A Phase II, open-label study evaluating pazopanib in patients with recurrent ovarian cancer." Gynecol Oncol 119(1): 32-37.

Frumovitz, M. and A. K. Sood (2007). "Vascular endothelial growth factor (VEGF) pathway as a therapeutic target in gynecologic malignancies." Gynecol Oncol 104(3): 768-778.

Gadducci, A., P. Viacava, S. Cosio, G. Fanelli, A. Fanucchi, D. Cecchetti, R. Cristofani and A. R. Genazzani (2003). "Intratumoral microvessel density, response to chemotherapy and clinical outcome of patients with advanced ovarian carcinoma." Anticancer Res 23(1B): 549-556.

Garcia, A. A., H. Hirte, G. Fleming, D. Yang, D. D. Tsao-Wei, L. Roman, S. Groshen, S. Swenson, F. Markland, D. Gandara, S. Scudder, R. Morgan, H. Chen, H. J. Lenz and A. M. Oza (2008). "Phase II clinical trial of bevacizumab and low-dose metronomic oral cyclophosphamide in recurrent ovarian cancer: a trial of the California, Chicago, and Princess Margaret Hospital phase II consortia." J Clin Oncol 26(1): 76-82.

Ghosh, S. and P. Maity (2010). "VEGF antibody plus cisplatin reduces angiogenesis and tumor growth in a xenograft model of ovarian cancer." J Environ Pathol Toxicol Oncol 29(1): 17-30.

Giantonio, B. J., P. J. Catalano, N. J. Meropol, P. J. O'Dwyer, E. P. Mitchell, S. R. Alberts, M. A. Schwartz and A. B. Benson, 3rd (2007). "Bevacizumab in combination with oxaliplatin, fluorouracil, and leucovorin (FOLFOX4) for previously treated metastatic colorectal cancer: results from the Eastern Cooperative Oncology Group Study E3200." J Clin Oncol 25(12): 1539-1544.

Gonzalez, R. R., S. Cherfils, M. Escobar, J. H. Yoo, C. Carino, A. K. Styer, B. T. Sullivan, H. Sakamoto, A. Olawaiye, T. Serikawa, M. P. Lynch and B. R. Rueda (2006). "Leptin signaling promotes the growth of mammary tumors and increases the expression of vascular endothelial growth factor (VEGF) and its receptor type two (VEGF-R2)." J Biol Chem 281(36): 26320-26328.

Goodheart, M. J., M. A. Vasef, A. K. Sood, C. S. Davis and R. E. Buller (2002). "Ovarian cancer p53 mutation is associated with tumor microvessel density." Gynecol Oncol 86(1): 85-90.

Gross, J., R. G. Azizkhan, C. Biswas, R. R. Bruns, D. S. Hsieh and J. Folkman (1981). "Inhibition of tumor growth, vascularization, and collagenolysis in the rabbit cornea by medroxyprogesterone." Proc Natl Acad Sci U S A 78(2): 1176-1180.

Guidi, A. J., G. Abu-Jawdeh, B. Berse, R. W. Jackman, K. Tognazzi, H. F. Dvorak and L. F. Brown (1995). "Vascular permeability factor (vascular endothelial growth factor) expression and angiogenesis in cervical neoplasia." J Natl Cancer Inst 87(16): 1237-1245.

Guidi, A. J., G. Abu-Jawdeh, K. Tognazzi, H. F. Dvorak and L. F. Brown (1996). "Expression of vascular permeability factor (vascular endothelial growth factor) and its receptors in endometrial carcinoma." Cancer 78(3): 454-460.

Gunderson, C. C., A. N. Fader, K. A. Carson and R. E. Bristow (2012). "Oncologic and Reproductive outcomes with progestin therapy in women with endometrial hyperplasia and grade 1 Adenocarcinoma: A systematic review." Gynecol Oncol. 125(2): 477-82.

Hamilton, C. A., G. L. Maxwell, M. R. Chernofsky, S. A. Bernstein, J. H. Farley and G. S. Rose (2008). "Intraperitoneal bevacizumab for the palliation of malignant ascites in refractory ovarian cancer." Gynecol Oncol 111(3): 530-532.

Han, E. S., R. A. Burger, K. M. Darcy, M. W. Sill, L. M. Randall, D. Chase, B. Parmakhtiar, B. J. Monk, B. E. Greer, P. Connelly, K. Degeest and J. P. Fruehauf (2010). "Predictive and prognostic angiogenic markers in a gynecologic oncology group phase II trial of bevacizumab in recurrent and persistent ovarian or peritoneal cancer." Gynecol Oncol 119(3): 484-490.

Heer, K., H. Kumar, V. Speirs, J. Greenman, P. J. Drew, J. N. Fox, P. J. Carleton, J. R. Monson and M. J. Kerin (1998). "Vascular endothelial growth factor in premenopausal women--indicator of the best time for breast cancer surgery?" Br J Cancer 78(9): 1203-1207.

Hollingsworth, H. C., E. C. Kohn, S. M. Steinberg, M. L. Rothenberg and M. J. Merino (1995). "Tumor angiogenesis in advanced stage ovarian carcinoma." Am J Pathol 147(1): 33-41.

Hopfl, G., O. Ogunshola and M. Gassmann (2004). "HIFs and tumors – causes and consequences." Am J Physiol Regul Integr Comp Physiol 286(4): R608-623.

Huang, L. and C. Li (2000). "Leptin: a multifunctional hormone." Cell Res 10(2): 81-92.

Hurwitz, H., L. Fehrenbacher, W. Novotny, T. Cartwright, J. Hainsworth, W. Heim, J. Berlin, A. Baron, S. Griffing, E. Holmgren, N. Ferrara, G. Fyfe, B. Rogers, R. Ross and F. Kabbinavar (2004). "Bevacizumab plus irinotecan, fluorouracil, and leucovorin for metastatic colorectal cancer." N Engl J Med 350(23): 2335-2342.

Jikihara, H., N. Terada, R. Yamamoto, Y. Nishikawa, O. Tanizawa, K. Matsumoto and N. Terakawa (1992). "Inhibitory effect of medroxyprogesterone acetate on angiogenesis induced by human endometrial cancer." Am J Obstet Gynecol 167(1): 207-211.

Kainz, C., P. Speiser, C. Wanner, A. Obermair, C. Tempfer, G. Sliutz, A. Reinthaller and G. Breitenecker (1995). "Prognostic value of tumour microvessel density in cancer of the uterine cervix stage IB to IIB." Anticancer Res 15(4): 1549-1551.

Kaku, T., T. Kamura, N. Kinukawa, H. Kobayashi, K. Sakai, N. Tsuruchi, T. Saito, S. Kawauchi, M. Tsuneyoshi and H. Nakano (1997). "Angiogenesis in endometrial carcinoma." Cancer 80(4): 741-747.

Kamat, B. R., L. F. Brown, E. J. Manseau, D. R. Senger and H. F. Dvorak (1995). "Expression of vascular permeability factor/vascular endothelial growth factor by human granulosa and theca lutein cells. Role in corpus luteum development." Am J Pathol 146(1): 157-165.

Kaye, S. B. (2007). "Bevacizumab for the treatment of epithelial ovarian cancer: will this be its finest hour?" J Clin Oncol 25(33): 5150-5152.

Kerbel, R. S. (2008). "Tumor angiogenesis." N Engl J Med 358(19): 2039-2049.

Kim, Y. H., M. A. Kim, I. A. Park, W. Y. Park, J. W. Kim, S. C. Kim, N. H. Park, Y. S. Song and S. B. Kang (2010). "VEGF polymorphisms in early cervical cancer susceptibility, angiogenesis, and survival." Gynecol Oncol 119(2): 232-236.

Kirschner, C. V., J. M. Alanis-Amezcua, V. G. Martin, N. Luna, E. Morgan, J. J. Yang and E. L. Yordan (1996). "Angiogenesis factor in endometrial carcinoma: a new prognostic indicator?" Am J Obstet Gynecol 174(6): 1879-1882; discussion 1882-1874.

Klopp, A. H., Y. Zhang, T. Solley, F. Amaya-Manzanares, F. Marini, M. Andreeff, B. Debeb, W. Woodward, R. Schmandt, R. Broaddus, K. Lu and M. G. Kolonin (2012). "Omental Adipose Tissue-Derived Stromal Cells Promote Vascularization and Growth of Endometrial Tumors." Clin Cancer Res 18(3): 771-82.

Kobold, S., S. Hegewisch-Becker, K. Oechsle, K. Jordan, C. Bokemeyer and D. Atanackovic (2009). "Intraperitoneal VEGF inhibition using bevacizumab: a potential approach for the symptomatic treatment of malignant ascites?" Oncologist 14(12): 1242-1251.

Kodama, J., N. Seki, K. Tokumo, A. Hongo, Y. Miyagi, M. Yoshinouchi, H. Okuda and T. Kudo (1999). "Vascular endothelial growth factor is implicated in early invasion in cervical cancer." Eur J Cancer 35(3): 485-489.

Krikun, G., F. Schatz and C. J. Lockwood (2004). "Endometrial angiogenesis: from physiology to pathology." Ann N Y Acad Sci 1034: 27-35.

Kudelka, A. P., T. Levy, C. F. Verschraegen, C. L. Edwards, S. Piamsomboon, W. Termrungruanglert, R. S. Freedman, A. L. Kaplan, D. G. Kieback, C. A. Meyers, K. A. Jaeckle, E. Loyer, M. Steger, R. Mante, G. Mavligit, A. Killian, R. A. Tang, J. U. Gutterman and J. J. Kavanagh (1997). "A phase I study of TNP-470 administered to patients with advanced squamous cell cancer of the cervix." Clin Cancer Res 3(9): 1501-1505.

Kumaran, G. C., G. C. Jayson and A. R. Clamp (2009). "Antiangiogenic drugs in ovarian cancer." Br J Cancer 100(1): 1-7.

Lee, J. S., H. S. Kim, J. J. Jung, M. C. Lee and C. S. Park (2002). "Angiogenesis, cell proliferation and apoptosis in progression of cervical neoplasia." Anal Quant Cytol Histol 24(2): 103-113.

Leung, K. M., W. Y. Chan and P. K. Hui (1994). "Invasive squamous cell carcinoma and cervical intraepithelial neoplasia III of uterine cervix. Morphologic differences other than stromal invasion." Am J Clin Pathol 101(4): 508-513.

Li, X. and U. Eriksson (2001). "Novel VEGF family members: VEGF-B, VEGF-C and VEGF-D." Int J Biochem Cell Biol 33(4): 421-426.

Loncaster, J. A., R. A. Cooper, J. P. Logue, S. E. Davidson, R. D. Hunter and C. M. West (2000). "Vascular endothelial growth factor (VEGF) expression is a prognostic factor for radiotherapy outcome in advanced carcinoma of the cervix." Br J Cancer 83(5): 620-625.

Mabuchi, S., Y. Terai, K. Morishige, A. Tanabe-Kimura, H. Sasaki, M. Kanemura, S. Tsunetoh, Y. Tanaka, M. Sakata, R. A. Burger, T. Kimura and M. Ohmichi (2008). "Maintenance treatment with bevacizumab prolongs survival in an in vivo ovarian cancer model." Clin Cancer Res 14(23): 7781-7789.

Mackay, H. J., A. Tinker, E. Winquist, G. Thomas, K. Swenerton, A. Oza, J. Sederias, P. Ivy and E. A. Eisenhauer (2010). "A phase II study of sunitinib in patients with locally advanced or metastatic cervical carcinoma: NCIC CTG Trial IND.184." Gynecol Oncol 116(2): 163-167.

Manenti, L., E. Riccardi, S. Marchini, E. Naumova, I. Floriani, A. Garofalo, R. Dossi, E. Marrazzo, D. Ribatti, E. Scanziani, M. Bani, D. Belotti, M. Broggini and R. Giavazzi (2005). "Circulating plasma vascular endothelial growth factor in mice bearing human ovarian carcinoma xenograft correlates with tumor progression and response to therapy." Mol Cancer Ther 4(5): 715-725.

Masood, R., J. Cai, T. Zheng, D. L. Smith, D. R. Hinton and P. S. Gill (2001). "Vascular endothelial growth factor (VEGF) is an autocrine growth factor for VEGF receptor-positive human tumors." Blood 98(6): 1904-1913.

Matei, D., M. W. Sill, H. A. Lankes, K. DeGeest, R. E. Bristow, D. Mutch, S. D. Yamada, D. Cohn, V. Calvert, J. Farley, E. F. Petricoin and M. J. Birrer (2011). "Activity of sorafenib in recurrent ovarian cancer and primary peritoneal carcinomatosis: a gynecologic oncology group trial." J Clin Oncol 29(1): 69-75.

Merritt, W. M., A. M. Nick, A. R. Carroll, C. Lu, K. Matsuo, M. Dumble, N. Jennings, S. Zhang, Y. G. Lin, W. A. Spannuth, A. A. Kamat, R. L. Stone, M. M. Shahzad, R. L. Coleman, R. Kumar and A. K. Sood (2010). "Bridging the gap between cytotoxic and biologic therapy with metronomic topotecan and pazopanib in ovarian cancer." Mol Cancer Ther 9(4): 985-995.

Mesiano, S., N. Ferrara and R. B. Jaffe (1998). "Role of vascular endothelial growth factor in ovarian cancer: inhibition of ascites formation by immunoneutralization." Am J Pathol 153(4): 1249-1256.

Micha, J. P., B. H. Goldstein, M. A. Rettenmaier, M. Genesen, C. Graham, K. Bader, K. L. Lopez, M. Nickle and J. V. Brown, 3rd (2007). "A phase II study of outpatient first-line paclitaxel, carboplatin, and bevacizumab for advanced-stage epithelial ovarian, peritoneal, and fallopian tube cancer." Int J Gynecol Cancer 17(4): 771-776.

Monk, B. J., D. C. Choi, G. Pugmire and R. A. Burger (2005). "Activity of bevacizumab (rhuMAB VEGF) in advanced refractory epithelial ovarian cancer." Gynecol Oncol 96(3): 902-905.

Monk, B. J., M. W. Sill, R. A. Burger, H. J. Gray, T. E. Buekers and L. D. Roman (2009). "Phase II trial of bevacizumab in the treatment of persistent or recurrent squamous cell carcinoma of the cervix: a gynecologic oncology group study." J Clin Oncol 27(7): 1069-1074.

Monk, B. J., L. J. Willmott and D. A. Sumner (2010). "Anti-angiogenesis agents in metastatic or recurrent cervical cancer." Gynecol Oncol 116(2): 181-186.

Motzer, R. J., T. E. Hutson, P. Tomczak, M. D. Michaelson, R. M. Bukowski, O. Rixe, S. Oudard, S. Negrier, C. Szczylik, S. T. Kim, I. Chen, P. W. Bycott, C. M. Baum and R. A. Figlin (2007). "Sunitinib versus interferon alfa in metastatic renal-cell carcinoma." N Engl J Med 356(2): 115-124.

Motzer, R. J., B. I. Rini, R. M. Bukowski, B. D. Curti, D. J. George, G. R. Hudes, B. G. Redman, K. A. Margolin, J. R. Merchan, G. Wilding, M. S. Ginsberg, J. Bacik, S. T. Kim, C. M. Baum and M. D. Michaelson (2006). "Sunitinib in patients with metastatic renal cell carcinoma." JAMA 295(21): 2516-2524.

Nakamizo, A., F. Marini, T. Amano, A. Khan, M. Studeny, J. Gumin, J. Chen, S. Hentschel, G. Vecil, J. Dembinski, M. Andreeff and F. F. Lang (2005). "Human bone marrow-derived mesenchymal stem cells in the treatment of gliomas." Cancer Res 65(8): 3307-3318.

Nakamura, M., J. M. Bodily, M. Beglin, S. Kyo, M. Inoue and L. A. Laimins (2009). "Hypoxia-specific stabilization of HIF-1alpha by human papillomaviruses." Virology 387(2): 442-448.

Nimeiri, H. S., A. M. Oza, R. J. Morgan, G. Friberg, K. Kasza, L. Faoro, R. Salgia, W. M. Stadler, E. E. Vokes and G. F. Fleming (2008). "Efficacy and safety of bevacizumab plus erlotinib for patients with recurrent ovarian, primary peritoneal, and fallopian tube cancer: a trial of the Chicago, PMH, and California Phase II Consortia." Gynecol Oncol 110(1): 49-55.

Nimeiri, H. S., A. M. Oza, R. J. Morgan, D. Huo, L. Elit, J. A. Knost, J. L. Wade, 3rd, E. Agamah, E. E. Vokes and G. F. Fleming (2010). "A phase II study of sorafenib in advanced uterine carcinoma/carcinosarcoma: a trial of the Chicago, PMH, and California Phase II Consortia." Gynecol Oncol 117(1): 37-40.

Norden, A. D., G. S. Young, K. Setayesh, A. Muzikansky, R. Klufas, G. L. Ross, A. S. Ciampa, L. G. Ebbeling, B. Levy, J. Drappatz, S. Kesari and P. Y. Wen (2008). "Bevacizumab for recurrent malignant gliomas: efficacy, toxicity, and patterns of recurrence." Neurology 70(10): 779-787.

Numnum, T. M., R. P. Rocconi, J. Whitworth and M. N. Barnes (2006). "The use of bevacizumab to palliate symptomatic ascites in patients with refractory ovarian carcinoma." Gynecol Oncol 102(3): 425-428.

Obermair, A., C. Tempfer, L. Hefler, O. Preyer, A. Kaider, R. Zeillinger, S. Leodolter and C. Kainz (1998). "Concentration of vascular endothelial growth factor (VEGF) in the serum of patients with suspected ovarian cancer." Br J Cancer 77(11): 1870-1874.

Olson, T. A., D. Mohanraj, L. F. Carson and S. Ramakrishnan (1994). "Vascular permeability factor gene expression in normal and neoplastic human ovaries." Cancer Res 54(1): 276-280.

Paley, P. J., B. A. Goff, A. M. Gown, B. E. Greer and E. H. Sage (2000). "Alterations in SPARC and VEGF immunoreactivity in epithelial ovarian cancer." Gynecol Oncol 78(3 Pt 1): 336-341.

Papetti, M. and I. M. Herman (2002). "Mechanisms of normal and tumor-derived angiogenesis." Am J Physiol Cell Physiol 282(5): C947-970.

Penson, R. T., D. S. Dizon, S. A. Cannistra, M. R. Roche, C. N. Krasner, S. T. Berlin, N. S. Horowitz, P. A. Disilvestro, U. A. Matulonis, H. Lee, M. A. King and S. M. Campos (2010). "Phase II study of carboplatin, paclitaxel, and bevacizumab with maintenance bevacizumab as first-line chemotherapy for advanced mullerian tumors." J Clin Oncol 28(1): 154-159.

Perren, T. J., A. M. Swart, J. Pfisterer, J. A. Ledermann, E. Pujade-Lauraine, G. Kristensen, M. S. Carey, P. Beale, A. Cervantes, C. Kurzeder, A. du Bois, J. Sehouli, R. Kimmig, A. Stahle, F. Collinson, S. Essapen, C. Gourley, A. Lortholary, F. Selle, M. R. Mirza, A. Leminen, M. Plante, D. Stark, W. Qian, M. K. Parmar and A. M. Oza (2011). "A phase 3 trial of bevacizumab in ovarian cancer." N Engl J Med 365(26): 2484-2496.

Perrot-Applanat, M., M. Ancelin, H. Buteau-Lozano, G. Meduri and P. Bausero (2000). "Ovarian steroids in endometrial angiogenesis." Steroids 65(10-11): 599-603.

Petridou, E., M. Belechri, N. Dessypris, P. Koukoulomatis, E. Diakomanolis, E. Spanos and D. Trichopoulos (2002). "Leptin and body mass index in relation to endometrial cancer risk." Ann Nutr Metab 46(3-4): 147-151.

Ranieri, G., R. Patruno, E. Ruggieri, S. Montemurro, P. Valerio and D. Ribatti (2006). "Vascular endothelial growth factor (VEGF) as a target of bevacizumab in cancer: from the biology to the clinic." Curr Med Chem 13(16): 1845-1857.

Rasila, K. K., R. A. Burger, H. Smith, F. C. Lee and C. Verschraegen (2005). "Angiogenesis in gynecological oncology-mechanism of tumor progression and therapeutic targets." Int J Gynecol Cancer 15(5): 710-726.

Reck, M., J. von Pawel, P. Zatloukal, R. Ramlau, V. Gorbounova, V. Hirsh, N. Leighl, J. Mezger, V. Archer, N. Moore and C. Manegold (2009). "Phase III trial of cisplatin plus gemcitabine with either placebo or bevacizumab as first-line therapy for nonsquamous non-small-cell lung cancer: AVAiL." J Clin Oncol 27(8): 1227-1234.

Reed, S. D., K. M. Newton, R. L. Garcia, K. H. Allison, L. F. Voigt, C. D. Jordan, M. Epplein, E. Swisher, K. Upson, K. J. Ehrlich and N. S. Weiss (2010). "Complex hyperplasia with and without atypia: clinical outcomes and implications of progestin therapy." Obstet Gynecol 116(2 Pt 1): 365-373.

Rose, D. P., E. M. Gilhooly and D. W. Nixon (2002). "Adverse effects of obesity on breast cancer prognosis, and the biological actions of leptin (review)." Int J Oncol 21(6): 1285-1292.

Rubatt, J. M., K. M. Darcy, A. Hutson, S. M. Bean, L. J. Havrilesky, L. A. Grace, A. Berchuck and A. A. Secord (2009). "Independent prognostic relevance of microvessel density in advanced epithelial ovarian cancer and associations between CD31, CD105, p53 status, and angiogenic marker expression: A Gynecologic Oncology Group study." Gynecol Oncol 112(3): 469-474.

Rutgers, J. L., T. F. Mattox and M. P. Vargas (1995). "Angiogenesis in uterine cervical squamous cell carcinoma." Int J Gynecol Pathol 14(2): 114-118.

Saaristo, A., T. Karpanen and K. Alitalo (2000). "Mechanisms of angiogenesis and their use in the inhibition of tumor growth and metastasis." Oncogene 19(53): 6122-6129.

Sandler, A., R. Gray, M. C. Perry, J. Brahmer, J. H. Schiller, A. Dowlati, R. Lilenbaum and D. H. Johnson (2006). "Paclitaxel-carboplatin alone or with bevacizumab for non-small-cell lung cancer." N Engl J Med 355(24): 2542-2550.

Santin, A. D., P. L. Hermonat, A. Ravaggi, S. Pecorelli, M. J. Cannon and G. P. Parham (1999). "Secretion of vascular endothelial growth factor in adenocarcinoma and squamous cell carcinoma of the uterine cervix." Obstet Gynecol 94(1): 78-82.

Schlenger, K., M. Hockel, M. Mitze, U. Schaffer, W. Weikel, P. G. Knapstein and A. Lambert (1995). "Tumor vascularity – a novel prognostic factor in advanced cervical carcinoma." Gynecol Oncol 59(1): 57-66.

Schultheis, A. M., G. Lurje, K. E. Rhodes, W. Zhang, D. Yang, A. A. Garcia, R. Morgan, D. Gandara, S. Scudder, A. Oza, H. Hirte, G. Fleming, L. Roman and H. J. Lenz (2008). "Polymorphisms and clinical outcome in recurrent ovarian cancer treated with cyclophosphamide and bevacizumab." Clin Cancer Res 14(22): 7554-7563.

Sillman, F., J. Boyce and R. Fruchter (1981). "The significance of atypical vessels and neovascularization in cervical neoplasia." Am J Obstet Gynecol 139(2): 154-159.

Smerdel, M. P., K. D. Steffensen, M. Waldstrom, I. Brandslund and A. Jakobsen (2010). "The predictive value of serum VEGF in multiresistant ovarian cancer patients treated with bevacizumab." Gynecol Oncol 118(2): 167-171.

Smith-McCune, K. (1997). "Angiogenesis in squamous cell carcinoma in situ and microinvasive carcinoma of the uterine cervix." Obstet Gynecol 89(3): 482-483.

Smith-McCune, K. K. and N. Weidner (1994). "Demonstration and characterization of the angiogenic properties of cervical dysplasia." Cancer Res 54(3): 800-804.

Spaeth, E. L., J. L. Dembinski, A. K. Sasser, K. Watson, A. Klopp, B. Hall, M. Andreeff and F. Marini (2009). "Mesenchymal stem cell transition to tumor-associated fibroblasts contributes to fibrovascular network expansion and tumor progression." PLoS One 4(4): e4992.

Spannuth, W. A., A. M. Nick, N. B. Jennings, G. N. Armaiz-Pena, L. S. Mangala, C. G. Danes, Y. G. Lin, W. M. Merritt, P. H. Thaker, A. A. Kamat, L. Y. Han, J. R. Tonra, R. L. Coleman, L. M. Ellis and A. K. Sood (2009). "Functional significance of VEGFR-2 on ovarian cancer cells." Int J Cancer 124(5): 1045-1053.

Spannuth, W. A., A. K. Sood and R. L. Coleman (2008). "Angiogenesis as a strategic target for ovarian cancer therapy." Nat Clin Pract Oncol 5(4): 194-204.

Stafl, A. and R. F. Mattingly (1975). "Angiogenesis of cervical neoplasia." Am J Obstet Gynecol 121(6): 845-852.

Tempfer, C., A. Obermair, L. Hefler, G. Haeusler, G. Gitsch and C. Kainz (1998). "Vascular endothelial growth factor serum concentrations in ovarian cancer." Obstet Gynecol 92(3): 360-363.

Tokumo, K., J. Kodama, N. Seki, Y. Nakanishi, Y. Miyagi, S. Kamimura, M. Yoshinouchi, H. Okuda and T. Kudo (1998). "Different angiogenic pathways in human cervical cancers." Gynecol Oncol 68(1): 38-44.

Torry, D. S., V. J. Holt, J. A. Keenan, G. Harris, M. R. Caudle and R. J. Torry (1996). "Vascular endothelial growth factor expression in cycling human endometrium." Fertil Steril 66(1): 72-80.

Vredenburgh, J. J., A. Desjardins, J. E. Herndon, 2nd, J. Marcello, D. A. Reardon, J. A. Quinn, J. N. Rich, S. Sathornsumetee, S. Gururangan, J. Sampson, M. Wagner, L. Bailey, D. D. Bigner, A. H. Friedman and H. S. Friedman (2007). "Bevacizumab plus irinotecan in recurrent glioblastoma multiforme." J Clin Oncol 25(30): 4722-4729.

Wiggins, D. L., C. O. Granai, M. M. Steinhoff and P. Calabresi (1995). "Tumor angiogenesis as a prognostic factor in cervical carcinoma." Gynecol Oncol 56(3): 353-356.

Willmott, L. J. and B. J. Monk (2009). "Cervical cancer therapy: current, future and anti-angiogensis targeted treatment." Expert Rev Anticancer Ther 9(7): 895-903.

Wright, J. D., D. Viviano, M. A. Powell, R. K. Gibb, D. G. Mutch, P. W. Grigsby and J. S. Rader (2006). "Bevacizumab combination therapy in heavily pretreated, recurrent cervical cancer." Gynecol Oncol 103(2): 489-493.

Yang, J. C., L. Haworth, R. M. Sherry, P. Hwu, D. J. Schwartzentruber, S. L. Topalian, S. M. Steinberg, H. X. Chen and S. A. Rosenberg (2003). "A randomized trial of bevacizumab, an anti-vascular endothelial growth factor antibody, for metastatic renal cancer." N Engl J Med 349(5): 427-434.

Yoneda, J., H. Kuniyasu, M. A. Crispens, J. E. Price, C. D. Bucana and I. J. Fidler (1998). "Expression of angiogenesis-related genes and progression of human ovarian carcinomas in nude mice." J Natl Cancer Inst 90(6): 447-454.

Yuan, S. S., K. B. Tsai, Y. F. Chung, T. F. Chan, Y. T. Yeh, L. Y. Tsai and J. H. Su (2004). "Aberrant expression and possible involvement of the leptin receptor in endometrial cancer." Gynecol Oncol 92(3): 769-775.

Zheng, Q., S. M. Dunlap, J. Zhu, E. Downs-Kelly, J. N. Rich, S. D. Hursting, N. A. Berger and O. Reizes (2012). "Leptin deficiency suppresses MMTV-Wnt-1 mammary tumor growth and abrogates tumor initiating cell survival." Endocr Relat Cancer. 2011 Jul 11; 18(4): 491-503.

Chapter **11**

Tumor Angiogenesis Regulators in Breast Cancer: Does Race Matter?

Laronna S. Colbert[1*] and LaTonia Taliaferro-Smith[2]

[1]*Assistant Professor of Clinical Medicine, Morehouse School of Medicine Hematology/Oncology Section, Atlanta, GA 30310*

[2]*Department of Hematology and Oncology, Winship Cancer Institute Emory University School of Medicine, Atlanta, GA*

ABSTRACT

There is a great deal of disparity regarding how race is measured and interpreted in medical literature. Despite these disparities new data is emerging which not only confirms the presence of specific polymorphisms within different ethnic groups but also confirms the relationship between the presence of these polymorphisms and

Correspondence Author: Morehouse School of Medicine, Department of Medicine, 720 Westview Drive, S.W., Atlanta, GA 30310. Email: lcolbert@msm.edu

gene expression and protein function. Importantly, the specific proteins involved in the process of angiogenesis are impacted by polymorphisms within their respective genes. Polymorphisms within these genes if they occur with greater frequency within certain ethnic groups can indeed influence and regulate the process of angiogenesis. Additionally polymorphisms within the genes encoding the proteins present in the tumor microenvironment impact on the process of angiogenesis and differ between ethnic groups. While acknowledging that epigenetic as well as post-translational changes within these proteins may also contribute to differences between ethnic groups we will focus on the specific pre-translational phenomenon of single nucleotide polymorphisms. The role of the adipocytokines: adiponectin and leptin, in mediating the process of angiogenesis is also the subject of ongoing studies. In this chapter we will review data on several of these molecules and other microenvironment molecules that could be responsible for differential tumor angiogenesis outcome ethnic groups.

Key words: Polymorphisms, Angiogenesis, Race, Ethnicity, Breast Cancer

1   Introduction

Tumor angiogenesis is a critical component of carcinogenesis and metastasis. Admittedly, many factors impact upon the process. We will not attempt to define race in this chapter but will review the current literature reporting the presence of polymorphisms found within certain ethnic groups and their influence on the process of angiogenesis.

As we review and analyze data relating to the differences in gene expression which affect angiogenesis pathways within specific racial/ethnic groups we will speculate on the importance of polymorphisms responsible for these genetic variations. We will review the limited data relating to the role of polymorphisms on breast cancer subtypes at diagnosis as well as on prognosis and response to treatment. The role of polymorphisms in differing responses to treatment between ethnic groups continues to be elucidated. We will also discuss findings from current literature analyzing the effect of polymorphisms on the expression and function of proteins constituting the tumor microenvironment. One important aspect that could influence breast cancer outcome between ethnic groups is the interaction of the adipocytokines with their receptors and receptor expression. The effect of the presence of polymorphisms on the expression of these components will be reviewed as well as the effect of polymorphisms on downstream events initiated by ligand/receptor binding including intracellular signaling pathways

2   Genetic Polymorphisms

Polymorphisms are gene variants that appear with a frequency of greater than 1% throughout the general population. They occur approximately once every 1,000 base pairs (bps). The most common and most simple polymorphism is the single nucleotide polymorphism (SNP). Additionally deletions, insertions, duplications and microsatellite repeats can occur with some measure of frequency. SNPs can be classified as either coding or non-coding depending on their placement within the gene. Polymorphisms can lead to changes in phenotype by altering protein structure directly, or they may be in linkage disequilibrium with a functional variant (Syvanen 2001). Polymorphisms in regulatory regions such as the promoter, the 5'/3' untranslated regions or the splice sites may have an influence on either transcription or mRNA stability (Wagner et al. 2007). The NCBI SNP database lists a total of 12,632,873 human SNPs (12,539,846 validated) (http://www.ncbi.nlm.nih.gov/SNP/snp_summary.cgi)

3   Role of Genetic Polymorphisms in Angiogenesis Regulation

Angiogenesis, the process of new blood vessel forming from pre-existing vessels, has become a much studied phenomenon in the pathogenesis of cancer. The relationship between the key components of angiogenesis must be exact in order for new vessels to grow. The effect of polymorphisms within the genes regulating angiogenesis and the tumor microenvironment including: hypoxia inducible factor-1 (*HIF-1*), vascular endothelial growth factor (*VEGF*), VEGF receptor (*VEGFR*), endothelial nitric oxide synthase (*eNOS*) (http://www.ncbi.nlm.nih.gov/SNP/snp_summary.cgi), Notch, matrix metalloproteinases, insulin-like growth factor (IGF), leptin and adiponectin are all key components of angiogenesis. The key polymorphisms of these genes reviewed in this chapter are listed in Table 1.

Table 1   Polymorphisms reviewed in this chapter

VEGF	HIF-1α	eNOS	MMP	ADIPONECTIN	LEP/LEPR
–2578C/A	C1772T	-786T/C	rs11644561 (G/A)	rs2241766	A19G
–1154G/A	G1790A	894G/T	rs11643630 (T/G)	rs1501299	Q223R
–634G/C			rs243865 (-1306 C/T)		K109R
+936C/T					K656N
–1498C/C					
–460T/C					

3.1   VEGF

The *VEGF* gene, which is located on chromosome 6p21.3 consists of eight exons and undergoes alternate splicing to form a family of proteins (Stevens et al. 2003). VEGF plays a role in various physiological as well as pathological processes. It functions mainly as a pro-angiogenic factor. However anti-angiogenic effects have been documented. Specific protein effects are dependent on the splice site choice in the terminal, eighth exon of chromosome 6. Proximal splice site selection (PSS) of exon 8 produces proangiogenic isoforms among which VEGF-A165 was first described (Nowak et al. 2010). Other isoforms include: VEGF-A121, VEGF-A145, VEGF-A148, VEGF-A183, VEGF-A189, and VEGF-A206. These additional isoforms are generated by alternative splicing of exons 6 and 7 (Harper and Bates 2008). Selection of a distal splice site selection (DSS) 66 bp further into exon 8 produces anti-angiogenic RNA isoforms that contain exon 8b, namely VEGF165b. Epithelial cells treated with insulin-like growth factor (IGF-1) increase PSS producing more VEGF165 and less VEGF165b (Nowak et al. 2010). IGF-1, as will be discussed later, plays a significant role within the tumor microenvironment and as such impacts greatly on the process of angiogenesis. Polymorphisms within the gene coding for IGF-1 will be reviewed.

A difference in the levels of circulating VEGF between different ethnic groups has been documented. It can be inferred that circulating levels of VEGF reflect gene activity. The studies of Ruggiero et al (Ruggiero et al. 2011) lend support to the hypothesis that distinct polymorphisms identified in discrete populations influence serum levels of VEGF within those populations. These researchers conducted a genome-wide linkage analysis in three isolated populations in Southern Italy. These groups were characterized by a large, unique genealogy with a small number of founders. Twenty-six different SNPs with a minor allele frequency (MAF) < 5% were identified. None of the *VEGF* gene SNPs which affected VEGF serum levels in one population was associated with the levels in the other two populations. The 6p21.1 *VEGF* gene region was identified as the main quantitative trait loci (QTL) for serum VEGF level variation (Ruggiero et al. 2011).

Conflicting evidence exists however regarding the role of VEGF in breast cancer susceptibility among ethnic groups. A population based study by Kataoka et al suggested that the *VEGF* C936T polymorphism was associated with a decreased risk of breast cancer among Chinese women (odds ratio, OR = 0.65) (Kataoka et al. 2006). Schneider et al. studied the association between the expression of *VEGF*-2578 AA and -1498 CC genotypes (alleles) and revealed an increased risk of breast cancer in patients expressing these alleles as compared to controls (P= 0.06 and P = 0.04, respectively). Within this study group of 1212 females, 89% were Caucasian. The study analysis was performed only on the Caucasian subjects (Schneider et al. 2008a).

The association between four *VEGF* polymorphisms (–2578C/A, –1154G/A, –634G/C, +936C/T) and breast cancer risk among postmenopausal women was examined in the Cancer Prevention Study II Nutrition Cohort (Jacobs et al. 2006). This large scale prospective cohort included 97,000 women, 97% of which were white and 1% black. Within the five-year follow-up 1683 women developed breast cancer. (Calle et al. 2002). A cohort comprised 509 of these women was studied by Jacobs et al (Jacobs et al. 2006). Their findings provided limited support for the correlation of *VEGF* polymorphisms and breast cancer risk. The –2578A and –1154G alleles were associated with increased risk of invasive disease but not in situ cancer (OR=1.46, 95%

confidence interval [CI] 1.00-2.14 for –2578CC vs AA and OR 1.64, 95% CI 1.02-2.64 for –1154GG vs AA). The +936C allele was associated with reduced risk of in situ disease (OR 0.59, 95% CI 0.37-0.93 for CC vs TT/CT) but was not clearly associated with invasive breast cancer (OR 1.21, 95% CI 0.88-1.67 for +936CC vs TT/CT) (Jacobs et al. 2006).

Null findings of race specific polymorphisms contributing to risk variation between ethnic groups have also been reported. Jin et al. studied the relationship between the genetic polymorphisms in the *VEGF* gene and the development of breast cancer in 571 familial breast cancer patients from Poland and Germany and 974 random breast cancer patients from Sweden (Jin et al. 2005). Of the four polymorphisms studied, –2578C/A, –1154G/A, +936C/T, and –634G/C none were found to significantly impact breast cancer susceptibility (OR 0.99, 95% CI, 0.85-1.15; p=0.93). Wang et al (Wang et al. 2011) performed a meta-analysis of 10 case-controlled studies which included 8175 cases and 8528 controls. The overall results did not show any association of the five *VEGF* polymorphisms (+936C/T, –1154A/G, –2578C/A, –634G/C, –460T/C) studied with breast cancer risk especially for Caucasians (Wang et al. 2011). However, the relationship between genetic variation of *VEGF* A rs833070 (–460C/T) and breast cancer risk among participants of the Shanghai Breast Cancer Genetics Study (SBCGS) found increased risk of breast cancer associated with TT genotype (OR = 1.26, 95% CI: 1.05–1.52, p = 0.016) (Beeghly-Fadiel et al. 2011a). Finally, in a case-control study including 804 Austrian breast cancer patients and 804 matched controls of the seven VEGF polymorphisms selected, only one (–634G/C) was associated with breast cancer risk with any degree of statistical significance (Langsenlehner et al. 2008).

3.2 VEGF Receptor

The vascular endothelial growth factor receptor family comprised three members in mammals: VEGFR-1 (Flt-1), VEGFR-2 (KDR/Flk-1) and VEGFR-3 (Flt-4). These receptors are transmembrane tyrosine kinase receptors which bind various members of the VEGF ligand family (Guo et al. 2010). VEGFR receptor is encoded by the kinase insert domain receptor gene (*KDR*) also known as fetal liver kinase 1 (*Flk-1*) and is a type III tyrosine kinase receptor.

The VEGFR-2 gene is located on chromosome 4 (4q11–q12) (Sait et al. 1995). It is organized into 30 exons separated by 29 introns (Jurado 2009). There are no studies to date analyzing SNPs of the genes encoding the VEGF receptor and their effects on breast cancer risk within specific ethnic groups.

3.3 Hypoxia Inducible Factor-1

The tumor microenvironment is characterized by a complex interplay between inflammation, hypoxia and angiogenesis. HIF-1 exists as a heterodimer of HIF-1α and HIF-1β. HIF-1 serves as a transcription activator to mediate cellular responses to hypoxia. Many hypoxia-related genes are mediated HIF. It has been postulated that perhaps up to 1% of the genome is hypoxia regulated (Semenza 2003). A hypoxic microenvironment initiates multiple cellular responses which result in tumor growth progression such as proliferation and angiogenesis. HIF-1α rapidly degrades in cells under normoxic conditions but is induced in the hypoxic microenvironment characteristic of tumors. Under hypoxic stabilization HIF-1α is translocated to the nucleus where heterodimerization occurs with HIF-1β. HIF-1 subsequently interacts with hypoxia response elements of target promoters (Tanimoto et al. 2003). Kim et al studied two missense polymorphisms, C1772T and G1790A located in the oxygen-dependent degradation domain (ODD) of the HIF-1α protein and revealed higher transcription activities than wild-type under normoxic or hypoxic conditions in Korean patients (Kim et al. 2008). Tanimoto et al have speculated on several mechanisms of the enhanced transactivation such as enhanced recruitment of transcriptional cofactors which interact with HIF-1α (Tanimoto et al. 2003). Nonetheless, a significant association was found between the C1772T polymorphism and expression of HIF-1α in the Korean patients studied by Kim et al (Kim et al. 2008). These patients tended to have poorer prognosis; possibly from more aggressive tumors. The presence of the G1790A polymorphism demonstrated no such association. These researchers further studied the frequency of these polymorphisms in Korean breast cancer patients and healthy controls. Specifically, this study evaluated the correlation between HIF-1α polymorphisms and clinicopathological features in these patients. The frequency of genotype, i.e. the allele distribution of the two polymorphisms did not differ significantly between patients and normal controls. HIF-1α levels however, as measured by immunostaining was positively correlated with high histological grade, lymph node metastasis, and Ki-67 levels in this group (Kim et al. 2008).

Ribeiro et al investigated the frequency of the C1772T and G1790A HIF-1 polymorphisms in four distinct populations representing three continents to further evaluate interethnic variability of these polymorphisms (Ribeiro et al. 2009). The populations studied were indigenous to Portugal, Mozambique, Colombia and Guinea-Bissau. The Mozambican population had an allelic frequency of 0.006 for the 1790A allele which was not detected among the other studied populations. The allelic frequency of the 1772T allele was 0.122 in Portugal, 0.151 in Columbia, 0.246 in Mozambique and 0.08 in Guinea-Bissau. These findings lend further support to the hypothesis that differences in the frequency of polymorphisms specifically those regulating genes involved in angiogenesis may be relevant in the etiology of the differences seen in tumor aggressiveness within various ethnic groups.

3.4   Nitric Oxide Synthase

Nitric oxide is synthesized from arginine by a family of three distinct NO synthase enzymes. Endothelial nitric oxide synthase-3 is encoded by the

NOS3 gene on chromosome 7. It is involved in a myriad of physiologic functions as well as pathologic functions. More than 800 polymorphisms have been identified in the *eNOS* gene as reported in the GenBank dbSNP database. (Available at URL: http://www.ncbi.nlm.nil.gov). Ghilardi et al found the –786T>C polymorphism to be associated with vascular invasion in a study involving Italian breast cancer patients (Ghilardi et al. 2003).

Schneider et al (Schneider et al. 2008b) demonstrated a greater likelihood of invasive breast cancer among women with *eNOS*–786 TT and *eNOS* 894 GG genotypes. The *eNOS*–786 T/G single nucleotide polymorphism (SNP) had a frequency of 0.39 among Caucasian patients studied and a frequency of 0.16 among African-American patients studied with an allele difference of <0.001 by race. The *eNOS* 894 G/T SNP had a Caucasian frequency of 0.32 and an African-American frequency of 0.10 with an allele difference of <0.001 by race. The *eNOS* 894 genotype was also associated with a greater likelihood of metastatic breast cancer (Schneider et al. 2008a).

Lu et al (Lu et al. 2006) conducted a case-control study of 421 non-Hispanic white women with a history of sporadic breast cancer and 423 controls to assess the association of specific polymorphisms with risk of breast cancer. Three polymorphisms were genotyped: –786T>C, 27 base pair (bp) variable number of tandem repeats (VNTR) in intron 4 and 894G>T which suggested an increased risk in sporadic breast cancer development within this ethnic group (Lu et al. 2006).

3.5   Notch

Notch signaling plays an important role in the processes of vasculogenesis and angiogenesis. Notch genes encode transmembrane proteins that serve as receptors for the Delta, Serrate, Lag-2 (DSL) family of ligands. There are four different Notch proteins (receptors) in mammals and five known ligands: Delta-like 1, Delta-like 3 and Delta-like 4, Jagged 1 and Jagged 2 (Brennan and Brown 2003). The effects of Notch signaling on angiogenesis can be traced back to its interaction with VEGF and HIF (Guo et al. 2011). Notch receptor binding to specific ligands initiates a series of intracellular processes which ultimately lead to the formation of a mobile cytoplasmic subunit (Notch intracellular domain or NICD). The Notch 1 NICD interacts with HIF-1α to recruit Notch-1 responsive promoters. Additional studies have demonstrated upregulation of DLL-4 and Notch expression. Through a feedback mechanism Notch signaling can in turn affect the expression of VEGF receptors.

Each of the four Notch receptor proteins are encoded by unique genes on various chromosomes. Few studies exist which have explored the effect of polymorphisms within these genes on receptor expression and function. Lee et al (Lee et al. 2007) performed an analysis of the four Notch genes to detect the presence of any mutations within specific cancer types. Their study revealed only one mutation within the cohort of 48 breast cancers. Fu et al (Fu et al. 2010) published the first study to demonstrate an association between SNP rs11249433 and expression of NOTCH2 gene located on in the 1p11.2 region. No sub-analysis was performed based on ethnicity (Fu et al. 2010).

4 Tumor Microenvironment and Gene Regulation

As previously discussed the tumor microenvironment serves as a critical milieu of many of the processes involved in breast carcinogenesis. Evidence supports the theory of cancer stem cells interacting with the components of the tumor microenvironment and cytokine networks to promote tumor growth and metastasis (Korkaya et al. 2011). Fibroblasts, differentiated cancer cells, cancer stem cells, mesenchymal cell, immune cells, macrophages, adipocytes, cytokines and growth factors all interact within this complex milieu. Additionally, the extracellular matrix molecules, i.e. collagen, glycosaminoglycans, elastin and glycoproteins influence the process of carcinogenesis (Luikart 1988).

4.1 Matrix Metalloproteinases

Matrix metalloproteinases (MMPs) are proteins secreted by cancer cells that modify the tumor microenvironment and influence angiogenesis.

MMPs comprised a family of upwards of 20 different endopeptidases which are categorized by substrate specificity, i.e. collagenases, stromelysins, gelatinases and membrane-type MMPs (Price et al. 2001). The MMPs contribute to angiogenesis through several different mechanisms including transcriptional regulation of VEGF, release of VEGF that is sequestered in the extracellular matrix (ECM) or bound to connective tissue growth factor and the post-translational processing of VEGF. Some MMPs act as positive regulators (MMP-1, MMP-2, MMP-9, MT-MMPs) while others act as negative regulators (MMP-19) mediating their effects via vessel regression (MMP-10) (Chabottaux and Noel 2007). Price et al (Price et al. 2001) functionally characterized three promoter variants of the human MMP-2 gene (at −1306, −790 and +220). The common C→T transition at −1306 (allele frequency 0.26) disrupts a Sp1-type promoter site (CCACC box) and subsequently inhibits promoter activity (Price et al. 2001). The product of MMP-2, is a 72 kDa gelatinase or type IV collagenase which has activity against many proteins including growth factor-binding proteins and growth factor receptors. MMP-2 can cleave insulin-like growth factor (IGF)-binding proteins and release IGFs (Zhou et al. 2004).

Conflicting data exists regarding the impact of MMP polymorphisms and breast cancer risk and survival. One of the more investigated polymorphisms of the MMPs is the −1306C→T (or C/T) within the MMP-2 gene. Data suggests either increased or decreased risk among individuals with the −1306C→T SNP. Beeghly-Fadiel et al studied promoter region polymorphisms of *MMP-2* in 6,066 Chinese women participating in the

Shanghai Breast Cancer Study (Beeghly-Fadiel et al. 2011b). Their findings revealed a decreased risk of breast cancer (odds ratio, OR 0.6) among those who were minor allele homozygotes for *rs11644561 (G/A)* compared to major allele homozygotes as well as among those who were minor allele homozygotes for *rs11643630 (T/G)* compared with major allele homozygotes (OR 0.8). Additionally a rare haplotype of both *rs11644561 A* and *rs11643630 G* was associated with a significantly reduced breast cancer risk (OR 0.6) while allele homozygotes for *rs243865 (–1306 C/T)* had an increased risk of breast cancer (OR 1.4) (Beeghly-Fadiel et al. 2009). Delgado-Enciso et al investigated the effect of the –1306CC genotype in the MMP-2 promoter on breast cancer risk in a Mexican population (Delgado-Enciso et al. 2008). Ninety patients were compared to 96 controls. The breast cancer patients exhibited a higher frequency of the MMP-2CC genotype (OR 2.15, 95% CI 1.1-4.1). The correlation was even more pronounced among younger patients (age $\leq$ 50 years) with an OR of 2.66 (95% CI 1.04–6.96). Zhou et al found the presence of SNPs within the promoter region of *MMP-2* (–1306C→T) associated with a reduced risk of breast cancer in an ethnically homogeneous population of Han Chinese women (OR 0.46, 95% CI 0.34-0.63) (Zhou et al. 2004). Roehe et al (Roehe et al. 2007) found no association between the –1306C→T polymorphism and breast cancer risk in a Brazilian population. Frequency of the CT genotype/allele was 0.24 among patients and 0.32 among controls (p = 0.22) (Roehe et al. 2007). Grieu et al found an association between SNP of the *MMP-2* promoter region (–1306 C→T) and breast cancer phenotype in an Australian population (>95% Caucasian descent) (Grieu et al. 2004). In patients with estrogen receptor negative (ER-) disease the *MMP-2* TT genotype was associated with poor survival (6/8, 75% deceased at end of study) compared to those with CC/CT genotype (p=0.002). In patients with estrogen receptor positive (ER+) disease the *MMP-2* TT genotype was associated with a trend towards good survival (10/10, 100% alive at end of study. The median follow-up time was 87 months (range, 2-116 months).

MMP-9 activity is needed to produce normal patterns of vascularization (Goto et al. 1993). Increased levels of MMP-9 are associated with highly vascularized and rapidly growing tumors (Chantrain et al. 2004). Przybylowska et al demonstrated a correlation between high level of MMP-9 and tumor size of breast cancers likely secondary to the regulatory role of MMP-9 in neovascularization and proliferation of breast cancer cells (Przybylowska et al. 2006). In this study conducted in Poland patients and age-matched healthy controls a correlation between the T allele and malignancy was observed (OR 2.61, 95% CI 1.33–4.87) The C→T substitution at the polymorphic site of the *MMP-9* gene promoter results in higher transcription activity due to the loss of a binding site for a repressor protein (Przybylowska et al. 2006).

Conversely, Beeghly-Fadiel et al did not find any significant association in Chinese patients with SNPs involving the promoter region of MMP-9 and breast cancer risk (OR 1.2, 95% CI 0.8–1.8) (Beeghly-Fadiel et al. 2011b).

4.2   Insulin-like Growth Factor-I

The IGF-I system plays an important role in normal human growth and development. It has also been heavily implicated in the development and progression of breast cancer (Chong et al. 2007). Its role in angiogenesis continues to be elucidated as the interplay between IGF-I, MMPs and other components of the tumor microenvironment to support tumor growth and metastasis. IGF-I is produced by stromal cells and can stimulate adjacent breast cancer cells via a paracrine route (Chong et al. 2007).

The IGF system includes IGF-I and II, their receptors, and six IGF binding proteins (IGFBP I-6) which serve as transporter proteins for IGF-I. Both IGF-I and IGF-II are encoded by single genes. The gene for IGF-I is located on the long arm of chromosome 12 while the gene for IGF-II is located on the short arm of chromosome 11. The IGFBP-1 gene is localized to chromosome 7, where it is contiguous with the gene encoding IGFBP-3. Harrela et al studied the magnitude of genetic contribution to the levels of circulating IGF-I (38%), IGF-II (66%), and IGFBP-3 (60%) by comparing levels in monozygotic and dizygotic twins (Harrela et al. 1996).

4.3   Adiponectin

Increasing evidence suggests that obesity, a condition characterized by an increase in adipocyte size and/or number, alters adipocytokine secretion and plays a major role in the metabolic syndrome. Obesity, a worldwide epidemic, is one of the primary risk factors for certain types of cancer, including post-menopausal breast cancer and endometrial cancer. Obesity significantly augments the risk of a number of morbidities. While the causes of these obesity-related cancers has been primarily attributed to excess estrogen production by adipose tissue, they have also been thought to be due, in part, to changes in the levels of various adipocytokines secreted by adipocytes. Although numerous adipocytokines have been identified, the effects of only a few in promoting or inhibiting mammary tumor growth have been extensively studied to date. These include leptin, hepatocyte growth factor (HGF), and adiponectin. Adiponectin is composed of 244 amino acids which generate a 30 kDa protein (18 amino acid long N-terminus long signal peptide followed by short hypervariable regions and collagen domain containing 22 repeated motifs). Adiponectin is produced primarily in white adipose tissue, but Viengehareun et al also showed that adiponectin is expressed in brown adipose T37i cell lines (derived from hibernomas and can undergo terminal differentiation into brown adipocytes) (Viengchareun et al. 2002). Circulating serum levels of adiponectin are approximately 5–10 μg/ml. Serum levels of adiponectin are significantly decreased in both obesity (body mass index greater than 30) and type 2 diabetes, indicative of a potential metabolic role in insulin resistance (Viengchareun et al. 2002).

Three isoforms of adiponectin are found in circulating plasma: low molecular weight (homotrimers), middle molecular weight (hexamers), high molecular weight (multimers, most biologically active form). Adiponectin trimer (LMW) associates, by disulfide-bonding, into medium molecular weight hexamers (MMW) and >300 kDa high molecular weight (HMW) oligomers. Two more distinct isoforms, full-length (fAd) and globular (gAd) adiponectin normally circulate in human serum at high concentrations and both forms are biologically active (Fruebis et al. 2001). gAd typically forms trimers that are homologous to tumor necrosis factor alpha (TNF-α) trimers. Waki et al (Waki et al. 2005) showed that globular adiponectin is generated in monocytes and/or neutrophils via cleavage of full-length adiponectin by leukocyte elastase (THP-1 monocytes were stimulated with phorbol 12-myristate 13-acetate, PMA) (Waki et al. 2005). Cleavage by this enzyme results in three major proteolytic fragments (18 kDA, 20 kDa and 25 kDA) and was inhibited by a leukocyte elastase-specific inhibitor (MeO-Suc-AAPV-CMK). Distinct biological actions and signal transduction properties have been reported for both the full-length and globular in various cell types, although the full significance of these events remains to be determined.

Two adiponectin receptors, AdipoR1 and AdipoR2, have been detected in human fat, muscle, liver, and a number of other cells. In mice, AdipoR1 is expressed primarily in lung, muscle, and spleen tissues, whereas AdipoR2 is predominantly detected in the liver. Humans express both adiponectin receptors in adipocytes, islet cells of the pancreas, macrophages, and vascular smooth muscle (Arita et al. 1999; Kharroubi et al. 2003; Chinetti et al. 2004). AdipoR1 has a high affinity for globular adiponectin and a low affinity for full-length adiponectin in muscle cells; conversely, AdipoR2 has an intermediate affinity for both isoforms of adiponectin. Both AdipoR1 and AdipoR2 are also detected in normal breast epithelial and breast cancer cells, where AdipoR1 appears to be the predominant functional receptor (Takahata et al. 2007). The biological effects of adiponectin are determined by the relative circulating levels, tissue-specificity, the presence of different adiponectin isoforms, and adiponectin receptor subtypes (Yamauchi et al. 2003).

Adiponectin was initially discovered as a hormone secreted by adipocytes in 3T3-L1 mice (Scherer et al. 1995). It was identified in humans one year later by three independent groups and termed AdipoQ by Hu et al, apM1 by Maeda et al and GBP28 by Nakano et al. (Hu et al. 1996; Nakano et al. 1996; Maeda et al. 2001). The adiponectin gene encodes a protein that is secreted by adipocytes and displays broad range of paracrine and endocrine effects on cellular metabolism and inflammation.

Both *in vivo* and *in vitro* studies have documented the effects of adiponectin on breast cancer pathogenesis. Dieudonne et al (Dieudonne et al. 2006) demonstrated the antiproliferative effects of adiponectin on human breast cancer MCF-7 cells. Upon exposure to recombinant adiponectin growth of the MCF-7 cells was inhibited, p42/p44 MAP kinase was inactivated, AMP kinase and apoptosis were activated (Dieudonne et al. 2006). Kaklamani et al (Kaklamani et al. 2008) measured total serum and HMW adiponectin in a case-control study of 74 female breast cancer patients and 76 controls. Patients with the highest adiponectin levels had a reduction in breast cancer risk of 65% (p = 0.04). Additionally, T47D cells exposed to adiponectin had a reduction in viable cells to 86% and proliferation to 66% (Kaklamani et al. 2008).

The adiponectin gene (ADIPOQ) is localized to chromosome 3q27. Teras et al (Teras et al. 2009) performed a nested case-control study of women enrolled in the American Cancer Society Center Prevention Study II examining the association between SNPs in ADIPOQ, ADIPOR1, and ADIPOR2. There were 648 cases and 659 controls. Ninety-nine percent of the participants were white. No statistically significant association was found between breast cancer risk and SNPs (p < 0.05) (Teras et al. 2009). Kaklamani et al (Kaklamani et al. 2008) genotyped 733 breast cancer patients and 839 controls for haplotyped-tagging SNPs of ADIPOQ and ADIPOR1. Two ADIPOQ SNPs (rs2241766 and rs1501299) were associated with breast cancer risk [rs2241766 T/G; OR 0.61, 95% CI 0.46-0.80 and rs1501299 G/G; OR 1.8, 95% CI 1.14-2.85). The study was conducted in New York City. Patients were 76.8% white and 9.3% black (Kaklamani et al. 2008).

5   Role of Leptin and Leptin Receptor Polymorphisms in Angiogenesis Regulation

Obesity is a strong risk factor for breast cancer in post-menopausal females and an adverse prognostic indicator in both pre- and post-menopausal females. Leptin, the product of the obesity (Ob) gene and it's interaction with its receptor (Ob-R) has been shown to impact the development of breast cancer in murine models (Gillespie et al., 2012). The interaction between leptin and VEGF/VEGF receptor in promoting angiogenesis has also been documented (Guo and Gonzalez-Perez 2011).

Leptin is a pro-proliferative, pro-inflammatory and pro-angiogenic factor. A complex crosstalk between leptin and the pro-inflammatory cytokine, IL-1, and the embryogenic/angiogenic signal, Notch, has been reported in breast cancer (Guo and Gonzalez-Perez 2011). Leptin is a 167-amino acid protein produced by the leptin gene (*LEP*). The *LEP* gene has been localized in humans to the 7q31.3 chromosome. It consists of three exons separated by two introns (Isse et al. 1995). Leptin is important in the regulation of adipose-tissue mass and body weight by regulating food intake and energy expenditure. Leptin is produced by the adipocytes of white adipose tissue. Its effects are mediated through binding to a specific receptor, encoded by *LEPR* located on human chromosome 1p31 (Paracchini et al. 2005). The leptin receptor, LEPR (Ob-R), has six isoforms produced by alternative RNA splicing of the ob gene. Based on these structural differences the receptor isoforms are classified as either long, short or soluble. The full length isoform (Ob-Rb) is the only isoform able to fully transduce activation signals into the recipient cell. The four short isoforms, Ob-Ra, Ob-Rc, Ob-Rd and Ob-Re are able to bind Janus kinases (JAKs). The soluble short isoform,

Ob-Re, can regulate serum leptin concentration, serve as a carrier protein, and transduce signals to recipient cells (Gorska et al. 2010). Isoforms of the leptin receptor have been identified in many tissues including the mammary gland, male and female reproductive organs, the gut, lung and hypothalamus. Several polymorphisms in the *LEP* gene as well as the *LEPR* gene have been identified. Paracchini et al. (Paracchini et al. 2005) recognized the ethnic variation in allelic frequencies of the polymorphisms *LEP* A19G, *LEPR* Q223R, K109R, and K656N. However, no evidence of an association between these genes and obesity was found through meta-analysis of available data (Paracchini et al. 2005). This conclusion was also supported by Heo et al. who previously analyzed the association and linkage of LEPR polymorphisms *LEPR* Q223R, K109R, and K656N to body mass index and waist circumference (Heo et al. 2002).

It must be acknowledged that there is an increased incidence of obesity among certain ethnic groups. Data from the National Health and Nutrition Examination Survey (NHANES) 2007-2008 reported obesity prevalence rates of 33% among Non-Hispanic white women ages 20 and above (64). Non-Hispanic black women had prevalence rates of 49.6% and Hispanics, including Mexican Americans had prevalence rates of 43%. Obesity was defined as having a body mass index (BMI) $\geq$ 30 (Flegal et al. 2010). It must also be acknowledged that obesity is a multifactorial phenomenon and positing a direct link between obesity and breast cancer pathogenesis is problematic. Increasing BMI is positively correlated with increased risk of death from cancer in general (Calle et al. 2003). Obese patients fare worse than lean patients in terms of recurrence rates and prognosis. Increasing BMI was been shown to be associated with increased death rates for breast cancer. Ewertz et al documented a 38% increase in the risk of breast cancer death over 30 years in patients with a BMI $\geq$ 30 kg/m^2 (Ewertz et al. 2011). Nonetheless, as leptin, the product of the obesity gene, is central in regulating food intake, it can be postulated that a change in its expression as brought about by a gene polymorphism may possibly impact on all the known functions mediated by leptin/Ob-R binding. Indeed, Avery et al performed a retrospective case-case study of 164 African-American and 172 White breast cancer patients (Avery 2011). Mean serum levels of leptin were higher in the African-American patients as compared to the White patients after adjusting for BMI. There was also a trend toward higher leptin levels in patients with triple negative tumors (p=0.06). Additionally, Cohen et al (Cohen et al. 2012) performed a cross-sectional study of 915 white and 892 black women enrolled in the Southern Community Cohort Study. Their findings revealed that leptin levels differ between races even after adjusting BMI. Blacks had higher leptin levels when compared to whites (mean 22.4 vs 19.0 ng/ml; p< 0.0001 unadjusted and 22.7 vs 18.8 ng/ml adjusted (Cohen et al. 2012). It has been mentioned previously in this chapter that African-American patients tend to have a higher incidence of triple negative breast cancers as compared to non-African-American patients (Carey et al. 2006).

A small increase in the risk of developing breast cancer was identified in European-American women carrying the *LEP* –2548AA genotype compared to those carrying the *LEP* –2548GG genotype (OR = 1.30) (Cleveland et al. 2010). This association was stronger among obese post-menopausal women with borderline statistical significance (OR = 1.86, P = 0.07). However there was no appreciable association between any studied polymorphism of either the *LEP* or *LEPR* in regards to breast cancer specific mortality or all-cause mortality among women with breast cancer.

Conversely, Snoussi et al documented a major increase in breast cancer risk associated with heterozygous LEP –2548AA (OR = 1.45) and homozygous LEP –2548AA (OR = 3.17) genotypes in Tunisian patients (Snoussi et al. 2006). A significant association was demonstrated between both the heterozygous LEPR 223QR genotype (OR = 1.68) and the homozygous LEPR 223RR genotype (OR = 2.26) and breast cancer. Additionally the LEP –2548A allele was significantly associated with a reduction in disease-free survival in breast cancer patients while the presence of the LEPR 223R allele was associated with a reduction in overall survival.

6   Gene Expression Profiling and Differences in Breast Cancer Subtype between African-Americans and European-Americans

The difference in incidence of breast cancer between ethnic groups has been well documented. Breast cancer rates among white women increased during the 1980s, stabilized from 1987-1994 then increased and peaked in 1999. Rates decreased from 1999 to 2002-2003 and have been stable since. This drop may be secondary to reductions in the use of menopausal hormone replacement therapy following the publications of findings from the Women's Health Initiative linking combined estrogen plus progestin to increased coronary heart disease and breast cancer. Additionally, declines in mammography use may have impacted the reduction in incidences of breast cancer among this group (American Cancer Society 2010). Conversely, among African-American women breast cancer rates have only stabilized since 1992 after an initial rise during the 1980s. African-American women are more like to be diagnosed with larger tumors and with more advanced disease at presentation. Overall, breast cancer death rates have decreased since 1997 however the percentage decrease has been less pronounced among African-Americans (1.6%) as compared Hispanics/Latinas (1.9%) and non-Hispanic whites (1.8%). African-Americans breast cancer death rates have declined at a greater rate when compared to Asian Americans/ Pacific Islanders (0.8%). In 2007 the death rates for African-Americans was 41% higher than in white women (American Cancer Society 2011-2012). Age-adjusted mortality in the United States from breast cancer in white women is 70 deaths per 100,000 compared to 85 deaths per 100,000 in African-American women. This difference is even more pronounced among women younger than age 50. In this subset mortality is 80% higher among African-American women compared to white women (9 vs 5 deaths per 100,000) (National Cancer Institute 2011).

Differences in outcome between African-American breast cancer patients and their European counterparts are well documented. In a series of consecutive trials conducted by the Southwest Oncology Group which included 6676 breast cancer patients a strong association between mortality and race was noted among premenopausal African-American patients with early-stage breast cancer (hazard ratio [HR] for death = 1.41) as well as postmenopausal early-stage African-American patients (HR for death = 1.61) (Albain et al. 2009). A meta-analysis of fourteen studies involving more than 10,000 African-American patients and 40,000 European-American patients conducted by Newman et al (Newman et al. 2002) also revealed African-American ethnicity as an independent predictor for worse breast cancer outcome. Many studies have also lent support to theories of biologic diversity as the etiology for these differences in outcome. These proposed differences must ultimately be linked to changes in gene expression and subsequent differences in protein production. While it must be acknowledged that these differences in incidence rates and mortality are multifactorial and can be partially attributed to socioeconomic factors, lifestyle, etc. the effect of biologic and genetic influences must also be recognized.

Gene expression profiling has allowed classification of breast cancer into distinct subtypes. "The genome wide expression patterns of tumors are a representation of the biology of the tumors; diversity in patterns reflects biological diversity." Sorlie et al (Sorlie et al. 2001) reported the subclassification of breast cancer in 2001 using hierarchical clustering of cDNA microarrays. An intrinsic gene set of 456 cDNA clones (427 unique genes) were selected to optimally identify the intrinsic characteristics of breast tumors. These were originally selected from 8,102 genes on the basis that they would show significantly greater expression variation between different tumors as compared to paired samples from the same tumor (Sorlie et al. 2001). Listed among the genes selected for inclusion on the cDNA clones were several which encode for components of the tumor microenvironment such as matrix metalloproteinase, collagen, keratin, integrin and insulin-like growth factor (Perou et al. 2000). The identified subtypes included: luminal A, luminal B, luminal C, basal-like and ERBB2 overexpressing (Sorlie et al. 2001). The subgroups were redefined in 2003 by the same author into: luminal A, luminal B, basal-like, ERBB2 overexpressing and the fifth group: "normal breast tissue-like" (Sorlie et al. 2003). The Carolina Breast Study determined the prevalence of five breast cancer subtypes based on gene expression: luminal A (ER+ and/ or progesterone receptor positive [PR+], HER2–), luminal B (ER+ and/or PR+, HER2+), basal-like (ER–, PR–, HER2–, cytokeratin 5/6 positive, and/ or HER1+), HER2+/ER– (ER–,PR–, and HER2+), and unclassified (negative for all 5 markers) among premenopausal and post-menopausal African-American and non-African-American patients (Carey et al. 2006). This study of 496 incident cases of invasive breast cancer was unique in that it oversampled African-American and premenopausal women to allow better representation of these two subpopulations. Carey et al (Carey et al. 2006) found the basal-like subtype to be more prevalent among premenopausal African-American women (39%) as compared to post-menopausal African-American women (14%) and non-African-American women (16%) of any age (p < 0.001). The luminal A subtype was less prevalent (36% vs 59% and 54%, respectively) (Carey et al. 2006).

As stated previously, differences in outcome between ethnic groups, disallowing for exogenous factors impacting incidence, diagnosis and treatment must ultimately be related to differences in gene expression and protein production. As evidenced by the above data there is a scientific basis to the characterization of breast cancer subtypes among ethnic groups. The link between the genes responsible for hormone receptor expression, ERBB2 (HER2/neu) expression and genes regulating angiogenesis and the tumor microenvironment will undoubtedly shed light on the complex pathogenesis of breast cancer initiation, proliferation and metastatic potential between ethnic groups. It is interesting to speculate whether more in-depth gene expression profiling to include detection of polymorphisms would lead to more precise subclassification of breast tumors and assignment to specific ethnic groups.

7   The Effect of Polymorphisms on Differing Responses to Treatment

Gene expression profiling has been used to make informed treatment decisions for breast cancer patients. Martin et al. established a two gene signature profile using genome-wide mRNA expression specific to tumor epithelium and stroma in African-American and European-American patients (Martin et al. 2009). Their findings suggest that angiogenesis plays a significant role in the differences in treatment response between these two groups. Increased microvessel density and macrophage infiltration was noted in the African-American patients.

In 2008, Schneider et al (Schneider et al. 2008b) published findings of a retrospective study examining the outcome of breast cancer patients with specific *VEGF* and *VEGFR-2* polymorphisms who were treated with paclitaxel and bevacizumab, a humanized monoclonal antibody against VEGF. Three hundred sixty-three samples were analyzed. The patients whose tumors expressed the *VEGF–2578 AA* genotype and who were treated with both paclitaxel and bevacizumab had a superior median overall survival (OS) (HR=0.58, 95% CI, 0.36 to 0.93; p=0.023) compared to those patients who had alternate genotypes but had also received the combination treatment. Additionally, those patients whose tumors expressed the *VEGF–1154 A* allele and were treated with bevacizumab in combination with paclitaxel also experienced a superior OS with an additive effect of each active allele in the combination treatment arm (HR=0.62; 95% CI, 0.46 to 0.83; p=0.001) These results were not seen in the control arm which comprised those patients receiving paclitaxel only. The candidate SNPs studied included: *VEGF* –2578 C/A, –1498 C/T, –1154 g/A, –634 G/C, 936 C/T and *VEGFR-2* -889 G/A (V2971) and 1416 A/T (Q472H). The frequency of these alleles among Caucasian and African-Americans was noted but no subgroup analysis was conducted based on ethnicity (Schneider et al. 2008b).

Jurado et al (Jurado 2009) performed genotype analysis for selected VEGFR-2 polymorphisms in 44 patients with non-curable solid tumors who had been treated with either bevacizumab, raf kinase inhibitors, vatalanib, sunitinib, sorafenib, ZD6474 or AMG706. Their analysis of VEGFR-2 polymorphisms identified the AA variant of intron-20 rs2219471with a significant difference in progression free survival (PFS) and overall survival (OS) compared to the AG ancestral variant. The studied polymorphisms were in promoter regions 5'UTR, 3"UTR; in exons 7, 8, 9, 11, 16, 17, 18, 21, 27, 30 and introns 9, 17, 20 (Jurado 2009).

Chabottaux et al (Chabottaux and Noel 2007) have suggested that abrogation of angiogenesis can rely on the production of protein fragments endowed with anti-angiogenic activities. Indeed angiogenic inhibitors are generated by degradation of extracellular matrix components such as collagen or plasminogen (Chabottaux and Noel 2007).

8   Conclusions and Perspectives

In answer to the question asked in the title of this chapter as to whether within the process of angiogenesis: Does race matter? We can confidently respond with: It depends. Admittedly, the data is scarce to support an unequivocal conclusion but nonetheless data does exist to suggest that polymorphisms within specific ethnic groups do indeed impact upon some aspects of the process of angiogenesis. For *VEGF* much conflicting data exists. For *VEGFR* the data is virtually nonexistent. However, compelling data does exist for the C1772T and G1790A polymorphisms of the *HIF* gene. Likewise the –786T/G and 894G/T polymorphisms of the *eNOS* gene seem to influence breast cancer risk among Caucasian patients. Polymorphisms within the genes coding for proteins constituting the tumor microenvironment: matrix metalloproteinases, IGF, adiponection and leptin have been identified and studied however compelling evidence for specific ethnic variability is lacking. The recently reported findings by Avery et al (Avery 2011) documenting elevated leptin levels within African-American as compared to White breast cancer patients regardless of BMI is provocative and opens the door for more genetic studies regarding this particular protein. Finally, as mentioned at the beginning of this chapter polymorphisms are just one type of mutation that can occur and the presence of deletions, insertions, duplications and microsatellite repeats may also affect incidence and outcome of breast patients if they too have specific ethnic variability within the genes that govern angiogenesis. Much work remains in exploring the magnitude of the effects of polymorphisms on angiogenesis regulation. The studies presented here simply reflect the tip of the iceberg in fully elucidating the impact of gene expression.

With the completion of the Human Genome Project and ongoing Cancer Genome Project comes the realization that genes regulate every aspect of cancer from initiation to progression, metastasis and response to treatment. Additionally, individual gene profiles are coming into play in clinical practice whereby genetic profiles are used to inform treatment decisions.

The Human Genome Project was initiated in 1990 with the goal of determining the sequence of chemical base pairs comprising DNA and of identifying and mapping the approximately 20,000–25,000 genes within the human genome. This $3 billion project was considered essentially complete in 2003 with documentation of the 3.3 billion base pairs comprising the human genome. Ongoing studies include the HapMap project whose objective is to catalog the common variants in European, East Asian and African genomes (Human Genome Project).

The Cancer Genome Atlas initiated in 2005 aims to catalog genetic mutations responsible for cancer. Sequence analysis will ultimately be obtained on 20-25 different tumor types using a myriad of techniques including gene expression profiling, copy number variation profiling, SNP genotyping, genome-wide DNA methylation profiling, RNA profiling and exon sequencing. Ultimately this data will be helpful in identifying common variants unique to specific ethnic groups and thus impact on diagnostic, treatment and prognostic strategies (The Cancer Genome Atlas). Finally, the 1000 Genomes Project is an international collaboration between China, Germany, the UK and the USA with a goal of deep sequencing at least 1000 individuals from different world wide populations in order to uncover the majority of genetic variations that occur at a population frequency greater than 1%. More than 9 million new SNPs as well as many novel indels (insertions/deletions), and a few large structural variants were identified in the initial phase of this study (The Cancer Genome Atlas).

9   Acknowledgements

This work supported in part by CREDO (MSCR) 2R25RR017694-06A1 to LSC.

REFERENCES

Albain, K. S., J. M. Unger, J. J. Crowley, C. A. Coltman and D. L. Hershman (2009). "Racial disparities in cancer survival among randomized clinical trials patients of the Southwest Oncology Group." J Natl Cancer Inst 101: 984-992.

American Cancer Society (2010). Cancer Facts & Figures.

American Cancer Society (2011-2012). Breast Cancer Facts & Figures.

Arita, Y., S. Kihara, N. Ouchi, M. Takahashi, K. Maeda, J. Miyagawa, K. Hotta, I. Shimomura, T. Nakamura, K. Miyaoka, H. Kuriyama, M. Nishida, S. Yamashita, K. Okubo, K. Matsubara, M. Muraguchi, Y. Ohmoto, T. Funahashi and Y. Matsuzawa (1999). "Paradoxical decrease of an adipose-specific protein, adiponectin, in obesity." Biochem Biophys Res Commun 257(1): 79-83.

Avery, T. P., K. R. Sexton, A. Brewster, R. El-Zein, M. Bondy (2011). "Racial variation of leptin levels in women with breast cancer." J Clin Oncol 29(suppl): abstr 1590.

Beeghly-Fadiel, A., W. Lu, J.-R. Long, X.-O. Shu, Y. Zheng, Q. Cai, Y.-T. Gao and W. Zheng (2009). "Matrix metalloproteinase-2 polymorphisms and breast cancer susceptibility." Cancer Epidemiol Biomarkers Prev 18: 1770-1776.

Beeghly-Fadiel, A., W. Lu, X.-O. Shu, J. Long, Q. Cai, Y. Xiang, Y.-T. Gao and W. Zheng (2011b). "MMP9 polymorphisms and breast cancer risk: a report from the Shanghai Breast Cancer Genetics Study." Breast Cancer Res Treat 126: 507-513.

Beeghly-Fadiel, A., X. O. Shu, W. Lu, J. Long, Q. Cai, Y. B. Xiang, Y. Zheng, Z. Zhao, K. Gu, Y. T. Gao and W. Zheng (2011a). "Genetic variation in VEGF family genes and breast cancer risk: a report from the Shanghai Breast Cancer Genetics Study." Cancer Epidemiol Biomarkers Prev 20(1): 33-41.

Brennan, K. and A. M. C. Brown (2003). "Is there a role for Notch signalling in human breast cancer?" Breast Cancer Res 5: 69-75.

Calle, E. E., C. Rodriguez, E. J. Jacobs, M. L. Almon, A. Chao, M. L. McCullough, H. S. Feigelson and M. J. Thun (2002). "The American Cancer Society Cancer Prevention Study II Nutrition Cohort: rationale, study design, and baseline characteristics." Cancer 94: 2490-2501.

Calle, E. E., C. Rodriguez, K. Walker-Thurmond and M. J. Thun (2003). "Overweight, obesity, and mortality from cancer in a prospectively studied cohort of U.S. adults." N Engl J Med 348(17): 1625-1638.

Carey, L. A., C. M. Perou, C. A. Livasy, L. G. Dressler, D. Cowan, K. Conway, G. Karaca, M. A. Troester, C. K. Tse, S. Edmiston, S. L. Deming, J. Geradts, M. C. Cheang, T. O. Nielsen, P. G. Moorman, H. S. Earp and R. C. Millikan (2006). "Race, breast cancer subtypes, and survival in the Carolina Breast Cancer Study." JAMA 295(21): 2492-2502.

Chabottaux, V. and A. Noel (2007). "Breast cancer progression: insights into multifaceted matrix metalloproteinases." Clin Exp Metastasis 24: 647-656.

Chantrain, C. F., H. Shimada, S. Jodele, S. Groshen, W. Ye, D. R. Shalinsky, Z. Werb, L. M. Coussens and Y. A. DeClerck (2004). "Stromal matrix metalloproteinase-9 regulates the vascular architecture in neuroblastoma by promoting pericyte recruitment." Cancer Res 64: 1675-1686.

Chinetti, G., C. Zawadski, J. C. Fruchart and B. Staels (2004). "Expression of adiponectin receptors in human macrophages and regulation by agonists of the nuclear receptors PPARalpha, PPARgamma, and LXR." Biochem Biophys Res Commun 314(1): 151-158.

Chong, Y. M., A. Subramanian, A. K. Sharma and K. Mokbel (2007). "The potential clinical applications of insulin-like growth factor-1 ligand in human breast cancer." Anticancer Res 27(3B): 1617-1624.

Cleveland, R. J., M. D. Gammon, C. M. Long, M. M. Gaudet, S. M. Eng, S. L. Teitelbaum, A. I. Neugut and R. M. Santella (2010). "Common genetic variations in the LEP and LEPR genes, obesity and breast cancer incidence and survival." Breast Cancer Res Treat 120(3): 745-752.

Cohen, S. S., J. H. Fowke, Q. Cai, M. S. Buchowski, L. B. Signorello, M. K. Hargreaves, W. Zheng, W. J. Blot and C. E. Matthews (2012). "Differences in the association between serum leptin levels and body mass index in black and white women: a report from the Southern Community Cohort Study." Ann Nutr Metab 60: 90-97.

Delgado-Enciso, I., F. R. Cepeda-Lopez, E. A. Monrroy-Guizar, J. R. Bautista-Lam, M. Andrade-Soto, G. Jonguitud-Olguin, A. Rodriguez-Hernandez, A. Anaya-Ventura, L. M. Baltazar-Rodriguez, M. Orozco-Ruiz, A. D. Soriano-Hernandez, I. P. Rodriguez-Sanchez, A. Lugo-Trampe, F. Espinoza-Gomez and M. L. Michel-Peregrina (2008). "Matrix metalloproteinase-2 promoter polymorphism is associated with breast cancer in a Mexican population." Gynecol Obstet Invest 65(1): 68-72.

Dieudonne, M. N., M. Bussiere, E. Dos Santos, M. C. Leneveu, Y. Giudicelli and R. Pecquery (2006). "Adiponectin mediates antiproliferative and apoptotic responses in human MCF7 breast cancer cells." Biochem Biophys Res Commun 345(1): 271-279.

Ewertz, M., M. B. Jensen, K. A. Gunnarsdottir, I. Hojris, E. H. Jakobsen, D. Nielsen, L. E. Stenbygaard, U. B. Tange and S. Cold (2011). "Effect of obesity on prognosis after early-stage breast cancer." J Clin Oncol 29(1): 25-31.

Flegal, K. M., M. D. Carroll, C. L. Ogden and L. R. Curtin (2010). "Prevalence and trends in obesity among US adults, 1999-2008." JAMA 303(3): 235-241.

Fruebis, J., T. S. Tsao, S. Javorschi, D. Ebbets-Reed, M. R. Erickson, F. T. Yen, B. E. Bihain and H. F. Lodish (2001). "Proteolytic cleavage product of 30-kDa adipocyte complement-related protein increases fatty acid oxidation in muscle and causes weight loss in mice." Proc Natl Acad Sci USA 98(4): 2005-2010.

Fu, Y.-P., H. Edvardsen, A. Kaushiva, J. P. Arhancet, T. M. Howe, I. Kohaar, P. Porter-Gill, A. Shah, H. Landmark-Hoyvik, S. D. Fossa, S. Ambs, B. Naume, A.-L. Borresen-Dale, V. N. Kristensen and L. Prokunina-Olsson (2010). "NOTCH2 in breast cancer: association of SNP rs11249433 with gene expression in ER-positive breast tumors without TP53 mutations." Mol Cancer 9: 113-113.

Ghilardi, G., M. L. Biondi, F. Cecchini, M. DeMonti, E. Guagnellini and R. Scorza (2003). "Vascular invasion in human breast cancer is correlated to T→786C polymorphism of NOS3 gene." Nitric Oxide 9: 118-122.

Gillespie, C., Guo S, Zhou W, Gonzalez.-Perez RR, (2011). "Leptin signaling disruption prevents DMBA-induced mammary tumors in lean and diet-induced-obesity (DIO) mice." 102nd Annual Meeting of the American Association for Cancer Research, AACR.

Gillespie C, Quarshie A, Penichet M, Gonzalez-Perez RR, (2012). Potential Role of Leptin Signaling in DMBA-induced Mammary Tumors by Non-Responsive C57BL/6J Mice Fed a High-Fat Diet. J Carcinogene Mutagene 2012, 3:132.

Gorska, E., K. Popko, A. Stelmaszczyk-Emmel, O. Ciepiela, A. Kucharska and M. Wasik (2010). "Leptin receptors." Eur J Med Res 15 Suppl 2: 50-54.

Goto, F., K. Goto, K. Weindel and J. Folkman (1993). "Synergistic effects of vascular endothelial growth factor and basic fibroblast growth factor on the proliferation and cord formation of bovine capillary endothelial cells within collagen gels." Lab Invest 69: 508-517.

Grieu, F., W. Q. Li and B. Iacopetta (2004). "Genetic polymorphisms in the MMP-2 and MMP-9 genes and breast cancer phenotype." Breast Cancer Res Treat 88: 197-204.

Guo, S., L. S. Colbert, M. Fuller, Y. Zhang and R. R. Gonzalez-Perez (2010). "Vascular endothelial growth factor receptor-2 in breast cancer." Biochim Biophys Acta 1806: 108-121.

Guo, S. and R. R. Gonzalez-Perez (2011). "Notch, IL-1 and leptin crosstalk outcome (NILCO) is critical for leptin-induced proliferation, migration and VEGF/VEGFR-2 expression in breast cancer." PLoS One 6(6): e21467.

Guo, S., M. Liu and R. R. Gonzalez-Perez (2011). "Role of Notch and its oncogenic signaling crosstalk in breast cancer." Biochim Biophys Acta 1815(21193018): 197-213.

Harper, S. J. and D. O. Bates (2008). "VEGF-A splicing: the key to anti-angiogenic therapeutics?" Nat Rev Cancer 8: 880-887.

Harrela, M., H. Koistinen, J. Kaprio, M. Lehtovirta, J. Tuomilehto, J. Eriksson, L. Toivanen, M. Koskenvuo, P. Leinonen, R. Koistinen and M. Seppala (1996). "Genetic and environmental components of interindividual variation in circulating levels of IGF-I, IGF-II, IGFBP-1, and IGFBP-3." J Clin Invest 98(11): 2612-2615.

Heo, M., R. L. Leibel, K. R. Fontaine, B. B. Boyer, W. K. Chung, M. Koulu, M. K. Karvonen, U. Pesonen, A. Rissanen, M. Laakso, M. I. J. Uusitupa, Y. Chagnon, C. Bouchard, P. A. Donohoue, T. L. Burns, A. R. Shuldiner, K. Silver, R. E. Andersen, O. Pedersen, S. Echwald, T. I. A. Sorensen, P. Behn, M. A. Permutt, K. B. Jacobs, R. C. Elston, D. J. Hoffman, E. Gropp and D. B. Allison (2002). "A meta-analytic investigation of linkage and association of common leptin receptor (LEPR) polymorphisms with body mass index and waist circumference." Int J Obes Relat Metab Disord 26: 640-646.

http://www.ncbi.nlm.nih.gov/SNP/snp_summary.cgi.

Hu, E., P. Liang and B. M. Spiegelman (1996). "AdipoQ is a novel adipose-specific gene dysregulated in obesity." J Biol Chem 271(18): 10697-10703.

Human Genome Project.

Isse, N., Y. Ogawa, N. Tamura, H. Masuzaki, K. Mori, T. Okazaki, N. Satoh, M. Shigemoto, Y. Yoshimasa, S. Nishi and et al. (1995). "Structural organization and chromosomal assignment of the human obese gene." J Biol Chem 270(46): 27728-27733.

Jacobs, E. J., H. S. Feigelson, E. B. Bain, K. A. Brady, C. Rodriguez, V. L. Stevens, A. V. Patel, M. J. Thun and E. E. Calle (2006). "Polymorphisms in the vascular endothelial growth factor gene and breast cancer in the Cancer Prevention Study II cohort." Breast Cancer Res 8.

Jin, Q., K. Hemminki, K. Enquist, P. Lenner, E. Grzybowska, R. Klaes, R. Henriksson, B. Chen, J. Pamula, W. Pekala, H. Zientek, J. Rogozinska-Szczepka, B. Utracka-Hutka, G. Hallmans and A. Forsti (2005). "Vascular endothelial growth factor polymorphisms in relation to breast cancer development and prognosis." Clin Cancer Res 11: 3647-3653.

Jurado, J., J.A. Ortega, P.Iglesias, J.L. Garcia-Puche, J. Belon (2009). "Vascular endothelial growth factor receptor-2 (VEGFr-2) genetic polymorphisms as predictors to antiangiogenic therapy." Clin Oncol 27 (suppl): abstr e14561.

Kaklamani, V. G., M. Sadim, A. Hsi, K. Offit, C. Oddoux, H. Ostrer, H. Ahsan, B. Pasche and C. Mantzoros (2008). "Variants of the adiponectin and adiponectin receptor 1 genes and breast cancer risk." Cancer Res 68(9): 3178-3184.

Kataoka, N., Q. Cai, W. Wen, X.-O. Shu, F. Jin, Y.-T. Gao and W. Zheng (2006). "Population-based case-control study of VEGF gene polymorphisms and breast cancer risk among Chinese women." Cancer Epidemiol Biomarkers Prev 15: 1148-1152.

Kharroubi, I., J. Rasschaert, D. L. Eizirik and M. Cnop (2003). "Expression of adiponectin receptors in pancreatic beta cells." Biochem Biophys Res Commun 312: 1118-1122.

Kim, H. O., Y. H. Jo, J. Lee, S. S. Lee and K.-S. Yoon (2008). "The C1772T genetic polymorphism in human HIF-1alpha gene associates with expression of HIF-1alpha protein in breast cancer." Oncol Rep 20: 1181-1187.

Korkaya, H., S. Liu and M. S. Wicha (2011). "Breast cancer stem cells, cytokine networks, and the tumor microenvironment." J Clin Invest 121: 3804-3809.

Langsenlehner, U., G. Wolf, T. Langsenlehner, A. Gerger, G. Hofmann, H. Clar, T. C. Wascher, B. Paulweber, H. Samonigg, P. Krippl and W. Renner (2008). "Genetic polymorphisms in the vascular endothelial growth factor gene and breast cancer risk. The Austrian "tumor of breast tissue: incidence, genetics, and environmental risk factors" study." Breast Cancer Res Treat 109: 297-304.

Lee, S. H., E. G. Jeong and N. J. Yoo (2007). "Mutational analysis of NOTCH1, 2, 3 and 4 genes in common solid cancers and acute leukemias." APMIS 115(12): 1357-1363.

Lu, J., Q. Wei, M. L. Bondy, T.-K. Yu, D. Li, A. Brewster, S. Shete, A. Sahin, F. Meric-Bernstam and L.-E. Wang (2006). "Promoter polymorphism (–786t>C) in the endothelial nitric oxide synthase gene is associated with risk of sporadic breast cancer in non-Hispanic white women age younger than 55 years." Cancer 107: 2245-2253.

Luikart, S. D. (1988). "Effects of extracellular matrix on the malignant phenotype." Yale J Biol Med 61: 35-38.

Maeda, N., M. Takahashi, T. Funahashi, S. Kihara, H. Nishizawa, K. Kishida, H. Nagaretani, M. Matsuda, R. Komuro, N. Ouchi, H. Kuriyama, K. Hotta, T. Nakamura, I. Shimomura and Y. Matsuzawa (2001). "PPARgamma ligands increase expression and plasma concentrations of adiponectin, an adipose-derived protein." Diabetes 50(9): 2094-2099.

Martin, D. N., B. J. Boersma, M. Yi, M. Reimers, T. M. Howe, H. G. Yfantis, Y. C. Tsai, E. H. Williams, D. H. Lee, R. M. Stephens, A. M. Weissman and S. Ambs (2009). "Differences in the tumor microenvironment between African-American and European-American breast cancer patients." PLoS One 4.

Nakano, Y., T. Tobe, N. H. Choi-Miura, T. Mazda and M. Tomita (1996). "Isolation and characterization of GBP28, a novel gelatin-binding protein purified from human plasma." J Biochem 120(4): 803-812.

National Cancer Institute (2011). "Surveillance, Epidemiology, and End Results Program (SEER) Seer Statistic Database: Mortality (2000-2008)." (October).

Newman, L. A., J. Mason, D. Cote, Y. Vin, K. Carolin, D. Bouwman and G. A. Colditz (2002). "African-American ethnicity, socioeconomic status, and breast cancer survival: a meta-analysis of 14 studies involving over 10,000 African-American and 40,000 White American patients with carcinoma of the breast." Cancer 94: 2844-2854.

Nowak, D. G., E. M. Amin, E. S. Rennel, C. Hoareau-Aveilla, M. Gammons, G. Damodaran, M. Hagiwara, S. J. Harper, J. Woolard, M. R. Ladomery and D. O. Bates (2010). "Regulation of vascular endothelial growth factor (VEGF) splicing from pro-angiogenic to anti-angiogenic isoforms: a novel therapeutic strategy for angiogenesis." J Biol Chem 285(8): 5532-5540.

Paracchini, V., P. Pedotti and E. Taioli (2005). "Genetics of leptin and obesity: a HuGE review." Am J Epidemiol 162: 101-114.

Perou, C. M., T. Sorlie, M. B. Eisen, M. van de Rijn, S. S. Jeffrey, C. A. Rees, J. R. Pollack, D. T. Ross, H. Johnsen, L. A. Akslen, O. Fluge, A. Pergamenschikov, C. Williams, S. X. Zhu, P. E. Lonning, A. L. Borresen-Dale, P. O. Brown and D. Botstein (2000). "Molecular portraits of human breast tumours." Nature 406: 747-752.

Price, S. J., D. R. Greaves and H. Watkins (2001). "Identification of novel, functional genetic variants in the human matrix metalloproteinase-2 gene: role of Sp1 in allele-specific transcriptional regulation." J Biol Chem 276: 7549-7558.

Przybylowska, K., A. Kluczna, M. Zadrozny, T. Krawczyk, A. Kulig, J. Rykala, A. Kolacinska, Z. Morawiec, J. Drzewoski and J. Blasiak (2006). "Polymorphisms of the promoter regions of matrix metalloproteinases genes MMP-1 and MMP-9 in breast cancer." Breast Cancer Res Treat 95: 65-72.

Ribeiro, A. L., J. Correia and V. Ribeiro (2009). "Ethnic variability of HIF-1alpha polymorphisms." Cancer Biomark 5(6): 273-277.

Roehe, A. V., A. P. Frazzon, G. Agnes, A. P. Damin, A. A. Hartman and M. S. Graudenz (2007). "Detection of polymorphisms in the promoters of matrix metalloproteinases 2 and 9 genes in breast cancer in South Brazil: preliminary results." Breast Cancer Res Treat 102(1): 123-124.

Ruggiero, D., C. Dalmasso, T. Nutile, R. Sorice, L. Dionisi, M. Aversano, P. Broet, A. L. Leutenegger, C. Bourgain and M. Ciullo (2011). "Genetics of VEGF serum variation in human isolated populations of cilento: importance of VEGF polymorphisms." PLoS One 6(2): e16982.

Sait, S. N., M. Dougher-Vermazen, T. B. Shows and B. I. Terman (1995). "The kinase insert domain receptor gene (KDR) has been relocated to chromosome 4q11→q12." Cytogenet Cell Genet 70: 145-146.

Scherer, P. E., S. Williams, M. Fogliano, G. Baldini and H. F. Lodish (1995). "A novel serum protein similar to C1q, produced exclusively in adipocytes." J Biol Chem 270(45): 26746-26749.

Schneider, B. P., M. Radovich, G. W. Sledge, J. D. Robarge, L. Li, A. M. Storniolo, S. Lemler, A. T. Nguyen, B. A. Hancock, M. Stout, T. Skaar and D. A. Flockhart (2008a). "Association of polymorphisms of angiogenesis genes with breast cancer." Breast Cancer Res Treat 111(1): 157-163.

Schneider, B. P., M. Wang, M. Radovich, G. W. Sledge, S. Badve, A. Thor, D. A. Flockhart, B. Hancock, N. Davidson, J. Gralow, M. Dickler, E. A. Perez, M. Cobleigh, T. Shenkier, S. Edgerton and K. D. Miller (2008b). "Association of vascular endothelial growth factor and vascular endothelial growth factor receptor-2 genetic polymorphisms with outcome in a trial of paclitaxel compared with paclitaxel plus bevacizumab in advanced breast cancer: ECOG 2100." J Clin Oncol 26(28): 4672-4678.

Semenza, G. L. (2003). "Targeting HIF-1 for cancer therapy." Nat Rev Cancer 3: 721-732.

Snoussi, K., A. D. Strosberg, N. Bouaouina, S. Ben Ahmed, A. N. Helal and L. Chouchane (2006). "Leptin and leptin receptor polymorphisms are associated with increased risk and poor prognosis of breast carcinoma." BMC Cancer 6: 38.

Sorlie, T., C. M. Perou, R. Tibshirani, T. Aas, S. Geisler, H. Johnsen, T. Hastie, M. B. Eisen, M. van de Rijn, S. S. Jeffrey, T. Thorsen, H. Quist, J. C. Matese, P. O. Brown, D. Botstein, P. E. Lonning and A. L. Borresen-Dale (2001). "Gene expression patterns of breast carcinomas distinguish tumor subclasses with clinical implications." Proc Natl Acad Sci USA 98: 10869-10874.

Sorlie, T., R. Tibshirani, J. Parker, T. Hastie, J. S. Marron, A. Nobel, S. Deng, H. Johnsen, R. Pesich, S. Geisler, J. Demeter, C. M. Perou, P. E. Lonning, P. O. Brown, A.-L. Borresen-Dale and D. Botstein (2003). "Repeated observation of breast tumor subtypes in independent gene expression data sets." Proc Natl Acad Sci USA 100: 8418-8423.

Stevens, A., J. Soden, P. E. Brenchley, S. Ralph and D. W. Ray (2003). "Haplotype analysis of the polymorphic human vascular endothelial growth factor gene promoter." Cancer Res 63: 812-816.

Syvanen, A. C. (2001). "Accessing genetic variation: genotyping single nucleotide polymorphisms." Nat Rev Genet 2: 930-942.

Takahata, C., Y. Miyoshi, N. Irahara, T. Taguchi, Y. Tamaki and S. Noguchi (2007). "Demonstration of adiponectin receptors 1 and 2 mRNA expression in human breast cancer cells." Cancer Lett 250: 229-236.

Tanimoto, K., K. Yoshiga, H. Eguchi, M. Kaneyasu, K. Ukon, T. Kumazaki, N. Oue, W. Yasui, K. Imai, K. Nakachi, L. Poellinger and M. Nishiyama (2003). "Hypoxia-inducible factor-1alpha polymorphisms associated with enhanced transactivation capacity, implying clinical significance." Carcinogenesis 24(11): 1779-1783.

Teras, L. R., M. Goodman, A. V. Patel, M. Bouzyk, W. Tang, W. R. Diver and H. S. Feigelson (2009). "No association between polymorphisms in LEP, LEPR, ADIPOQ, ADIPOR1, or ADIPOR2 and postmenopausal breast cancer risk." Cancer Epidemiol Biomarkers Prev 18: 2553-2557.

http://en.wikipedia.org/w/index.php?title=The_Cancer_Genome_ Atlas&printable=yes.

Viengchareun, S., M. C. Zennaro, L. Pascual-Le Tallec and M. Lombes (2002). "Brown adipocytes are novel sites of expression and regulation of adiponectin and resistin." FEBS Lett 532(3): 345-350.

Wagner, K., K. Hemminki and A. Forsti (2007). "The GH1/IGF-1 axis polymorphisms and their impact on breast cancer development." Breast Cancer Res Treat 104: 233-248.

Waki, H., T. Yamauchi, J. Kamon, S. Kita, Y. Ito, Y. Hada, S. Uchida, A. Tsuchida, S. Takekawa and T. Kadowaki (2005). "Generation of globular fragment of adiponectin by leukocyte elastase secreted by monocytic cell line THP-1." Endocrinology 146(2): 790-796.

Wang, K., L. Liu, Z.-M. Zhu, J.-H. Shao and L. Xin (2011). "Five polymorphisms of vascular endothelial growth factor (VEGF) and risk of breast cancer: a meta-analysis involving 16,703 individuals." Cytokine 56: 167-173.

Yamauchi, T., J. Kamon, Y. Ito, A. Tsuchida, T. Yokomizo, S. Kita, T. Sugiyama, M. Miyagishi, K. Hara, M. Tsunoda, K. Murakami, T. Ohteki, S. Uchida, S. Takekawa, H. Waki, N. H. Tsuno, Y. Shibata, Y. Terauchi, P. Froguel, K. Tobe, S. Koyasu, K. Taira, T. Kitamura, T. Shimizu, R. Nagai and T. Kadowaki (2003). "Cloning of adiponectin receptors that mediate antidiabetic metabolic effects." Nature 423(6941): 762-769.

Zhou, Y., C. Yu, X. Miao, W. Tan, G. Liang, P. Xiong, T. Sun and D. Lin (2004). "Substantial reduction in risk of breast cancer associated with genetic polymorphisms in the promoters of the matrix metalloproteinase-2 and tissue inhibitor of metalloproteinase-2 genes." Carcinogenesis 25: 399-404.

Chapter 12

Obesity, Diabetes and Metabolic Syndrome Impact on Tumor Angiogenesis

João Incio[1] and Raquel Soares[2]*

[1]Edwin L. Steele Laboratory of Tumor Biology, Dept of Radiation Oncology
Massachusetts General Hospital, Harvard Medical School
100 Blossom St, Cox 7, Boston, MA 02114 USA. Email: jincio@steele.mgh.harvard.edu

[2]Dept of Biochemistry (U38-FCT), Faculty of Medicine, University of Porto
Al Prof Hernâni Monteiro, 4200-319, Portugal. Email: raqsoa@med.up.pt

CHAPTER OUTLINE

- ▶ Introduction
- ▶ Association between systemic metabolic disorders and cancer
- ▶ Metabolic disorders and tumor angiogenesis
- ▶ Metabolic disorders may modulate response to anti-angiogenic therapies
- ▶ From obesity to cancer and back
- ▶ Therapeutic strategies to target tumor angiogenesis in metabolic disorders
- ▶ Conclusions and perspectives
- ▶ References

ABSTRACT

In the last few decades, the world has been challenged with a cluster of metabolic disturbances that induce susceptibility to cancer and cardiovascular diseases, two major pathological conditions that contributes to the increasing morbidity and mortality rates in the 21st century. Among these disturbances are obesity, diabetes and metabolic syndrome. Metabolic syndrome co-aggregates established cardiovascular risk factors, including insulin resistance, hypertension, dislipidemia and central obesity itself. Interestingly, adipose tissue is no longer considered a merely energy storage tissue. Rather, the release of distinct hormones, as well as of inflammatory cytokines,

*Correspondence Author: Dept of Biochemistry (U38-FCT), Faculty of Medicine, University of Porto, Al Prof Hernâni Monteiro, 4200-319, Portugal. Email: raqsoa@med.up.pt

rendered it an established endocrine and inflammatory organ with a potential to modulate angiogenesis. Accordingly, obesity, diabetes and metabolic syndrome are characterized by distinct features that enhance angiogenesis, such as hypoxia, low grade inflammation, oxidative stress, hormone imbalance or hyperglycemia, which are also key players in cancer development and progression. Corroborating these findings, metabolic disorders have been associated with increased incidence and poor outcome of several types of neoplasia, including the highly prevalent breast, prostate and colorectal cancer. Therefore, angiogenesis is very likely a key player in this association between these metabolic conditions and cancer.

Given the epidemic character of obesity, diabetes and metabolic syndrome, which are already considered to be major public health threats, and elucidating the angiogenic mechanisms implicated in these metabolic disorders predisposing and promoting cancer, becomes of paramount importance. The present chapter focuses on the pro-angiogenic environment that is primarily developed in these metabolic disorders, and its relevance to tumor progression.

Key words: Obesity, metabolic syndrome, diabetes, angiogenesis, adipokines, cancer, inflammation

1   Introduction

The metabolic syndrome (MS) encompasses a cluster of risk factors for diabetes, cardiovascular disease, which also associates with the development of certain types of cancer. Altogether, these pathological conditions are major causes of morbidity and mortality in the western world. The ethiopathogenesis of MS is complex, comprising distinct factors. Increasing evidence indicates that angiogenesis plays important roles in the pathogenesis of the MS. In certain conditions, angiogenesis is excessive, whereas in other pathological situations there is an impairment of this process. A huge amount of secreted growth factors and cytokines are known to play a role in MS. Nevertheless, the completely understanding on the association of these factors and the development and progression of MS needs more investigation. The current chapter focuses on the impact of various MS risk factors, among them obesity and diabetes play relevant roles. One of the processes imbalanced by MS condition is the vascular component, which provides tumour growth advantage. The current chapter offers an overview of tumor angiogenesis and MS relationships. We present data supporting the need to develop approach towards new preventive and therapeutic strategies in cancer.

2   Association between Systemic Metabolic Disorders and Cancer

2.1   Metabolic Disorders

Obesity has become a major worldwide public health problem, and its incidence is increasing at an alarming rate (Monteiro et al. 2009). The prevalence of obese and overweight people in the United States has been rising continuously since the late 1970's, with over 30% of the population being obese and 60% overweight in year 2000 (Flegal et al. 2002). Obesity, MS and type 2 diabetes (T2D) each share a common characteristic of insulin resistance. It has already been established that obesity plays a major role in the onset and progression in MS and T2D. Therefore, the pandemic of obesity has additional detrimental effects and strong probability of increasing the incidence of insulin resistance. (Zimmet et al. 1997; Fox et al. 2006). MS is a complex group of biochemical abnormalities and associated clinical conditions, and is characterized by abdominal obesity, impaired glucose tolerance, dyslipidemia, and high blood pressure (Vona-Davis et al. 2007). T2D is a condition normally associated with obesity, where peripheral resistance to insulin leads to glucose intolerance. From 1980 through 2004 (Zimmet et al. 1997; Fox et al. 2006) the number of Americans with diabetes increased from 5.8 million to 14.7 million (CDC 2007). Currently, obesity, T2D and the metabolic syndrome are more prevalent in developed countries as oppose to the less developed countries. Obesity, T2D and MS rates are higher in developed populations where a high intake of refined carbohydrates and saturated fats are more prevalent, although the rate of developing countries are increasing rapidly due to the adoption of many of the lifestyle characteristics of the Western world (Xue and Michels 2007).

Obesity and its related conditions are associated with the two most prevalent and deadly diseases in the Western world, as each displays high morbidity and mortality rates, namely cancer and cardiovascular diseases. Obese patients have substantially decreased life expectancies, with obesity-associated diseases accounting for 400,000 deaths per year in the US, triggering the largest growth in mortality over the past decade, and an explosion of medical costs (Peeters et al. 2003; Mokdad et al. 2004). In fact, obesity is a direct cause for the decrease in life expectancy of Americans for the first time in generations (Olshansky et al. 2005).

2.2 From Metabolic Disorders to Cancer

The focus of this chapter is the role that obesity and related conditions play in the development and progression of cancer. The high incidence of cancer in North America has been correlated to a major environmental influence consisting of a "western lifestyle", which is mainly attributed to a positive energy balance, a combination of excessive dietary energy intake, and lack of exercise. As a result, these factors eventually contribute and promotes to the complications manifested through MS and/or obesity and T2D. Cancer thrives best in a pro-tumor environment. This is the case of increased cancer cell proliferation via a systemic metabolic derangement, i.e., obesity. It is not surprising, therefore, that adiposity is associated with increased incidence of colon, breast, kidney, esophagus, endometrial, gallbladder, and other forms of cancer (Calle and Kaaks 2004a). A recent study found that women who gained 55 pounds or more after age 18 had almost 50% greater risk of breast cancer after menopause, compared to those who maintained their weight, making obesity the most important risk and prognostic factor for post-menopausal breast cancer (Eliassen et al. 2006). Furthermore, a number of large-scale studies have demonstrated that obesity leads to an increase in not only incidence, but also cancer-related mortality. In fact, the American Cancer Society estimated that in the United States up to 14 percent of all deaths from cancer in men and 20 percent of those in women could be caused by overweight and obesity (Calle et al. 2003). Moreover, obesity appears to be a negative prognostic factor for several cancer related events, including disease free interval, overall survival, cancer outcome, and an independent predictive variable of cancer stage (Majed et al. 2008; Maccio et al. 2010). Multivariate analysis supports that obesity is an independent prognostic factor for obese patients who present more advanced tumors at diagnosis (Eliassen et al. 2006). Obesity was also shown to be associated with metastasis recurrence, in particular with bone, visceral and lymph node metastatic involvement, suggesting that it may potentiate the metastatic spread of tumors (Daling et al. 2001).

The correlation between several types of cancer with obesity and T2D has also been recaptured in animal models (Gordon et al. 2008; Nunez et al. 2008), but the mechanistic causes of this association have yet to be determined. In a spontaneous model of breast cancer, mammary tumor incidence and weight were increased in obese vs. normal weighted mouse, and the latency of development of mammary tumors decreased (Dogan et al. 2007; Khalid et al. 2009). Furthermore, experiments using several xenograph models have shown that a high fat diet, which leads to an obese or diabetic phenotype, induces higher rate of primary tumor growth and metastasis (Rose et al. 1991; Gordon et al. 2008). In a particular study, changes in the hepatic microenvironment associated with obesity sustained the presence of tumor cells in the liver, and increased the incidence of hepatic metastases after intrasplenic/portal inoculation of colon carcinoma cells (Wu et al. 2010).

Angiogenesis has been suggested among the various pathophysiological mechanisms to explain the link between obesity, T2D and MS with cancer. Therefore, the dysfunctional and pro-angiogenic adipose tissue may be an unifying and underlying factor in these relationships.

3 Metabolic Disorders and Tumor Angiogenesis

One of the main mechanisms that lead to tumor initiation and progression is the stimulation of angiogenesis – the formation of new vessels from pre-existing ones. It was previously shown in different models of carcinogenesis that angiogenesis is an important process even during early neoplastic transformation, with pre-neoplastic lesions presenting increased vessel density, dilation, tortuosity and hyperpermeability (Hagendoorn et al. 2006; Kim et al. 2010). Furthermore, new vessels allow the established tumor to grow, and also guarantee a good supply of nutrients and oxygen from the circulation to support the rapidly dividing cells and a way to export metabolic products. These processes also enable metastasis (Costa et al. 2004; Incio et al. 2006; Costa et al. 2007; Negrao et al. 2007). In the clinical setting, high level of angiogenic activity in tumors can predict a poor prognosis (Uzzan et al. 2004). Angiogenesis is thus critical for tumor development and progression.

Recent data suggest that obesity correlates with increased angiogenesis in established breast tumors, which is assumed to result from increased systemic and local levels of pro-angiogenic molecules (Gu et al. 2011). However, no study has established causality, nor dissected the mechanisms underlying this association. Obesity and angiogenesis are known to be interdependent processes (Fukumura et al. 2003; Tam et al. 2009). However, in cancer, it is unclear what particular obesity-associated factors play a major role in regulating the angiogenic process in pre-neoplastic and neoplastic lesions. Macrophages express leptin receptor. Therefore, it has been suggested that leptin could attract macrophages into breast adipose tissue leading to the development of a pro-angiogenic tumor microenvironment (Rose et al. 2004). However, causality between these events has not been established yet.

In this chapter, we will explore the angiogenic process as the link between the high incidence metabolic disorders and cancer (Fig. 1)

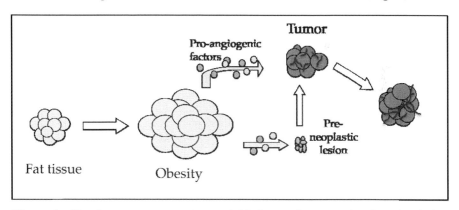

Figure 1 Obesity may promote tumor initiation and progression by stimulating angiogenesis.

Color image of this figure appears in the color plate section at the end of the book.

3.1   Adipose Tissue as an Angiogenic Promoter

Contrary to what was thought for many years, adipose tissue is not a passive deposit of triglycerides, but rather an active tissue with many biological and endocrine functions (Monteiro et al. 2009). It was after the discovery of leptin that the adipose tissue was formally accepted and recognized as a new endocrine organ that releases cytokine-like polypeptides responsible for widespread biological effects (adipocytokines or adipokines) (Monteiro et al. 2009). The altered adipocytokine and growth factor profile in centrally obese patients is believed to be an important contributing factor to the pathogenesis of many co-morbidities associated with obesity, including cancer (Lysaght et al. 2011). Obesity, particularly central obesity, results in the establishment of a state of chronic systemic low-grade inflammation associated with pro-tumorigenic and pro-angiogenic factors secreted by the adipose tissue in proportion to the tissue mass (Costa et al. 2007).

The importance of angiogenesis for the growth of solid tumors as well as for the development of obesity is well recognized (Costa et al. 2007). The VEGF family and the major promoters of angiogenesis, are known to play a crucial role in adipose tissue expansion which occurs in obesity; hypoxia is also a powerful stimulus to its production (Fukumura et al. 2003; Tam et al. 2009). Adipose tissue has a normal level of oxygenation in lean animals, but becomes hypoxic during obesity (Hosogai et al. 2007; Rausch et al. 2008), presumably because as fat tissue expands in volume it exceeds the oxygenation capacity of the existing local vasculature (Bouloumie et al. 2002). Hypoxia has profound effects on many different aspects of adipose tissue biology. Moreover, hypoxia and angiogenesis are crucial for tumor growth, and angiogenic factors and hypoxia-induced cytokines derived from adipocytes are actively involved in such processes (Park et al. 2011). Adipocytes in culture respond to hypoxia by increasing the production of pro-angiogenic and inflammatory molecules (Lolmede et al. 2003; Wang et al. 2007). In fact, proliferating adipocytes present with a deregulated production of a wide variety of pro-angiogenic molecules besides VEGF, including leptin, hepatocyte growth factor (HGF), insulin-like growth factor I (IGF-I), tumor necrosis factor α (TNF-α), angiopoietin-2, transforming growth factor β (TGF-β) interleukin (IL)-6, IL-8, IL1-α and tissue factor (TF) (Catalan et al. 2011; Lysaght et al. 2011; Park et al. 2011). Highly hypoxic regions have been identified at the tips of epididymal fat tissue in mice, and these regions are also characterized by increased expression of a host of pro-angiogenic molecules, dense vascular networks, and substantial macrophage infiltration (Cho et al. 2007). Of note, inflammatory cells infiltrating the adipose tissue and adipose stromal cells contribute significantly to VEGF production (Cho et al. 2007). When conditioned media (CM) from visceral adipose tissue was added to esophageal and colorectal tumor cells *in vitro*, cells proliferated more if the source of the CM was from obese patients. In addition, neutralizing VEGF in the CM significantly decreased tumor cell proliferation. These findings suggest that in addition to a potential local effect in the omentum, adipose tissue accumulation in the visceral cavity may regulate cancer development and growth at distant sites. Furthermore, whenever the adipose tissue is a substantial part of the tumor microenvironment, as in breast cancer, adipocyte abundance may also influence tumor growth.

In preclinical animal models, mice fed a Western-type or high-fat diet displayed increased VEGF in circulation, visceral fat and also in implanted tumors, together with increased tumor angiogenesis (Llaverias et al. 2010; Gu et al. 2011). Furthermore, in one study the increase in the pro-angiogenic VEGF and monocyte chemoattractant protein (MCP)-1 in tumors of obese mice associated with HIF-1 overexpression (Kimura and Sumiyoshi 2007). Elevated serum levels of pro-angiogenic VEGF, IGF-1, leptin and several other adipokines, and reduced levels of the anti-angiogenic adiponectin are found in overweight or obese patients (Lysaght et al. 2011). In addition, several molecules secreted by the adipose tissue, such as VEGF, leptin, and TNFα have been related to the risk of developing cancer. In contrast, other molecules, such as adiponectin (which is lower in obese persons) seem to have a protective role in carcinogenesis (see section 3.1.1.2., Costa et al. 2007). This suggests that the elevated levels of pro-angiogenic molecules in overweight and obese patients may contribute to the increased risk of developing cancer in obese patients and of having poor prognosis when it is diagnosed. Nevertheless, despite all the observations from clinical studies and pre-clinical models that support a role of angiogenesis in bridging metabolic disorders with cancer, research is still ongoing to establish causality.

3.1.1 Role of Adipose Tissue Derived Hormones

Obesity, T2D and the MS each has in common, among other abnormalities, an increased production of leptin and a decreased production of adiponectin by adipose tissue (Cao et al. 2001). These changes in plasma leptin and adiponectin have been associated with an increase in cancer risk and with more aggressive tumors. Leptin stimulates, and adiponectin inhibits, tumor cell proliferation and the microvessel angiogenesis, which is essential for cancer development and progression (Cao et al. 2001; Vona-Davis et al. 2007).

3.1.1.1 Leptin

Circulating levels of pro-inflammatory adipokines such as leptin have been shown to correlate with visceral adipose tissue. The principal sources of leptin are the pre-adipocytes and adipocytes. The plasma levels of leptin, a hormone released by adipose tissue, are increased in obesity, T2D and MS (Fischer et al. 2002; Rose et al. 2004). Furthermore, data from *in vitro* and *in vivo* studies confirm that leptin not only promotes cellular proliferation, but also exerts direct angiogenic effects, inducing endothelial cell (EC) proliferation and migration (89, 90). Human ECs stimulated with leptin form tubules in three-dimensional collagen gels that are similar to capillaries. In addition, interaction of leptin with its receptor on EC leads to activation of the signal transducer and activator of transcription 3 (Stat3) pathway, which plays a key role in many cellular processes such as cell growth and apoptosis (Sierra-Honigmann et al. 1998). Similar to VEGF-A, leptin induces the formation of fenestrated capillaries, as confirmed by the absence of fenestrations in leptin deficient *ob/ob* mice. Moreover, leptin has a synergistic effect on stimulation by VEGF and FGF-2 (Cao et al. 2001). Besides a direct pro-angiogenic activity, leptin upregulates VEGF expression in EC and cancer cells via activation of the JAK/Stat3 signaling pathway, and induces VEGF-driven tumor angiogenesis in mice (Suganami et al. 2004; Gonzalez et al. 2006).

In addition, leptin is also implicated in inflammation and oxidative stress, two conditions that are associated with angiogenesis (Costa et al. 2007). Through activation of its receptor, leptin induces endothelin-1 and nitric oxide synthase (NOS) in EC, enhancing the expression of endothelial adhesion molecules essential for the angiogenic process (Catalan et al. 2011). All these reports are in agreement with the finding that leptin and its receptor are both overexpressed in high grade breast tumors (Artac and Altundag 2012).

3.1.1.2 Adiponectin

Adiponectin is synthesized and secreted almost exclusively by adipocytes. Adiponectin is the most abundant adipokine and its level is inversely correlated with central obesity, T2D and various components of the MS. The degree of insulin resistance appears to contribute more significantly to levels of adiponectin than the degree of adiposity (Cao et al. 2001; Matsuzawa 2006). Not only has adiponectin been shown to inhibit tumor growth in animal models, but it also acts as an anti-angiogenic and anti-inflammatory cytokine (Cao et al. 2001). Adiponectin inhibits EC migration and proliferation *in vitro* and neo-angiogenesis *in vivo* in the chick chorioallantoic membrane and cornea assays. Adiponectin also reduces angiogenesis in T241 fibrosarcoma cells in mice (Brakenhielm et al. 2004a). However, some reports suggest that adiponectin may also exert pro-angiogenic effects. Adiponectin has been reported to activate adenosine monophosphate kinase in EC, leading to enhanced *in vivo* angiogenesis in murine matrigel plug and rabbit cornea assays, and inhibits apoptosis in HUVECs cultured *in vitro* (Kobayashi et al. 2004; Ouchi et al. 2004). Research is ongoing to clarify the role of adiponectin in the angiogenic process.

3.1.2 Chronic Inflammation

Interestingly, obesity and metabolic syndrome have been considered low-grade inflammatory states, given the fact that the inflammation that accompanies these conditions is not significant to cause massive tissue injury. As mentioned before, highly hypoxic regions in fat tissue in mice are characterized by increased expression of a host of pro-angiogenic molecules and dense vascular networks, but also present with substantial macrophage infiltration (Cho et al. 2007). Two seminal studies published in 2003 showed that bone marrow-derived macrophages infiltrate fat tissue during obesity, and the amount of macrophage infiltration increases linearly with the degree of obesity (Weisberg et al. 2003; Xu et al. 2003). Furthermore, significantly higher proportions of CD8+ T cells and NKT cells have been found in visceral adipose tissue (Lysaght et al. 2011).

These inflammatory cells infiltrating the adipose tissue and other adipose stromal cells can contribute significantly to VEGF production, ultimately promoting angiogenesis (Cho et al. 2007). In addition, adipose tissue macrophages were found to be a major source of cytokine production, responsible for almost all TNFα and significant portions of iNOS and IL-6 expression by fat tissue (Weisberg et al. 2003; Xu et al. 2003). In addition, these pro-inflammatory cytokines may activate neighboring EC to further attract inflammatory cells, enhancing their adhesion and extravasation (Soares and Costa 2009). Moreover, leptin produced by tumor cells and adipocytes may play a role in attracting macrophages into breast adipose tissue, since these cells express leptin receptor, and this could ultimately lead to a pro-angiogenic breast tumor microenvironment (Gruen et al. 2007). One study (Park et al. 2011) proposed that the influence of obesity on tumor progression and survival in colorectal cancer is due to obesity-related inflammation, rather than factors associated with obesity *per se*. The chronic inflammation that characterizes obesity is associated with increased plasma levels of C-Reactive protein (CRP), IL-6 and plasminogen activator inhibitor (Park et al. 2011). CRP has been shown to be strongly angiogenic in both *in vitro* and *in vivo* models (Polverini and Leibovich 1984), but some controversy still exists – in one *in vitro* assay CRP mediated inhibition of angiogenesis (Yang et al. 2005). In conclusion, it seems that inflammation and eventually immune cell infiltration in the adipose tissue surrounding tumors could play a major role in tumor progression.

3.1.3 Insulin Resistance and Hyperglycemia: Angiogenesis Stimulators in Cancer

3.1.3.1 Impact on Normal Vasculature

Metabolic changes seen in obesity, diabetes, and metabolic syndrome induce alterations in multiple vascular beds that result in inflammation around the vessel wall, endothelial dysfunction and ultimately propensity to vascular injury (Burke et al. 2004; Paoletti et al. 2006). Hyperglycemia, insulin resistance, dyslipidemia, hypertension, and advanced glycation end products have all been implicated in the pathogenesis of accelerated arterial disease in patients with diabetes (Creager et al. 2003). Hyperglycemia, which is a common feature in diabetic patients even under treatment, frequently results in glycation of many circulating factors, such as FGF, HGF, PIGF and PDGF (Duraisamy et al. 2001). These events imply that a substantial number of angiogenic factors may be affected in this complex picture. The identification of these factors and their signaling pathways is essential for the development of novel therapeutic strategies. Furthermore, experimentally induced hyperglycemia and hyperinsulinemia can reduce endothelium-dependent vasodilation, which suggests that these metabolic parameters, normally present in obesity, T2D and MS can have direct effects on the arterial wall.

3.1.3.2 Insulin and IGF-1: Tumor Promoters in Obesity and Related Conditions

As mentioned before, the correlation between obesity and cancer is often regarded as a result of increased secretion of pro-tumorigenic adipokines (e.g. Leptin). However, that may not always be the case. Studies in fatless A-Zip/F-1 mice, which have undetectable adipokine levels, but display accelerated tumor formation, suggest that adipokines may not be required for enhanced tumor development (Nunez et al. 2006). The A-Zip/F-1 mice are diabetic and display elevated circulating levels of insulin, insulin-like growth factor-1 (IGF-1), and inflammatory cytokines that are frequently associated with obesity and carcinogenesis. Therefore, pathways associated with insulin resistance and inflammation, rather than adipocyte-derived factors, may represent key prevention or therapeutic targets for disrupting the link between obesity and cancer. IGF-1 receptor (IGF-1R) activation not only leads to increased proliferation and apoptosis evasion of breast cancer cells, but may also promote tumor angiogenesis (Ryan and Goss 2008; Gallagher and LeRoith 2010). The frequent state of insulin resistance present in obesity, T2D and MS commonly present with an increased production of IGF-1 by the liver and a decrease in IGFBP-1 and IGFBP-2 (IGF- binding proteins), leading to increased bioavailable IGF-1 in circulation (Boyd 2003; Nunez et al. 2006). Mice that have low IGF-1 levels are less susceptible to mammary cancer, and IGF-1 supplementation restores cancer susceptibility (Bianchini et al. 2002; Calle and Thun 2004b). Furthermore, in a genetic mouse model with liver specific IGF-1 deficiency, it was concluded that IGF-1 not only increases colon tumor growth, but also increases expression of VEGF, and leads to the development of neovascularization and metastases (Wu et al. 2002). In this same model, IGF-1 was further described as mediating the obesity-induced increase in tumor cell growth and hepatic metastasis. In humans, obesity and related metabolic conditions are associated with increased levels of IGF-1 in circulation, and increased IGF-1 is related to the development of some forms of cancer, including pre- and post-menopausal breast cancer (Murphy et al. 1995; Godsland 2009). In addition, breast cancer patients frequently present with IGF-1R overexpression in tumors, and IGF-1R is known to correlate with disease progression, recurrence and metastasis (Ryan and Goss 2008). As mentioned before, adipose tissue may also mediate tumor growth through local effects. In fact, the mammary gland is surrounded by adipose tissue, and adipose tissue is the second major source of IGF-1 (Ryan and Goss 2008), suggesting that the mammary epithelium of obese women will be exposed to higher levels of IGF-1 than that of lean women. Thus, local breast adipose tissue may be a relevant contributor to breast cancer.

3.1.3.3 Insulin/IGF-1 and Angiogenesis

IGF-1 is a potent angiogenic factor. It stimulates neovascularization by inducing expression and secretion of VEGF by cancer cells via HIF-1-

dependent and -independent pathways, and it was shown in a colon cancer model that blockade of IGF-1R function inhibits growth and angiogenesis (Reinmuth et al. 2002). Vascular components such as endothelial cells (ECs) and pericytes / vascular smooth muscle cells express high levels of Insulin receptor and IGF-1R, potentially becoming exposed to increased insulin and IGF-1 stimulation in the obese setting. IGF-1 is an important factor promoting retinal angiogenesis, and an IGF-1R antagonist was able to suppress retinal neovascularization *in vivo* by inhibiting vascular endothelial growth factor signaling (Smith et al. 1999). Another study demonstrated that both IGF-1 and VEGF, during choroidal neovascularization stimulate SDF-1-induced angiogenesis (Sengupta et al. 2010). On ECs, activation of the IGF-1R stimulates proliferation, migration and tube formation (Delafontaine et al. 2004).

3.2 Pathological Difference between Adipose Tissue Depots

Subcutaneous and visceral adipose tissue depots are pathologically distinct, with visceral adipose tissue being more metabolically active, and often the fat component is associated with cancer incidence and progression (Ibrahim 2010). In one study, human visceral adipose tissue expression of IL-6 and VEGF mRNA was greater than subcutaneous adipose tissue (Fain 2006), and circulating levels of VEGF and pro-inflammatory adipokines, such as leptin, have been shown to correlate with central adiposity as opposed to subcutaneous fat in non-cancer cohorts (Lysaght et al. 2011). In addition, circulating levels of leptin have been reported to correlate solely with omental (visceral) mRNA expression (Moses et al. 2001). This may explain why body mass index (BMI), which takes into account the total weight of the patient, does not always correlate with proangiogenic markers. In one study, neither VEGF-A nor Ang-2 were associated with BMI or serum adiponectin levels, despite the positive correlations observed between serum Ang-2 and serum C-peptide (a stable marker of circulating insulin levels), as well as between VEGF-A and IGF-1 (Volkova et al. 2011). This and other studies are raising awareness that BMI may not be an accurate marker for visceral adiposity, which as mentioned may be where most pro-angiogenic adipokines are produced in the obese setting. In fact, a recent study found that waist circumference, but not BMI, was associated with cancer survival (Haydon et al. 2006).

On the other hand, in an additional study, it was found that morbidly obese patients with colorectal cancer stayed free of disease after therapy for a shorter period of time on average, but no change in overall survival was observed, suggesting that severe obesity, rather than a continuum of BMI, impacts negatively on survival (Meyerhardt et al. 2008). In a study using a syngeneic tumor model, the release of VEGF from subcutaneous fat of postmenopausal female mice was higher than visceral fat, which suggests that increased VEGF expression in subcutaneous fat may also locally promote breast cancer progression in postmenopausal obesity (Gu et al.

2011). Taken together, these data suggests that although the overwhelming majority of studies point to visceral fat, rather than overall fat, as the culprit in promoting tumor growth and angiogenesis, in some cases local production of pro-angiogenic factors by adipose tissue may also contribute to tumor progression.

4 Metabolic Disorders may Modulate Response to Anti-angiogenic Therapies

4.1 VEGF-A And Resistance to Anti-angiogenic Therapy

Adipose tissue, as mentioned before, secretes a wide array of pro-angiogenic molecules and elevated serum VEGF levels have been found in overweight or obese patients. In patients with breast or colorectal cancer, high tumor levels of VEGF were associated with poor prognosis (Protani et al. 2010)). Interestingly, non tumor-bearing obese animals proved resistant to anti-VEGF treatment (as an anti-obesity therapy) drug in the long term (Rupnick et al. 2002). The efficacy of anti-angiogenic agents in the clinical setting has been disappointing, This is underlined by the recent removal of bevacizumab from the restricted group of approved molecularly targeted therapy for breast cancer. Rapid development of resistance to anti-angiogenic therapy occurs at least, in part, due to the switching to alternative pro-angiogenic pathways by the tumor, and to increased mobilization of bone marrow derived myeloid cells (BMDC) to the tumor microenvironment, promoting angiogenesis (Carmeliet and Jain 2011a; Carmeliet and Jain 2011b). There are currently no predictive biomarkers of response to VEGF-targeted therapy, but in patients with metastatic kidney or colon cancer treated with chemotherapy or chemo plus bevacizumab (anti-VEGF antibody), obesity predicted worse outcome only in the group of patients that received bevacizumab (Guiu et al. 2010; Ladoire et al. 2011). It was concluded that patients with high visceral fat might either not benefit from bevacizumab or require a higher dosage. This data could support the hypothesis that a large amount of visceral fat may be associated with high VEGF levels. Therefore, the accumulation of fat could be linked to the resistance to bevacizumab-based regimens found in obese patients. Nevertheless, this may still be the case when only chemotherapy is provided, as shown in a study with breast cancer patients, where high tumor levels of VEGF associated with greater resistance to conventional chemotherapy (Protani et al. 2010). However, a recent study has raised some controversy and questioned this notion that obese patients may be more resistant to anti-angiogenic therapy. In fact, Steffens and colleagues have shown that in metastatic renal cell carcinoma, contrary to what had been previously described, visceral obesity was a positive predictive biomarker when patients received anti-angiogenic therapy (Steffens et al. 2011). Thus, larger studies are necessary to clarify this matter.

4.2 Immune Cells and Resistance to Anti-angiogenic Therapy

Another reason for the failure of anti-agiogenic therapy is believed to be due to infiltration of BMDC. Inherent anti-VEGF refractoriness is associated with infiltration of the tumor tissue by CD11b+Gr1+ myeloid cells, and when anti-VEGF treatment was combined with a mAb that targets myeloid cells, growth of refractory tumors was more effectively inhibited than with anti-VEGF alone (Shojaei et al. 2007). Immune cell infiltration may be one of the mechanisms of obesity induced-tumor progression. Therefore, it is possible that this process could also induce resistance to anti-angiogenic therapies.

5   From Obesity to Cancer and Back

Cancer itself may be promoting a more pro-angiogenic environment particularly in the obese setting. In one recent study, real-time PCR analysis indicated that the expression of HIF1α and VEGFA mRNAs in visceral adipose tissue was significantly higher in colorectal patients compared to control volunteers (Pischon et al. 2006). Furthermore, it was observed pro-angiogenic IGF-1 mRNA followed the same trend as VEGF and HIF1α, whereas expression of IGFBP3 mRNA (IGF-1 regulatory protein), was downregulated in colorectal patients.

6   Therapeutic Strategies to Target Tumor Angiogenesis in Metabolic Disorders

6.1   Targeting Angiogenesis with Metformin

Metformin is the most frequently prescribed drug for T2D, but its potential as a cancer drug has recently emerged. Provocative studies found approximately 50% decrease in incidence of breast cancer. This was parallel to higher pathological complete response rates (three fold) in established tumors among diabetics treated with metformin compared with diabetics under other therapies (Goodwin and Stambolic 2011). It has been shown that metformin has effects in different tumor cell lines, and can act directly on cancer cell metabolism, proliferation and apoptosis via activation of AMPK. Furthermore, it also has systemic effects by lowering the pro-tumorigenic and angiogenic insulin and IGF-1 in circulation, thus reducing the binding of these growth factors to insulin/IGF-1 receptors (IR/IGF-1R) that are almost ubiquitously present, and frequently overexpressed, on human breast cancers (Furundzija et al. 2010). In mouse studies, metformin inhibits tumorigenesis in some models (*MMTV-neu*), and is able to reduce tumor growth most effectively in the context of insulin resistance and high IGF-1 levels typical of obesity (Anisimov et al. 2005). Furthermore, some studies have shown that metformin could inhibit angiogenesis *in vitro*, and has activity *in vivo* on inflammatory angiogenesis using the murine sponge model of angiogenesis, frequently used to study inflammation (Isoda et al. 2006). Metformin was shown to exert inhibitory action on various cell types including vascular (anti-proliferative effect on normal vascular smooth muscle cells) and inflammatory cells, particularly macrophages and neutrophils (Isoda et al. 2006). In this way, metformin could also be considered an antiangiogenic therapy, most effective in the obese setting. However, in recent studies metformin, under certain conditions (e.g. high dose) and in some mouse models, unexpectedly lead to an apparent increase in tumor angiogenesis (Phoenix et al. 2009; Martin et al. 2012). This controversy is being debated among researchers, and thus the utility of metformin as an anti-angiogenic agent must be carefully evaluated. Metformin has long been used widely as a T2D treatment with few adverse effects. It has well known pharmacodynamic and pharmacokinetic profiles. In addition, metformin is an inexpensive drug that can be delivered orally. Therefore, the clinical applicability of metformin as an anti-cancer therapy would be relatively easy and fast compared to other anti-tumor therapeutics. Indeed, the use of metformin could be advantageous over antibodies targeting IGF-1 or its receptor, whose clinical toxicity profiles are not yet established. In fact, clinical studies for the neoadjuvant and adjuvant uses of metformin are ongoing. Moreover, Phase 2 trials using metformin in metastatic setting are being planned (Furundzija et al. 2010).

6.2 Targeting or Not Angiogenesis in Metabolic Disorders?

As mentioned before, during adipose tissue expansion, although adipocytes can in principle induce vascularization through the secretion of angiogenic factors, obese adipose tissue is frequently in a hypoxic state due to insufficient angiogenesis, causing a direct deregulation of adipokine secretion. Thus, therapies that promote angiogenesis are being considered for obesity and it is related to metabolic disorders (Brakenhielm et al. 2004a). However, this poses major therapeutic challenges: angiogenesis may be one missing link between obesity and cancer, thus therapies that promote this process may eventually increase the risk of cancer in obese patients.

7 Conclusions and Perspectives

Considering the epidemic character of obesity, diabetes and metabolic syndrome, elucidating the cancer promoting mechanisms implicated in these metabolic disorders is critical. The present chapter focused on angiogenesis as the potential missing link, and discussed pre-clinical and clinical evidence supporting this mechanism, as well as the therapeutic implications of these findings (Fig. 2).

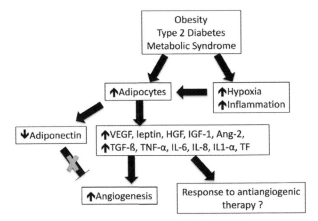

Figure 2 Conditions like Obesity, T2D and MS enlarge the pool of pro-angiogenic growth factors and pro-inflammatory cytokines. These factors stimulate angiogenesis, and eventually modulate response to anti-angiogenic therapies.

REFERENCES

Anisimov, V. N., L. M. Berstein, P. A. Egormin, T. S. Piskunova, I. G. Popovich, M. A. Zabezhinski, I. G. Kovalenko, T. E. Poroshina, A. V. Semenchenko, M. Provinciali, F. Re and C. Franceschi (2005). "Effect of metformin on life span and on the development of spontaneous mammary tumors in HER-2/neu transgenic mice." Exp Gerontol 40(8-9): 685-693.

Artac, M. and K. Altundag (2012). "Leptin and breast cancer: an overview." Med Oncol 29(3): 1510-1514.

Bianchini, F., R. Kaaks and H. Vainio (2002). "Overweight, obesity, and cancer risk." Lancet Oncol 3(9): 565-574.

Bouloumie, A., K. Lolmede, C. Sengenes, J. Galitzky and M. Lafontan (2002). "Angiogenesis in adipose tissue." Ann Endocrinol (Paris) 63(2 Pt 1): 91-95.

Boyd, D. B. (2003). "Insulin and cancer." Integr Cancer Ther 2(4): 315-329.

Brakenhielm, E., R. Cao, B. Gao, B. Angelin, B. Cannon, P. Parini and Y. Cao (2004a). "Angiogenesis inhibitor, TNP-470, prevents diet-induced and genetic obesity in mice." Circ Res 94(12): 1579-1588.

Burke, A. P., F. D. Kolodgie, A. Zieske, D. R. Fowler, D. K. Weber, P. J. Varghese, A. Farb and R. Virmani (2004). "Morphologic findings of coronary atherosclerotic plaques in diabetics: a postmortem study." Arterioscler Thromb Vasc Biol 24(7): 1266-1271.

Calle, E. E. and R. Kaaks (2004a). "Overweight, obesity and cancer: epidemiological evidence and proposed mechanisms." Nat Rev Cancer 4(8): 579-591.

Calle, E. E., C. Rodriguez, K. Walker-Thurmond and M. J. Thun (2003). "Overweight, obesity, and mortality from cancer in a prospectively studied cohort of U.S. adults." N Engl J Med 348(17): 1625-1638.

Calle, E. E. and M. J. Thun (2004b). "Obesity and cancer." Oncogene 23(38): 6365-6378.

Cao, R., E. Brakenhielm, C. Wahlestedt, J. Thyberg and Y. Cao (2001). "Leptin induces vascular permeability and synergistically stimulates angiogenesis with FGF-2 and VEGF." Proc Natl Acad Sci USA 98(11): 6390-6395.

Carmeliet, P. and R. K. Jain (2011a). "Molecular mechanisms and clinical applications of angiogenesis." Nature 473(7347): 298-307.

Carmeliet, P. and R. K. Jain (2011b). "Principles and mechanisms of vessel normalization for cancer and other angiogenic diseases." Nat Rev Drug Discov 10(6): 417-427.

Catalan, V., J. Gomez-Ambrosi, A. Rodriguez, B. Ramirez, C. Silva, F. Rotellar, J. L. Hernandez-Lizoain, J. Baixauli, V. Valenti, F. Pardo, J. Salvador and G. Fruhbeck (2011). "Up-regulation of the novel proinflammatory adipokines lipocalin-2, chitinase-3 like-1 and osteopontin as well as angiogenic-related factors in visceral adipose tissue of patients with colon cancer." J Nutr Biochem 22(7): 634-641.

CDC (2007). "Centers for Disease Control and Prevention, National Diabetes Surveillance System: prevalence of diabetes." Version current 6 October 2005 nternet: http://www.cdc.gov/diabetes/statistics/prev/national/ figpersons.htm

Cho, C. H., Y. J. Koh, J. Han, H. K. Sung, H. Jong Lee, T. Morisada, R. A. Schwendener, R. A. Brekken, G. Kang, Y. Oike, T. S. Choi, T. Suda, O. J. Yoo and G. Y. Koh (2007). "Angiogenic role of LYVE-1-positive macrophages in adipose tissue." Circ Res 100(4): e47-57.

Costa, C., J. Incio and R. Soares (2007). "Angiogenesis and chronic inflammation: cause or consequence?" Angiogenesis 10(3): 149-166.

Costa, C., R. Soares and F. Schmitt (2004). "Angiogenesis: now and then." APMIS 112(7-8): 402-412.

Creager, M. A., T. F. Luscher, F. Cosentino and J. A. Beckman (2003). "Diabetes and vascular disease: pathophysiology, clinical consequences, and medical therapy: Part I." Circulation 108(12): 1527-1532.

Daling, J. R., K. E. Malone, D. R. Doody, L. G. Johnson, J. R. Gralow and P. L. Porter (2001). "Relation of body mass index to tumor markers and survival among young women with invasive ductal breast carcinoma." Cancer 92(4): 720-729.

Delafontaine, P., Y. H. Song and Y. Li (2004). "Expression, regulation, and function of IGF-1, IGF-1R, and IGF-1 binding proteins in blood vessels." Arterioscler Thromb Vasc Biol 24(3): 435-444.

Dogan, S., X. Hu, Y. Zhang, N. J. Maihle, J. P. Grande and M. P. Cleary (2007). "Effects of high-fat diet and/or body weight on mammary tumor leptin and apoptosis signaling pathways in MMTV-TGF-alpha mice." Breast Cancer Res 9(6): R91.

Duraisamy, Y., M. Slevin, N. Smith, J. Bailey, J. Zweit, C. Smith, N. Ahmed and J. Gaffney (2001). "Effect of glycation on basic fibroblast growth factor induced angiogenesis and activation of associated signal transduction pathways in vascular endothelial cells: possible relevance to wound healing in diabetes." Angiogenesis 4(4): 277-288.

Eliassen, A. H., G. A. Colditz, B. Rosner, W. C. Willett and S. E. Hankinson (2006). "Adult weight change and risk of postmenopausal breast cancer." JAMA 296(2): 193-201.

Fain, J. N. (2006). "Release of interleukins and other inflammatory cytokines by human adipose tissue is enhanced in obesity and primarily due to the nonfat cells." Vitam Horm 74: 443-477.

Fischer, S., M. Hanefeld, S. M. Haffner, C. Fusch, U. Schwanebeck, C. Kohler, K. Fucker and U. Julius (2002). "Insulin-resistant patients with type 2 diabetes mellitus have higher serum leptin levels independently of body fat mass." Acta Diabetol 39(3): 105-110.

Flegal, K. M., M. D. Carroll, C. L. Ogden and C. L. Johnson (2002). "Prevalence and trends in obesity among US adults, 1999-2000." JAMA 288(14): 1723-1727.

Fox, C. S., M. J. Pencina, J. B. Meigs, R. S. Vasan, Y. S. Levitzky and R. B. D'Agostino, Sr. (2006). "Trends in the incidence of type 2 diabetes mellitus from the 1970s to the 1990s: the Framingham Heart Study." Circulation 113(25): 2914-2918.

Fukumura, D., A. Ushiyama, D. G. Duda, L. Xu, J. Tam, V. Krishna, K. Chatterjee, I. Garkavtsev and R. K. Jain (2003). "Paracrine regulation of angiogenesis and adipocyte differentiation during in vivo adipogenesis." Circ Res 93(9): e88-97.

Furundzija, V., J. Fritzsche, J. Kaufmann, H. Meyborg, E. Fleck, K. Kappert and P. Stawowy (2010). "IGF-1 increases macrophage motility via PKC/p38-dependent alphavbeta3-integrin inside-out signaling." Biochem Biophys Res Commun 394(3): 786-791.

Gallagher, E. J. and D. LeRoith (2010). "The proliferating role of insulin and insulin-like growth factors in cancer." Trends Endocrinol Metab 21(10): 610-618.

Godsland, I. F. (2009). "Insulin resistance and hyperinsulinaemia in the development and progression of cancer." Clin Sci (Lond) 118(5): 315-332.

Gonzalez, R. R., S. Cherfils, M. Escobar, J. H. Yoo, C. Carino, A. K. Styer, B. T. Sullivan, H. Sakamoto, A. Olawaiye, T. Serikawa, M. P. Lynch and B. R. Rueda (2006). "Leptin signaling promotes the growth of mammary tumors and increases the expression of vascular endothelial growth factor (VEGF) and its receptor type two (VEGF-R2)." J Biol Chem 281(36): 26320-26328.

Goodwin, P. J. and V. Stambolic (2011). "Obesity and insulin resistance in breast cancer - Chemoprevention strategies with a focus on metformin." Breast 20 Suppl 3: S31-35.

Gordon, R. R., K. W. Hunter, P. Sorensen and D. Pomp (2008). "Genotype X diet interactions in mice predisposed to mammary cancer. I. Body weight and fat." Mamm Genome 19(3): 163-178.

Gruen, M. L., M. Hao, D. W. Piston and A. H. Hasty (2007). "Leptin requires canonical migratory signaling pathways for induction of monocyte and macrophage chemotaxis." Am J Physiol Cell Physiol 293(5): C1481-1488.

Gu, J. W., E. Young, S. G. Patterson, K. L. Makey, J. Wells, M. Huang, K. B. Tucker and L. Miele (2011). "Postmenopausal obesity promotes tumor angiogenesis and breast cancer progression in mice." Cancer Biol Ther 11(10): 910-917.

Guiu, B., J. M. Petit, F. Bonnetain, S. Ladoire, S. Guiu, J. P. Cercueil, D. Krause, P. Hillon, C. Borg, B. Chauffert and F. Ghiringhelli (2010). "Visceral fat area is an independent predictive biomarker of outcome after first-line bevacizumab-based treatment in metastatic colorectal cancer." Gut 59(3): 341-347.

Hagendoorn, J., R. Tong, D. Fukumura, Q. Lin, J. Lobo, T. P. Padera, L. Xu, R. Kucherlapati and R. K. Jain (2006). "Onset of abnormal blood and lymphatic vessel function and interstitial hypertension in early stages of carcinogenesis." Cancer Res 66(7): 3360-3364.

Haydon, A. M., R. J. Macinnis, D. R. English and G. G. Giles (2006). "Effect of physical activity and body size on survival after diagnosis with colorectal cancer." Gut 55(1): 62-67.

Hosogai, N., A. Fukuhara, K. Oshima, Y. Miyata, S. Tanaka, K. Segawa, S. Furukawa, Y. Tochino, R. Komuro, M. Matsuda and I. Shimomura (2007). "Adipose tissue hypoxia in obesity and its impact on adipocytokine dysregulation." Diabetes 56(4): 901-911.

Ibrahim, M. M. (2010). "Subcutaneous and visceral adipose tissue: structural and functional differences." Obes Rev 11(1): 11-18.

Incio, J., R. Lopes, I. Azevedo and S. Soares (2006). "Prevention of both angio and atherogenesis: inhibitory properties of polyphenols (Xanthumol) in smooth muscle cells." European Journal of Medical Research 11:122(Suppl II).

Isoda, K., J. L. Young, A. Zirlik, L. A. MacFarlane, N. Tsuboi, N. Gerdes, U. Schonbeck and P. Libby (2006). "Metformin inhibits proinflammatory responses and nuclear factor-kappaB in human vascular wall cells." Arterioscler Thromb Vasc Biol 26(3): 611-617.

Khalid, S., D. Hwang, Y. Babichev, R. Kolli, S. Altamentova, S. Koren, P. J. Goodwin, M. Ennis, M. Pollak, N. Sonenberg and I. G. Fantus (2009). "Evidence for a tumor promoting effect of high-fat diet independent of insulin resistance in HER2/Neu mammary carcinogenesis." Breast Cancer Res Treat 122(3): 647-659.

Kim, P., E. Chung, H. Yamashita, K. E. Hung, A. Mizoguchi, R. Kucherlapati, D. Fukumura, R. K. Jain and S. H. Yun (2010). "In vivo wide-area cellular imaging by side-view endomicroscopy." Nat Methods 7(4): 303-305.

Kimura, Y. and M. Sumiyoshi (2007). "High-fat, high-sucrose, and high-cholesterol diets accelerate tumor growth and metastasis in tumor-bearing mice." Nutr Cancer 59(2): 207-216.

Kobayashi, H., N. Ouchi, S. Kihara, K. Walsh, M. Kumada, Y. Abe, T. Funahashi and Y. Matsuzawa (2004). "Selective suppression of endothelial cell apoptosis by the high molecular weight form of adiponectin." Circ Res 94(4): e27-31.

Ladoire, S., F. Bonnetain, M. Gauthier, S. Zanetta, J. M. Petit, S. Guiu, I. Kermarrec, E. Mourey, F. Michel, D. Krause, P. Hillon, L. Cormier, F. Ghiringhelli and B. Guiu (2011). "Visceral fat area as a new independent predictive factor of survival in patients with metastatic renal cell carcinoma treated with antiangiogenic agents." Oncologist 16(1): 71-81.

Llaverias, G., C. Danilo, Y. Wang, A. K. Witkiewicz, K. Daumer, M. P. Lisanti and P. G. Frank (2010). "A Western-type diet accelerates tumor progression in an autochthonous mouse model of prostate cancer." Am J Pathol 177(6): 3180-3191.

Lolmede, K., V. Durand de Saint Front, J. Galitzky, M. Lafontan and A. Bouloumie (2003). "Effects of hypoxia on the expression of proangiogenic factors in differentiated 3T3-F442A adipocytes." Int J Obes Relat Metab Disord 27(10): 1187-1195.

Lysaght, J., E. P. van der Stok, E. H. Allott, R. Casey, C. L. Donohoe, J. M. Howard, S. A. McGarrigle, N. Ravi, J. V. Reynolds and G. P. Pidgeon (2011). "Pro-inflammatory and tumour proliferative properties of excess visceral adipose tissue." Cancer Lett 312(1): 62-72.

Maccio, A., C. Madeddu, G. Gramignano, C. Mulas, C. Floris, D. Massa, G. Astara, P. Chessa and G. Mantovani (2010). "Correlation of body mass index and leptin with tumor size and stage of disease in hormone-dependent postmenopausal breast cancer: preliminary results and therapeutic implications." J Mol Med (Berl) 88(7): 677-686.

Majed, B., T. Moreau, K. Senouci, R. m. Salmon, A. Fourquet and B. Asselain (2008). "Is obesity an independent prognosis factor in woman breast cancer?" Breast Cancer Research and Treatment 111(2): 329-342.

Martin, M., R. Hayward, A. Viros and R. Marais (2012). "Metformin Accelerates the Growth of BRAFV600E -Driven Melanoma by Upregulating VEGF-A." Cancer Discovery 2:344-355.

Matsuzawa, Y. (2006). "Therapy Insight: adipocytokines in metabolic syndrome and related cardiovascular disease." Nat Clin Pract Cardiovasc Med 3(1): 35-42.

Meyerhardt, J. A., D. Niedzwiecki, D. Hollis, L. B. Saltz, R. J. Mayer, H. Nelson, R. Whittom, A. Hantel, J. Thomas and C. S. Fuchs (2008). "Impact of body mass index and weight change after treatment on cancer recurrence and survival in patients with stage III colon cancer: findings from Cancer and Leukemia Group B 89803." J Clin Oncol 26(25): 4109-4115.

Mokdad, A. H., J. S. Marks, D. F. Stroup and J. L. Gerberding (2004). "Actual causes of death in the United States, 2000." Jama 291(10): 1238-1245.

Monteiro, R., R. Soares, S. Guerreiro, D. Pestana, C. Calhau and I. Azevedo (2009). "Red wine increases adipose tissue aromatase expression and regulates body weight and adipocyte size." Nutrition 25(6): 699-705.

Moses, A. G., N. Dowidar, B. Holloway, I. Waddell, K. C. Fearon and J. A. Ross (2001). "Leptin and its relation to weight loss, ob gene expression and the acute-phase response in surgical patients." Br J Surg 88(4): 588-593.

Murphy, L. J., P. Molnar, X. Lu and H. Huang (1995). "Expression of human insulin-like growth factor-binding protein-3 in transgenic mice." J Mol Endocrinol 15(3): 293-303.

Negrao, M. R., J. Incio, R. Lopes, I. Azevedo and R. Scares (2007). "Xanthohumol inhibits angiogenic vessels but not stable vessels." Faseb Journal 21(6): A1095-A1095.

Nunez, N. P., W. J. Oh, J. Rozenberg, C. Perella, M. Anver, J. C. Barrett, S. N. Perkins, D. Berrigan, J. Moitra, L. Varticovski, S. D. Hursting and C. Vinson (2006). "Accelerated tumor formation in a fatless mouse with type 2 diabetes and inflammation." Cancer Res 66(10): 5469-5476.

Nunez, N. P., S. N. Perkins, N. C. Smith, D. Berrigan, D. M. Berendes, L. Varticovski, J. C. Barrett and S. D. Hursting (2008). "Obesity accelerates mouse mammary tumor growth in the absence of ovarian hormones." Nutr Cancer 60(4): 534-541.

Olshansky, S. J., D. J. Passaro, R. C. Hershow, J. Layden, B. A. Carnes, J. Brody, L. Hayflick, R. N. Butler, D. B. Allison and D. S. Ludwig (2005). "A potential decline in life expectancy in the United States in the 21st century." N Engl J Med 352(11): 1138-1145.

Ouchi, N., H. Kobayashi, S. Kihara, M. Kumada, K. Sato, T. Inoue, T. Funahashi and K. Walsh (2004). "Adiponectin stimulates angiogenesis by promoting cross-talk between AMP-activated protein kinase and Akt signaling in endothelial cells." J Biol Chem 279(2): 1304-1309.

Paoletti, R., C. Bolego, A. Poli and A. Cignarella (2006). "Metabolic syndrome, inflammation and atherosclerosis." Vasc Health Risk Manag 2(2): 145-152.

Park, J., D. M. Euhus and P. E. Scherer (2011). "Paracrine and endocrine effects of adipose tissue on cancer development and progression." Endocr Rev 32(4): 550-570.

Peeters, A., J. J. Barendregt, F. Willekens, J. P. Mackenbach, A. Al Mamun and L. Bonneux (2003). "Obesity in adulthood and its consequences for life expectancy: a life-table analysis." Ann Intern Med 138(1): 24-32.

Phoenix, K. N., F. Vumbaca and K. P. Claffey (2009). "Therapeutic metformin/ AMPK activation promotes the angiogenic phenotype in the ERalpha negative MDA-MB-435 breast cancer model." Breast Cancer Res Treat 113(1): 101-111.

Pischon, T., P. H. Lahmann, H. Boeing, C. Friedenreich, T. Norat, A. Tjonneland, J. Halkjaer, K. Overvad, F. Clavel-Chapelon, M. C. Boutron-Ruault, G. Guernec, M. M. Bergmann, J. Linseisen, N. Becker, A. Trichopoulou, D. Trichopoulos, S. Sieri, D. Palli, R. Tumino, P. Vineis, S. Panico, P. H. Peeters, H. B. Bueno-de-Mesquita, H. C. Boshuizen, B. Van Guelpen, R. Palmqvist, G. Berglund, C. A. Gonzalez, M. Dorronsoro, A. Barricarte, C. Navarro, C. Martinez, J. R. Quiros, A. Roddam,

N. Allen, S. Bingham, K. T. Khaw, P. Ferrari, R. Kaaks, N. Slimani and E. Riboli (2006). "Body size and risk of colon and rectal cancer in the European Prospective Investigation Into Cancer and Nutrition (EPIC)." J Natl Cancer Inst 98(13): 920-931.

Polverini, P. J. and S. J. Leibovich (1984). "Induction of neovascularization in vivo and endothelial proliferation in vitro by tumor-associated macrophages." Lab Invest 51(6): 635-642.

Protani, M., M. Coory and J. H. Martin (2010). "Effect of obesity on survival of women with breast cancer: systematic review and meta-analysis." Breast Cancer Res Treat 123(3): 627-635.

Rausch, M. E., S. Weisberg, P. Vardhana and D. V. Tortoriello (2008). "Obesity in C57BL/6J mice is characterized by adipose tissue hypoxia and cytotoxic T-cell infiltration." Int J Obes (Lond) 32(3): 451-463.

Reinmuth, N., W. Liu, F. Fan, Y. D. Jung, S. A. Ahmad, O. Stoeltzing, C. D. Bucana, R. Radinsky and L. M. Ellis (2002). "Blockade of insulin-like growth factor I receptor function inhibits growth and angiogenesis of colon cancer." Clin Cancer Res 8(10): 3259-3269.

Rose, D. P., J. M. Connolly and C. L. Meschter (1991). "Effect of dietary fat on human breast cancer growth and lung metastasis in nude mice." J Natl Cancer Inst 83(20): 1491-1495.

Rose, D. P., D. Komninou and G. D. Stephenson (2004). "Obesity, adipocytokines, and insulin resistance in breast cancer." Obes Rev 5(3): 153-165.

Rupnick, M. A., D. Panigrahy, C. Y. Zhang, S. M. Dallabrida, B. B. Lowell, R. Langer and M. J. Folkman (2002). "Adipose tissue mass can be regulated through the vasculature." Proc Natl Acad Sci USA 99(16): 10730-10735.

Ryan, P. D. and P. E. Goss (2008). "The emerging role of the insulin-like growth factor pathway as a therapeutic target in cancer." Oncologist 13(1): 16-24.

Sengupta, N., A. Afzal, S. Caballero, K. H. Chang, L. C. Shaw, J. J. Pang, V. C. Bond, I. Bhutto, T. Baba, G. A. Lutty and M. B. Grant (2010). "Paracrine modulation of CXCR4 by IGF-1 and VEGF: implications for choroidal neovascularization." Invest Ophthalmol Vis Sci 51(5): 2697-2704.

Shojaei, F., X. Wu, A. K. Malik, C. Zhong, M. E. Baldwin, S. Schanz, G. Fuh, H. P. Gerber and N. Ferrara (2007). "Tumor refractoriness to anti-VEGF treatment is mediated by CD11b+Gr1+ myeloid cells." Nat Biotechnol 25(8): 911-920.

Sierra-Honigmann, M. R., A. K. Nath, C. Murakami, G. Garcia-Cardena, A. Papapetropoulos, W. C. Sessa, L. A. Madge, J. S. Schechner, M. B. Schwabb, P. J. Polverini and J. R. Flores-Riveros (1998). "Biological action of leptin as an angiogenic factor." Science 281(5383): 1683-1686.

Smith, L. E., W. Shen, C. Perruzzi, S. Soker, F. Kinose, X. Xu, G. Robinson, S. Driver, J. Bischoff, B. Zhang, J. M. Schaeffer and D. R. Senger (1999). "Regulation of vascular endothelial growth factor-dependent retinal neovascularization by insulin-like growth factor-1 receptor." Nat Med 5(12): 1390-1395.

Soares, R. and C. Costa (2009). "Angiogenesis in metabolic syndrome (book chapter)." Oxidative stress, inflammation and angiogenesis in metabolic syndrome (Springer-Verlag, The Netherlands): 85-99.

Steffens, S., V. Grunwald, K. I. Ringe, C. Seidel, H. Eggers, M. Schrader, F. Wacker, M. A. Kuczyk and A. J. Schrader (2011). "Does obesity influence the prognosis of metastatic renal cell carcinoma in patients treated with vascular endothelial growth factor-targeted therapy?" Oncologist 16(11): 1565-1571.

Suganami, E., H. Takagi, H. Ohashi, K. Suzuma, I. Suzuma, H. Oh, D. Watanabe, T. Ojima, T. Suganami, Y. Fujio, K. Nakao, Y. Ogawa and N. Yoshimura (2004). "Leptin stimulates ischemia-induced retinal neovascularization: possible role of vascular endothelial growth factor expressed in retinal endothelial cells." Diabetes 53(9): 2443-2448.

Tam, J., D. G. Duda, J. Y. Perentes, R. S. Quadri, D. Fukumura and R. K. Jain (2009). "Blockade of VEGFR2 and not VEGFR1 can limit diet-induced fat tissue expansion: role of local versus bone marrow-derived endothelial cells." PLoS One 4(3): e4974.

Uzzan, B., P. Nicolas, M. Cucherat and G. Y. Perret (2004). "Microvessel density as a prognostic factor in women with breast cancer: a systematic review of the literature and meta-analysis." Cancer Res 64(9): 2941-2955.

Volkova, E., J. A. Willis, J. E. Wells, B. A. Robinson, G. U. Dachs and M. J. Currie (2011). "Association of angiopoietin-2, C-reactive protein and markers of obesity and insulin resistance with survival outcome in colorectal cancer." Br J Cancer 104(1): 51-59.

Vona-Davis, L., M. Howard-McNatt and D. P. Rose (2007). "Adiposity, type 2 diabetes and the metabolic syndrome in breast cancer." Obes Rev 8(5): 395-408.

Wang, B., I. S. Wood and P. Trayhurn (2007). "Dysregulation of the expression and secretion of inflammation-related adipokines by hypoxia in human adipocytes." Pflugers Arch 455(3): 479-492.

Weisberg, S. P., D. McCann, M. Desai, M. Rosenbaum, R. L. Leibel and A. W. Ferrante, Jr. (2003). "Obesity is associated with macrophage accumulation in adipose tissue." J Clin Invest 112(12): 1796-1808.

Wu, Y., P. Brodt, H. Sun, W. Mejia, R. Novosyadlyy, N. Nunez, X. Chen, A. Mendoza, S. H. Hong, C. Khanna and S. Yakar (2010). "Insulin-like growth factor-I regulates the liver microenvironment in obese mice and promotes liver metastasis." Cancer Res 70(1): 57-67.

Wu, Y., S. Yakar, L. Zhao, L. Hennighausen and D. LeRoith (2002). "Circulating insulin-like growth factor-I levels regulate colon cancer growth and metastasis." Cancer Res 62(4): 1030-1035.

Xu, H., G. T. Barnes, Q. Yang, G. Tan, D. Yang, C. J. Chou, J. Sole, A. Nichols, J. S. Ross, L. A. Tartaglia and H. Chen (2003). "Chronic inflammation in fat plays a crucial role in the development of obesity-related insulin resistance." J Clin Invest 112(12): 1821-1830.

Xue, F. and K. B. Michels (2007). "Diabetes, metabolic syndrome, and breast cancer: a review of the current evidence." Am J Clin Nutr 86(3): s823-835.

Yang, H., B. Nan, S. Yan, M. Li, Q. Yao and C. Chen (2005). "C-reactive protein decreases expression of VEGF receptors and neuropilins and inhibits VEGF165-induced cell proliferation in human endothelial cells." Biochem Biophys Res Commun 333(3): 1003-1010.

Zimmet, P. Z., D. J. McCarty and M. P. de Courten (1997). "The global epidemiology of non-insulin-dependent diabetes mellitus and the metabolic syndrome." J Diabetes Complications 11(2): 60-68.

Index

Color Plate Section

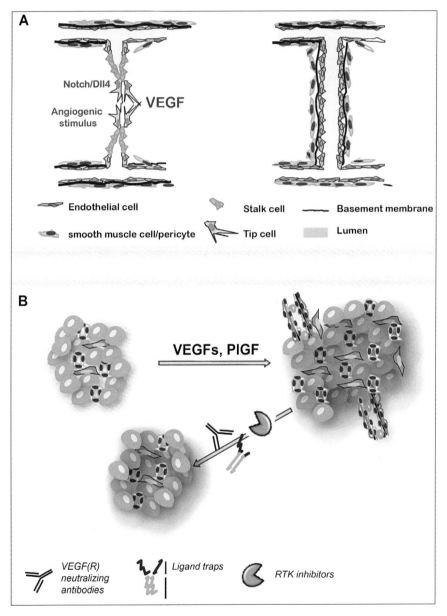

Figure 1 (Chapter 1) See text page 4 for caption.

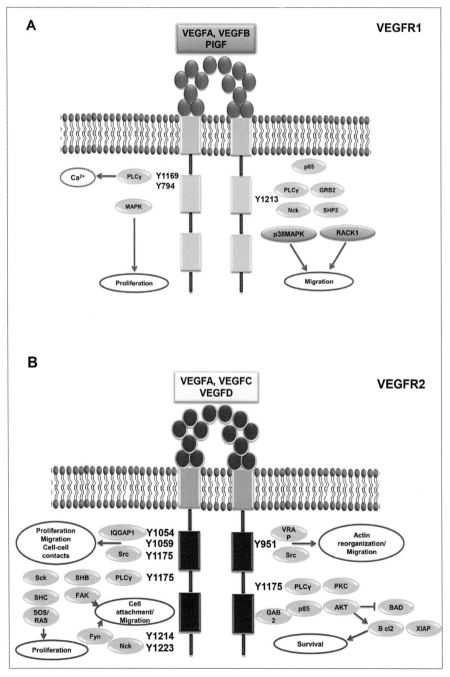

Figure 2 (Chapter 1)   See text page 8 for caption.

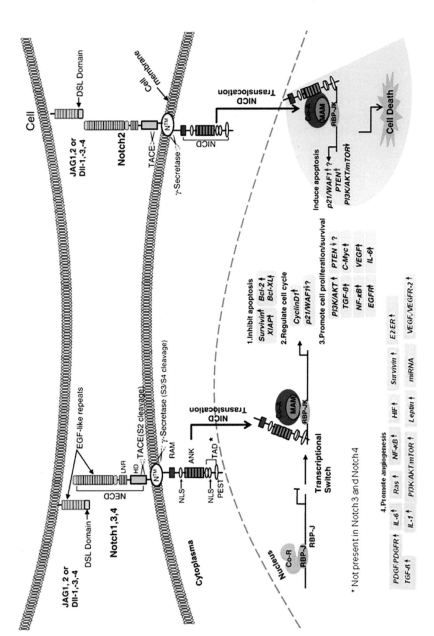

Figure 1 (Chapter 2) See text page **42** for caption.

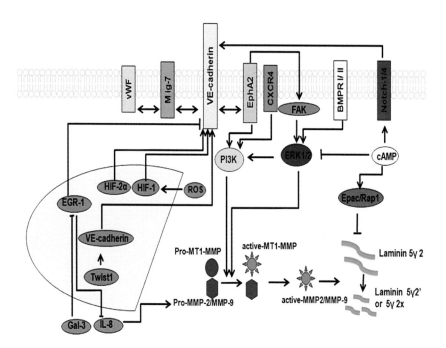

Figure 1 (Chapter 3)   See text page **95** for caption.

Chapter 4

Figure 1

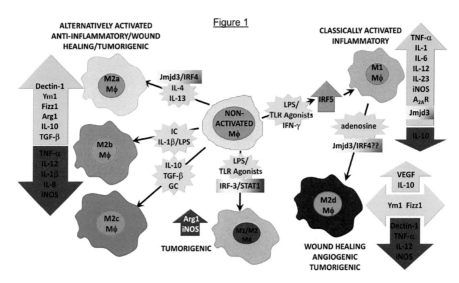

Figure 1 (Chapter 4)   See text page **117** for caption.

Figure 2

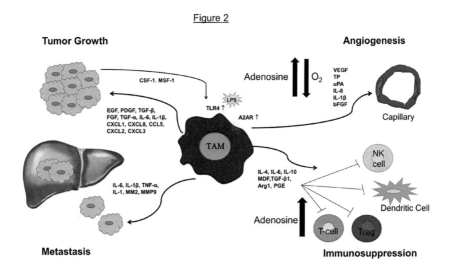

Figure 2 (Chapter 4)   See text page **118** for caption.

Chapter 5

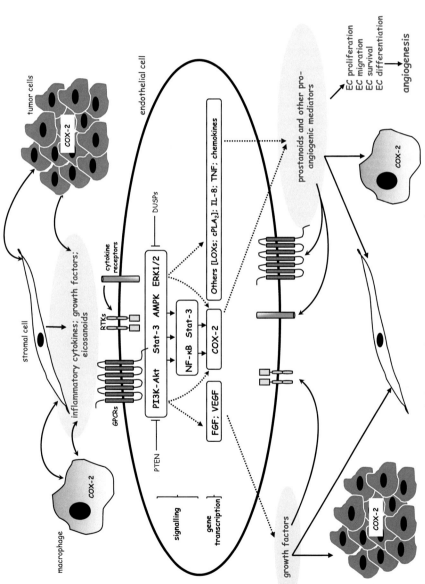

Figure 1 (Chapter 5)   See text page 154 for caption.

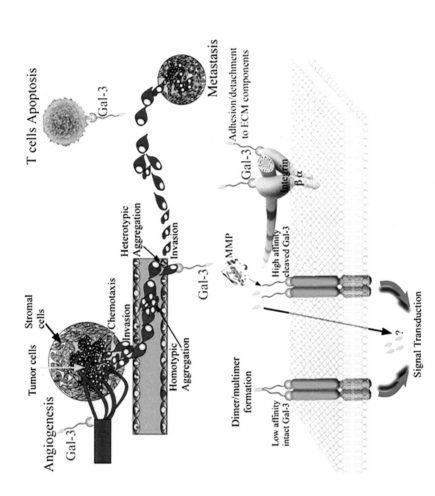

Figure 4 (Chapter 6) See text page **223** for caption.

Chapter 7

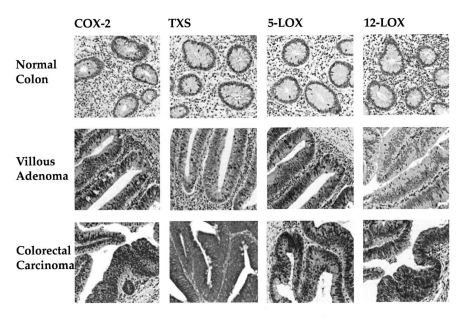

COX-2   **TXS**   **5-LOX**   **12-LOX**

Normal
Colon

Villous
Adenoma

Colorectal
Carcinoma

Figure 2 (Chapter 7)   See text page **248** for caption.

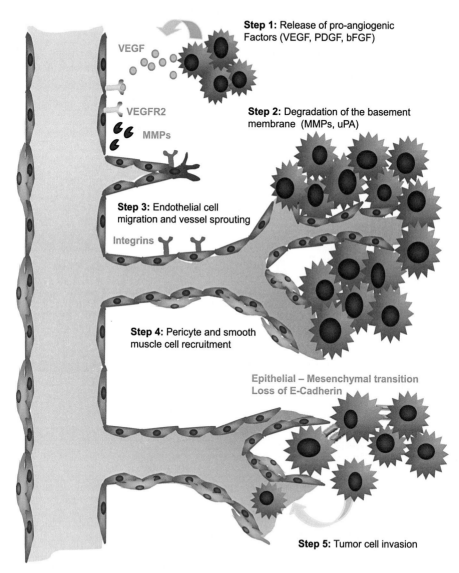

Step 1: Release of pro-angiogenic Factors (VEGF, PDGF, bFGF)

VEGF

VEGFR2

MMPs

Step 2: Degradation of the basement membrane (MMPs, uPA)

Step 3: Endothelial cell migration and vessel sprouting

Integrins

Step 4: Pericyte and smooth muscle cell recruitment

Epithelial – Mesenchymal transition Loss of E-Cadherin

Step 5: Tumor cell invasion

Figure 3 (Chapter 7)   See text page 253 for caption.

A) Intersegmental Vessel Formation

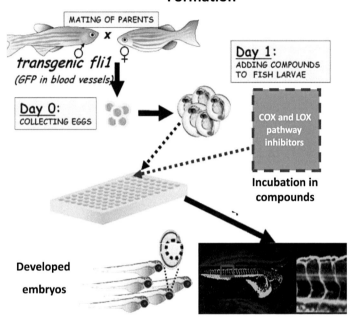

B)

Untreated Control	Untreated Control

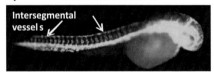
Intersegmental vessels

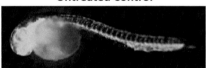

TXS Inhibitor (ozagrel)	TXS/5-LOX Inhibitor (ketoconazole)

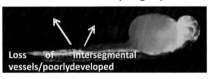
Loss of intersegmental vessels/poorly developed

TP Antagonist (seratrodast)	TP/EP4 Antagonist (AH-23848)

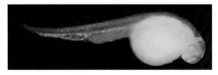

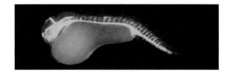

Figure 5A&B (Chapter 7)   See text page **274** for caption.

Chapter 8

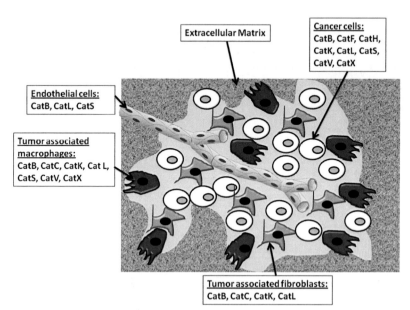

Figure 1 (Chapter 8)   See text page **303** for caption.

Figure 2 (Chapter 8)   See text page **304** for caption.

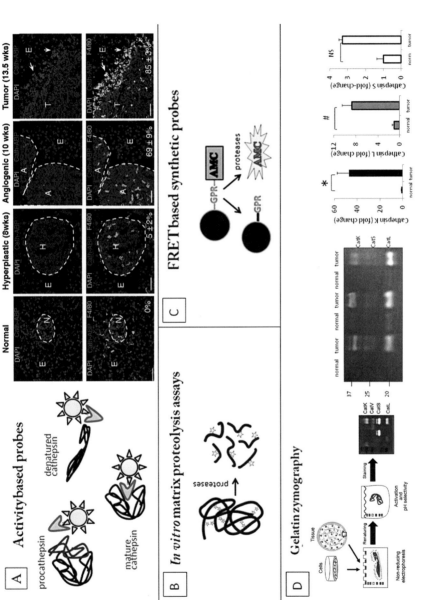

Figure 3 (Chapter 8)   See text page **315** for caption.

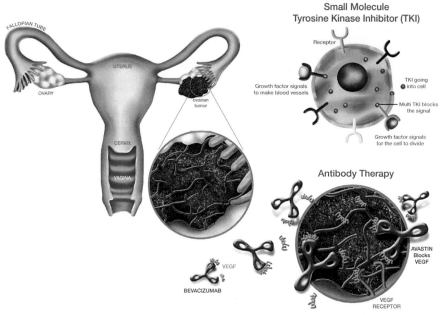

Figure 1 (Chapter 10)   See text page **368** for caption.

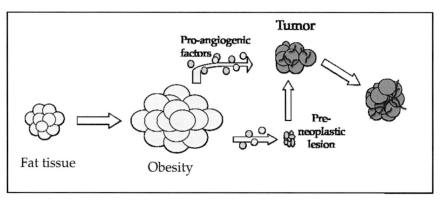

Figure 1 (Chapter 12)   See text page **417** for caption.